# GROUP THEORY IN CHEMISTRY AND SPECTROSCOPY

## A Simple Guide to Advanced Usage

BORIS S. TSUKERBLAT
*Department of Chemistry*
*Ben-Gurion University of the Negev, Israel*



DOVER PUBLICATIONS, INC.
Mineola, New York

*This book is dedicated to the memory of my parents*



*Copyright*



*Bibliographical Note*

This Dover edition, first published in 2006, is an unabridged republication of the edition published by Academic Press Ltd., London, 1994. An errata list, an additional appendix, and some updated material have been specially prepared for this edition.

*Library of Congress Cataloging-in-Publication Data*

Tsukerblat, B. S. (Boris Samoilovich)
Group theory in chemistry and spectroscopy : a simple guide to advanced usage / Boris S. Tsukerblat.
p. cm.
Originally published: London ; San Diego : Academic Press, c1994, in series: Theoretical chemistry.
Includes bibliographical references and index.
ISBN 0-486-45035-X (pbk.)
1. Group theory. 2. Chemistry, Physical and theoretical. I. Title.

QD455.3.G75T76 2006
541'.2—dc22

2006040233

Manufactured in the United States of America
Dover Publications, Inc., 31 East 2nd Street, Mineola, N.Y. 11501

# Contents

# Preface

The physical and chemical applications of group theory are usually based on the geometric symmetries of atoms, molecules and crystals, as well as on the symmetries of the equations describing the properties and behaviour of the physical system under consideration. For a contemporary chemist, group theory is not only a key element of the quantum mechanical methods of investigating the electronic structure of matter—knowledge of symmetry and its group theoretical implications is also widely applied in analysing the results of practically all spectroscopic techniques currently employed in organic and inorganic chemistry. In fact, group theory has become a working tool of all present-day chemists concerned not only with the synthesis of new substances but also with their electronic structure and properties.

The present work is intended as a handbook on group theory for chemists and experimental physicists who use spectroscopy and require knowledge of the electronic structures of materials under investigation. The book will be readily understood by students with the background in physics and mathematics provided by an up-to-date degree course in chemistry. Many of the key concepts are introduced in a simple way as the material is presented. Indeed, the first three chapters, covering concepts of geometric symmetry and point groups, might well be assimilated at a pre-university level.

This book's salient features are as follows:

(1) The deductive approach characteristic of many books on mathematics and theoretical physics has been avoided, since it would have been a barrier to approaching the physical and chemical applications that are the main topic of this book. All the fundamental concepts are introduced by way of simple examples, relating to particular chemical and physical problems;

(2) Since neither chemists nor spectroscopists require proofs of the theorems used, very few are supplied. Most attention is paid to explanation of the

principal conclusions, their meaning and the ways in which they are to be used. Proofs are given only when they are necessary for a proper understanding of the principles involved.

(3) In view of the practical bias of the book, the main results of group theory are presented in all sections as procedures, making possible their systematic and step-by-step application. The same approach is adopted in providing a detailed description of the main tables [1–9], whose correct use constitutes a considerable part of the practical application of the theory.

(4) The book is constructed around examples that are analysed in detail. Each chapter contains problems whose solution develops practical skill and provides a valuable supplement to the material presented in the rest of the text.

The structure of this book is as follows. Chapters 1 and 2 give a detailed description of symmetry operations, enumeration of point groups and rules for the determination of molecular symmetry. In Chapter 3, using simple examples, the fundamental concepts of reducible and irreducible representations are introduced, and character tables are described. It is shown in Chapter 4 how the splitting of the central-ion energy levels in a coordination compound leads to the notion of the reduction of a representation. The description of many-electron ions (Chapter 5) is associated with the notions of the reduction of spherical group representations, the direct product of irreducible representations, and Wigner and Clebsch–Gordan coefficients. The projection operator technique, the use of spherical tensors in point groups, and the generation of basis functions are described in Chapter 6, which deals with the semi-empirical theory of the crystal field. The subsequent development of the projection operator technique and the notion of site symmetry are used in the directed valence theory (Chapter 7) and the molecular orbital method (Chapter 8). Chapter 9 is concerned with optical spectra. Here the notion of a selection rule is introduced, and the Wigner–Eckart theorem and the method of the effective dipole moment are described. Because of the wide application of laser spectroscopy in present-day chemistry, a detailed treatment of the selection rules and polarization dependence of two-photon spectra is given. The theory of double groups is considered in connection with the study of spin–orbit interaction (Chapters 10 and 11). The widely used method of invariants is described in Chapter 12 in connection with electron paramagnetic resonance. Through consideration of examples with a gradually increasing degree of sophistication, the analysis of electric field effects in paramagnetic resonance is presented in an easily understandable form. Problems involving polynuclear coordination compounds (Chapter 13) are solved using irreducible tensor methods and through group-theoretical classification of exchange multiplets. Finally, although Chapter 14 does not contain any new group-theoretical concepts, the examples presented earlier are used to illustrate many of the fundamental concepts at a deeper level.

The author's own understanding of group theory owes much to the books [9–26], which we recommended for more detailed study of particular topics.

Finally, I appreciate very much the highly important contribution of Professor McWeeny in the final version of this book. It is a special pleasure for me to convey to him my deepest thanks.

BORIS S. TSUKERBLAT

# ERRATA

| | | |
|---|---|---|
| Page xiii | Line 4 | For: [9–26] Read: [9–26], [A.1–A.6] |
| Page 11 | Table 1.1 | Line 1, Column 2<br>For: Reflection within the plane ($\hat{\sigma}$)<br>Read: Reflection in the plane ($\hat{\sigma}$) |
| Page 33 | Line –5 | For: belong to the same class,<br>Read: belong to different classes, |
| Page 33 | Line –4 | For: are equivalent<br>Read: are not equivalent |
| Page 33 | Line –3 | For: three classes: $\hat{E}$, $\hat{C}_2$, $2\hat{\sigma}_V$<br>Read: four classes: $\hat{E}$, $\hat{C}_2$, $\hat{\sigma}_V$, $\hat{\sigma}'_V$ |
| Page 47 | Table 2.3 | Column IV, Line 2 For: $\mathbf{D}_6$ Read: $\mathbf{D}_3$ |
| Page 47 | Table 2.3 | Column IV, Add Line 4: $\mathbf{C}_{3h}$, $\mathbf{D}_{3d}$ |
| Page 47 | Table 2.3 | Column VI, Line 3 For: $\mathbf{C}_{3h}$, $\mathbf{D}_{3d}$ Read: $\mathbf{C}_6$ |
| Page 83 | Line –13 | For: $g_1 + g_2 = 1$ Read: $g_1 = g_2 = 1$ |
| Page 92 | Line 3 | Add: Characters of point groups are given in Appendix I. |
| Page 132 | Table 5.5 | Column $2\hat{\sigma}_d$ , Line $A_1 \times A_2$ For: 1 Read: $-1$ |
| Page 168 | Line 11 | For: $(\kappa + m' - l)!$ Read: $(\kappa + m' - m)!$ |
| Page 168 | Line 12 | For: $(\cos \frac{1}{2}\beta)^{2l-m-m'-2\kappa}$<br>Read: $(\cos \frac{1}{2}\beta)^{2l+m-m'-2\kappa}$ |
| Page 169 | Line –1 | For: Appendix I. Read: Appendix II. |
| Page 170 | Line –16 | For: point the groups are<br>Read: the point groups are |
| Page 172 | Line –1 | For: Appendix II. Read: Appendix III. |
| Page 186 | Line –12 | For: six Read: five |
| Page 186 | Line –11 | Delete: (1) $s, p_z, p_x, p_y$ $(\mathrm{sp}^3)$; |
| Page 186 | Line –10 | For: (2) Read: (1) |
| Page 186 | Line –9 | For: (3) Read: (2) |
| Page 186 | Line –8 | For: (4) Read: (3) |
| Page 186 | Line –7 | For: (5) Read: (4) |
| Page 186 | Line –6 | For: (6) Read: (5) |
| Page 199 | Line 7 | Add: [A.9, A.10] after *molecular orbital method.* |
| Page 207 | Line –13 | For: where $\omega = e^{2\pi i/3} = \cos \frac{2}{3}\pi + \mathrm{i} \sin \frac{2}{3}\pi = \frac{1}{2} + \frac{1}{2}\sqrt{3}\mathrm{i}$.<br>Read: where $\omega = e^{\pi i/3} = \cos \frac{\pi}{3} + \mathrm{i} \sin \frac{\pi}{3}$. |
| Page 207 | Line –9 | For: $\Psi(\mathrm{E}_1 y) = \frac{1}{2}(p_1 + p_2 - p_4 - p_5)$,<br>Read: $\Psi(\mathrm{E}_1 y) = \frac{1}{2}(p_2 + p_3 - p_5 - p_6)$, |

# ERRATA

| | | |
|---|---|---|
| Page 207 | Line -7 | For: $\Psi(\mathrm{E}_2 b) = \frac{1}{2}(p_1 - p_2 + p_3 - p_4)$.<br>Read: $\Psi(\mathrm{E}_2 b) = \frac{1}{2}(p_2 - p_3 + p_5 - p_6)$. |
| Page 240 | Line 6 | For: [72] Read: [72], [A.11]: |
| Page 267 | Line 17 | Add: Such kind of calculation is given in [A.12]. |
| Page 300 | Line -4 | For: (see [62]) Read: (see [62] and [A.13, A.14]) |
| Page 308 | Line 9 | For: [13.36] Read: (13.36) |
| Page 308 | Line -1 | For: [2, 41, 42, 61, 84] Read: [2, 41, 42, 62, 84, 87]. |
| Page 309 | Line 12 | For: transformations connecting sets<br>Read: transformations (recoupling) connecting sets |
| Page 314 | Line -4 | For: $= (-1)^{s_3+s_2+s_4+s_7}[(2s_3+1)(2s_7+1)]^{-1/2}$<br>Read: $= \delta_{s_3 s_6}\delta_{s_7 s_8}(-1)^{s_3+s_2+s_4+s_7}[(2s_3+1)(2s_7+1)]^{-1/2}$ |
| Page 315 | Line -12, -11 | For: already too elaborate Read: rather complicated even |
| Page 316 | Line 14 | For: *recoupling procedure,* Read: *decoupling procedure* |
| Page 316 | Line -5 | For: $= (1/\sqrt{6}[2s(2s+3)(2s+2)$<br>Read: $= (1/\sqrt{6}[s(2s+3)(s+1)$ |
| Page 317 | Line -16 | For: *system* (*recoupling*). Read: *system* (*decoupling*). |
| Page 318 | Line 3 | For: Applying the recoupling procedure<br>Read: Applying the decoupling procedure |
| Page 319 | Line 6 | For: $\hat{H}^{(4)} = \sum_{\kappa}(2\kappa+1)$ Read: $\hat{H}^{(4)} = \sum_{\kappa}(2\kappa+1)^{1/2}$ |
| Page 319 | Line 13 | For: $(S'_{12})s_3 s_4(S_{34})S'\rangle$ Read: $(S'_{12})s_3 s_4(S'_{34})S'\rangle$ |
| Page 319 | Line 15 | For: $\times(2\kappa_{34}+1)(2\kappa'_{34}+1)]^{1/2}$<br>Read: $\times(2\kappa_{34}+1)(2S_{34}+1)(2S'_{34}+1)$ |
| Page 319 | Line -10 | Equation (13.80) At the end add: $-2J_{24}\mathbf{s}_3\mathbf{s}_4 - 2J\mathbf{s}_3\mathbf{s}_4$ |
| Page 319 | Line -8 | For: The recoupling procedure<br>Read: The decoupling procedure |
| Page 334 | Figure 13.8 | In caption: For: [94] Read: [95] |
| Page 351 | Line -3 | For: [106] Read: [106], [A.47, A.48]. |
| Page 364 | Line -7 | For: [111–113], Read: [111–113], [A.61], |
| Page 364 | Line -6 | For: [114] Read: [114], [A.62] |
| Page 393 | Line 4 | Add: For further development of the vibronic models for mixed-valence systems see [A.63, A.64] and references therein. |
| Page 395 | Line 1 | For: Appendix I Read: Appendix II |
| Page 395 | Line 6 | For: AI.1 Read: AII.1 |
| Page 395 | Line -2 | For: AI.1 Read: AII.1 |

# ADDITIONAL NOTES

Note concerning page-301

When the ground terms of the ions are orbitally degenerate the form of the exchange interaction is dramatically complicated. The problem of degeneracy is out of the scope of this book and we give to the reader only some guiding references. First ideas in the field were proposed by Van Vleck [A.15] and worked out by Levy [A.16-A.18]. Kugel and Khomskii [A.19-A.20] considered the kinetic exchange (for a review of this mechanism see articles [A.13], [A.14]) between the orbitally degenerate ions and developed on this basis the theory of orbital and structural ordering in Jahn-Teller crystals [A.21]. The underlying ideas proposed in cited works have been applied to the field of the molecular magnetism with their subsequent developing toward the more detailed description of the crystal field states and structure of the electronic shells including *ab-initio* evaluation of the exchange parameters for the exchange coupled metal ions [A.22-A.28]. A new approach to the problem of the kinetic exchange between orbitally degenerate ions has been developed in [A.29-A.37]. The orbitally dependent exchange Hamiltonian has been deduced in a general form, that is, for the arbitrary terms and electronic configurations of the constituent metal ions in crystal fields, for a given overall symmetry of the system and with the account for spin-orbital interaction. The usage of the point (and more general) symmetry group arguments allowed to express the effective exchange Hamiltonian in terms of the irreducible tensor operators, spin operators, the parameters of the constituent moieties (cubic crystal field splitting $Dq$ and Racah parameters), and the set of the intercenter transfer integrals specific for each overall symmetry of the pair. This Hamiltonian proved to be an efficient tool for the description of the magnetic properties of the clusters and especially their magnetic anisotropy [A.38-A.39] and gives rise to a new development in molecular magnetism of metal clusters containing orbitally degenerate ions ( see [A.40-A.43] and references therein).

Note concerning page 319

Application of the irreducible tensor operator technique to the calculation of the thermodynamic properties of exchange clusters ( in the model that includes isotropic and anisotropic interactions) and the efficient program MAGPACK is given in [A.44-A.45]. A new general promising approach based on the additional symmetry related to permutation of spin sites has been proposed by Waldmann [A.46].

Note concerning page 353

Site-symmetry approach for the delocalized electronic pairs has been developed in refs. [A.49-A.51] (see also review articles [A.52-A.53]) The development of the angular momentum based approach to the problem of mixed valency is given in refs. {A.54.-A600] (see also references therein).

# 1 Symmetry Transformations and Groups

## 1.1 SYMMETRY TRANSFORMATIONS

### 1.1.1 Definition of a symmetry operation

The symmetry of an object (e.g. a molecule or a crystal) is determined by the set of transformations that maintain the distance between any two points and bring the object into self-coincidence. Such transformations change the geometrical configuration of the object to one that cannot be distinguished from the initial configuration, and are called *symmetry operations* (or *symmetry transformations*). Every possible symmetry operation can be reduced to one of the following three operations or a combination of them:

(1) rotation through a definite angle about some axis;
(2) mirror reflection in a plane;
(3) parallel transport (translation).

The last operation can be considered only with reference to a crystal lattice of infinite extent. In this book we shall deal only with systems of finite dimensions (molecules, complexes, impurity ions or atoms in crystals), whose symmetry transformations may be realized only by operations of the first two types.

### 1.1.2 Rotation operation. Symmetry axes

If a body is brought into self-coincidence on rotation through an angle $2\pi/n = 360°/n$ around some axis, this axis is called an $n$th order *symmetry axis*, and is denoted by $C_n$. The corresponding rotation *operation* is denoted by $\hat{C}_n$ (this notation is similar to that used for operators in quantum mechanics). The following symbols are used to denote axes in the figures [27–29]:

| Axis: | $C_2$ | $C_3$ | $C_4$ | $C_5$ | $C_6$ |
|---|---|---|---|---|---|
| Symbol: | 0 | △ | □ | ⬠ | ⬡ |

The influence of the operation $\hat{C}_3$ on the molecule $BH_3$ is shown in Fig. 1.1. Configurations II and I cannot be distinguished the atoms being identical (the enumeration does not imply any physical features that distinguish one configuration from another).

**Example 1.1** Water molecules have the symmetry axis $C_2$ coinciding with the bisectrix of the angle HOH (Fig. 1.2a).

**Example 1.2** Molecules of $BH_3$ (Fig. 1.2b) have three axes of the 2nd order ($C_2$) lying in the molecular plane and possess an axis of 3rd order ($C_3$) that is perpendicular to this plane.

**Example 1.3** The complex anions $[PtCl_4]^{2-}$ have the symmetry axes of a square (Fig. 1.2c). Of the four axes $C_2$, two (the diagonals) intersect the chlorine ions, while the other two bisect the ClPtCl angles.

**Example 1.4** Molecules of benzene (Fig. 1.2d) possess a $C_6$ axis and six $C_2$ axes. Similarly to the Example 1.2c, here there are two types of $C_2$ axes—the diagonal and the bisecting ones (only one axis of each type is shown in the figure).

For $n = 1$ the operation $C_1$ implies rotation through an angle $2\pi$, which always results in self-coincidence of the object. If there is only one $C_1$ axis, there is no symmetry as such. An example is the FClSO molecule (Fig. 1.2e). It is obvious that there are no axes of higher order ($C_2$, $C_3$, etc.), since the molecule consists of different atoms. The operation $\hat{C}_1$ leaves all atoms in their original places. This transformation is called the *identity*, and is denoted by $\hat{E}$ (thus $\hat{C}_1 = \hat{E}$).

Linear molecules ($H_2$, HCl etc., Fig. 1.2f, g) have $C_\infty$ axes. The corresponding operation is rotation through any angle. It is clear that the existence of a $C_\infty$ axis does not exclude other symmetry axes (Fig. 1.2f).

In Fig. 1.3 the successive use of two rotation operations $\hat{C}_n$ (with $n = 3$) changes the molecule $BH_3$ from configuration I to III, which is indistinguishable from I. Therefore the resulting rotation is also a symmetry operation,

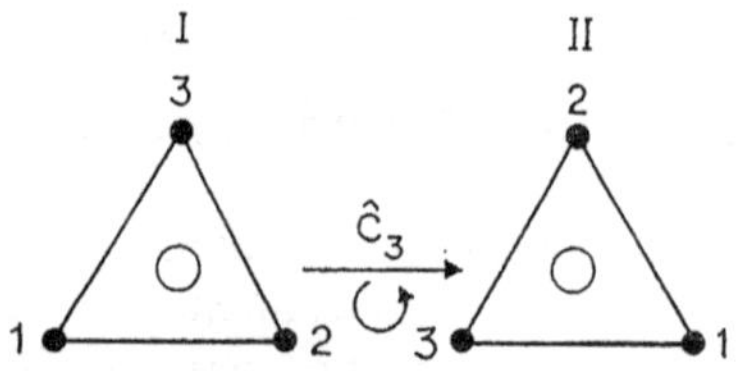


**Fig. 1.1** The action of the operation $\hat{C}_3$ on the $BH_3$ molecule (●—H, ○—B).

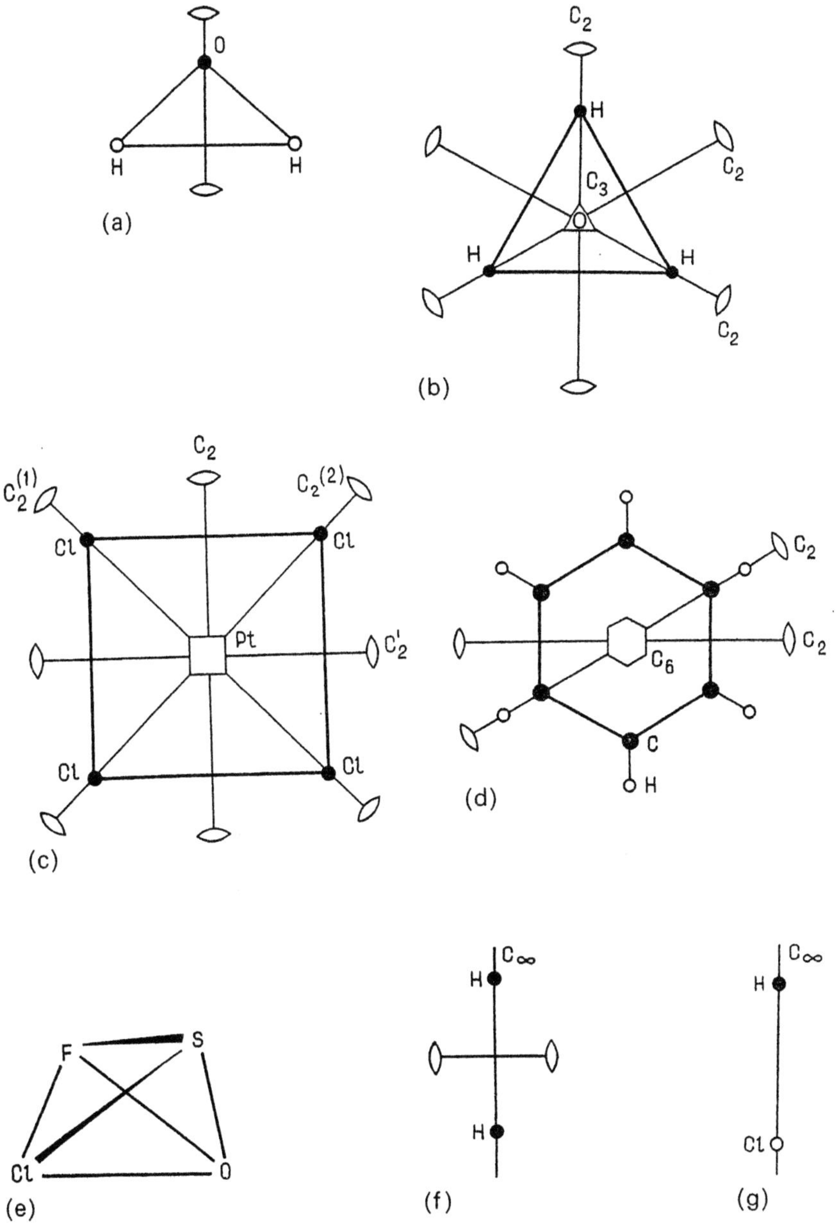


**Fig. 1.2** The arrangement of symmetry axes: ,(a) water; (b) $BH_3$ molecule; (c) the complex ion $[PtCl_4]^{2-}$; (d) benzene; (e) FClSO; (f) the $C_\infty$ and $C_2$ axes in the hydrogen molecule; (g) the $C_\infty$ axis in HCl.

denoted by $\hat{C}_n^2$. In general, rotations $\hat{C}_n^2, \hat{C}_n^3, \ldots, \hat{C}_n^m$ are also symmetry operations, corresponding to rotations through angles $2(2\pi/n)$, $3(2\pi/n), \ldots, m(2\pi/n)$ respectively. It is clear that $\hat{C}_n^n = \hat{E}$, since $n(2\pi/n) = 2\pi$ (configurations I and IV in Fig. 1.3). Thus the fact that a molecule has a $C_n$ axis leads to the existence of a definite number of symmetry operations, not just one. For example, in the case of a $C_4$ axis the symmetry operations are

$$\hat{C}_4, \quad \hat{C}_4^2, \quad \hat{C}_4^3, \quad \hat{C}_4^4 = \hat{E}, \quad \hat{C}_4^5, \ldots \ .$$

However, since $5(2\pi/4) = 2\pi + 2\pi/4$, $\hat{C}_4^5 = \hat{C}_4$; i.e. the rotation $\hat{C}_5 = \hat{E}\hat{C}_4$ is not considered as distinct from $\hat{C}_4$. Thus a $C_4$ axis gives rise to only four operations. It should be noted that $\hat{C}_4^2 = \hat{C}_2$, since this operation is a rotation through 180°. Thus the presence in a molecule of a $C_4$ axis implies the existence of a $C_2$ axis.

Let us consider the sequence of operations related to an axis $C_6$:

$$\hat{C}_6, \quad \hat{C}_6^2, \quad \hat{C}_6^3, \quad \hat{C}_6^4, \quad \hat{C}_6^5, \quad \hat{C}_6^6,$$

or

$$\hat{C}_6, \quad \hat{C}_3, \quad \hat{C}_2, \quad \hat{C}_3^2, \quad \hat{C}_6^5, \quad \hat{E}.$$

We see that the six operations of $C_6$ include the set of three operations $\hat{C}_3$, $\hat{C}_3^2$ and $\hat{E}$, associated with a $C_3$ axis, and two rotations $\hat{C}_2$, $\hat{E}$ around the $C_2$ axis. In other words, a molecule with a $C_6$ axis always has $C_3$ and $C_2$ axes. In the general case the presence of a $C_n$ axis implies the presence of axes $C_{n/p}$, where $p$ is a divisor of $n$.

It should be noted that a molecule with a $C_n$ axis must have $n$ identical atoms in a plane perpendicular to the axis (Figs 1.2a–d). It is clear that any number of atoms may be located *on* the symmetry axis.

A number of examples are given later of organic molecules and inorganic and coordination compounds having symmetry axes $C_2$, $C_3$, $C_4$ and $C_6$. Systems with $C_5$ axes are much less common: an example is the cyclopentadienyl anion $C_5H_5^-$. One example of a molecule with a $C_7$ axis is the tropilium ion $C_7H_7^+$ [13]. Finally, the ion $[C_8H_8]^{2-}$ has a $C_8$ axis; this is perhaps the only case of this type of symmetry [13].

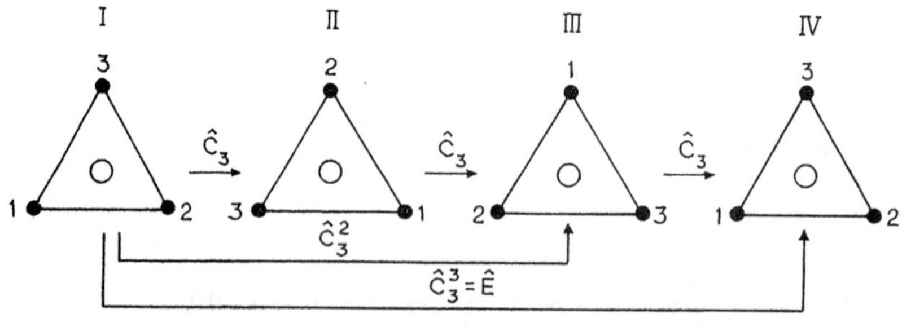


**Fig. 1.3** The action of the operation $\hat{C}_3^m$.

### 1.1.3 Reflection operation. Symmetry planes

Suppose that we have a point $P$ in space and a plane $\sigma$ (Fig. 1.4). We choose the coordinate system shown in Fig. 1.4. A *mirror reflection* in the plane $\sigma$ (called for simplicity a "*reflection*") is the operation $\hat{\sigma}$ that results in the point $P(x_0, y_0, z_0)$ being transformed into its mirror image $P'$. In this case the coordinates $x_0$ and $y_0$ remain unchanged, while $z_0$ changes its sign. This may be written as follows:

$$\hat{\sigma}P(x_0, y_0, z_0) = P'(x_0, y_0, -z_0).$$

If such an operation takes an object into self-coincidence, the object is said to possess a *symmetry plane.*

A symmetry plane, if its exists, always intersects the segments connecting the identical atoms at the mid-point, and is perpendicular to these segments. A planar molecule always has a symmetry plane. Examples include the chloroethylene molecule (Fig. 1.5a) and some other planar chlorosubstituted ethylenes $C_2H_{4-n}Cl_n$. As well as the axes, all the planar molecules referred to above also have symmetry planes (Figs 1.2a–d). It is obvious that the molecular plane need not be the only symmetry plane. Systems with symmetry planes $(\sigma, \sigma_1)$ and $(\sigma, \sigma_1, \sigma_2, \sigma_3)$, namely the water and $BH_3$ molecules, are shown in Figs 1.5(b, c). The FClSO molecule has no symmetry planes. Any linear molecule has an infinite number of planes containing the symmetry axis.

The reflection operation, when repeated, brings an object back to its initial configuration, so that $\hat{\sigma}^2 = \hat{E}$, $\hat{\sigma}^3 = \hat{\sigma}$ etc. The presence of a symmetry plane implies only one operation, $\hat{\sigma}$.

The existence of a symmetry plane imposes some restrictions on the number and position of atoms in a molecule: the number of identical atoms not located within the plane must be even, and a unique atom of any type must always lie in the symmetry plane.

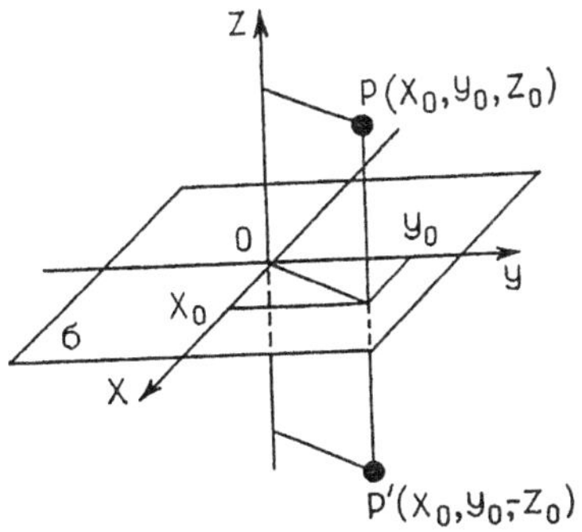


**Fig. 1.4** The mirror reflection operation.

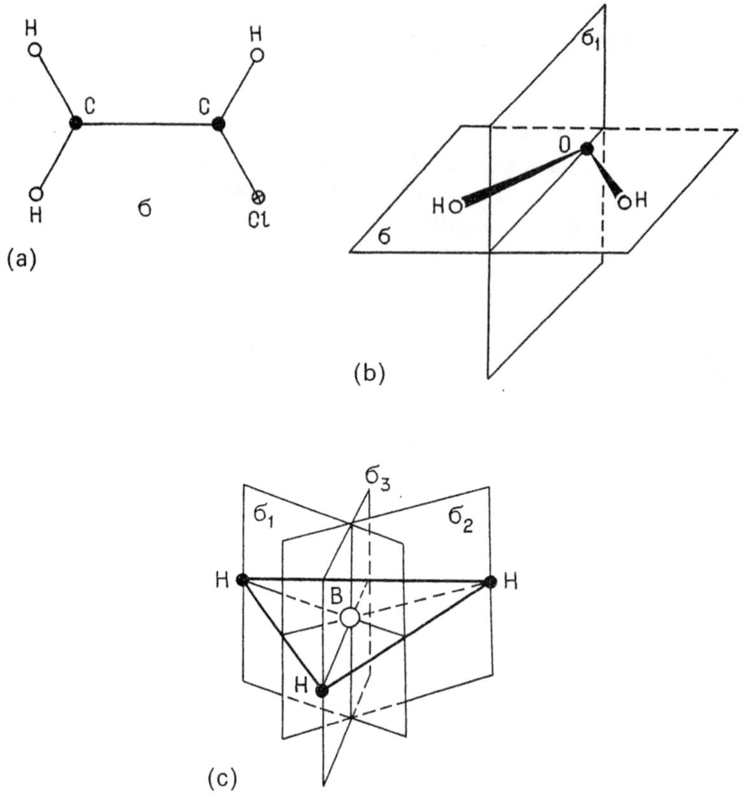


**Fig. 1.5** Symmetry planes of (a) chloroethylene, (b) water and (c) $BH_3$.

### 1.1.4 Improper rotation. Rotoflection axes

*Improper rotation* is an operation that can be achieved in two stages: a rotation $\hat{C}(\varphi)$ about the axis $C$ through a certain angle $\varphi$ combined with reflection $\hat{\sigma}$ in a plane perpendicular to the axis. The shift of a point $A$ due to the operation of improper rotation is shown in Fig. 1.6. In the first stage (the rotation $\hat{C}(\varphi)$) $A$ is shifted to the position $A'$; the reflection $\hat{\sigma}$ then takes it to the position $A_2$:

$$\hat{C}(\varphi)A_1 = A', \quad \hat{\sigma}A' = A_2.$$

If we change the sequence of the operations $\hat{C}$ and $\hat{\sigma}$, the result remains the same (Fig. 1.6):

$$\hat{\sigma}A_1 = A'', \quad \hat{C}(\varphi)A'' = A_2.$$

The operation of improper rotation through an angle $\varphi$ is denoted by $\hat{S}(\varphi)$. An improper rotation through an angle $2\pi/n$ is denoted by $\hat{S}_n$, and $S_n$ is a *rotoflection axis* of $n$th order.

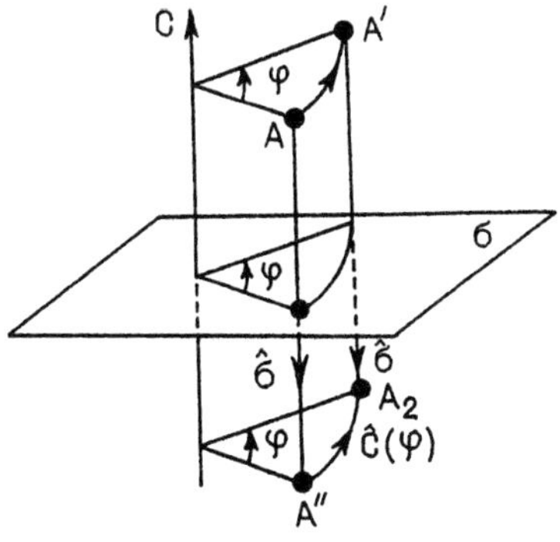


**Fig. 1.6** Improper rotation operation.

In those cases where the axis $C_n$ and the plane $\sigma$ are the symmetry elements, the molecule also has a rotoflection axis $S_n$. It should be emphasized, however, that $C_n$ and $\sigma$ *individually* may not be symmetry elements, even when the molecule has a rotoflection axis $S_n$.

**Example 1.5** Figure 1.7 shows the effect of the operation $S_n$ on the so-called "twisted" $C_2H_4$ ethylene configuration (where the $CH_2$ planes are perpendicular: Fig. 1.7a. Comparison of configurations II and III with I (Fig. 1.7b), shows that $\hat{C}_4$ and $\hat{\sigma}$ are not symmetry operations. Configurations IV and I are indistinguishable, so that the $C_2H_4$ molecule has a rotoflection axis $S_4$. It should be noted that Fig. 1.7 again shows that the final result (IV) does not depend on the sequence of the operations $\hat{C}_4$ and $\hat{\sigma}$

**Example 1.6** A tetrahedral complex of the type $[CuCl_4]^{2-}$ is shown in Fig. 1.8. A Cu ion is located centrally in the cube while ligands occupy the four non-adjacent vertices. Such a representation of a tetrahedron makes it clear that there are three axes $S_4$ intersecting at the origin and bisecting the opposite sides of the cube. It should be noted that the axes $C_4$ and planes $\sigma$ are not independently symmetry elements.

To denote the improper rotation axes $S_4$, the following symbols are used:

| | | | | |
|---|---|---|---|---|
| Axis: | $S_3$ | $S_4$ | $S_5$ | $S_6$ |
| Symbol: |  |  |  |  |

Figures 1.7 and 1.8 show that in these cases, as well as the improper rotation axes, there are also $C_2$ axes. This is the reason for the use of the particular geometric symbols for the $S_4, S_6, \ldots$ even-order axes.

As a rule, it is easy to understand when improper rotation is a new symmetry operation. It is clear that, in addition to $\hat{S}_n$, there are also symmetry operations $\hat{S}_n^2$, $\hat{S}_n^3$ etc., with

$$\hat{S}_n^m = \underbrace{\hat{C}_n\hat{\sigma}\ldots\hat{C}_n\sigma}_{m}.$$

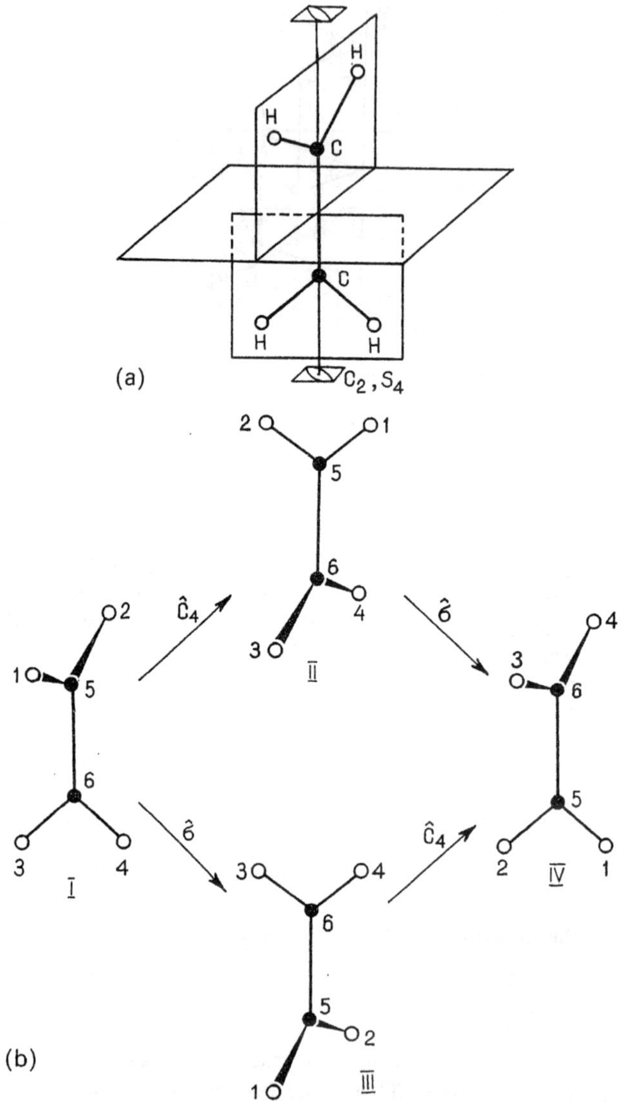


**Fig. 1.7** (a) Ethylene molecule. (b) The effect of the operation $\hat{S}_4$.

Again, this notation signifies that the operation $\hat{S}_n$ is repeated $m$ times. Since the order of rotations and reflections does not matter,

$$\hat{S}_n^m = \underbrace{\hat{C}_n \ldots \hat{C}_n}_{m \text{ times}} \underbrace{\hat{\sigma} \ldots \hat{\sigma}}_{m \text{ times}} = \hat{C}_n^m \cdot \underbrace{\hat{\sigma} \ldots \hat{\sigma}}_{m \text{ times}} \equiv \hat{C}_n^m \hat{\sigma}^m.$$

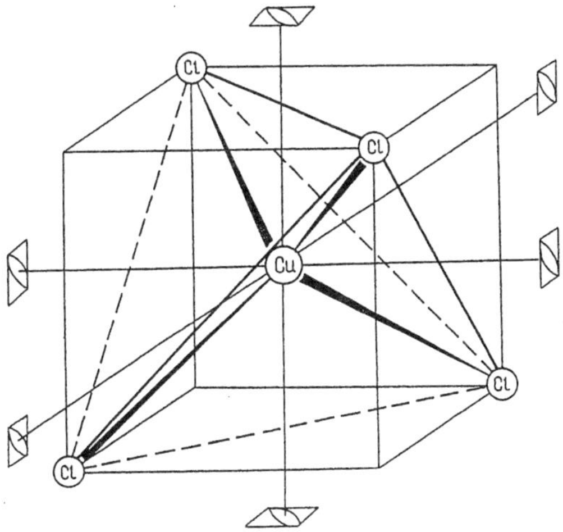


**Fig. 1.8** Improper rotation symmetry axes $S_4$ of the tetrahedral complex anion $[CuCl_4]^{2-}$.

Since an even number of reflections returns any point to its initial position,

$$\hat{\sigma}^m = \begin{cases} \hat{E} & \text{for } even \text{ m}, \\ \hat{\sigma} & \text{for } odd \text{ m}. \end{cases}$$

It follows that when $m$ is odd, $\hat{S}_n^n = \hat{C}_n^n\hat{\sigma} = \hat{E}\hat{\sigma} = \hat{\sigma}$ and the molecule has a symmetry plane $\sigma$. If the sequence of symmetry operations $\hat{S}_n^m$ is continued, we find $\hat{S}_n^{n+1}$:

$$\hat{S}_n^{n+1} = \hat{S}_n\hat{S}_n^n = \hat{S}_n\hat{C}_n^n\hat{\sigma}^n = \hat{S}_n\hat{E}\hat{\sigma} = \hat{S}_n\hat{\sigma} = \hat{C}_n\hat{\sigma}\hat{\sigma} = \hat{C}_n\hat{\sigma}^2 = \hat{C}_n\hat{E} = \hat{C}_n.$$

Consequently, an improper rotation axis of odd order implies the presence of a $C_n$ axis and a $\sigma$ symmetry plane.

The minimum number of rotations $\hat{S}_n$ by which the molecule returns to its initial configuration is $2n$:

$$\hat{S}_n^{2n} = \hat{C}_n^{2n}\hat{\sigma}^{2n} = \hat{E}.$$

Thus for odd $n$ an $S_n$ axis is associated with $2n$ mirror rotations. For example, an axis $S_3$ gives rise to six operations:

$$\hat{S}_3, \quad \hat{S}_3^2 = \hat{C}_3^2, \quad \hat{S}_3^3 = \hat{\sigma},$$
$$\hat{S}_3^4 = \hat{C}_3, \quad \hat{S}_3^5 = \hat{C}_3^2\hat{\sigma}, \quad \hat{S}_3^6 = \hat{E}.$$

Now, let us consider the $\hat{S}_n$ axes of even order. In this case $\hat{S}_n^n = \hat{E}$, while $C_n$ and $\sigma$ are not symmetry elements. For example, the $S_6$ axis gives rise to six operations:

$$\hat{S}_6, \quad \hat{S}_6^2 = \hat{C}_6^2\hat{\sigma}^2 = \hat{C}_3, \quad \hat{S}_6^3 = \hat{C}_6^3\hat{\sigma}^3 = \hat{C}_2\hat{\sigma} = \hat{S}_2,$$
$$\hat{S}_6^4 = \hat{C}_3^2, \quad \hat{S}_6^5, \quad \hat{S}_6^6 = \hat{E}.$$

It should be noted that the sequence obtained includes operations $\hat{C}_3$, $\hat{C}_3^2$ and $\hat{E}$ associated with the $C_3$ axis. This is the case for any even $n$: the existence of an $S_n$ axis also requires a $C_{n/2}$ axis.

### 1.1.5 Inversion

A very important particular case is the second-order rotoflection axis $S_2$. The transformation $\hat{S}_2 = \hat{C}_2\hat{\sigma}_2$ is shown in Fig. 1.9. The point $A_2$ lies on the straight line $A_1O$ intersecting the origin of coordinates and $A_2O = A_1O$, so that the operation $\hat{S}_2$ changes the signs of the coordinates:

$$\hat{S}_2 A_1(x_0, y_0, z_0) = A_2(-x_0, -y_0, -z_0).$$

The improper rotation $\hat{S}_2$ is called the *inversion* at the point $O$ (or, briefly

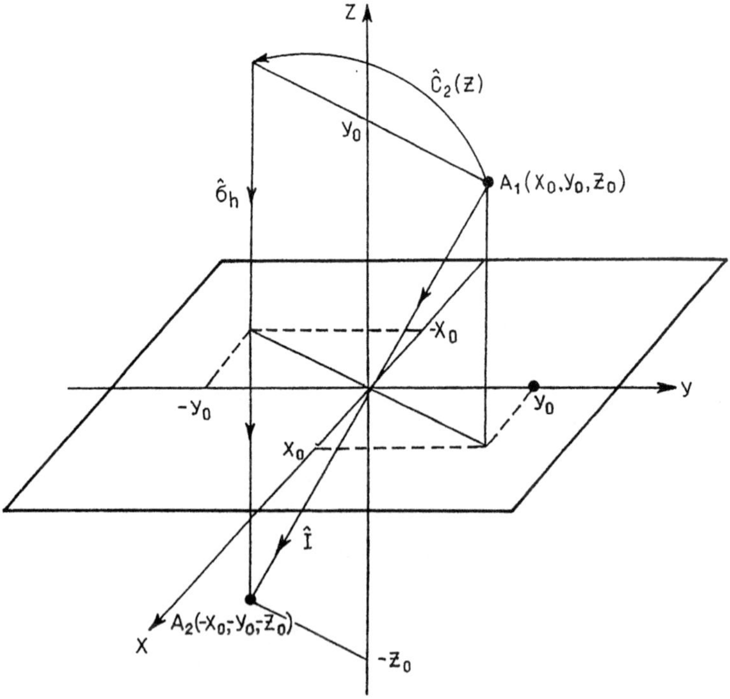


**Fig. 1.9** The inversion operation.

"*inversion*") and a special notation $\hat{I}$ is introduced for this operation (similarly, the $S_2$ axis is denoted by $I$). The point $O$ is called the *centre of inversion*. The complex anion $[PtCl_4]^{2+}$ (Fig. 1.2c) and the benzene molecule (Fig. 1.2d) are examples of systems with inversion symmetry. It is obvious that such systems have a common feature: they must contain an even number of atoms if there is no atom at the centre of inversion, and an odd number if there is such an atom.

### 1.1.6 Summary

The preceding sections have introduced all the symmetry elements and symmetry operations needed in discussing molecular symmetry. The main results are given in Table 1.1.

**Table 1.1** Four types of symmetry elements and operations.

| Symmetry elements | Symmetry operations |
|---|---|
| 1. Symmetry plane ($\sigma$) | Reflection within the plane ($\hat{\sigma}$) |
| 2. Rotation axis ($C_n$) of $n$th order | Rotations $\hat{C}_n^m$ |
| 3. Rotoflection axis $S_n$ | Improper rotations $\hat{S}_n^m$ |
| 4. Inversion centre $I$ | Inversion $\hat{I}$ |

## 1.2 MULTIPLICATION OF SYMMETRY OPERATIONS. COMMUTATIVITY

Since, in general, molecules have several symmetry elements, the question arises concerning the results of successive repetition of the corresponding operations. Let us assume $X$ and $Y$ to be two symmetry elements. We denote by $\hat{Y}\hat{X}$ the result of the successive action of the operations $\hat{X}$ and then $\hat{Y}$.† The notation $\hat{X}\hat{Y}$ represents the reverse sequence of operations: $\hat{Y}$ followed by $\hat{X}$. The operations $\hat{X}\hat{Y}$ and $\hat{Y}\hat{X}$ are called the *products* of the operations $\hat{X}$ and $\hat{Y}$, and the successive action of operations is called *multiplication*. Raising $\hat{X}$ to the $m$th power ($\hat{X}^m$) has a similar meaning, as anticipated in the discussion of rotation operations, where $\hat{C}_n^m = (\hat{C}_n)^m$. We have met other examples of multiplication: bearing in mind the definition, it is easily seen that an improper rotation $\hat{S}_n$ is a product of a rotation $\hat{C}_n$ and a reflection $\hat{\sigma}$. The resulting operation $\hat{S}_n$ is independent of the order of the factors:

$$\hat{C}_n\hat{\sigma} = \hat{\sigma}\hat{C}_n = \hat{S}_n.$$

† The convention of writing the first operation on the right is similar to that used in dealing with, for example, differential operators.

Let us now consider another example. We assume two planes $\sigma_1$ and $\sigma_2$, perpendicular to the plane of the page and intersecting at an angle $\varphi$. It can be seen from Fig. 1.10a that the operation $\hat{Z}_{21} = \hat{\sigma}_2\hat{\sigma}_1$ transforms $A$ into $A_2$:

$$\hat{\sigma}A = A_1, \quad \hat{\sigma}_2 A_1 = A_2, \quad \hat{\sigma}_2\hat{\sigma}_1 A = A_2.$$

The operation $Z_{12} = \hat{\sigma}_1\hat{\sigma}_2$ yields a different result: $\hat{\sigma}_1\hat{\sigma}_2 A = B_2$. This example shows that the result of multiplication may depend upon the order of the factors: $\hat{\sigma}_1\hat{\sigma}_2 \neq \hat{\sigma}_2\hat{\sigma}_1$. If $\hat{X}\hat{Y} = \hat{Y}\hat{X}$, the operations $\hat{X}$ and $\hat{Y}$ are said to be *commutative* or *commuting* but when $\hat{X}\hat{Y} \neq \hat{Y}\hat{X}$ they are said to be *non-commutative* or *non-commuting*. In the particular case of two perpendicular planes $\sigma_1$ and $\sigma_2$ ($\varphi = \frac{1}{2}\pi$), the operations $\hat{\sigma}_1$ and $\hat{\sigma}_2$ are commutative (Fig. 1.10b).

In a similar fashion, it is easy to establish all the possible pairs of commutative symmetry operations. They will be enumerated without proof:

(1) two rotations $\hat{C}(\varphi_1)$ and $\hat{C}_2(\varphi_2)$ around one axis:

$$\hat{C}(\varphi_1)\hat{C}(\varphi_2) = \hat{C}(\varphi_2)\hat{C}(\varphi_1) = \hat{C}(\varphi_1 + \varphi_2);$$

(2) two successive rotations around mutually perpendicular $C_2$ axes;

(3) two reflections in mutually perpendicular planes;

(4) rotation and reflection within a plane perpendicular to the rotation axis (a rotoflexive transformation);

(5) a rotation $\hat{C}(\varphi)$ around the $C$ axis and inversion through a point on the axis:

$$\hat{C}(\varphi)\hat{I} \equiv \hat{C}(\varphi)\hat{\sigma}\hat{C}_2 = \hat{I}\hat{C}(\varphi);$$

(6) reflection in a plane and inversion at a point on the plane.

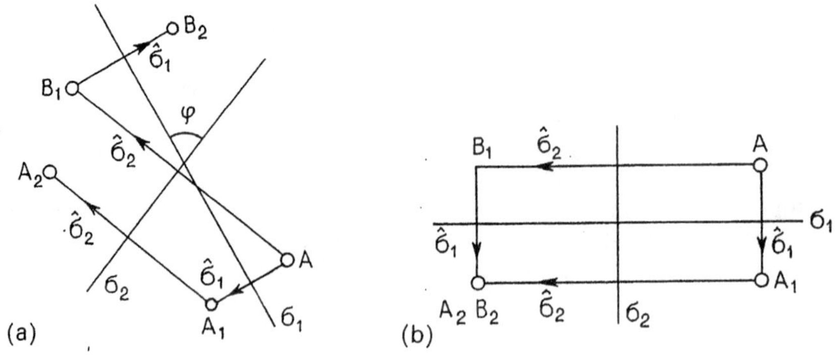


**Fig. 1.10** Reflection in two planes: (a) those intersecting at an angle $\varphi$; (b) those that are mutually perpendicular.

Let us now introduce the important notion of an *inverse* operation. The operation $\hat{Y}$ is an inverse operation to $\hat{X}$ if $\hat{Y}\hat{X} = \hat{E}$; it is denoted by $\hat{X}^{-1}$. For example, $\hat{C}_3^{-1} = \hat{C}_3^2$, since $\hat{C}_3^2\hat{C}_3 = \hat{C}_3^3 = \hat{E}$. It is easy to find the operation $(\hat{C}_n^m)^{-1}$ inverse to the rotation $\hat{C}_n^m$: $(\hat{C}_n^m)^{-1} = \hat{C}_n^{n-m}$, since $\hat{C}_n^{n-m}\hat{C}_n^m = \hat{C}_n^{n-m+m} = \hat{C}_n^n = \hat{E}$. Sometimes $\hat{X}^{-1} = \hat{X}$; an example is $\hat{X} = \hat{\sigma}$, for which $\hat{\sigma}^{-1} = \hat{\sigma}$, since $\hat{\sigma}\hat{\sigma} = \hat{E}$.

The following obvious property should be noted:

$$(\hat{X}\hat{Y})^{-1} = \hat{Y}^{-1}\hat{X}^{-1},$$

which is verified directly $(\hat{Y}^{-1}\hat{X}^{-1}\hat{X}\hat{Y} = \hat{E})$. In general, for any product of $n$ operations

$$(\hat{X}_1\hat{X}_2\ldots\hat{X}_n)^{-1} = \hat{X}_n^{-1}\ldots\hat{X}_2^{-1}\hat{X}_1^{-1}.$$

The above conclusions will be applied in the next two sections to establish the symmetry groups.

## 1.3 INTERRELATION BETWEEN SYMMETRY ELEMENTS

The examples given so far illustrate the properties of particular symmetry elements for specific molecular systems. We now turn to the problem of finding *all* symmetry elements for a given molecule. A direct approach to this problem consists in finding successively all those operations that bring the molecule into self-coincidence. There is, however, another approach based on the fact that many symmetry elements are interrelated. It is obvious that if $X$ and $Y$ are symmetry elements then, as well as the operations $\hat{X}$ and $\hat{Y}$, there are at least two other symmetry operations: $\hat{Z}_1 = \hat{Y}\hat{X}$ and $\hat{Z}_2 = \hat{X}\hat{Y}$. Some new symmetry elements may correspond to these operations. We have already discussed some particular cases when considering the improper rotations $\hat{S}_n^m$. Let us now extend this discussion.

### 1.3.1 Symmetry axes

We first consider the following example. Assume that a molecule has a $C_3(z)$ axis and a $C_2 \perp C_3$ axis within the plane $\sigma(xy)$ (Fig. 1.11a). The operation $\hat{C}_2$ takes the point $A$ into $A_1$. The point $A$ is located "above" the $xy$ plane (symbol ○), while $A_1$ is located "below" the plane (symbol ⊗). The operation $\hat{C}_3$ transforms $A_1$ into $A_2$; in other words, $\hat{C}_3(z)\hat{C}_2A = A_2$. It should be noted (Fig. 1.11a) that the rotation operation $\hat{C}_2'$ around the $C_2'$ axis, which makes an angle $\frac{2}{3}\pi$ to the $C_2$ axis, leads to the same result $\hat{C}_2'A = A_2$. In terms of operators, this can be expressed as:

$$\hat{C}_2' = \hat{C}_3(z)\hat{C}_2.$$

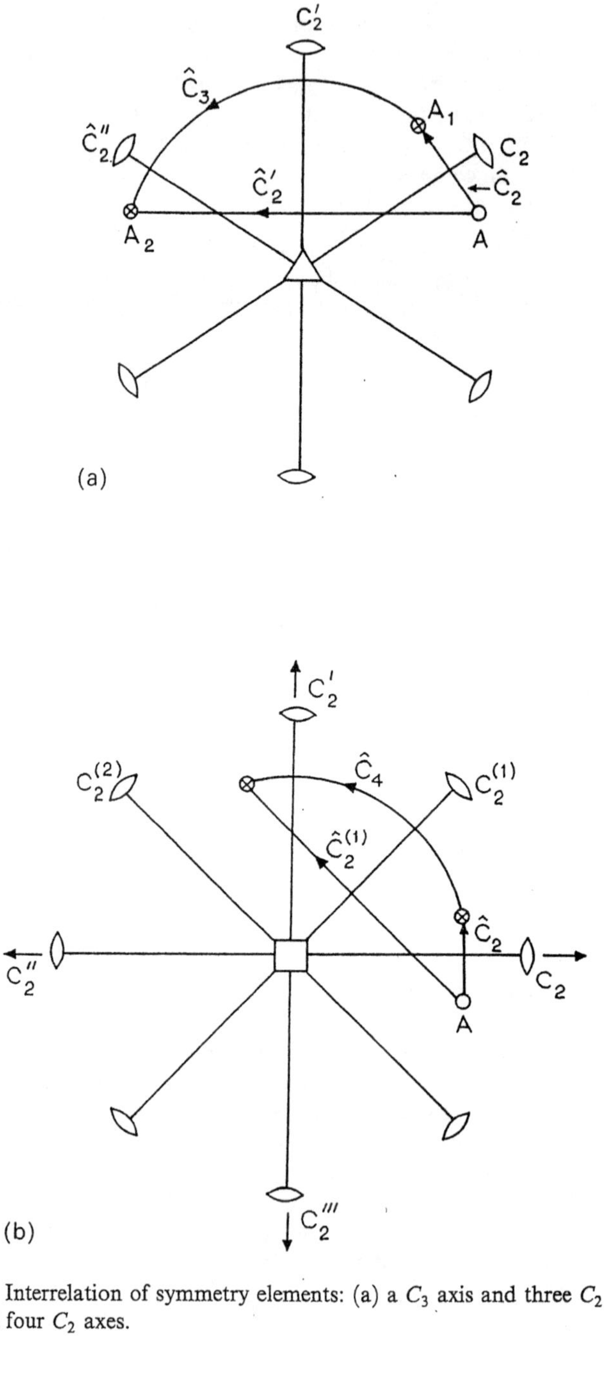


**Fig. 1.11** Interrelation of symmetry elements: (a) a $C_3$ axis and three $C_2$ axes; (b) a $C_4$ axis and four $C_2$ axes.

Expressed differently, the symmetry elements $C_3(z)$ and $C_2$ require an axis $C_2'$. The existence of a $C_2''$ axis is proved in the same way.

In the general case an $n$th order axis $C_n$ and a $C_2 \perp C_n$ axis give rise to $n-1$ axes $C_2$ in the plane $\sigma$ ($\sigma \perp C_n$), with the $C_2$ axes intersecting one another at angles $\pi/n$. Each of the $C_2$ axes can be obtained applying the operation $C_n$ to one of the others. Symbolically, this may be written as

$$\hat{C}_3 C_2 = C_2'', \quad \hat{C}_3 C_2' = C_2, \quad \hat{C}_3 C_2'' = C_2'.$$

Some explanation is required in the case of even $n$. Consider, for example, the $C_4(z)$ and $C_2 \perp C_4(z)$ axes (Fig. 1.11b). The rotation $\hat{C}_4(z)$ transforms $C_2$ into $C_2'$, and a second rotation $\hat{C}_4(z)$ transforms the $C_2$ axis into the position $C_2''$ with the opposite rotation direction (indicated by arrows in Fig. 1.11b). The axes $C_4$, $C_2$ and $C_2'$ do not comprise the complete set of symmetry elements. It is easy to see that the existence of $C_4$ and $C_2$ axes implies the existence of $C_2^{(1)}$ and $C_2^{(2)}$ axes. Indeed, it can be seen from Fig. 1.11b that $\hat{C}_4\hat{C}_2 = \hat{C}_2^{(1)}$. It is impossible to transform the sets of axes $C_2$, $C_2'$ and $C_2^{(1)}$, $C_2^{(2)}$ into one another by rotations $\hat{C}_4(z)$. This example illustrates a general assertion, whose importance will become clear later: when $n$ is *odd*, the operation $\hat{C}_n$ transforms *all* the $n$ horizontal axes $C_2$ into one another. When $n$ is *even*, the $C_2$ axes are divided into *two sets*; it is impossible to bring the $C_2$ axes belonging to different sets to coincidence with each other using the operations $\hat{C}_n^m$.

Thus not every set of symmetry elements can play the role of a set of molecular symmetry elements. For example, there are no molecules having only $C_3$ and $C_2 \perp C_3$ axes, since the presence of these axes implies that there are also two more $C_2$ axes.

### 1.3.2 Axes and planes

Let us consider some important cases of interrelated symmetry elements.

(1) A $C_n$ axis and a plane that intersects it result in the appearance of $n-1$ planes intersecting each other at angles $\pi/n$.

**Example 1.7** Consider a $C_3$ axis and a plane $\sigma_1$. It can be seen from Fig. 1.12a that $\hat{C}_3 A = A'$ and $\hat{\sigma}_1 A' = A''$. On the other hand, $\hat{\sigma}_2 A = A''$; hence $\hat{\sigma}_1 \hat{C}_3 = \hat{\sigma}_2$. This shows the existence of the symmetry plane $\sigma_2$. Similarly, we come to the conclusion that there is a plane $\sigma_3$. The planes $\sigma_1$, $\sigma_2$, $\sigma_3$ are brought into coincidence by rotations $\hat{C}_3$. This conclusion holds for a $C_n$ axis of any odd order $n$. When $n$ is even, the planes $\sigma$ are divided into two sets of different type, and it is impossible to bring into coincidence planes of different type by means of operations $\hat{C}_n^m$.

(2) Let us take a set consisting of a $C_n(z)$ axis and $n$ axes $C_2$ perpendicular to

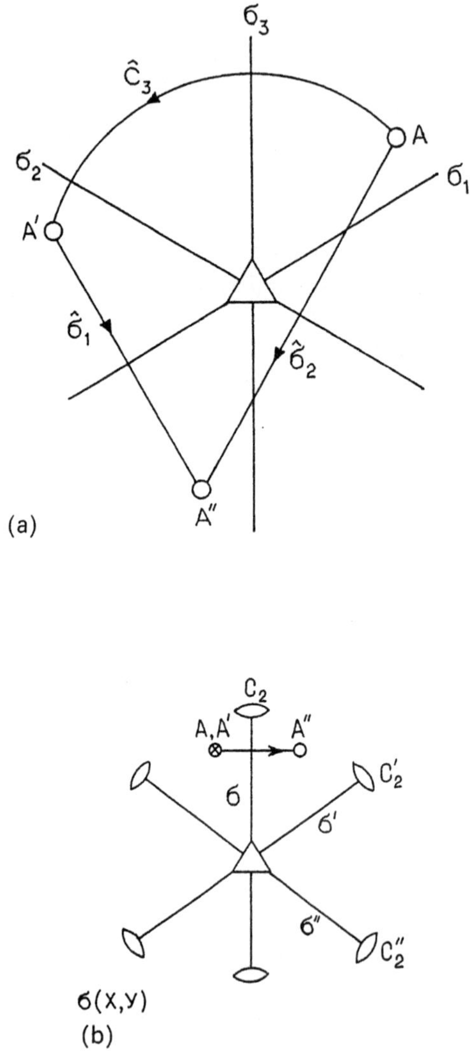


**Fig. 1.12** Symmetry axes and planes: (a) a $C_3(z)$ axis and three $\sigma$ planes passing through the $C_3$ axis; (b) a $C_3(z)$ axis, three $C_2$ ($C_2 \perp C_3$) axes and planes $\sigma$, $\sigma'$, $\sigma''$, $\sigma(xy)$.

$C_n$. Suppose we have established that a molecule has a horizontal $\sigma(xy)$ symmetry plane containing $C_2$ axes. Can all the above-mentioned symmetry elements constitute a complete set? To answer this question, we examine all possible products of the forms $\hat{C}_n\hat{\sigma}$ and $\hat{C}_2\hat{\sigma}$ and determine whether they include new operations. As an example, we take the case $n = 3$.

**Example 1.8** Consider a $C_3(z)$ axis, three $C_2$ axes and a plane $\sigma(xy)$. We shall examine the product $C_2\hat{\sigma}(xy)$. The operation $\hat{C}_2\hat{\sigma}(xy)$ transforms the point $A$ into $A''$ (Fig. 1.12b, where $A$ is "above" the $\sigma(xy)$ plane while $A'$ is "below" this plane). On the other hand, a reflection $\hat{\sigma}$ within the vertical plane $\sigma$, which comprises the $C_2$ and $C_3$ axes, leads to the same result:

$$\hat{\sigma}A = A'', \qquad \hat{\sigma} = \hat{C}_2\hat{\sigma}(xy).$$

The existence of $\sigma'$ and $\sigma''$ symmetry planes is proved in the same way. Thus the presence of a $C_n$ axis and a plane perpendicular to this axis and having $n$ axes $C_2$ implies the existence of $n$ extra $\sigma$ planes containing both the $C_n$ and $C_2$ axes.

The above results are important when determining the full set of symmetry elements for any given molecule.

## 1.4 DEFINITION OF A GROUP

In mathematics a *group* is defined as a set of elements satisfying the following four requirements.

(1) There is a law of combination, called *multiplication*, such that, given any two elements $\hat{X}$ and $\hat{Y}$ of the set, there is a third element $\hat{Z} = \hat{X}\hat{Y}$ that also belongs to the set.

It is essential that the element $\hat{Z}$ belong to the same set; in other words, the multiplication of two elements must not give an object that is not contained in the set. To prevent misunderstanding, it is necessary to be clear about the meaning of the terms being used here. A group is an abstract mathematical concept, and its "elements" may have a very general interpretation. For example, they may be integers (both positive and negative, including zero), and the group "multiplication" may correspond to simple addition. It has been shown that the set of symmetry operations for a given molecule satisfies this requirement when "multiplication" is taken to be the successive application of two symmetry operations. The symmetry operations are then the group elements (note that the term "elements" used here should not be confused with the geometric symmetry elements such as axes and planes).

(2) The set contains a unit element $\hat{E}$ such that $\hat{E}\hat{X} = \hat{X}\hat{E}$ for any $\hat{X}$ contained in the set.

In a molecular symmetry group this is the identity transformation $\hat{E}$.

(3) The law of combination (1) for the elements of the set is *associative*; that is, for any $\hat{X}$, $\hat{Y}$, $\hat{Z}$ in the set

$$\hat{Z}(\hat{Y}\hat{X}) = (\hat{Z}\hat{Y})\hat{X}.$$

More generally,

$$\hat{X}_1\hat{X}_2\ldots\hat{X}_n = (\hat{X}_1\hat{X}_2)(\hat{X}_3\hat{X}_4)\hat{X}_5\ldots\hat{X}_n = (\hat{X}_1\hat{X}_2\hat{X}_3)\hat{X}_4\ldots\hat{X}_n \quad \text{etc.}$$

(4) Any element $\hat{X}$ has an *inverse* $\hat{X}^{-1} = \hat{Y}$, such that

$$\hat{X}\hat{Y} = \hat{Y}\hat{X} = \hat{E}.$$

All the symmetry operations discussed so far satisfy this requirement ($\hat{\sigma}^{-1} = \hat{\sigma}$, $\hat{C}_3^{-1} = \hat{C}_3^2$ etc.).

In this book we are concerned only with the *symmetry groups*, whose elements are *symmetry operations*. In this context, the following definitions should be noted:

(i) The number of elements in the group is called the *group order*. There are *finite* and *infinite* groups. The order of the symmetry group of a nonlinear molecule is finite. The order of the symmetry group of a linear molecule is infinite, since its elements include the operations of rotation through any angle about the molecular axis.

(ii) If all the group operations are commutative, the group is called an *Abelian group*. An example is the group of $\hat{C}_n^m$ rotations, whose commutativity is obvious:

$$\hat{C}_n^m\hat{C}_n^k = \hat{C}_n^k\hat{C}_n^m = \hat{C}_n^{m+k}.$$

(iii) a *cyclic group* is one whose elements are obtained by raising one operation to various powers: $\hat{X}, \hat{X}^2, \hat{X}^3, \ldots \hat{X}^n$. If a group is finite then, for some $n$, $\hat{X}^n = \hat{E}$, as is the case for the group of rotations (where $\hat{C}_n^n = \hat{E}$). Any cyclic group is an Abelian group, since $\hat{X}^m$ and $\hat{X}^k$ are commutative. The order of a cyclic group is called its *period* of operation, and is denoted by $\{X\}$.

(iv) A *subgroup* is a subset of the elements of the given group that itself forms a group. The cyclic group $\hat{X}, \hat{X}^2, \ldots, \hat{X}^n$ is a commonly occurring subgroup (of order $n$). Another group that often appears as a subgroup comprises a reflection $\hat{\sigma}$ and the identity $\hat{E}$ ($= \hat{\sigma}^2$); clearly it is of second order.

## PROBLEMS

**1.1** Find the symmetry planes of the tetrahedral molecule in Fig. 1.8.

**1.2** Find the product of the operations $\hat{C}_6$ and $\hat{C}_2$ (Fig. 1.2d).

# 2 Point Groups and Their Classes

## 2.1 EQUIVALENT SYMMETRY ELEMENTS AND ATOMS

In Section 1.3 it was shown that under the influence of symmetry operations some symmetry elements coincide with each other. For example, under the action of $\hat{C}_3$ ($C_3 \perp C_2$) operations three $C_2$ axes are transformed into each other. If a symmetry element $A$ is transformed into $B$ by means of an operation $\hat{X}$ belonging to the same group as the operations $\hat{A}$ and $\hat{B}$ then it is said that the element $A$ is *conjugated* to the element $B$:$B = \hat{X}A$.

It is obvious that $B$ is also conjugated to $A$, since $\hat{X}^{-1}B = A$, and according to the definition in Section 1.4, $\hat{X}^{-1}$ belongs to the same group. The conjugation is therefore *mutual*. Mutually conjugate symmetry elements are called *equivalent*.

Suppose we have three symmetry elements: $A$, $B$ and $C$. Let $A$ be conjugate to $B$ and $C$: $\hat{X}A = B$, $\hat{Y}A = C$, where $\hat{X}$ and $\hat{Y}$ are group operations. Then the operation $\hat{Z} = \hat{Y}\hat{X}^{-1}$ from the same group transforms $B$ into $C$:

$$\hat{Z}B = \hat{Y}\hat{X}^{-1}B = \hat{Y}A = C.$$

Therefore all three symmetry elements $A$, $B$ and $C$ are conjugate with each other in pairs. Equivalent symmetry elements can thus be collected together in sets.

**Example 2.1** The three $C_2$ axes located within the plane $\sigma(xy)$ and intersecting at the angles of $\frac{2}{3}\pi$ are equivalent, if there is a $C_3(z)$ axis (Fig. 1.2b).

**Example 2.2** The four $C_2$ axes (Fig. 1.2c) are divided into two sets of equivalent axes if there exists a $C_4$ axis:

$$\{C_2, C_2'\} \quad \text{and} \quad \{C_2^{(1)}, C_2^{(2)}\}.$$

**Example 2.3** The symmetry axes of the octahedral complex $MX_6$ are shown in Fig. 2.1. A metal ion M is located at the origin of the coordinate system, and the ligands X

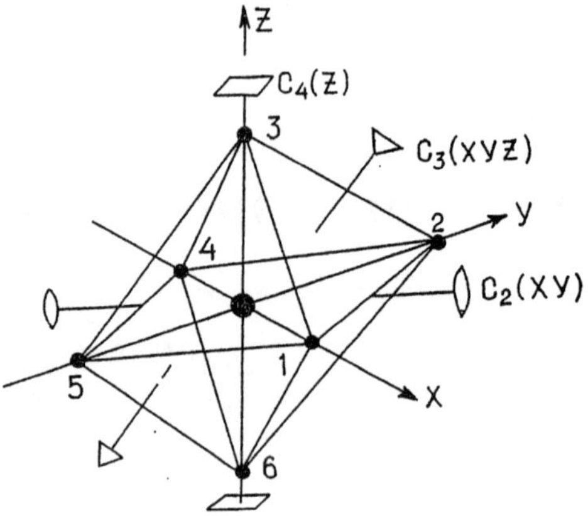


**Fig. 2.1** The symmetry axes of an octahedral molecule (one axis of each type is represented).

are located on the axes. The three $C_4$ axes coincide with the [100], [010] and [001] directions, four $C_3$ axes coincide with the [111], [$\bar{1}$11], [1$\bar{1}$1] and [11$\bar{1}$] directions, and six $C_2$ axes are located along the [110], [1$\bar{1}$0], [101], [10$\bar{1}$], [011] and [01$\bar{1}$] directions (Table 2.1). The $C_4$ axes are equivalent, since they coincide with each other through rotations $\hat{C}_3(xyz) : \hat{C}_3(xyz)C_4(x) = C_4(y)$ etc. The $C_3$ axes are also equivalent, since they are transformed into each other by operations $\hat{C}_4$ $(C_4(z)C_3(xyz) = C_3(\bar{x}yz)$ etc.). The equivalence of the six $C_2$ axes is proved similarly.

**Table 2.1** Symmetry axes of the octahedral complex $MX_6$ ($x, y, z$ are the positive and $\bar{x}, \bar{y}, \bar{z}$ the negative directions of the coordinate axes).

| Axes | Directions |
|---|---|
| $3C_4$ | $C_4(x)$, $C_4(y)$, $C_4(z)$ |
| $4C_3$ | $C_3(xyz)$, $C_3(\bar{x}yz)$, $C_3(x\bar{y}z)$, $C_3(xy\bar{z})$ |
| $6C_2$ | $C_2(xy)$, $C_2(yz)$, $C_2(zx)$, $C_2(\bar{x}y)$, $C_2(\bar{y}z)$, $C_2(\bar{z}x)$ |

Atoms of the same kind, that can be brought into coincidence with each other by symmetry operations, are said to be *equivalent*.

**Example 2.4** In a $PF_5$ molecule (Fig. 2.2), the equatorial fluorine atoms constitute a set of equivalent atoms: they can be transformed into each other by reflection in the symmetry planes.

Thus a set of symmetry elements of any molecule can be divided into sets of equivalent elements, and the atoms comprising the molecule can be divided into sets of equivalent atoms.

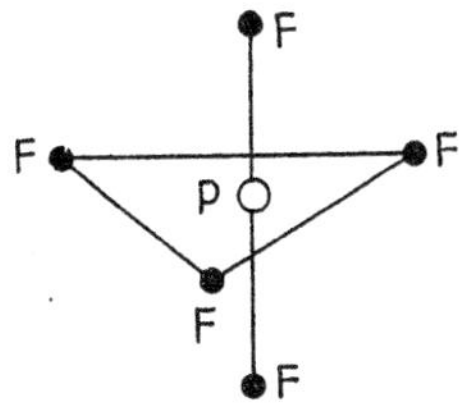


**Fig. 2.2** Equivalent and inequivalent atoms in the $PF_5$ molecule.

## 2.2 CLASSES OF CONJUGATE SYMMETRY OPERATIONS

Suppose $\hat{A}$ and $\hat{B}$ are two operations of a symmetry group. We form all possible products of the type $\hat{C}\hat{B}\hat{C}^{-1}$, where $\hat{C}$ is an operation of the same group. If we can find a $\hat{C}$ for which $\hat{A} = \hat{C}\hat{B}\hat{C}^{-1}$ then the operation $\hat{A}$ is said to be *conjugate* to $\hat{B}$. The use of the same terminology as for the symmetry elements (Section 2.1) is deliberate. We shall show below that conjugate symmetry elements give rise to conjugate operations. It is clear that $\hat{B}$ is also conjugated with $\hat{A}$, since

$$\hat{B} = \hat{C}^{-1}\hat{A}\hat{C} = \hat{D}\hat{A}\hat{D}^{-1},$$

where $\hat{D} = \hat{C}^{-1}$. Mutually conjugate operations have the following properties:

(1) Each operation is self-conjugate ($\hat{A} = \hat{B}$ for $\hat{C} = \hat{E}$).

(2) If $\hat{A}$ is conjugate to $B$, and $\hat{B}$ is conjugate to $\hat{C}$, then $\hat{A}$ is conjugate to $\hat{C}$.

This latter property makes it possible to speak about sets of mutually conjugate group operations. The complete collections of mutually conjugate group operations are called *classes*.

(3) Conjugate operations have the same orders. Let us assume that the order of the operation $\hat{B}$ is equal to $n$ (that is, $\hat{B}^n = \hat{E}$). Then

$$\hat{A}^n = (\hat{C}\hat{B}\hat{C}^{-1})^n = \underbrace{\hat{C}\hat{B}\hat{C}^{-1}\hat{C}\hat{B}\hat{C}^{-1}\ldots\hat{C}\hat{B}\hat{C}^{-1}}_{n \text{ times}} = \hat{C}\hat{E}\hat{C}^{-1} = \hat{E}.$$

The fact that $\hat{A}^n = \hat{E}$ implies that the orders of $\hat{A}$ and $\hat{B}$ are equal.

To split the set of group operations into classes, we take any operation $\hat{A}$ and form all possible products of the form $\hat{X}\hat{A}\hat{X}^{-1}$; we then repeat this for another operation $\hat{B}$. Let us illustrate how this rule works for a specific symmetry group.

**Example 2.5** We consider the following group of operations:

$$\hat{E}, \quad \hat{C}_4, \quad \hat{C}_4^2 = \hat{C}_2, \quad \hat{C}_4^3, \quad 2\hat{\sigma}_v, \quad 2\hat{\sigma}_d.$$

This is known as the $\mathbf{C}_{4v}$ group; its symmetry elements (Fig. 2.3) are the $C_4$ axis

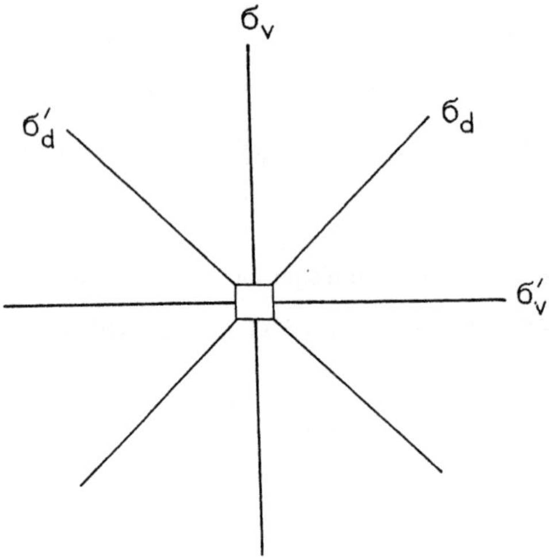


**Fig. 2.3** The symmetry elements of $\mathbf{C}_{4v}$ (projection onto a plane perpendicular to the $C_4$ axis).

and four vertical planes, $2\sigma_v$ and $2\sigma_d$ (the diagonals). In defining classes, it is convenient (using the methods given in Section 1.3) to produce a multiplication table for the group operations, as in Table 2.2, which is for the $\mathbf{C}_{4v}$ group. (Note that the entries in the table are $YX$; thus $\hat{\sigma}_d\hat{\sigma}_v = \hat{C}_4^3, \hat{\sigma}_v\hat{\sigma}_d = \hat{C}_4$ etc.).

The identity operation produces a class of first order. Let us take a $\hat{C}_4$ operation and obtain all the products $\hat{X}\hat{C}_4\hat{X}^{-1}$:

$$\hat{E}\hat{C}_4\hat{E}^{-1} = \hat{C}_4,$$

$$\hat{C}_4\hat{C}_4\hat{C}_4^{-1} = \hat{C}_4\hat{C}_4\hat{C}_4^3 = \hat{C}_4,$$

**Table 2.2** Multiplication table for the $\mathbf{C}_{4v}$ group.

| $\hat{Y}$ \ $\hat{X}$ | $\hat{E}$ | $\hat{C}_4$ | $\hat{C}_2$ | $\hat{C}_4^3$ | $\hat{\sigma}_v$ | $\hat{\sigma}'_v$ | $\hat{\sigma}_d$ | $\hat{\sigma}'_d$ |
|---|---|---|---|---|---|---|---|---|
| $\hat{E}$ | $\hat{E}$ | $\hat{C}_4$ | $\hat{C}_2$ | $\hat{C}_4^3$ | $\hat{\sigma}_v$ | $\hat{\sigma}'_v$ | $\hat{\sigma}_d$ | $\hat{\sigma}'_d$ |
| $\hat{C}_4$ | $\hat{C}_4$ | $\hat{C}_2$ | $\hat{C}_4^3$ | $\hat{E}$ | $\sigma'_d$ | $\hat{\sigma}_d$ | $\sigma_v$ | $\sigma'_v$ |
| $\hat{C}_2$ | $\hat{C}_2$ | $\hat{C}_4^3$ | $\hat{E}$ | $\hat{C}_4$ | $\hat{\sigma}'_v$ | $\hat{\sigma}_v$ | $\hat{\sigma}'_d$ | $\hat{\sigma}_d$ |
| $\hat{C}_4^3$ | $\hat{C}_4^3$ | $\hat{E}$ | $\hat{C}_4$ | $\hat{C}_2$ | $\hat{\sigma}_d$ | $\hat{\sigma}'_d$ | $\hat{\sigma}'_v$ | $\hat{\sigma}_v$ |
| $\hat{\sigma}_v$ | $\hat{\sigma}_v$ | $\hat{\sigma}_d$ | $\hat{\sigma}'_v$ | $\hat{\sigma}'_d$ | $\hat{E}$ | $\hat{C}_2$ | $\hat{C}_4$ | $\hat{C}_4^3$ |
| $\hat{\sigma}'_v$ | $\hat{\sigma}'_v$ | $\hat{\sigma}'_d$ | $\hat{\sigma}_v$ | $\hat{\sigma}_d$ | $\hat{C}_2$ | $\hat{E}$ | $\hat{C}_4^3$ | $\hat{C}_4$ |
| $\hat{\sigma}_d$ | $\hat{\sigma}_d$ | $\hat{\sigma}'_v$ | $\hat{\sigma}'_d$ | $\hat{\sigma}_v$ | $\hat{C}_4^3$ | $\hat{C}_4$ | $\hat{E}$ | $\hat{C}_2$ |
| $\hat{\sigma}'_d$ | $\sigma'_d$ | $\hat{\sigma}_v$ | $\hat{\sigma}_d$ | $\hat{\sigma}'_v$ | $\hat{C}_4$ | $\hat{C}_4^3$ | $\hat{C}_2$ | $\hat{E}$ |

$$\hat{C}_2\hat{C}_4\hat{C}_2^{-1} = \hat{C}_4,$$
$$\hat{C}_4^3\hat{C}_4(\hat{C}_4^3)^{-1} = \hat{C}_4,$$
$$\hat{\sigma}_v\hat{C}_4(\hat{\sigma}_v)^{-1} = \hat{\sigma}_v\hat{C}_4\hat{\sigma}_v = \hat{\sigma}_v\hat{\sigma}_d' = \hat{C}_4^3,$$
$$\hat{\sigma}_v'\hat{C}_4(\hat{\sigma}_v')^{-1} = \hat{C}_4^3,$$
$$\hat{\sigma}_d\hat{C}_4(\hat{\sigma}_d)^{-1} = \hat{C}_4^3,$$
$$\hat{\sigma}_d'\hat{C}_4(\hat{\sigma}_d')^{-1} = \hat{C}_4^3.$$

Thus the operations $\hat{C}_4$ and $\hat{C}_4^3$ form a class of second order. Continuing this process, we obtain the following classes:

$$\hat{E};\quad \hat{C}_4,\ \hat{C}_4^3;\quad \hat{C}_2;\quad \hat{\sigma}_v,\ \hat{\sigma}_v';\quad \hat{\sigma}_d,\ \hat{\sigma}_d'.$$

In an Abelian group each operation forms a class. It should be emphasized that a class is not a subgroup.

Finally it should be pointed out that the full collection of the operations of a group may be generated from a smaller number of operations by means of the successive multiplication procedure. In the case of the $\mathbf{C}_{4v}$ group one can see that successive multiplication of the two operations, $\hat{C}_4$ and $\hat{\sigma}_v$, gives rise to all eight operations. Using Table 2.2 we can obtain the following products:

$$\hat{\sigma}_v^2 = \hat{E},\qquad \hat{C}_4\hat{\sigma}_v = \hat{\sigma}_d,$$
$$\hat{\sigma}_d\hat{\sigma}_v = \hat{C}_4^3,\qquad \hat{\sigma}_d\hat{\sigma}_d' = \hat{C}_2,$$
$$\hat{\sigma}_v\hat{C}_4^3 = \hat{\sigma}_d',\qquad \hat{\sigma}_v\hat{C}_2 = \hat{\sigma}_v'.$$

The set of operations obtained in such a way exhausts the $\mathbf{C}_{4v}$ group. The operations whose successive multiplications generate the full set of group operations constitute a set of *generating operations* or *generators of the group.*

The two rotations $\hat{C}_4(y)$ and $\hat{C}_4(z)$ generate all twenty-four operations of the **O** group. For example

$$\hat{C}_4(y)\hat{C}_4(z) = \hat{C}_3(xyz),$$
$$\hat{C}_4(z)\hat{C}_4(y) = \hat{C}_3(\bar{x}yz),\ \text{etc.}$$

Generally there are several ways of choosing the generators. In the case of **O** group this may be for example $\hat{C}_3(xyz)$ and $\hat{C}_4(z)$.

## 2.3 RULES FOR ESTABLISHING CLASSES

The method discussed in the previous section is applicable to all groups. For symmetry groups there is a simpler and easy-to-grasp rule for forming the classes, which is always applied. In order to determine the operations $\hat{B} = \hat{X}\hat{A}\hat{X}^{-1}$, it is not necessary to consider all operations $\hat{X}$. The operation $\hat{X}$ can be found according to the following rule:

$\hat{X}$ is an operation that transforms the symmetry element $A$ into $B$.

Let us check this rule, using the group $\mathbf{C}_{4v}$ as an example. Two reflections $\hat{\sigma}_v$ and $\hat{\sigma}'_v$ constitute a class, since

$$\hat{C}_4\hat{\sigma}_v\hat{C}_4^{-1} = \hat{\sigma}'_v.$$

On the other hand, the relation $\hat{C}_4\sigma_v = \sigma'_v$ holds for planes $\sigma_v$ and $\sigma'_v$. Therefore, if symmetry planes are brought into coincidence by rotation, both reflections belong to the same class. Since there is no operation in the $\mathbf{C}_{4v}$ group taking the planes $\sigma_v$ and $\sigma_d$ into coincidence, reflections in these planes belong to different classes. There are similar rules for rotational transformations. Rotations through the same angle about equivalent axes will belong to the same class (the order of conjugate operations being the same).

If

$$\hat{X}C(1) = C(2)$$

then

$$\hat{C}(2) = \hat{X}\hat{C}(1)\hat{X}^{-1},$$

where $\hat{C}(1)$ and $\hat{C}(2)$ are rotations about the 1 and 2 axes through an angle $\varphi$. In order to verify that this is true, we apply the operation $\hat{X}\hat{C}(1)\hat{X}^{-1}$ to the $C(2)$ axis, i.e. we transform all points lying on the axis: $\hat{X}\hat{C}(1)\hat{X}^{-1}C(2)$. The operation $\hat{X}^{-1}$ transforms the $C(2)$ axis into $C(1)$, the operation $\hat{C}(1)$ is a rotation about $C(1)$, and finally $\hat{X}$ transforms the $C(1)$ axis into $C(2)$. Therefore atoms situated on the $C(2)$ axis retain their initial positions. Hence $\hat{X}\hat{C}(1)\hat{X}^{-1}$ is a rotation about the $C(2)$ axis through the same angle $\varphi$. We have thus shown that two rotations through the same angle about equivalent axes belong to the same class.

The problem of dividing the group operations into classes thus reduces to establishing the equivalent symmetry elements. These equivalent elements are related to operations belonging to the same class. The rotations $\hat{C}_n^m$ and $\hat{C}_n^{-m}$ through the same angle $(2\pi/n)m$ in different directions require a special treatment. These rotations belong to the same class only if the group contains an operation that reverses the axis direction, i.e. the sense of rotation. Such operations are a reflection $\hat{\sigma}_v$ in the plane containing the $C_n$ axis, and a rotation $\hat{C}_2$ about the perpendicular $C_2$ axis. In other words, for rotations and reflections of the above types the following relationships hold:

$$\hat{C}_n^{-m} = \hat{\sigma}_v\hat{C}_n^m\hat{\sigma}_v^{-1}, \qquad \hat{C}_n^{-m} = \hat{C}_2\hat{C}_n^m\hat{C}_2^{-1}.$$

The first of these relationships is illustrated by Fig. 2.4. The reflection $\hat{\sigma}_v^{-1}$, the rotation $\hat{C}_6$ and the reflection $\hat{\sigma}_v$ transform the point $A$ into $A_1$, $A_2$ and $A_3$

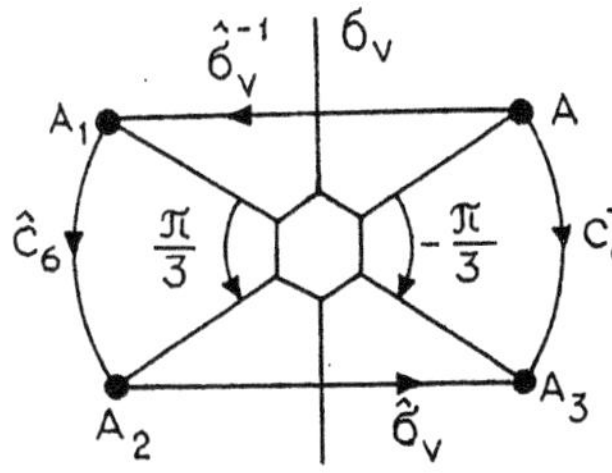


**Fig. 2.4** Conjugate $\hat{C}_6$ and $\hat{C}_6^{-1}$ rotations.

consecutively, which is achieved in one step by the operation $\hat{C}_6^{-1}$. It should be noted that the reflection $\hat{\sigma}_h$ in the horizontal plane ($\sigma_h \perp C_6$) changes simultaneously both the axis direction and the rotation direction. Hence the existence of the $\sigma_h$ plane does not ensure conjugation of the rotations $\hat{C}_n^m$ and $\hat{C}_n^{-m}$.

The above arguments make it possible to formulate the following rules for dividing symmetry groups into classes:

(1) Two rotations through the same angle about non-coincident axes belong to the same class provided that the group includes a symmetry operation that takes the axes into coincidence.

(2) Two rotations $\hat{C}_n^m$ and $\hat{C}_n^{-m}$ about the same axis belong to the same class provided that there is a vertical $\sigma_v$ plane or a horizontal $C_2$ axis. In this case the $C_n$ axis is called a *two-sided axis*.

(3) Two reflections belong to the same class if the group includes an operation that takes the planes into coincidence.

(4) The operations $\hat{E}$ and $\hat{I}$ constitute classes ($\hat{I}$ being the inversion).

(5) There is a simple rule for establishing the classes for groups that include inversion. Suppose **G** is a group containing no inversion, for example one consisting of three operations $\hat{E}$, $\hat{X}$ and $\hat{Y}$, while $\mathbf{G}_i$ is a group consisting of two operations $\hat{E}$ and $\hat{I}$. Then the set of six operations $\hat{E}\hat{E} = \hat{E}$, $\hat{X}\hat{E} = \hat{X}$, $\hat{Y}\hat{I} = \hat{Y}$, $\hat{E}\hat{I} = \hat{I}$, $\hat{X}\hat{I}$, $\hat{Y}\hat{I}$ is called the *direct product* of the groups **G** and $\mathbf{G}_i$ and is denoted by $\mathbf{G} \times \mathbf{G}_i$. To each class of the group **G** there correspond two classes of the direct product group.

The above properties of symmetry operations and rules for establishing classes allow the construction of symmetry groups and their classes. From the examples given, one may note a common feature of the symmetry transformations. All these transformations, as well as their products, leave one point of the molecule unchanged (note that it is not necessary that an atom occupy this point). In the $[PtCl_4]^{2-}$ complex this point is the platinum ion (Fig. 1.2c) and in the water molecule it is the oxygen atom (Fig. 1.5b), while in benzene (Fig. 1.2d) and $BH_3$ (Fig. 1.2b) the fixed points are the centres of the hexagon and the triangle respectively. This is a feature common to all molecular symmetry groups. (If such a point did not exist, the product of symmetry operations would result in a translational movement of the molecule.) Molecular symmetry groups are therefore called *point groups*.

In order to use group theory in chemical and spectroscopic applications, it is necessary first of all to determine the symmetry group of the object being investigated (a molecule, a complex or an impurity centre in a crystal). The following sections therefore deal with elements and symmetry operations of point groups.

## 2.4 POINT GROUPS

### 2.4.1 The rotation groups $\mathbf{C}_n$

The $\mathbf{C}_n$ group consists of rotations $\hat{C}_n^m$ about the $n$th order $C_n$ axis. Molecules with $\mathbf{C}_n$ symmetry have no other symmetry elements (such as planes, other axes or an inversion centre). The group $\mathbf{C}_n$ is a cyclic group (Section 1.4), and all operations $\hat{C}_n^m$ are commutative; therefore each of these groups constitutes a class. (Note that henceforth point groups will be denoted by bold sans serif capital letters, **C**, **D** etc.)

*The group* $\mathbf{C}_1$ This contains only the identity operation $\hat{C}_1 = \hat{E}$. This symmetry is of the lowest possible order; in other words, a molecule belonging to the $C_1$ group has no symmetry at all.

**Example 2.6** The FCl SO molecule (Fig. 1.2e).

**Example 2.7** The asymmetric conformation of the five-membered ring envelope of the chelate shown in Fig. 2.5(a) [30].

**Example 2.8** The asymmetric conformation of the bath form of six-membered ring in the chelate is shown in Fig. 2.5(b) [30].

**Example 2.9** Methy ethyl ketone (Fig. 2.5c).

**Example 2.10** Two enantiomeric forms of the *cis*-isomer of $MA_2B_2C_2$, where M is a metal ion and A, B, C are ligands (Figs. 2.5d, e). Such systems include the complexes $Pt(NH_3)_2(NO_2)_2Cl_2$, *cis*-$[Coen(NH_3)_2Cl_2]^{2+}$, $Pten(NH_3)_2Cl_2]^{2-}$ and *cis*-$[PtenCl_2(NH_3)py]^{2+}$.

*The group* $\mathbf{C}_2$ This consists of two operations, $\hat{C}_2$ and $\hat{E}$, constituting two classes.

**Example 2.11** The symmetrically tapered five-membered ring of the chelate shown in Fig. 2.6a.

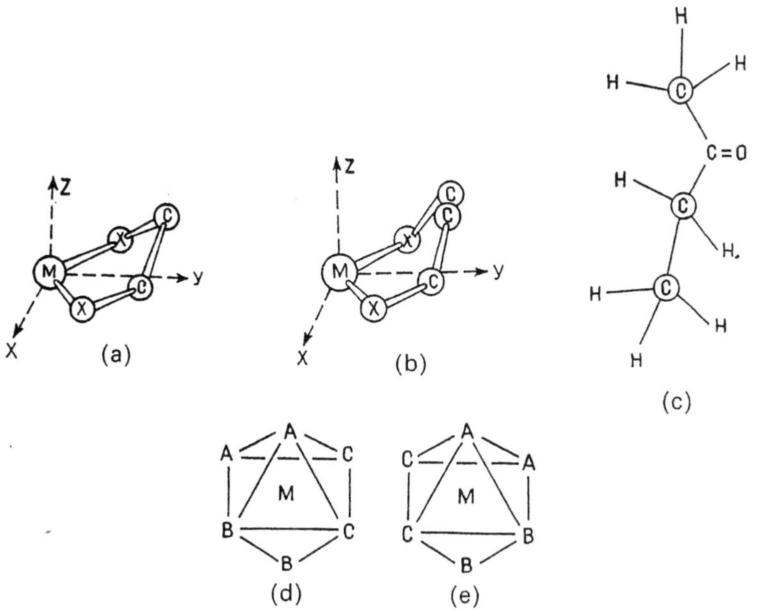


**Fig. 2.5** $\mathbf{C}_1$ symmetry molecules: asymmetric conformations of five-membered (a) and six-membered (b) chelate rings, methyl ethyl ketone (c), and the enantiomeric *cis* forms of an $MA_2B_2C_2$ complex (d, e). (After [30].)

**Example 2.12** The symmetrically tapered bath conformation of the six-membered ring in the chelate shown in Fig. 2.6b.

**Example 2.13** *Trans*-1,2-dichlorcyclopropane (Fig. 2.6c).

**Example 2.14** The $H_2C{=}CII_2$ molecule in a configuration in which the $CH_2$ and $CI_2$ planes are rotated through an angle $\alpha \neq \frac{1}{2}\pi$ (Figs 2.6d, e). (When the $CH_2$ and $CI_2$ planes are perpendicular, they become symmetry planes $\sigma$ and the molecule belongs to the group $\mathbf{C}_{2v}$.

*The group* $\mathbf{C}_3$ A $C_3$ axis gives rise to the three operations $\hat{C}_3$, $\hat{C}_3^2$ and $\hat{E}$. Each operation constitutes a class.

**Example 2.15** The $H_3C{-}Cl_3$ molecule with the configuration shown in Figs 2.7(a, b); the $H_3$ and $Cl_3$ triangles are rotated through an angle $\alpha \neq \frac{1}{6}\pi$ relative to each other (at $\alpha = \frac{1}{6}\pi$ the symmetry becomes higher).

**Example 2.16** Triphenylarsine (Fig. 2.7c); the benzene rings are inclined with respect to the plane of the page.

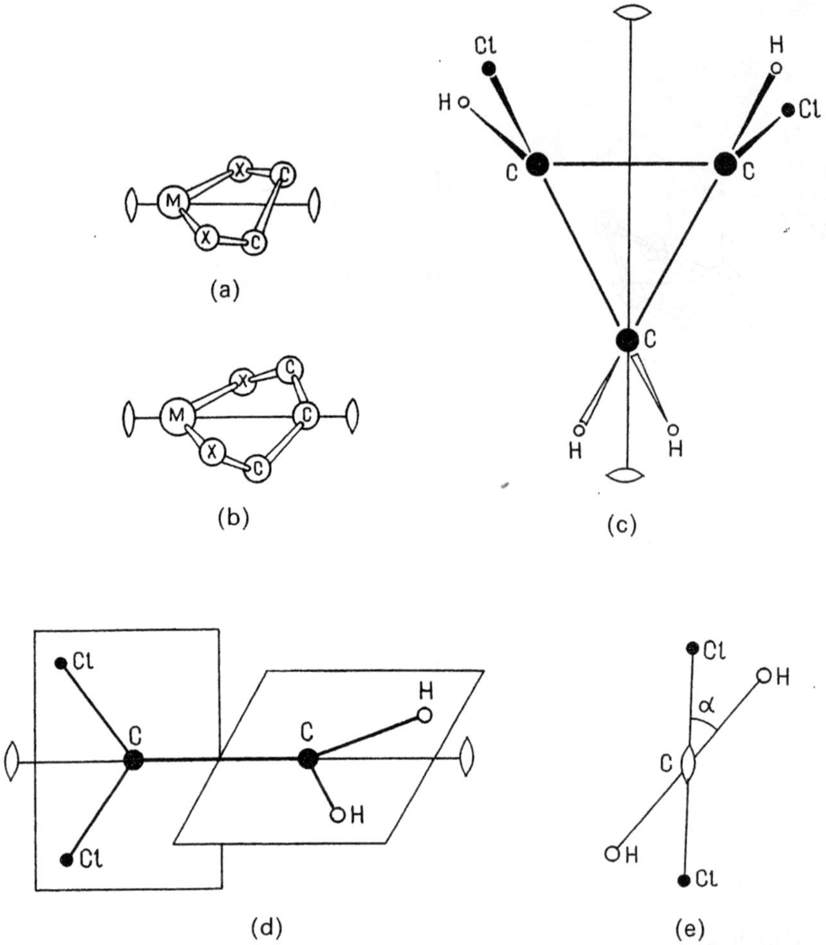


**Fig. 2.6** $\mathbf{C}_2$ symmetry molecules: (a) a symmetrically tapered five-membered chelate ring; (b) a symmetrically tapered bath conformation of a six-membered chelate ring (after [30]), (c) *trans*-1,2-dichlocyclopropane; (d, e) dichlorethylene.

*The group* $\mathbf{C}_4$ This consists of the four rotations $\hat{C}_4$, $\hat{C}_4^2$, $\hat{C}_4^3$ and $\hat{C}_4^4 = \hat{E}$ about the $C_4$ axis, which constitute four classes.

**Example 2.17** Tetraphenylcyclobutadiene (Fig. 2.8); the benzene rings are inclined with respect to the plane of the page.

*The group* $\mathbf{C}_6$ This includes six rotations $\hat{C}_6^m$, each of which constitutes a class.

**Example 2.18** Hexaphenylbenzene (Fig. 2.9).

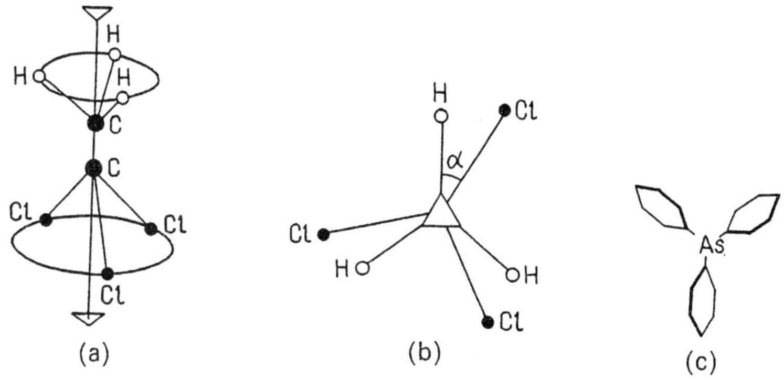


**Fig. 2.7** $\mathbf{C}_3$ symmetry molecules: (a, b) 1,1,1-trichlorethane; (c) triphenylarsine.

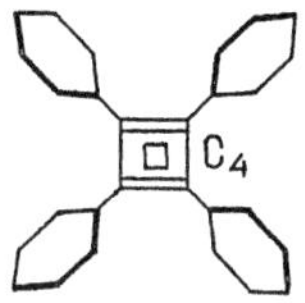


**Fig. 2.8** Tetraphenylcyclobutadiene: point group $\mathbf{C}_4$.

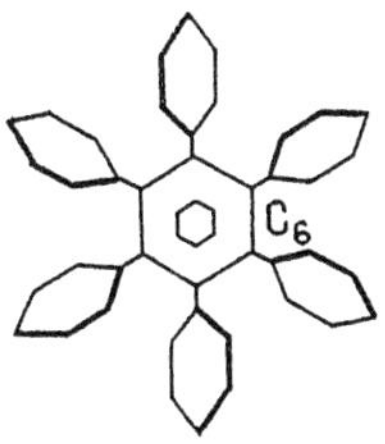


**Fig. 2.9** Hexaphenylbenzene: point group $\mathbf{C}_6$.

## 2.4.2 Groups of rotoflection transformations $\mathbf{S}_{2n}$

The set of rotoflection transformations $S_{2n}$ about an even-order improper rotation axis constitutes the group $\mathbf{S}_{2n}$. There is no point in discussing groups $\mathbf{S}_n$ of odd order, since an improper rotation $\hat{S}_n$ in this case is not a new symmetry operation (Section 1.1.4). All the groups $\mathbf{S}_{2n}$ are cyclic. Thus each operation over these groups constitutes a class.

*The group* $\mathbf{S}_2$ This consists of two operations: $\hat{E}$ and $\hat{I}$. It is called an inversion group and is denoted by $\mathbf{C}_i$.

**Example 2.18** 1,2-difluoro-1,2-dichloroethane (Fig. 2.10); the mid-point of the C–C segment is the centre of inversion.

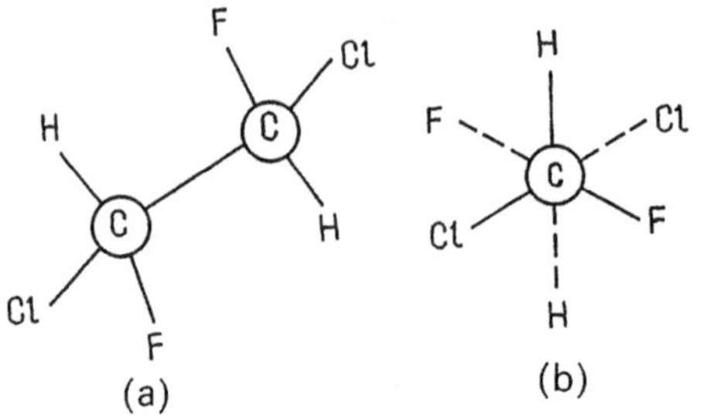


**Fig. 2.10** (a) 1,2-Difluoro-1,2-dichlorethane. (b) Projection along the axis connecting the carbon atoms. The point group is $\mathbf{C}_i$ (after [17]).

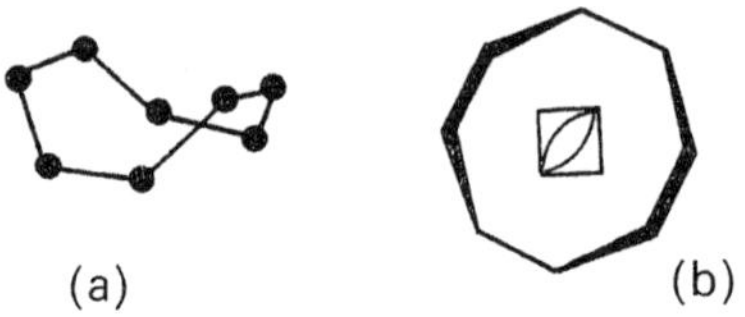


**Fig. 2.11** (a) The $\mathbf{S}_4$ symmetry cyclooctane molecule. (b) Projection onto a plane perpendicular to the $S_4$ axis. (After [17].)

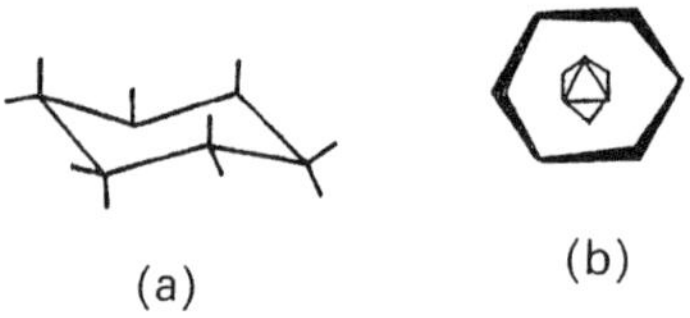


**Fig. 2.12** Cyclohexane: (a) space structure; (b) projection onto a plane perpendicular to the $S_6$ axis (after [17]).

*The group* $\mathbf{S}_4$ This consists of four operations, $\hat{E}_1$, $\hat{C}_2$, $\hat{S}_4$, $\hat{S}_4^3$ which constitute four classes.

**Example 2.19** Cyclooctane (Fig. 2.11).

*The group* $\mathbf{S}_6$ This consists of six operations: $\hat{E}$, $\hat{C}_3$, $\hat{C}_3^2$, $\hat{I}$, $\hat{C}_3^2\hat{\sigma}_h \equiv \hat{S}_6^{-1}$ and $\hat{S}_6$.

**Example 2.20** Cyclohexane in the "armchair" conformation (Fig. 2.12).

## 2.4.3 The groups $\mathbf{C}_{nh}$

Now let us consider a collection of symmetry elements including a $C_n$ axis and a plane of symmetry $\sigma_h$ perpendicular to this axis. The set of rotations $\hat{C}_n^m$, reflections $\hat{\sigma}_h$ and their products $\hat{C}_n^m\hat{\sigma}_h$ and $\hat{\sigma}_h\hat{C}_n^m$ satisfy the four requirements determining a symmetry group. Such groups are denoted by $\mathbf{C}_{nh}$. We shall discuss them for the simple cases $n = 1, \ldots, 4$, summarizing the results afterwards.

*The group* $\mathbf{C}_{1h}$ This consists of two operations, $\hat{C}_1 = \hat{E}$ and $\hat{\sigma}_h$, and is denoted by $\mathbf{C}_s$. Molecules belonging to $\mathbf{C}_s$ have only a symmetry plane.

**Example 2.21** Formamide (Fig. 2.13a) has only one symmetry plane, which is the molecular plane.

**Example 2.22** 1,1-Dibromo-2,2-dichlorocyclopropane (Fig. 2.13b); the symmetry plane contains the three carbon atoms.

**Example 2.23** *o*-Chlorobromobenzene (Fig. 2.13c).

*The group* $\mathbf{C}_{2h}$ The symmetry elements—the $C_2$ axis and the $\sigma_h$ plane—give rise to the operations $\hat{E}$, $\hat{C}_2$, $\hat{\sigma}_h$ and the product $\hat{\sigma}_h\hat{C}_2 = \hat{C}_2\hat{\sigma}_h = \hat{I}$. All the operations of $\mathbf{C}_{2h}$ ($\hat{E}$, $\hat{C}_2$, $\hat{\sigma}_h$, $\hat{I}$) are commutative, and constitute four classes.

**Example 2.24** The planar *trans*-ClHC–CHCl molecule (Fig. 2.14); the $\sigma_h$ plane

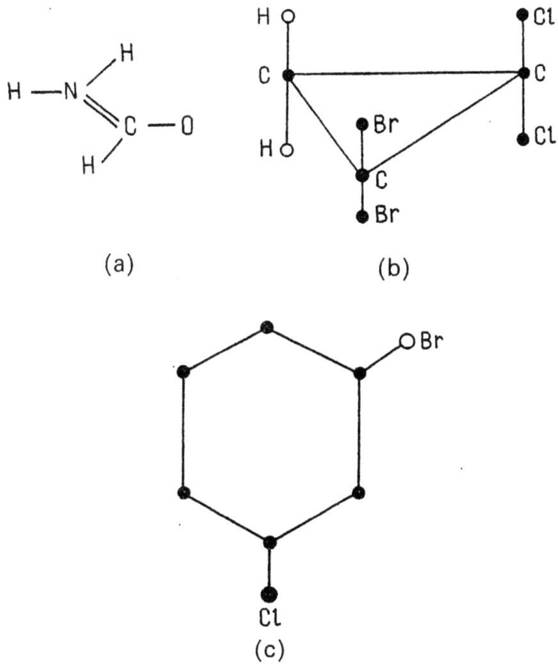


**Fig. 2.13** $\mathbf{C}_s$ symmetry molecules: (a) formamide; (b) 1,1-dibromo-2,2-dichlorcyclopropane; (c) *o*-chlorobromobenzene.

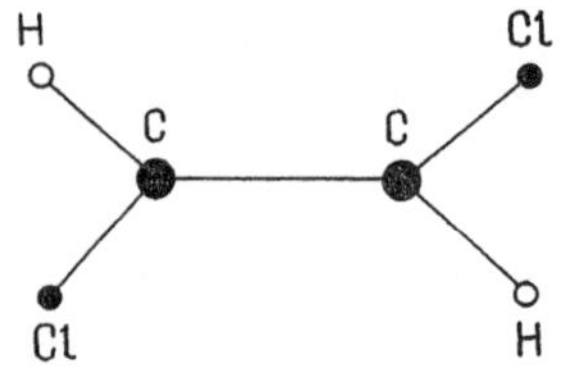


**Fig. 2.14** A $\mathbf{C}_{2h}$ symmetry molecule: *trans*-ClHC–CHCl.

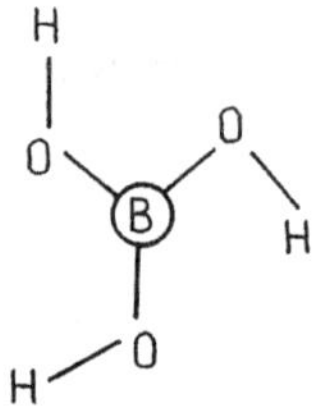


**Fig. 2.15** The $B(OH)_3$ molecule: $\mathbf{C}_{3h}$ symmetry.

coincides with the plane of the page; the $C_2$ axis is perpendicular to $\sigma_h$ and passes through the mid-point of the C–C bond.

*The group* $\mathbf{C}_{3h}$ The symmetry elements are $C_3$ and $\sigma_h$. Since the rotations $\hat{C}_n^m$ and the reflections $\hat{\sigma}_h$ are commutative, (Section 1.2) the operations $\hat{C}_3^m\hat{\sigma}_h$ and $\hat{\sigma}_h\hat{C}_3^m$ coincide: $\hat{\sigma}_h\hat{C}_3 = \hat{S}_3$ while $\hat{\sigma}_h\hat{C}_3^2 = \hat{S}_3^5 = \hat{S}_3^{-1}$. Thus we have obtained a collection of the group operations:

$$\hat{E}, \quad \hat{C}_3, \quad \hat{C}_3^2, \quad \hat{\sigma}_h, \quad \hat{S}_3, \quad \hat{S}_3^5.$$

The group is Abelian (Section 1.4) and each operation produces a class.

**Example 2.25** The $B(OH)_3$ molecule (Fig. 2.15).

*The group* $\mathbf{C}_{4h}$ This is also an Abelian group, and consists of eight operations, including inversion:

$$\hat{E}, \quad \hat{C}_4, \quad \hat{C}_2 \equiv \hat{C}_4^2, \quad \hat{C}_4^3, \quad \hat{I}, \quad \hat{S}_4^3, \quad \hat{\sigma}_h, \quad \hat{S}_4.$$

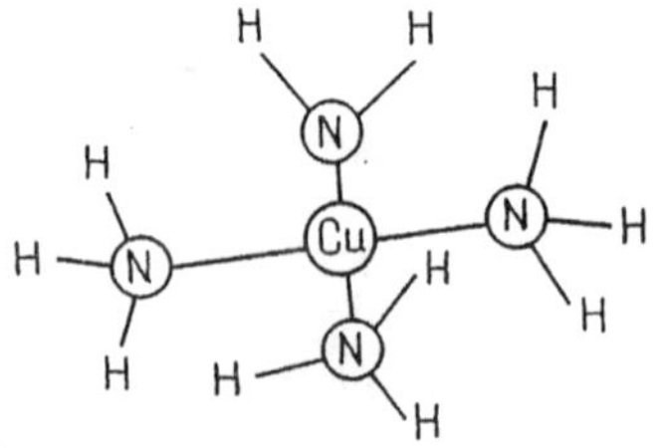


**Fig. 2.16** The $Cu(NH_3)_4$ ($\mathbf{C}_{4h}$) complex.

**Example 2.26** The complex $Cu(NH_3)_4$ (Fig. 2.16) [17]; the copper and four nitrogen atoms form a square.

Thus the group $\mathbf{C}_{nh}$ contains $2n$ operations: $n$ rotations (that is, operations of $\mathbf{C}_n$ and $n$ improper rotations $\hat{C}_n^m \hat{\sigma}_h (m = 1, 2, \ldots, n)$. $\mathbf{C}_{nh}$ are Abelian, and the number of classes is equal to the number of operations. If $n$ is even, $\mathbf{C}_{nh}$ includes an inversion, and the molecule is centrosymmetric.

### 2.4.4 The groups $\mathbf{C}_{nv}$

A $C_n$ axis and a vertical plane $\sigma_v$ in which it lies cannot on their own be the symmetry elements of a group, since (Section 1.3.2) this axis and plane give rise to a further $n - 1$ planes $\sigma_v$, intersecting at angles $\pi/n$. Any group having $C_n$ and $\sigma_v$ as symmetry elements must also have these additional $\sigma_v$ planes.

*The group* $\mathbf{C}_{2v}$ The symmetry elements of this group are the $C_2$ axis and two mutually perpendicular planes, $\sigma_v$ and $\sigma_v'$. Therefore the operations include a rotation $\hat{C}_2$, the identity $\hat{E}$ and two reflections $\hat{\sigma}_v$ and $\hat{\sigma}_v'$. The products $\hat{\sigma}_v \hat{C}_2$ and $\hat{C}_2 \hat{\sigma}_v$ do not give rise to new symmetry operations. Indeed, it can be seen from Fig. 2.17 that $\hat{C}_2 \hat{\sigma}_v = \hat{\sigma}_v'$. Thus $\mathbf{C}_{2v}$ consists of four operations: $\hat{E}$, $\hat{C}_2$, $\hat{\sigma}_v$, $\hat{\sigma}_v'$.

Now let us apply the rules for establishing classes that we formulated in Section 2.3. The reflections $\hat{\sigma}_v$ and $\hat{\sigma}_v'$ belong to the same class, since the planes $\sigma_v$ and $\sigma_v'$ are equivalent. The rotation $\hat{C}_2$ itself constitutes a class. Hence $\mathbf{C}_{2v}$ has three classes: $\hat{E}$, $\hat{C}_2$, $2\hat{\sigma}_v$.

**Example 2.27** Formaldehyde (Fig. 2.18a): the $C_2$ axis intersects the C=O bond, and one of the $\sigma_v$ planes is the molecular plane while the other is perpendicular to it.

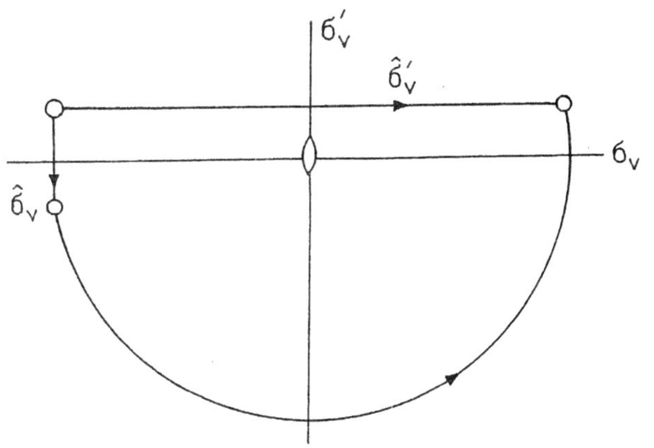


**Fig. 2.17** Multiplication of the operations $\hat{C}_2$ and $\hat{\sigma}_v$.

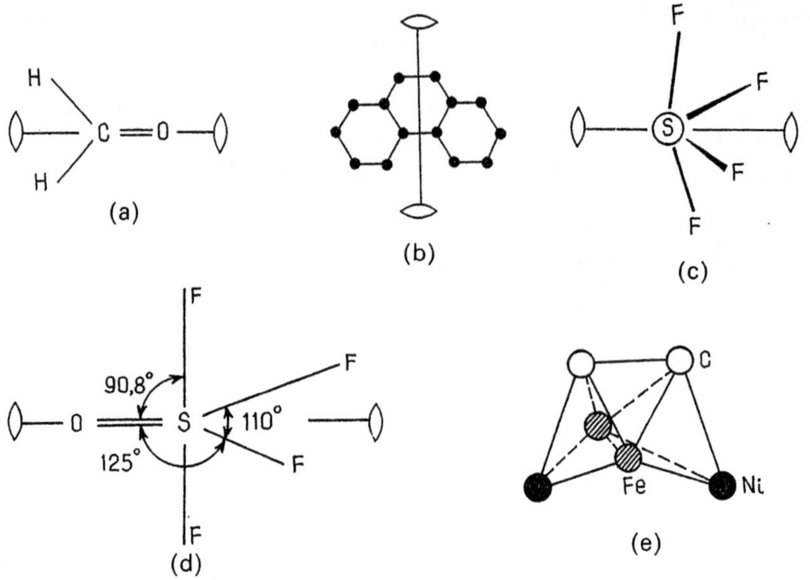


**Fig 2.18** $\mathbf{C}_{2v}$ symmetry molecules: (a) formaldehyde; (b) phenanthroline; (c) sulphur tetrafluoride; (d) $SOF_4$; (e) the cluster $F_2Ni_2C_2$.

**Example 2.28** Phenanthroline (Fig. 2.18b).

**Example 2.29** The $SF_4$ molecule, with the configuration showed in Fig. 2.18c; the $SF_2(\sigma_v)$ planes are mutually perpendicular. Other tetrahalides ($SeF_4$, $SeCl_4$ and $TeCl_4$) have similar structures [31].

**Example 2.30** The $SOF_4$ molecule (Fig. 2.18d) belongs to $\mathbf{C}_{2v}$, if we neglect the small difference between the FSO angle (90.8°) and a right-angle [31].

**Example 2.31** The heteroatomic cluster $Fe_2Ni_2C_2$ [32] (Fig. 2.18e); the $C_2$ axis intersects the Fe–Fe and C–C bonds at their mid-points, and one of the $\sigma_v$ planes

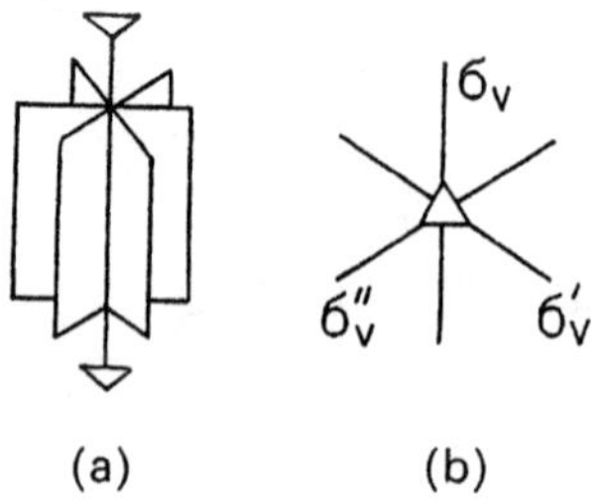


**Fig. 2.19** (a) Symmetry elements of $\mathbf{C}_{3v}$. (b) Projection onto a plane perpendicular to the $C_3$ axis.

contains the C and Ni atoms while the other passes through the Fe–Fe bond and the mid-point of the C–C bond.

*The group* $\mathbf{C}_{3v}$ The $C_3$ axis and three planes $\sigma_v$ are shown in Fig. 2.19. Six symmetry operations, $\hat{E}$, $\hat{C}_3$, $\hat{C}_3^2$, $\hat{\sigma}_v$, $\hat{\sigma}'_v$, $\hat{\sigma}''_v$, are divided into three classes: $\hat{E}$, $2\hat{C}_3$, $3\hat{\sigma}_v$. The $C_3$ axis belonging to $\mathbf{C}_{3v}$ appears to be two-sided (in contrast with $\mathbf{C}_3$ and $\mathbf{C}_{3h}$ owing to the presence of the $\sigma_v$ planes; therefore the rotations $C_3$ and $\hat{C}_3^2 = \hat{C}_3^{-1}$ belong to the same class (Section 2.3). The three $\sigma_v$ planes are equivalent, since they are brought into coincidence by $\hat{C}_3$ rotations.

**Example 2.32** Methyl chloride (Fig. 2.20a); the $C_3$ axis is shown in the figure; the three planes $\sigma_v$ contain ClCH atoms.

**Example 2.33** The trinuclear copper(II) cluster $Cu_3(C_6H_5N_2O)_3(OH)SO_3$ (Figs. 2.20b, including three copper and axial ligands (Fig. 2.20.c).

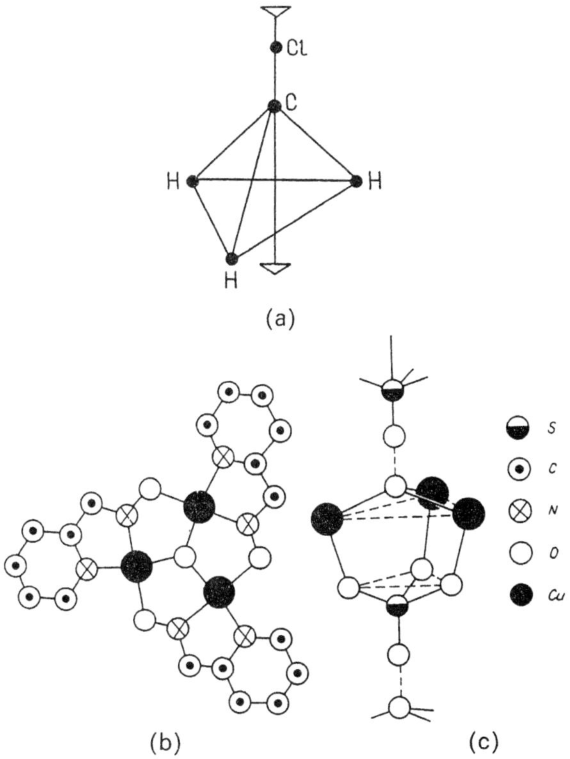


**Fig. 2.20** $\mathbf{C}_{3v}$ symmetry molecules: (a) methyl chloride; (b, c) the trinuclear copper(II) cluster $Cu_3(C_6H_5N_2O)_3(OH)SO_3$—projection onto a plane perpendicular to the $C_3$ axis; (b) and a schematic view (c).

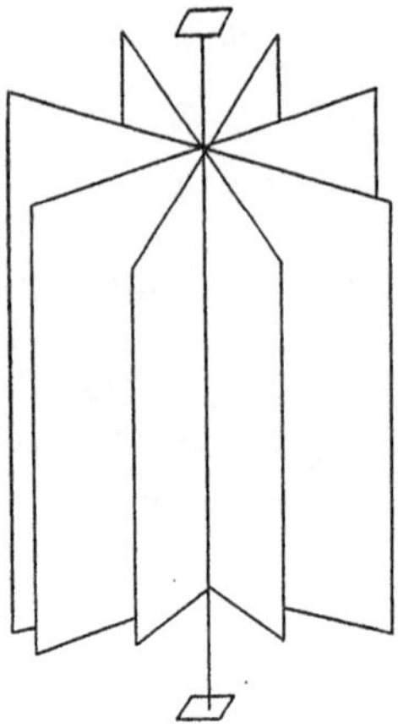

**Fig. 2.21** Symmetry elements of $\mathbf{C}_{4v}$.

*The group* $\mathbf{C}_{4v}$ This group (Fig. 2.21) contains eight operations (Section 2.2), which are divided into five classes;

$$\hat{E},\quad 2\hat{C}_4,\quad \hat{C}_2 \equiv \hat{C}_4^2,\quad 2\hat{\sigma}_v,\quad 2\hat{\sigma}_d.$$

The four reflections in the vertical planes are divided into two classes ($2\hat{\sigma}_v$ and $2\hat{\sigma}_d$, Fig. 2.21), with the two $\hat{C}_4$ rotations bringing those in each class into coincidence; the $C_4$ axis is two-sided.

**Example 2.34** The $IOF_5$ molecule (Fig. 2.22a).

**Example 2.35** The $BrF_5$ molecule (Fig. 2.22b).

It follows from the above discussion that $\mathbf{C}_{nv}$ contains $2n$ operations: $n$ rotations $\hat{C}_n^m$ and $n$ reflections $\hat{\sigma}_v$. The $C_n$ axis is two-sided. For odd $n = 2k + 1$

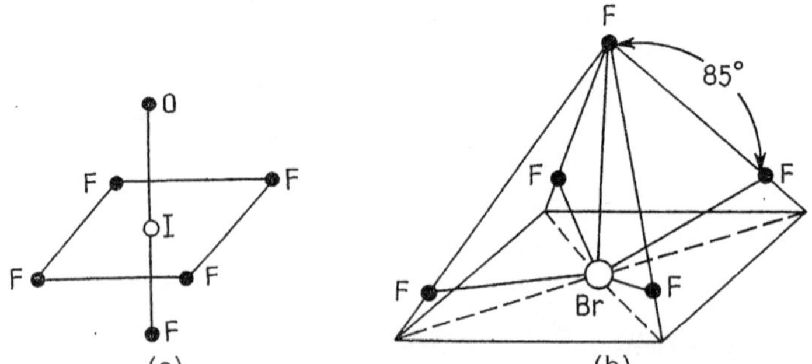


**Fig. 2.22** $\mathbf{C}_{4v}$ symmetry molecules: (a) $IOF_5$; (b) $BrF_5$.

all the reflections form a single class. The $\hat{C}_n^m$ and $\hat{C}_n^{-m}$ rotations are conjugate and form $k$ classes, $\hat{E}$ being a single class. Thus $C_{2k+1,v}$ has $k+2$ classes. When $n$ is even $(n = 2k)$, the reflections are divided into two classes: $\mathbf{C}_{2k,v}$ has $k+3$ classes.

### 2.4.5 The dihedral groups $\mathbf{D}_n$

The groups $\mathbf{D}_n$ are called *dihedral* and contain as symmetry elements the vertical $C_n$ axis and $n$ horizontal $C_2$ axes intersecting at angles $\pi/n$. The symbols for these axes are primed: $C_2'$, $C_2''$ etc. The symmetry elements of $\mathbf{D}_2$, $\mathbf{D}_3$, and $\mathbf{D}_4$ are shown in Fig. 2.23. The $\mathbf{D}_n$ have no symmetry planes. $\mathbf{D}_1(\hat{E}, \hat{C}_2)$ is similar to $\mathbf{C}_2$.

*The group* $\mathbf{D}_2$ This includes three operations of rotation around the three $C_2$ axes: $\hat{E}$, $\hat{C}_2$, $\hat{C}_2'$, $\hat{C}_2''$ (Fig. 2.23a). The group has no operations that bring the axes into coincidence. Therefore each operation constitutes a separate class.

**Example 2.36** Biphenyl (Fig. 2.24a) [17].

**Example 2.37** Ethylene in the tapered (twisted) configuration (Fig. 2.24b); the lines of CHC bonds are not perpendicular.

*The group* $\mathbf{D}_3$ The system of axes is shown in Fig. 2.23b. The rotations $\hat{C}_2'$, $\hat{C}_2''$, $\hat{C}_2'''$ belong to the same class, since $\hat{C}_3$ rotations bring the horizontal axes into coincidence. Because of these rotations, the $C_3$ axis is two-sided, so that the operations $\hat{C}_3$ and $\hat{C}_3^2$ form a class. Hence the six operations of $\mathbf{D}_3$ are divided into three classes:

$$\hat{E}, \quad 3\hat{C}_2, \quad 2\hat{C}_3.$$

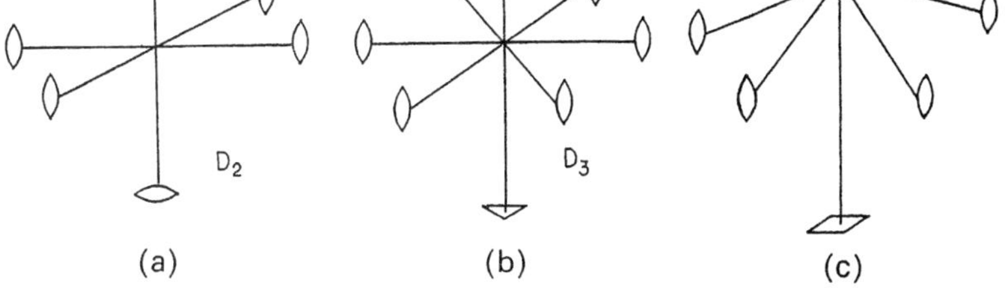

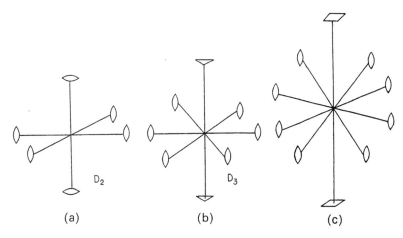

**Fig. 2.23** Symmetry elements of $\mathbf{D}_n$: (a) $\mathbf{D}_2$; (b) $\mathbf{D}_3$; (c) $\mathbf{D}_4$.

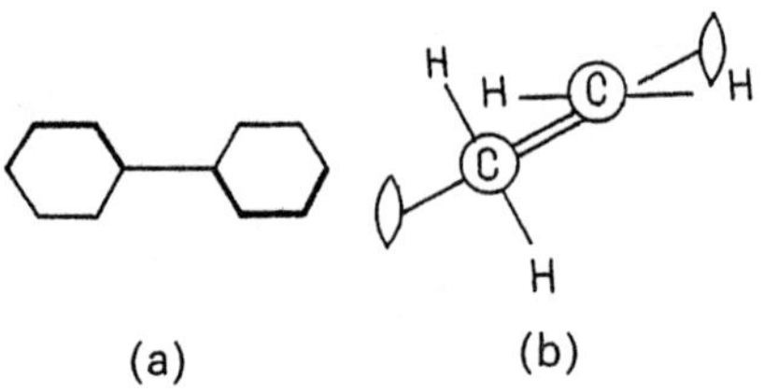


**Fig. 2.24** $\mathbf{D}_2$ symmetry molecules: (a) biphenyl; (b) "twisted" conformation of ethylene.

**Example 2.38** Ethane with a tapered configuration (Fig. 2.25a, b); here the minimum angle between the projections of the C–H bonds in the two methyl groups differs from 60°. If this angle is 60° (Fig. 2.25c), the symmetry is higher.

*The group* $\mathbf{D}_4$ The symmetry elements are the $C_4$ axis and the four horizontal $C_2$ axes (Fig. 2.23a). The rotations $C_4$ bring each of the latter into coincidence (similarly to the $\sigma_v$ planes in $\mathbf{C}_{4v}$). Therefore the $\hat{C}'_2$ operations are divided into two classes. The $C_4$ axis is two-sided, so that the operations $\hat{C}_4$ and $\hat{C}_4^3$ form a class. Thus the eight operations of $\mathbf{D}_4$ are divided into five classes:

$$\hat{E}, \quad 2\hat{C}_4, \quad \hat{C}_2 = \hat{C}_4^2, \quad 2\hat{C}'_2, \quad 2\hat{C}''_2.$$

In all $\mathbf{D}_n$ the $C_n$ axis is two-sided and the horizontal axes are equivalent for odd $n = 2k + 1$ and form two sets for even $n = 2k$. $\mathbf{D}_{2k}$ has $k + 3$ classes ($\hat{E}$, two classes with $k$ rotations $\hat{C}_2$ in each class, and operation $C_2$) and $k - 1$ classes with two rotations about the $C_n$ axis in each class. $\mathbf{D}_{2k+1}$ has $k + 2$ classes: $\hat{E}$, $2k + 1$ rotations $\hat{C}'_2$, and $k$ classes with two rotations around the $C_{2n}$ axis.

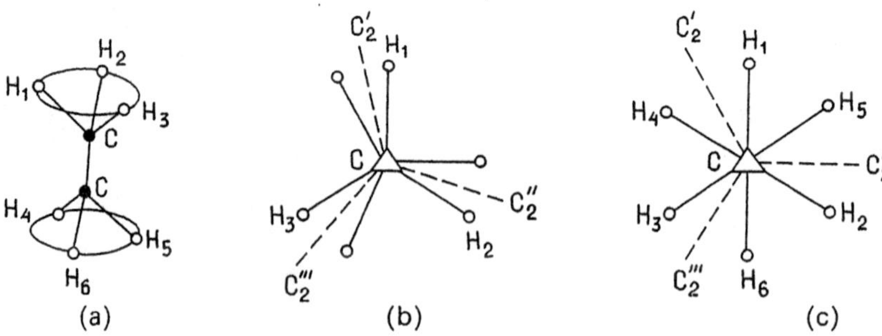


**Fig. 2.25** (a) Ethane in the tapered conformation. (b, c) Projections onto a plane perpendicular to the C–C bond.

## 2.4.6 The groups $\mathbf{D}_{nh}$

Let us add a horizontal plane to the system of axes of $\mathbf{D}_n$. The set of operations produced by this set of symmetry elements does not form a group (Section 1.3.2). The $C_n$ axis and the plane containing $n$ axes $C_2'$ perpendicular to this plane give rise to a further $n$ vertical planes $\sigma_v$ passing through the $C_n$ and $C_2'$ axes. In Section 1.3.2 this was discussed in detail using as an example the $C_3(z)$, $3C_2$ axes and the $\sigma(xy)$ planes. The resulting group $\mathbf{D}_{nh}$ includes $4n$ operations: $2n$ operations of $\mathbf{D}_n$, $n$ reflections $\sigma_h$ and $n$ improper rotations $\hat{C}_n^m\hat{\sigma}_h$. The symmetry elements of $\mathbf{D}_{2h}$ and $\mathbf{D}_{3h}$ are shown in Fig. 2.26.

*The group* $\mathbf{D}_{2h}$ The symmetry elements are shown in Fig. 2.26(a). The $\mathbf{D}_{2h}$ group operations can be obtained by multiplying the $\mathbf{D}_2$ group operations $(\hat{E}, \hat{C}_2(x), \hat{C}_2(y), \hat{C}_2(z))$ by $\hat{\sigma}_h$ and combining them with $\mathbf{D}_2$:

$$\hat{\sigma}_h\hat{E} = \hat{\sigma}_h, \quad \hat{\sigma}_h\hat{C}_2(z) = \hat{I}, \quad \hat{\sigma}_h\hat{C}_2(x) = \hat{\sigma}_v(xz), \quad \hat{\sigma}_h\hat{C}_2(y) = \sigma_v(yz).$$

Thus the $\sigma_h$ plane doubles the number of operations of $\mathbf{D}_2$. Let us divide the eight operations of $\mathbf{D}_2$ $(\hat{E}, 3\hat{C}_2, \hat{I}, \hat{\sigma}_h, 2\hat{\sigma}_v)$ into classes. It should be noted that $\hat{\sigma}_h$ commutates with all the other operations. $\mathbf{D}_{2h}$ may be written as the direct product (Section 2.3) of $\mathbf{D}_2$ and $C_s$: $\mathbf{D}_{2h} = \mathbf{D}_2 \times \mathbf{C}_s$. Therefore, half of the $\mathbf{D}_{nh}$ group classes coincide with the $\mathbf{D}_n$ classes. The remainder are obtained by multiplying the $\mathbf{D}_n$ classes by $\hat{\sigma}_h$, resulting in eight classes of the group $\mathbf{D}_{2h}$:

$$\hat{E}, \quad \hat{C}_2(x), \quad C_2(y), \quad \hat{C}_2(z), \quad \hat{\sigma}_h, \quad \hat{I}, \quad \hat{\sigma}_v(xz), \quad \hat{\sigma}_v(yz).$$

**Example 2.39** The planar $I_2Cl_6$ molecule (Fig. 2.27a).

**Example 2.40** *p*-Dichlorobenzene (Fig. 2.27b).

**Example 2.41** The coordination compound bis(dimethylglyoximato)nickel(II) (Fig. 2.27c); this complex has the $\mathbf{D}_{2h}$ group symmetry elements, ignoring the hydrogen atoms.

*The group* $\mathbf{D}_{3h}$ This is the direct product $\mathbf{D}_3 \times \mathbf{C}_s$ (Fig. 2.26b). Using the geometric approach to the multiplication of operations (Section 1.2), it is possible to obtain $\mathbf{D}_{3h}$ as

$$\hat{E}, \quad \hat{\sigma}_h, \quad 2\hat{C}_3, \quad 2\hat{S}_3, \quad 3\hat{C}_2', \quad 3\hat{\sigma}_v.$$

This notation corresponds to the distribution of operations in six classes.

**Example 2.42** Planar triangular molecules $BH_3$ (Fig. 1.2b), the $NO_3^-$ ion and $BF_3$ (where N and B are located in the centre of the $H_3$ and $F_3$ triangles respectively), the "screened" ethane conformation and the bipyramidal $PF_5$ molecule (Fig. 2.2).

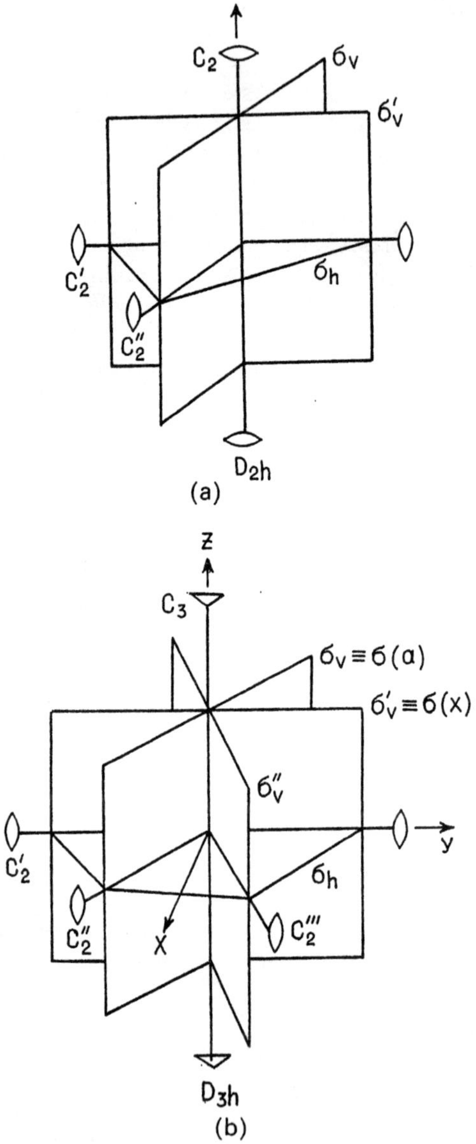


**Fig. 2.26** Symmetry elements of (a) $\mathbf{D}_{2h}$ and (b) $\mathbf{D}_{3h}$.

*The group* $\mathbf{D}_{4h}$ This includes 16 operations of the direct product $\mathbf{D}_4 \times \mathbf{C}_s$, which are distributed in 10 classes:

$$\hat{E},\quad 2\hat{C}_4,\quad \hat{C}_2 = \hat{C}_4^2,\quad 2\hat{C}'_2,\quad 2\hat{C}''_2,\quad \hat{\sigma}_h,\quad 2\hat{\sigma}_v,\quad 2\hat{\sigma}_d,\quad 2\hat{S}_4,\quad \hat{S}_2 \equiv \hat{I}.$$

Similarly to $\mathbf{D}_4$, the horizontal axes here form two inequivalent collections,

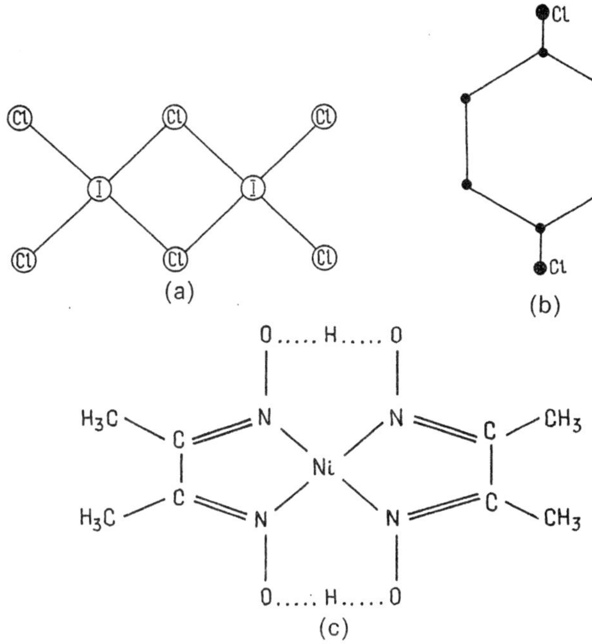


**Fig. 2.27** $\mathbf{D}_{nh}$ symmetry molecules: (a) $I_2Cl_6$; (b) *p*-dichlorobenzene; (c) the coordination compound bis(dimethylglyoximato)nickel(II).

$2C_2'$ and $2C_2''$, with the $2\hat{C}_2'$ and $2\hat{C}_2''$ operations belonging to different classes. In the same way, the reflections in the vertical planes are divided into two classes. It should be noted that $\mathbf{D}_{4h}$ includes an inversion, and the molecules of this group are centrosymmetric.

**Example 2.43** Square planar complexes of the $[PtCl_4]^-$ type (Fig. 1.2).

**Example 2.44** *Trans-* complexes of the $MA_4X_2$ type, where M is a metal, and A and X are monodentate ligands (Fig. 2.28a).

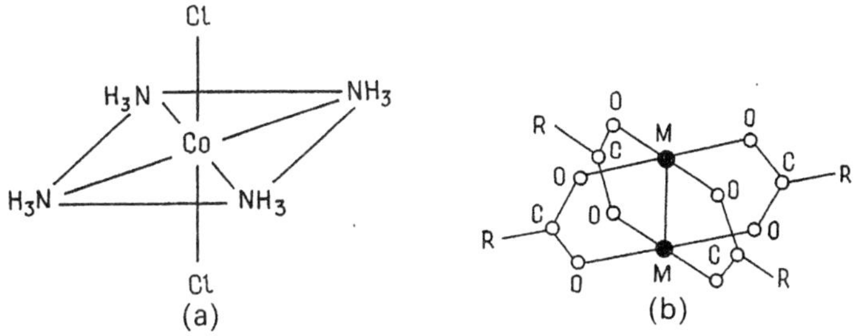


**Fig. 2.28** $\mathbf{D}_{4h}$ symmetry molecules: (a) the complex $[Co(NH_3)_4Cl_2]$; (b) a binuclear cluster $M_2(O_2CR)_4$, where M is a metal ion.

**Example 2.45** Binuclear carboxylate clusters of the $M_2(O_2CR)_4$ type involving transition metals (Fig. 2.28b) [33], with a structure of "lantern" form.

From the above analysis of the groups $\mathbf{D}_{2h}$, $\mathbf{D}_{3h}$ and $\mathbf{D}_{4h}$, the laws for general $n$ are clear. $\mathbf{D}_{nh}$ has twice as many classes as $\mathbf{D}_n$. For odd $n$ and $\sigma_v$ reflections belong to one class, while for even $n$ they form two classes. The improper rotations $\hat{\sigma}_h \hat{C}_n^k$ and $\hat{\sigma}_h \hat{C}_n^{\hat{k}}$ are conjugate in pairs. The $\mathbf{D}_{nh}$ of even order contain an inversion. All planar molecules possessing a $C_n$ axis belong to $\mathbf{D}_{nh}$. We have already discussed some typical examples. Further examples of systems of this type, with a higher degree of symmetry, include ferrocene, with a "screened" conformation ($\mathbf{D}_{5h}$), benzene and coronene ($\mathbf{D}_{6h}$).

### 2.4.7 The groups $\mathbf{D}_{nd}$

Let us attach vertical diagonal planes to the axis system of $\mathbf{D}_n$ (Fig. 2.29). Each of the $n$ planes bisects the angle between the horizontal $C_2'$ axes. The resulting system of symmetry elements of $\mathbf{D}_{nd}$ leads to $4n$ operations: $2n$ rotations of $\mathbf{D}_n$, $n$ reflections in the diagonal planes and $n$ transformations of the $\hat{C}_2'\hat{\sigma}_d$ type. Let us determine the nature of these transformations, taking $\mathbf{D}_{2d}$ as an example. From Fig. 2.30 it is obvious that successive application of the operations $\hat{\sigma}_d'$ and $\hat{C}_2''$ transforms the point $A$ into $B$. The same result is obtained by applying the $\hat{C}_4$ operation with subsequent reflection in the horizontal plane. Hence $\hat{C}_2\hat{\sigma}_d = \hat{\sigma}_d\hat{C}_2 = \hat{S}_4$. The existence of the $\sigma_d$ planes thus transforms the $C_2$ axis into the $S_4$ rotoflection axis. This result also holds for the general case of $C_n$ axes.

*The group* $\mathbf{D}_{2d}$ This is obtained by adding improper rotations of the $\hat{C}_2'\hat{\sigma}_d$ type to $\mathbf{D}_2$; the operations $\hat{S}_4$ and $\hat{S}_4^3$ belong to the same class. The eight operations are thus divided into five classes:

$$\hat{E},\quad \hat{C}_2,\quad 2\hat{S}_4,\quad 2\hat{C}_2',\quad 2\hat{\sigma}_d.$$

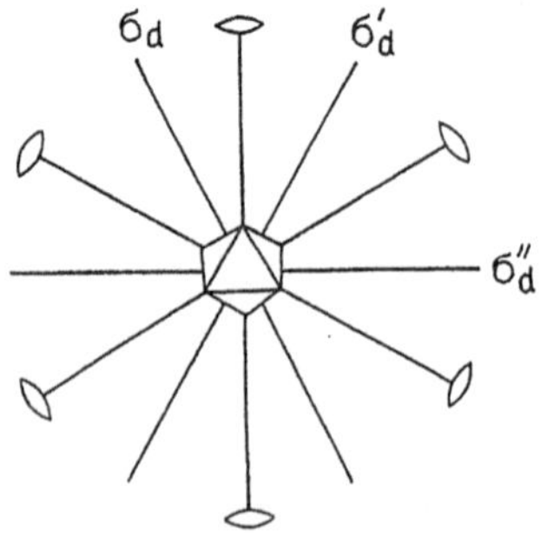


**Fig. 2.29** Symmetry elements of the $\mathbf{D}_{3d}$ group (projection onto a plane perpendicular to the $S_6$ axis).

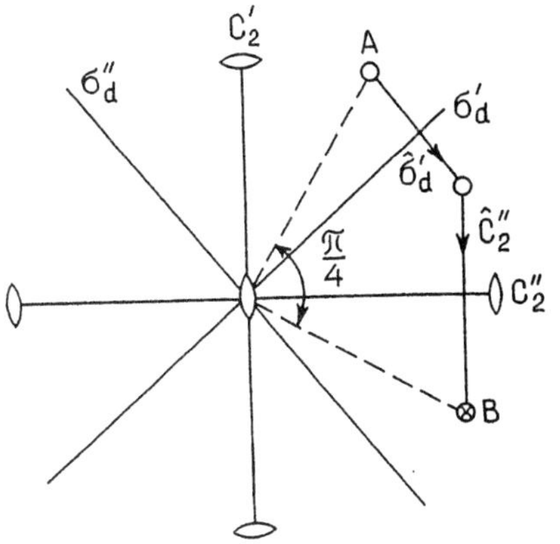


**Fig. 2.30** The improper rotation $\hat{S}_4 = \hat{C}_2\hat{\sigma}_d$ in the $\mathbf{D}_{2d}$ group.

**Example 2.46** The tapered ("chessboard") ethylene conformation (Fig. 2.24b), for which the lines of the bonds HCH are mutually perpendicular.

**Example 2.47** The spirane shown in Fig. 2.31a.

**Example 2.48** The binuclear cluster $W_2Cl_4[P(CH_3)_3]_4$ (Fig. 2.31b).

*The group* $\mathbf{D}_{3d}$ This group (Fig. 2.29) is treated similarly. The 12 operations are distributed into six classes as follows:

$$\hat{E}, \quad 2\hat{S}_6, \quad 2\hat{C}_3, \quad \hat{I}, \quad 3\hat{C}_2', \quad 3\hat{\sigma}_d.$$

Among its operations, $\mathbf{D}_{3d}$ includes the inversion.

**Example 2.49** The tapered ("chessboard") ethane conformation (Fig. 2.25c); the HCH angles projected onto the plane perpendicular to the $C_3$ axis are equal to 60°.

**Example 2.50** The binuclear cobalt carbonyl shown in Fig. 2.32a.

**Example 2.51** The heteronuclear trimeric cluster of cobalt and zinc $Zn[Co(CO_4)_4]_2$ (Fig. 2.32b) [31].

In the general case, $\mathbf{D}_{nd}$ of even order $n = 2k$ consists of $2k + 3$ classes: $\hat{E}, \hat{C}_2 \equiv \hat{C}_{2k}^k$, $k - 1$ classes each having two conjugate rotations $\hat{C}_k^m$ and $\hat{C}_k^{-m}$, a class of $2k$ rotations $C_2'$, a class of $2k$ reflections $\sigma_d$, and $k$ classes having two improper rotations each. For odd $n = 2k + 1$ the group contains an inversion and is the direct product $\mathbf{D}_{2k+1} \times \mathbf{C}_i$. According to the rule given in Section 2.3, $\mathbf{D}_{2k+1,d}$ contains $2k + 4$ classes, half of which coincide with the $\mathbf{D}_{2k+1}$ classes.

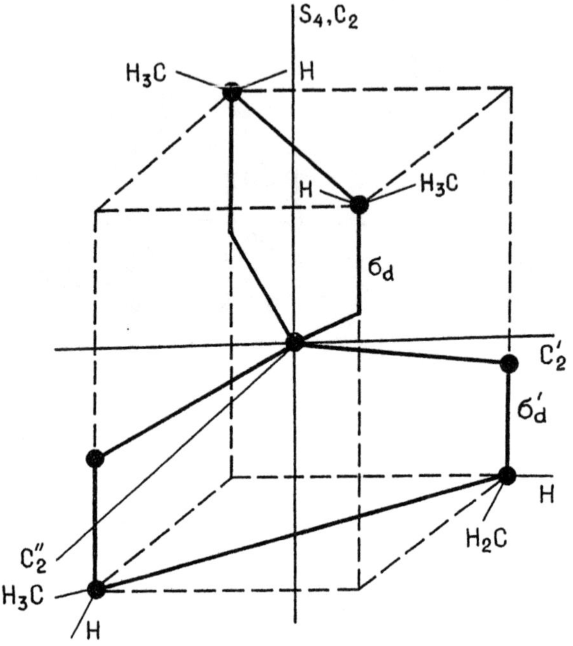


(a)

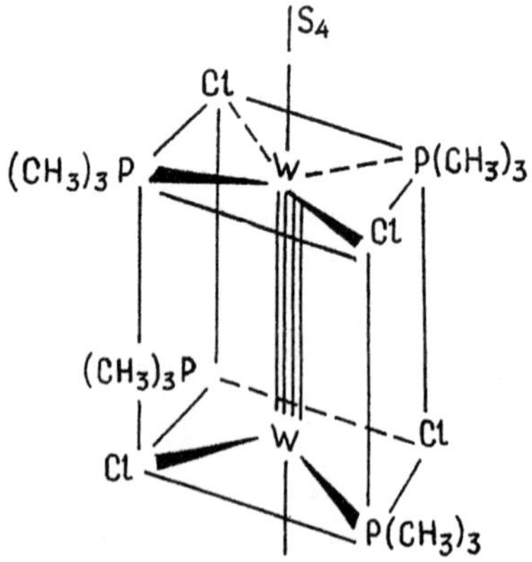


(b)

**Fig. 2.31** $\mathbf{D_{2d}}$ symmetry molecules: (a) a spiran; (b) the binuclear tungsten cluster $[W_2Cl_4P(CH_3)_3]_4$.

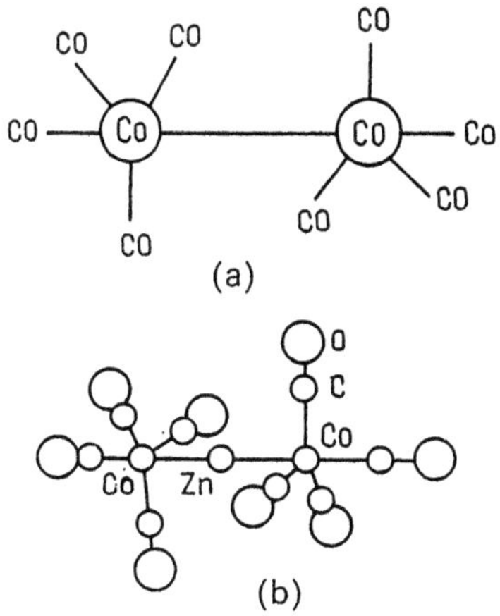


**Fig. 2.32** $\mathbf{D}_{3d}$ symmetry molecules: (a) a binuclear cobalt carbonyl; (b) the heteronuclear cluster $Zn[Co(CO_4)_4]_2$.

## 2.4.8 The cubic groups ($\mathbf{T}$, $\mathbf{T}_d$, $\mathbf{T}_h$, $\mathbf{O}$, $\mathbf{O}_h$)

Cubic groups include either all or some of the symmetry operations of a cube. The groups $\mathbf{T}$, $\mathbf{T}_d$, $\mathbf{T}_h$ are called *tetrahedral*, while $\mathbf{O}$, $\mathbf{O}_h$ are *octahedral.*

*The Group* $\mathbf{T}$ The group symmetry elements are the axes of a tetrahedron. The positions of these axes are shown in Fig. 2.33(a). The vertices of the tetrahedron are positioned on the non-adjacent vertices of the cube; the four $C_3$ axes coincide with the spatial diagonals of the cube, while the $C_2$ and $S_4$ axes coincide with the $C_4$ axes of the cube. The group $\mathbf{T}$ contains 12 operations, which form four classes:

$$\hat{E}, \quad 4\hat{C}_3, \quad 4\hat{C}_3^2, \quad 3\hat{C}_2.$$

The group $\mathbf{T}_h$ is the direct product $\mathbf{T} \times \mathbf{C}_i$.

*The group* $\mathbf{T}_d$ This is produced by attaching to $\mathbf{T}$ six $\sigma_d$ planes (each plane passing through the two axes $C_3$ and $C_2$) as well as three $S_4$ axes, coinciding with the $C_2$ axes. $\mathbf{T}_d$ has 24 operations and 5 classes: $\hat{E}, 8\hat{C}_3, 6\hat{\sigma}_d, 6\hat{S}_4$ and $3\hat{C}_2$. Tetrahedral molecules such as $CH_4$ and $CCl_4$ have $\mathbf{T}_d$ symmetry.

*The group* $\mathbf{O}$ The symmetry elements are the axes of the cube (Fig. 2.33b) or of the octahedron (Fig. 2.1); these are the four $C_3$ axes (spatial diagonals), the six $C_2$ axes passing through the mid-points of the opposite edges, and the three $C_4$ axes passing through the centres of the opposite faces. The 24 operations are divided into 5 classes:

$$\hat{E}, \quad 8\hat{C}_3, \quad 6\hat{C}_2, \quad 6\hat{C}_4, \quad 3\hat{C}_4^2.$$

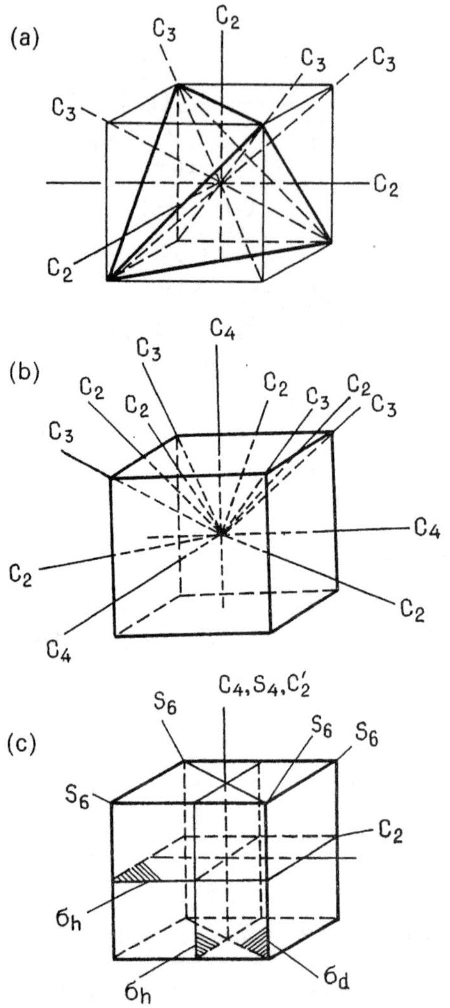


**Fig. 2.33** Symmetry elements of the cubic groups (a) **T**, (b) **O** and (c) $\mathbf{O}_h$.

*The group* $\mathbf{O}_h$ This is the direct product $\mathbf{O} \times \mathbf{C}_i$. Such a symmetry is characteristic of a cube or an octahedron (Fig. 2.33c). Half of the operations of **O** and $\mathbf{O}_h$ coincide, the remaining 24 of the $\mathbf{O}_h$ operations are obtained by multiplying the **O** operations by $\hat{I}$. In comparison with **O**, the number of classes is doubled. The group $\mathbf{O}_h$ has 10 classes:

$$\hat{E}, \quad 8\hat{C}_3, \quad 6\hat{C}_2, \quad 6\hat{C}_4, \quad 3\hat{C}_2', \quad \hat{I} \quad 6\hat{S}_4, \quad 8\hat{S}_6, \quad 3\hat{\sigma}_h, \quad 6\hat{\sigma}_d.$$

The group $\mathbf{O}_h$ possesses the highest degree of symmetry that is possible in crystals (Section 2.5).

As well as the molecular symmetry groups described above, there are also the *icosahedral* groups **I** and $\mathbf{I}_h$, which will not be considered in this book. Suffice it to say that an icosahedron is a regular polyhedron with 20 triangular facets. **I** also contains a pentagonal dodecahedron, i.e. a regular polyhedron with 12 pentagonal facets.

### 2.4.9 Continuous groups

Linear molecules possess a special symmetry called *axial symmetry*. If a molecule is not symmetric relative to its centre, it belongs to the axial group $\mathbf{C}_{\infty v}$. This group includes such operations as rotations about the axis through an arbitrary angle and reflections in any plane passing through the axis. $\mathbf{C}_{\infty v}$ is continuous, since it has an infinite number of operations. A linear molecule that is symmetrical relative to its centre is described by $\mathbf{D}_{\infty h}$. This group consists of an infinite number of arbitrary rotations about the $\mathbf{C}_\infty$ axis, an infinite number of $\hat{C}_2$ rotations about the $C_2 \perp C_\infty$ axis, and the reflections in the $\sigma_h$ and the $\sigma_v$ planes. To $\mathbf{D}_{\infty h}$ belong diatomic molecules of the $X_2$ type and triatomic XAX molecules.

A free atom possesses spherical symmetry. A spherical group **K** or a *group of rotations* consists of an infinite number of arbitrary rotations about any axis, passing through the atom, as well as reflections in the planes containing the atom.

## 2.5 CRYSTALLOGRAPHIC POINT GROUPS

In contrast to molecules, crystals have no $C_5$, $C_7$, $C_8$ etc. axes. This follows from the so-called "law of rational indices" [20]. Only 32 point groups satisfy this empirical law. These groups form the possible crystalline systems (Table 2.3).

The symbols used so far for point groups belong to the Schoenflies system.

**Table 2.3** Crystalline systems.

| | Systems | | | | | | |
|---|---|---|---|---|---|---|---|
| | I Triclinic | II Monoclinic | III Rhombic | IV Trigonal | V Tetragonal | VI Hexagonal | VII Cubic |
| Point groups | $C_1$<br>$C_i$ | $C_{2h}$, $C_2$<br>$C_s$ | $D_{2h}$, $D_2$<br>$C_{2v}$ | $D_{3h}$<br>$D_6$, $C_{3v}$<br>$S_6$, $C_3$ | $D_{4h}$, $D_4$<br>$C_{4v}$, $C_{4h}$<br>$C_4$, $S_4$<br>$D_{2d}$ | $D_{6h}$, $D_6$<br>$C_{6v}$, $C_{6h}$<br>$C_{3h}$, $D_{3d}$ | $O_h$, O<br>$T_d$, $T_h$<br>T |

**Table 2.4** Three systems of notation for point groups.

| Schoenflies | Shubnikov | International | Schoenflies | Shubnikov | International |
|---|---|---|---|---|---|
| $\mathbf{C}_1$ | 1 | 1 | $\mathbf{C}_{4h}$ | $4:m$ | $4/m$ |
| $\mathbf{C}_i$ | $\bar{2}$ | $\bar{1}$ | $\mathbf{D}_{2d}$ | $\bar{4}\cdot m$ | $\bar{4}2m$ |
| $\mathbf{C}_{2h}$ | $2:m$ | $2/m$ | $\mathbf{S}_4$ | $\bar{4}$ | $\bar{4}$ |
| $\mathbf{C}_2$ | 2 | 2 | $\mathbf{C}_4$ | 4 | 4 |
| $\mathbf{C}_s$ | $m$ | $m$ | $\mathbf{D}_{6h}$ | $m\cdot 6:m$ | $6/mmm$ |
| $\mathbf{D}_{2h}$ | $m\cdot 2:m$ | $mmm$ | $\mathbf{D}_6$ | $6:2$ | 622 |
| $\mathbf{D}_2$ | $2:2$ | 222 | $\mathbf{C}_{6v}$ | $6\cdot m$ | $6mm$ |
| $\mathbf{C}_{2v}$ | $2\cdot m$ | $mm2$ | $\mathbf{C}_{6h}$ | $6:m$ | $6/m$ |
| $\mathbf{D}_{3h}$ | $m\cdot 3:m$ | $\bar{6}m2$ | $\mathbf{D}_{3d}$ | $\bar{6}\cdot m$ | $\bar{3}m$ |
| $\mathbf{D}_3$ | $3:2$ | 32 | $\mathbf{C}_{3h}$ | $3:m$ | $\bar{6}$ |
| $\mathbf{C}_{3v}$ | $3\cdot m$ | $3m$ | $\mathbf{C}_6$ | 6 | 6 |
| $\mathbf{S}_6$ | $\bar{6}$ | $\bar{3}$ | $\mathbf{O}_h$ | $\bar{6}/4$ | $m3m$ |
| $\mathbf{C}_3$ | 3 | 3 | $\mathbf{O}$ | $3/4$ | 432 |
| $\mathbf{D}_{4h}$ | $m\cdot 4:m$ | $4/mmm$ | $\mathbf{T}_d$ | $3/\bar{4}$ | $\bar{4}3m$ |
| $\mathbf{D}_4$ | $4:2$ | 422 | $\mathbf{T}_h$ | $\bar{6}/2$ | $m/3$ |
| $\mathbf{C}_{4v}$ | $4\cdot m$ | $4mm$ | $\mathbf{T}$ | $3/2$ | 23 |

In the literature on crystal chemistry two systems of notation are generally used: the Shubnikov system and the International system. We give for reference the notations for the 32 point groups in the three systems (Table 2.4) without explaining the meaning of the symbols.

Finally, it should be mentioned that in order to describe symmetries in magnetic systems, so-called ‘colour groups’ are used [20]. However, the study of crystalline magnetic materials is beyond the scope of this book.

## 2.6 RULES FOR THE DETERMINATION OF MOLECULAR SYMMETRY

In this section rules are formulated that facilitate the deterination of molecular symmetry point groups. A set of symmetry elements is found as follows:

(1) First we determine whether the molecule under consideration belongs to the axial groups $\mathbf{D}_{\infty h}$ or $\mathbf{C}_{\infty v}$. Only linear molecules possess such symmetries.

(2) If the molecule is nonlinear, we seek a symmetry axis $C_n$ of higher order. If this is the $C_5$ axis, it is necessary to find whether it is unique. A molecule with several $C_5$ axes belongs to $\mathbf{I}_h$ or $\mathbf{I}$. If there are other $C_5$ axes we may proceed to Step (3). If there is only one $C_4$ axis, the analysis is also carried out according to Step (3). A molecule with three $C_4$ axes belongs to $\mathbf{O}_h$ or $\mathbf{O}$. Finally, if a molecule has four $C_3$ axes, it has $\mathbf{T}$

symmetry (there are neither planes nor a symmetry centre), $\mathbf{T}_h$ symmetry (there is a symmetry centre) or $\mathbf{T}_d$ symmetry (there are six planes and three $S_4$ axes). In general, a high degree of icosahedral or cubic symmetry is immediately obvious. The choice between different symmetry groups of the above-mentioned types is a little more complicated.

(3) If a molecule has a single $C_n$ axis, it is necessary to determine whether there are $n$ axes $C'_2$ perpendicular to the $C_n$ axis. If such axes do exist, the molecule belongs to one of the dihedral groups. If there are no other symmetry elements, this is $\mathbf{D}_n$; if there is a horizontal plane, it is $\mathbf{D}_{nh}$; if there are no horizontal planes but $n$ vertical planes, it is $\mathbf{D}_{nd}$.

(4) If the molecule has a vertical $C_n$ axis but no horizontal $C'_2$ axes, it is necessary to find the rotoflection axis $S_{2n}$ or to prove its absence. In the latter case the molecule belongs to $\mathbf{C}_n$ if the symmetry elements include no horizontal or vertical planes, to $\mathbf{C}_{nh}$ if there is a horizontal plane, and to $\mathbf{C}_{nv}$ if there are $n$ vertical planes.

(5) The low-symmetry groups $\mathbf{C}_s$ and $\mathbf{C}_i$ are determined with ease: a molecule that has only a symmetry plane belongs to $\mathbf{C}_s$, while a molecule with a centre of inversion belongs to $\mathbf{C}_i$.

We now consider some examples of the use of Steps (1)–(5).

**Example 2.52** The allene molecule $H_2C{=}C{=}CH_2$ (Fig. 2.34) with a "chessboard" conformation [13].

(1) The molecule is nonlinear; hence it does not belong to any of the continuous groups.

(2) There are no $C_5$ or $C_4$ axes.

(3) The symmetry axis $C_2$ intersects line of bonds $C{=}C{=}C$; there are no higher-order axes. From Fig. 2.34 it can be seen that there are $C'_2$ and $C''_2$ axes perpendicular to the $C_2$ axis. Therefore the molecule belongs to one of the

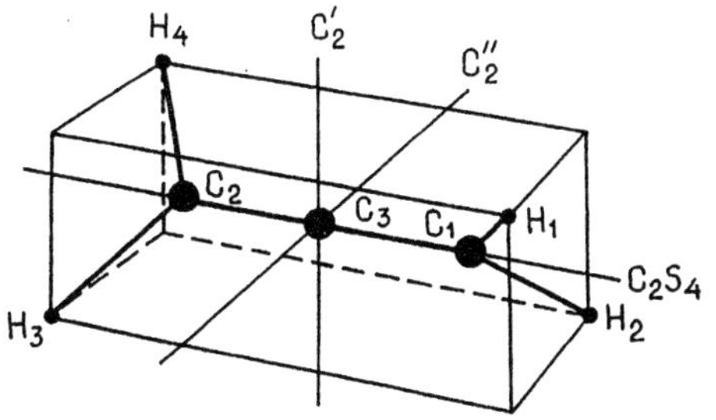


**Fig. 2.34** The allene molecule and its symmetry elements ($\mathbf{D}_{2d}$).

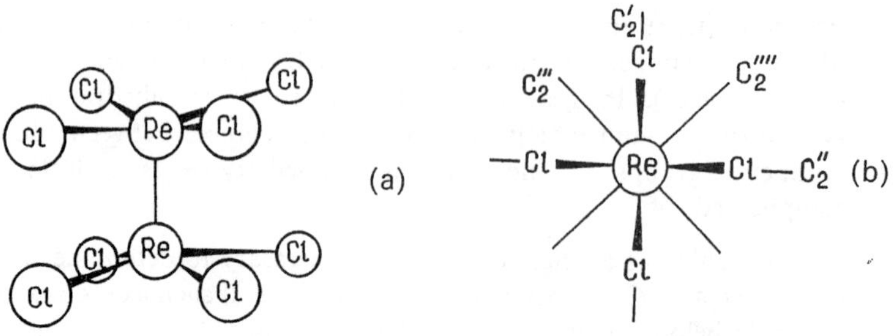


**Fig. 2.35** A binuclear rhenium cluster: (a) structure of the $[Re_2Cl_8]^{2-}$ anion; (b) the view along the Re–Re bond.

dihedral groups. There is no horizontal plane passing through the $C_2'$ and $C_2''$ axes, but there are two vertical planes ($C_1H_1H_2$ and $C_2H_3H_4$). Hence the $C_2$ axis is also the rotoflection axis $S_4$, and thus the allene molecule belongs to the point group $\mathbf{D}_{2d}$.

**Example 2.53** The binuclear rhenium cluster $[Re_2Cl_8]^{2-}$ [31] (Fig. 2.35), which has a "screened" conformation.

(1) The molecule is nonlinear.

(2) It is obvious that there is only one $C_4$ axis, coinciding with the Re–Re bond.

(3) The cluster possesses four $C_2'$ axes perpendicular to the $C_4$ axis and passing through the mid-point of the Re–Re bond. These four $C_2'$ axes are located in the symmetry plane $\sigma_h$ (the reflection $\hat{\sigma}_h$ changes the positions of the $ReCl_4$ groups). Therefore the point symmetry of the $[Re_2Cl_8]^{2-}$ anion is $\mathbf{D}_{4h}$.

**Example 2.54** The $B_4Cl_4$ molecule (Fig. 2.36), with the B–Cl bonds radiating from the vertices of the tetrahedron formed by the boron atoms.

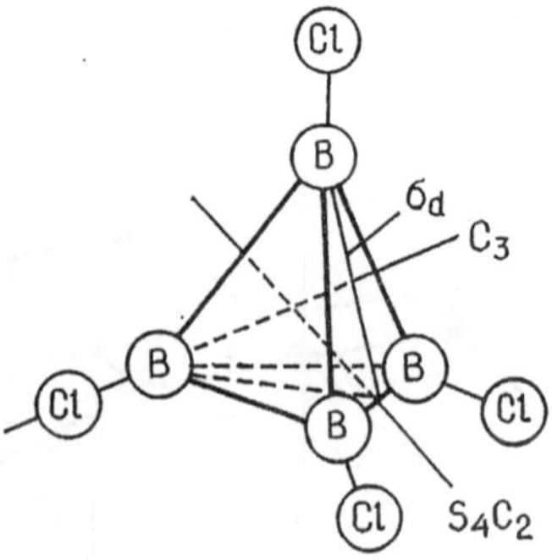


**Fig. 2.36** The $B_4Cl_4$ molecule and its symmetry elements ($\mathbf{T}_d$).

(1) The molecule is nonlinear.

(2) There are no $C_4$ axes. There are four $C_3$ axes, indicating **T** symmetry. From Fig. 2.36 it can be seen that the molecule has six $\sigma_d$ planes and the three $S_4$ axes; thus it has $\mathbf{T}_d$ symmetry.

**Example 2.55** The $[Bi_2Cl_8]^{2-}$ ion [31] (Fig. 2.37). The ligands form square pyramids around each Bi atom, with the bridging chlorine atoms in common; the Bi atoms are located slightly below the basal planes of the pyramids.

(1) The molecule is nonlinear.

(2) There are no $C_5$, $C_4$ or $C_3$ axes.

(3) There is a $C_2$ axis coinciding with the line joining the chorine bridge atoms. There are no $C_2' \perp C_2$ axes; this excludes the possibility of dihedral symmetry.

(4) The symmetry plane $\sigma_h$ perpendicular to the $C_2$ axis passes through the Bi–Bi line. The molecule has $\mathbf{C}_{2h}$ symmetry.

In later chapters it will become clear that symmetry concepts are most effectively used for systems with a high degree of symmetry. Therefore the problem sometimes arises concerning the comparison of different point groups—in other words, the question of which of two groups is "more symmetrical". It is obvious that $\mathbf{O}_h$ is more symmetrical than $\mathbf{D}_{4h}$, since the latter lacks several symmetry elements of the former (four $C_3$ axes etc.). This conclusion is consistent with intuitive concepts regarding symmetry: a regular octahedron ($\mathbf{O}_h$) is more symmetrical than one that is deformed (compressed or stretched) along the $C_4$ axis ($\mathbf{D}_{4h}$). Nevertheless, it is not always possible to make an unambiguous comparison between the symmetries of two molecules. Consider, for example, $\mathbf{C}_2$, $\mathbf{C}_s$ and $\mathbf{C}_i$. Each contains one non-trivial symmetry element. In molecules belonging to these groups the atoms not located on the symmetry elements constitute equivalent pairs, and comparing the symmetries of the three given groups is of no use. However, such a comparison is possible

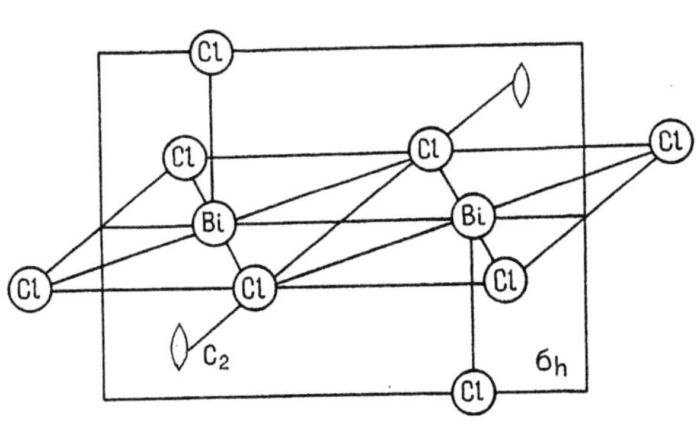


**Fig. 2.37** The $[BiCl_8]^{2-}$ ion and its symmetry elements ($\mathbf{C}_{2h}$).

**Table 2.5** Subgroups of the 32 point groups [1].

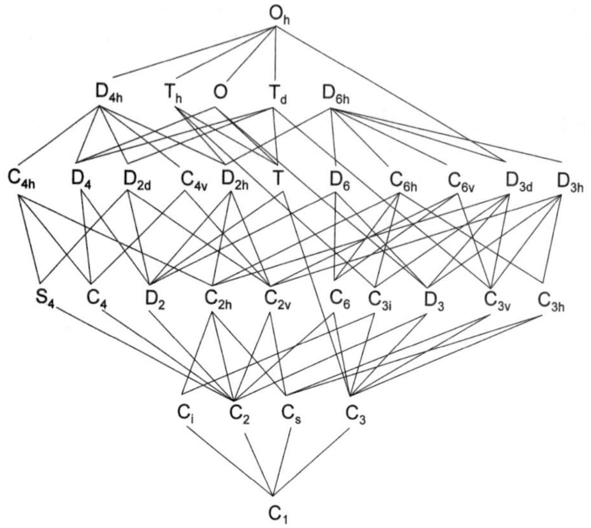


for a chain of groups, each of which is a subgroup of the preceding one, for example

$$\mathbf{O}_h \rightarrow \mathbf{D}_{4h} \rightarrow \mathbf{C}_{4v} \rightarrow \mathbf{C}_4 \rightarrow \mathbf{C}_2 \rightarrow \mathbf{C}_1.$$

Each successive group in this chain is less symmetric than the preceding one. This decrease in symmetry is called *reduction*. It can result from substitution of ligands in coordination compounds or the application of external electric or magnetic fields; the formation of complexes can also lead to reduction. Manifestations of symmetry reduction are discussed below; here we show in Table 2.5 the subgroups of all the 32 point groups [1]. Using this table, it is easy to follow the reduction chains of any point group.

## PROBLEMS

**2.1** Find the equivalent symmetry elements of a tetrahedral molecule.

**2.2** Check the multiplication table of $\mathbf{C}_{4v}$ (Table 2.2).

**2.3** Using the multiplication table, prove that for $\mathbf{C}_{4v}$, $\hat{C}_4\hat{\sigma}_v\hat{C}_4^{-1} = \hat{\sigma}'_v$.

**2.4** Using the multiplication table divide the $\mathbf{C}_{4v}$ group operations into classes.

**2.5** Prove that the $\sigma_h \perp C_3$ plane does not provide conjugation of the rotations $\hat{C}_3$ and $\hat{C}_3^{-1}$.

**2.6** Find the products $\hat{C}_3\hat{\sigma}_v$, $\hat{\sigma}_v\hat{C}_3$ in $\mathbf{C}_{3v}$.

**2.7** Analyse the $\hat{C}_3\hat{\sigma}_d$ transformation of $\mathbf{D}_{3d}$ and prove that the $C_3$ axis is also the rotoflection $S_6$ axis.

**2.8** Prove that the rotoflection transformations $\hat{S}_n^k$ and $\hat{S}_n^{-k}$ in $\mathbf{D}_{nd}$ for even $n$ are conjugate in pairs.

**2.9** Determine the symmetry point groups of the following molecules: the hexameric molybdenum cluster $[Mo_6Cl_8]^{4+}$ (Fig. 2.38); the binuclear complex anion $[Tl_2Cl_9]^{3-}$ formed from two octahedral thallium fragments $TlCl_6$ with a common facet (Fig. 2.39); and an $IF_7$ molecule that is a pentagonal bipyramid (Fig. 2.40).

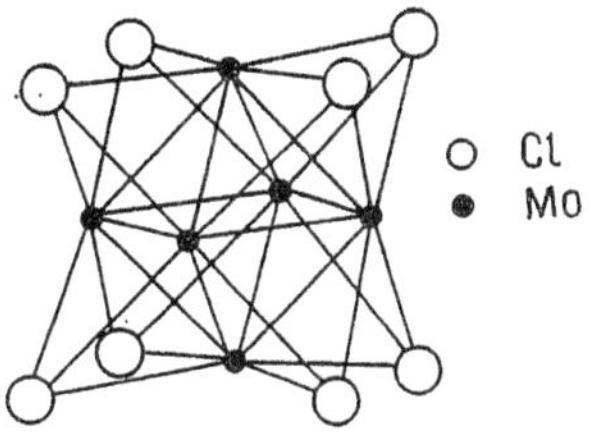


**Fig. 2.38** The hexameric cluster $[Mo_6Cl_8]^{4+}$.

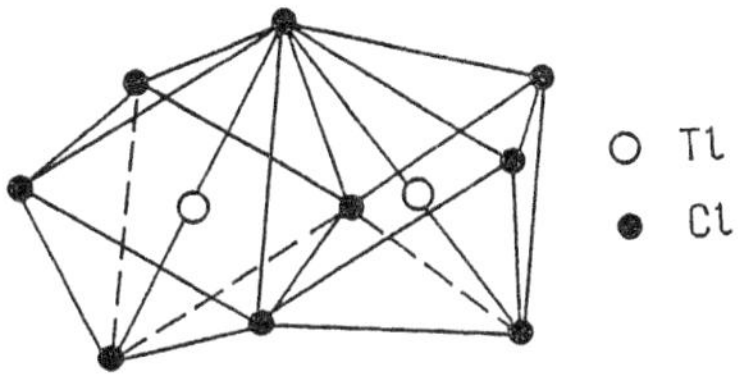


**Fig. 2.39** The binuclear complex anion $[Tl_2Cl_9]^{3-}$.

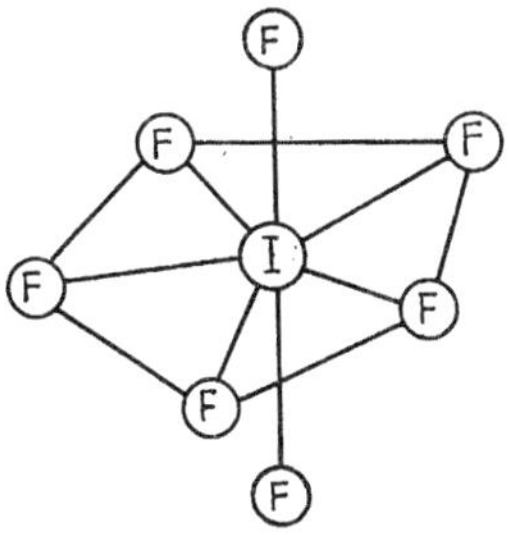


**Fig. 2.40** The $IF_7$ molecule.

**2.10** Prove that the point groups of isomeric forms of the substituted octahedral complexes $MA_4B_2$, $MA_3B_3$ and $MA_2B_2C_2$ correspond to those shown in Fig. 2.41 [30].

**2.11** Prove that the point groups of the three isomers of octahedral complexes $M(AA)B_2C_2$ (Fig. 2.42a) with a symmetric bidentate ligand AA and of the four isomers of complexes $M(AB)C_2D_2$ (Fig. 2.42b) with a non-symmetric planar chelate ligand AB correspond to those shown in the figures.

**2.12** Analyse the point symmetry of isomeric forms of the bischelate complexes with symmetrical AA (Fig. 2.43a) and non-symmetrical AB (Fig. 2.43b) ligands.

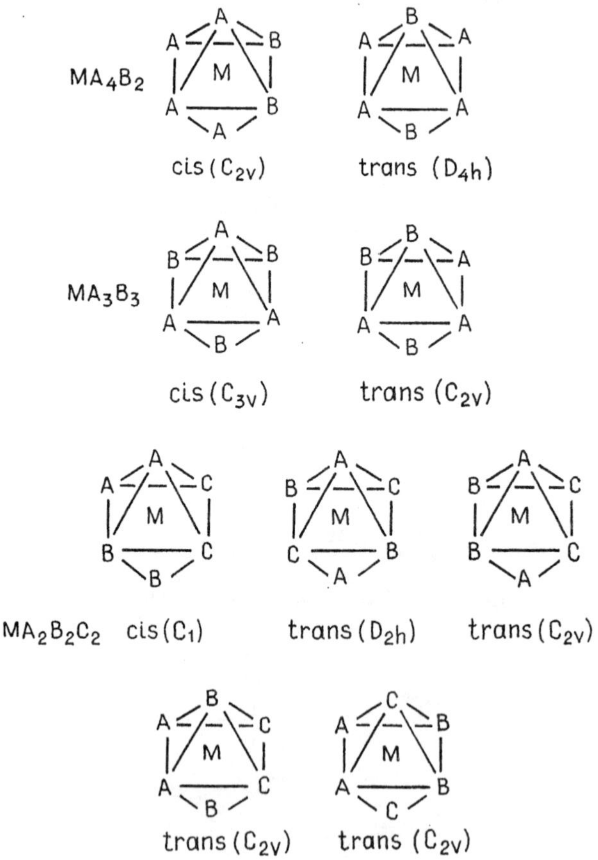


**Fig. 2.41** Isomeric forms of the octahedral complexes $MA_4B_2$, $MA_3B_3$ and $MA_2B_2C_2$.

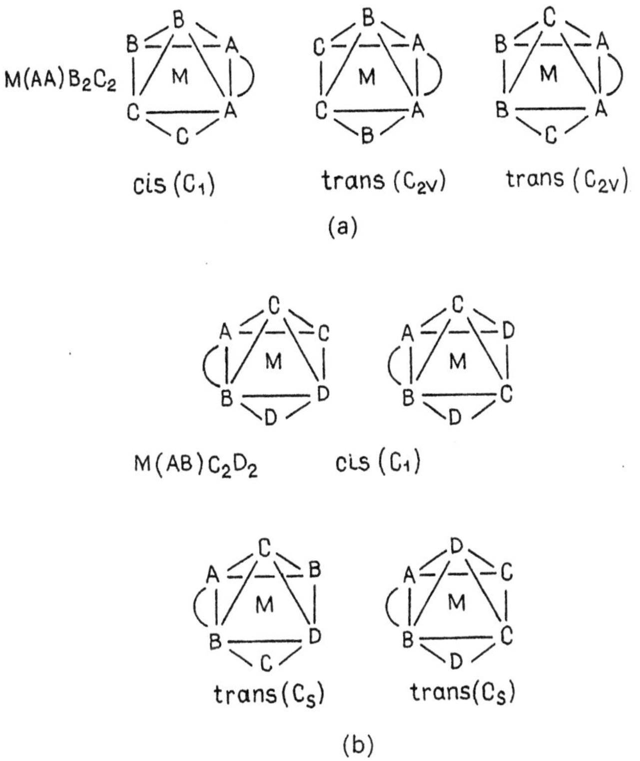


**Fig. 2.42** Isomeric forms of the octahedral complexes (a) $M(AA)B_2C_2$ and (b) $MC(AB)C_2D$ with bidentate ligands (after [30]).

**2.13** Prove that $\hat{C}_4(y)$ and $\hat{C}_4(z)$ are the generating operations of the group **O**.

**2.14** Find the generating operations of the $\mathbf{D}_{3d}$ and $\mathbf{D}_{4h}$ groups.

**2.15** Prove that $\mathbf{O}_h$ and $\mathbf{T}_d$ possess three generating operations each. Find these generating operations and construct the full sets of operations.

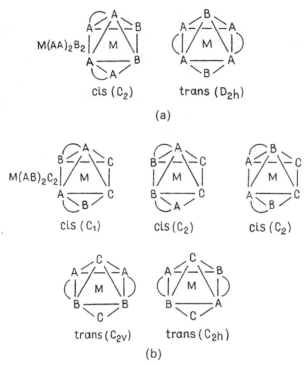

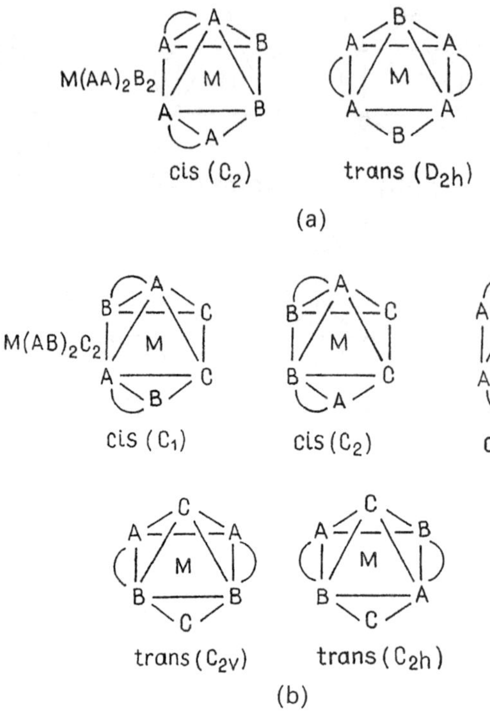


**Fig. 2.43** Isomeric forms of the bis-chelate octahedral complexes of type $M(AA)_2B_2$ with (a) symmetric and (b) nonsymmetric ligands (after [30]).

# 3 Representations of Point Groups

## 3.1 MATRICES AND VECTORS

### 3.1.1 Definition of a matrix

This chapter deals with the important concept of a *group representation*, which forms the basis for all of the chemical and physical applications of group theory. Before proceeding with group representation theory, let us recall some details regarding matrices and vectors, which will be needed in the following.

Matrices are arrays of numbers of symbols obeying specific algebraic rules. They are usually denoted with square brackets, for example

$$\begin{bmatrix} 1 & 3 & 2 & 1 \\ 2 & 7 & 3 & 5 \\ 1 & 0 & 0 & 4 \\ 2 & 2 & 3 & 1 \end{bmatrix}, \qquad \begin{bmatrix} 3 & 5 & 2 & 7 \\ 1 & 3 & 4 & 0 \end{bmatrix}.$$

The first of these is a *square matrix*, with number of rows equal to number of columns—this number is called the *dimension* or *order* of the matrix. The second is a *rectangular matrix*, consisting of two rows and four columns (a $2 \times 4$ matrix). Using the notation $\mathbf{A} \equiv [A]$, a matrix can be represented as

$$\mathbf{A} \equiv \begin{bmatrix} A_{11} & A_{12} & \cdots & A_{1n} \\ A_{21} & A_{22} & \cdots & A_{2n} \\ \vdots & \vdots & & \vdots \\ A_{m1} & A_{m2} & \cdots & A_{mn} \end{bmatrix}. \tag{3.1}$$

The first subscript is the row number of the $m \times n$ matrix $\mathbf{A}$, while the second indicates the column number. $A_{ij}$ are the matrix *elements*; $A_{11}, A_{22}, \ldots, A_{ii}$ are the *diagonal elements* of a square matrix (forming the *main diagonal*). A square matrix with all diagonal elements unity and all nondiagonal ones zero is called a *unit matrix* and is denoted by **1**.

Important particular cases are column and row matrices:

$$\begin{bmatrix} A_1 \\ A_2 \\ \vdots \\ A_n \end{bmatrix}, \qquad [A_1 \quad A_2 \quad \ldots \quad A_n]. \tag{3.2}$$

By convention, the components of a vector $(x, y, z)$ are usually collected into a column, while the sets of unit vectors $(\mathbf{i}\,\mathbf{j}\,\mathbf{k})$ (and, later, "basis functions") are written as row matrices:

$$\begin{bmatrix} x \\ y \\ z \end{bmatrix}, \qquad [\mathbf{i} \; \mathbf{j} \; \mathbf{k}]. \tag{3.3}$$

We note that the columns of the matrix **A** in (3.1) may be regarded as columns of vector components.

### 3.1.2 Matrix multiplication

We now consider some rules for manipulating matrices, i.e. *matrix algebra*. The sum of matrices **A** and **B** is matrix **C** whose elements are formed by addition:

$$C_{ij} = A_{ij} + B_{ij}. \tag{3.4}$$

Multiplication of a matrix by a number $\alpha$ obeys a simple rule: $\alpha\mathbf{A}$ is the matrix with the elements $\alpha A_{ij}$. The rule for matrix multiplication is more complex. We shall consider it first with reference to square matrices **A** and **B**. Suppose that **A** and **B** are of second order. Then their product $\mathbf{C} = \mathbf{AB}$ is expressed as

$$\begin{bmatrix} A_{11} & A_{12} \\ A_{21} & A_{22} \end{bmatrix} \begin{bmatrix} B_{11} & B_{12} \\ B_{21} & B_{22} \end{bmatrix} = \begin{bmatrix} A_{11}B_{11} + A_{12}B_{21} & A_{11}B_{12} + A_{12}B_{22} \\ A_{21}B_{11} + A_{22}B_{21} & A_{21}B_{12} + A_{22}B_{22} \end{bmatrix}. \tag{3.5}$$

The elements $C_{11}$ and $C_{12}$ of the first row of the product are obtained by multiplying the first row of **A** by the first and second columns of **B** respectively; the elements of the second row, $C_{21}$ and $C_{22}$, are obtained similarly. For matrices of arbitrary dimension $n$ it is possible to write

$$C_{ij} = A_{i1}B_{1j} + A_{i2}B_{2j} + \cdots + A_{in}B_{nj} \equiv \sum_{k=1}^{n} A_{ik}B_{kj}, \tag{3.6}$$

where the summation is over row $i$ of **A** and column $j$ of **B**. Thus the multiplication rule is as follows: the $ij$ element of the product is obtained by multiplying row $i$ of **A** by column $j$ of **B**.

Rectangular matrices are multiplied in a similar way; their product **AB**, however, has a meaning only when the number of columns of **A** (i.e. the number of elements in each of its rows) is equal to the number of rows of **B** (i.e. the number of elements in each of its columns). On multiplying an $n \times p$ matrix **A** by a $p \times m$ matrix **B**, we obtain an $n \times m$ matrix **C**. Of course, one matrix may be square. Let us apply the rule (3.6) for multiplying the following $3 \times 2$ and $2 \times 4$ matrices:

$$\begin{bmatrix} A_{11} & A_{12} \\ A_{21} & A_{22} \\ A_{31} & A_{32} \end{bmatrix} \begin{bmatrix} B_{11} & B_{12} & B_{13} & B_{14} \\ B_{21} & B_{22} & B_{23} & B_{24} \end{bmatrix} = \begin{bmatrix} C_{11} & C_{12} & C_{13} & C_{14} \\ C_{21} & C_{22} & C_{23} & C_{24} \\ C_{31} & C_{32} & C_{33} & C_{34} \end{bmatrix}, \tag{3.7}$$

$$C_{11} = A_{11}B_{11} + A_{12}B_{21}, \qquad C_{21} = A_{21}B_{11} + A_{22}B_{21},$$

$$C_{12} = A_{11}B_{12} + A_{12}B_{22}, \qquad C_{22} = A_{21}B_{12} + A_{22}B_{22}.$$

The other elements of the $3 \times 4$ matrix **C** are obtained in the same way.

It should be noted that, in general, matrix multiplication is noncommutative. Consider, for example, the following products:

$$\begin{bmatrix} 1 & 0 \\ 0 & -1 \end{bmatrix} \begin{bmatrix} 0 & 1 \\ 1 & 0 \end{bmatrix} = \begin{bmatrix} 0 & 1 \\ -1 & 0 \end{bmatrix},$$

$$\begin{bmatrix} 0 & 1 \\ 1 & 0 \end{bmatrix} \begin{bmatrix} 1 & 0 \\ 0 & -1 \end{bmatrix} = \begin{bmatrix} 0 & -1 \\ 1 & 0 \end{bmatrix}.$$

However, matrix multiplication is associative, i.e.

$$\mathbf{ABC} = \mathbf{A}(\mathbf{BC}) = (\mathbf{AB})\mathbf{C}.$$

The notion of an inverse matrix $\mathbf{A}^{-1}$ must also be introduced. This matrix satisfies the relation $\mathbf{A}^{-1}\mathbf{A} = \mathbf{1}$. Only so-called *nonsingular matrices*, i.e. those whose determinants are nonzero, can have inverses. Rules for obtaining inverse matrices are given in most textbooks on linear algebra.

### 3.1.3 Multiplication of block-diagonal matrices

A matrix all of whose elements are zero except for those on the main diagonal ($A_{ii}$), is called a *diagonal matrix*. Of great importance later will be matrices that consist of *blocks* or *submatrices*, all other elements being zero. For example, the matrix

$$\left[\begin{array}{cc|ccc|c} 1 & 2 & 0 & 0 & 0 & 0 \\ 0 & 1 & 0 & 0 & 0 & 0 \\ \hline 0 & 0 & 1 & 3 & 2 & 0 \\ 0 & 0 & 2 & 7 & 4 & 0 \\ 0 & 0 & 4 & 2 & 1 & 0 \\ \hline 0 & 0 & 0 & 0 & 0 & 3 \end{array}\right]$$

consists of second-, third- and first-order blocks. Such matrices are called *block-diagonal matrices*. Let us multiply (according to the rule (3.6)) two block-diagonal matrices having the same dimensions and block positions:

$$\left[\begin{array}{cc|ccc|c} 1 & 2 & 0 & 0 & 0 & 0 \\ 0 & 1 & 0 & 0 & 0 & 0 \\ \hline 0 & 0 & 1 & 3 & 2 & 0 \\ 0 & 0 & 2 & 7 & 4 & 0 \\ 0 & 0 & 4 & 2 & 1 & 0 \\ \hline 0 & 0 & 0 & 0 & 0 & 3 \end{array}\right] \left[\begin{array}{cc|ccc|c} 2 & 3 & 0 & 0 & 0 & 0 \\ 1 & 4 & 0 & 0 & 0 & 0 \\ \hline 0 & 0 & 3 & 1 & 2 & 0 \\ 0 & 0 & 2 & 4 & 1 & 0 \\ 0 & 0 & 1 & 3 & 5 & 0 \\ \hline 0 & 0 & 0 & 0 & 0 & 2 \end{array}\right]$$

$$= \left[\begin{array}{cc|ccc|c} 4 & 11 & 0 & 0 & 0 & 0 \\ 1 & 4 & 0 & 0 & 0 & 0 \\ \hline 0 & 0 & 11 & 19 & 15 & 0 \\ 0 & 0 & 24 & 42 & 31 & 0 \\ 0 & 0 & 17 & 15 & 15 & 0 \\ \hline 0 & 0 & 0 & 0 & 0 & 6 \end{array}\right].$$

We can see that multiplication of block-diagonal matrices gives a matrix with similar structure. In other words, blocks of equal dimensions can be multiplied independently,

$$\begin{bmatrix} 1 & 2 \\ 0 & 1 \end{bmatrix} \begin{bmatrix} 2 & 3 \\ 1 & 4 \end{bmatrix} = \begin{bmatrix} 4 & 11 \\ 1 & 4 \end{bmatrix},$$

$$\begin{bmatrix} 1 & 3 & 2 \\ 2 & 7 & 4 \\ 4 & 2 & 1 \end{bmatrix} \begin{bmatrix} 3 & 1 & 2 \\ 2 & 4 & 1 \\ 1 & 3 & 5 \end{bmatrix} = \begin{bmatrix} 11 & 19 & 15 \\ 24 & 42 & 48 \\ 17 & 15 & 15 \end{bmatrix},$$

$$[3][2] = [6],$$

and positioned in the same sequence as in the multiplied matrices. This can be expressed as follows:

$$\begin{bmatrix} \mathbf{A}_1 & & & \\ & \mathbf{A}_2 & & \\ & & \ddots & \\ & & & \mathbf{A}_n \end{bmatrix} \begin{bmatrix} \mathbf{B}_1 & & & \\ & \mathbf{B}_2 & & \\ & & \ddots & \\ & & & \mathbf{B}_n \end{bmatrix} = \begin{bmatrix} \mathbf{A}_1\mathbf{B}_1 & & & \\ & \mathbf{A}_2\mathbf{B}_2 & & \\ & & \ddots & \\ & & & \mathbf{A}_n\mathbf{B}_n \end{bmatrix} \tag{3.8}$$

where $\mathbf{A}_i$ and $\mathbf{B}_i$ are submatrices of the same dimension, and $\mathbf{A}_i\mathbf{B}_i$ are their products (the zero elements of the matrices have been omitted for clarity).

### 3.1.4 Matrix characters

The *trace* or *character* $\chi$ of a matrix is the sum of its diagonal elements:

$$\chi(\mathbf{A}) = \sum_i A_{ii}. \tag{3.9}$$

If, for example, $\mathbf{A} = \begin{bmatrix} 1 & 2 \\ 3 & 5 \end{bmatrix}$ then $\chi(\mathbf{A}) = 6$. The character is an important quantity, used frequently later.

We shall need the following properties of characters.

(1) The character of a product of two (not necessarily commuting) matrices does not depend upon the sequence of multiplication. In other words, for two matrices $\mathbf{A}$ and $\mathbf{B}$,

$$\chi(\mathbf{AB}) = \chi(\mathbf{BA}), \tag{3.10}$$

even if $\mathbf{AB} \neq \mathbf{BA}$. This can be proved as follows:

$$\left.\begin{aligned} \chi(\mathbf{AB}) &= \sum_i (\mathbf{AB})_{ii} = \sum_i \sum_k A_{ik}B_{ki}, \\ \chi(\mathbf{BA}) &= \sum_k (\mathbf{BA})_{kk} = \sum_{ik} B_{ki}A_{ik} \\ &= \sum_i \sum_k A_{ik}B_{ki} \equiv \chi(\mathbf{AB}). \end{aligned}\right\} \tag{3.11}$$

(2) The characters of conjugate matrices coincide. *Conjugate matrices* (similar to conjugate operations) $\mathbf{A}$ and $\mathbf{B}$ are matrices of the same order that are related by

$$\mathbf{A} = \mathbf{Q}^{-1}\mathbf{B}\mathbf{Q}, \tag{3.12}$$

where $\mathbf{Q}$ is another matrix of the same order. A transformation of the type (3.12) is called a *similarity transformation*. The equality of the characters of $\mathbf{A}$ and $\mathbf{B}$ can be proved using the associativity of matrix multiplication and the property (1):

$$\chi(\mathbf{A}) = \chi(\mathbf{Q}^{-1}\mathbf{BQ}) = \chi(\mathbf{BQQ}^{-1}) = \chi(\mathbf{B1}) = \chi(\mathbf{B}). \qquad (3.13)$$

## 3.2 MATRIX FORM OF GEOMETRICAL TRANSFORMATIONS

Let $\mathbf{r}$ be a general vector representing the position of a point P on the $xy$-plane, with cartesian coordinates $x$, $y$. In terms of the unit vectors $\mathbf{i}$, $\mathbf{j}$ along the $x$ and $y$ axes we write

$$\mathbf{r} = x\mathbf{i} + y\mathbf{j} \qquad (3.14)$$

where $x$, $y$ are the components of the vector $\mathbf{r}$. If we now rotate all vectors through the angle $\vartheta$ so that they become $\mathbf{r}'$, $\mathbf{i}'$, $\mathbf{j}'$ we see that

$$\mathbf{i}' = \mathbf{i}\cos\vartheta + \mathbf{j}\sin\vartheta,$$
$$\mathbf{j}' = -\mathbf{i}\sin\vartheta + \mathbf{j}\cos\vartheta, \qquad (3.15)$$

while $\mathbf{r}'$ (the new vector, related to $\mathbf{i}'$ and $\mathbf{j}'$ just as $\mathbf{r}$ was related to $\mathbf{i}$ and $\mathbf{j}$) will be

$$\mathbf{r}' = x\mathbf{i}' + y\mathbf{j}' = (x\cos\vartheta - y\sin\vartheta)\mathbf{i} + (x\sin\vartheta + y\cos\vartheta)\mathbf{j} = x'\mathbf{i} + y'\mathbf{j}.$$

In this way the components of the new vector are related to those of the original vector by

$$x' = x\cos\vartheta - y\sin\vartheta,$$
$$y' = x\sin\vartheta + y\cos\vartheta. \qquad (3.16)$$

On writing the components of a vector as a column matrix, as in (3.3), the rotation of $\mathbf{r}$ into $\mathbf{r}'$ is described by the matrix equation

$$\begin{bmatrix} x' \\ y' \end{bmatrix} = \mathbf{D}(\vartheta)\begin{bmatrix} x \\ y \end{bmatrix}, \qquad (3.17)$$

where

$$\mathbf{D}(\vartheta) = \begin{bmatrix} \cos\vartheta & -\sin\vartheta \\ \sin\vartheta & \cos\vartheta \end{bmatrix}, \qquad (3.18)$$

is a rotation matrix.

It is important to note that the unit vectors in (3.14) do not follow the same transformation law as the components: in fact, (3.15) may be written

$$[\mathbf{i}' \ \mathbf{j}'] = [\mathbf{i} \ \mathbf{j}]\,\mathbf{D}(\vartheta), \tag{3.19}$$

where the unit vectors are collected in a row matrix, while the rotation matrix appears on the right. With this notation the vector $\mathbf{r}$ in (3.14) becomes a row-column product:

$$\mathbf{r} = [\mathbf{i} \ \mathbf{j}] \begin{bmatrix} x \\ y \end{bmatrix}. \tag{3.20}$$

In applying the rotation operator we have defined the rotated vector as $\mathbf{r}' = x\mathbf{i}' + y\mathbf{j}'$ (only vectors being rotated, not numbers) and thus

$$\mathbf{r}' = [\mathbf{i}' \ \mathbf{j}'] \begin{bmatrix} x \\ y \end{bmatrix} = [\mathbf{i} \ \mathbf{j}]\,\mathbf{D}(\vartheta) \begin{bmatrix} x \\ y \end{bmatrix} = [\mathbf{i} \ \mathbf{j}] \begin{bmatrix} x' \\ y' \end{bmatrix}.$$

This equation confirms that the components of the new vector are indeed given by (3.17). It is convenient to indicate the effect of a general operation $\hat{R}$ by writing, for example,

$$\hat{R}\mathbf{i} - \mathbf{i}', \ \hat{R}\mathbf{j} = \mathbf{j}', \ \hat{R}[\mathbf{i} \ \mathbf{j}] = [\mathbf{i}' \ \mathbf{j}'], \tag{3.21}$$

and with this notation (3.19) becomes

$$\hat{R}[\mathbf{i} \ \mathbf{j}] = [\mathbf{i} \ \mathbf{j}]\mathbf{D}(\hat{R}), \tag{3.22}$$

where $\mathbf{D}(\hat{R})$ is the matrix whose columns in the components of $\mathbf{i}'$ and $\mathbf{j}'$ in terms of $\mathbf{i}$ and $\mathbf{j}$, just as in (3.15). This basic equation defines the matrix $\mathbf{D}(\hat{R})$ that represents the operation $\hat{R}$.

Assume $\mathbf{r}_1$ is a vector representing the position of the point with the coordinates $x_1$ and $y_1$ (Fig. 3.1). Let us rotate $\mathbf{r}_1$ through an angle $\vartheta$ so that in its new position, $\mathbf{r}_2$, the coordinates are $(x_2 y_2)$:

$$\begin{bmatrix} x_2 \\ y_2 \end{bmatrix} = \mathbf{D}(\vartheta) \begin{bmatrix} x_1 \\ y_1 \end{bmatrix}, \tag{3.23}$$

where the rotation matrix is given by the equation (3.18).

Let us now rotate the vector $\mathbf{r}_2$ through an angle $\varphi$ so that it becomes $\mathbf{r}_3$. In matrix form this rotation is expressed as

$$\begin{bmatrix} x_3 \\ y_3 \end{bmatrix} = \mathbf{D}(\varphi) \begin{bmatrix} x_2 \\ y_2 \end{bmatrix}. \tag{3.24}$$

The transformation from $\mathbf{r}_1$ to $\mathbf{r}_3$ can also be performed directly by the rotation matrix $\mathbf{D}(\vartheta + \varphi)$:

$$\begin{bmatrix} x_3 \\ y_3 \end{bmatrix} = \mathbf{D}(\vartheta + \varphi) \begin{bmatrix} x_1 \\ y_1 \end{bmatrix}, \qquad \mathbf{D}(\vartheta + \varphi) = \begin{bmatrix} \cos(\vartheta + \varphi) & -\sin(\vartheta + \varphi) \\ \sin(\vartheta + \varphi) & \cos(\vartheta + \varphi) \end{bmatrix}. \tag{3.25}$$

Let us obtain the $\mathbf{D}(\vartheta)\mathbf{D}(\varphi)$ product according to the matrix multiplication rule:

$$\mathbf{D}(\vartheta)\mathbf{D}(\varphi) = \begin{bmatrix} (\cos\vartheta \cos\varphi - \sin\vartheta \sin\varphi) & (-\cos\vartheta \sin\varphi - \sin\vartheta \sin\varphi) \\ (\sin\vartheta \cos\varphi + \cos\vartheta \sin\varphi) & (-\sin\vartheta \sin\varphi + \cos\vartheta \cos\varphi) \end{bmatrix}$$

$$= \begin{bmatrix} \cos(\vartheta+\varphi) & -\sin(\vartheta+\varphi) \\ \sin(\vartheta+\varphi) & \cos(\vartheta+\varphi) \end{bmatrix}$$

$$= \mathbf{D}(\vartheta+\varphi). \tag{3.26}$$

The matrices $\mathbf{D}(\vartheta)$ and $\mathbf{D}(\varphi)$ correspond to the rotation operations $\hat{C}(\vartheta)$ and $\hat{C}(\varphi)$, while $\mathbf{D}(\vartheta+\varphi)$ corresponds to the rotation operation $\hat{\mathbf{C}}(\vartheta+\varphi)$. For rotation operations the multiplication rule is $\hat{C}(\vartheta+\varphi) = \hat{C}(\vartheta)\hat{C}(\varphi)$, and it follows from (3.26) that the same rule is valid for rotation matrices:

$$\begin{array}{ccc} \hat{C}(\vartheta)\hat{C}(\varphi) & = & \hat{C}(\vartheta+\varphi) \\ \updownarrow \quad \updownarrow & & \updownarrow \\ \mathbf{D}(\vartheta)\mathbf{D}(\varphi) & = & \mathbf{D}(\vartheta+\varphi). \end{array} \tag{3.27}$$

Let us write in matrix form the reflection operation $\hat{\sigma}(xz)$ in the vertical $(x, z)$ plane. It follows from Fig. 3.1 that

$$\left.\begin{aligned} x_4 &= 1 \cdot x_1 + 0 \cdot y_1, \\ y_4 &= 0 \cdot x_1 - 1 \cdot y_1, \\ \text{or} \quad \begin{bmatrix} x_4 \\ y_4 \end{bmatrix} &= \begin{bmatrix} 1 & 0 \\ 0 & -1 \end{bmatrix} \begin{bmatrix} x_1 \\ y_1 \end{bmatrix} \end{aligned}\right\} \tag{3.28}$$

Let us denote the *reflection matrix* by $\mathbf{D}(\hat{\sigma}(xz))$, emphasizing its relation to the operation $\hat{\sigma}(xz)$. On applying a reflection to the vector $\mathbf{r}_2$ with components

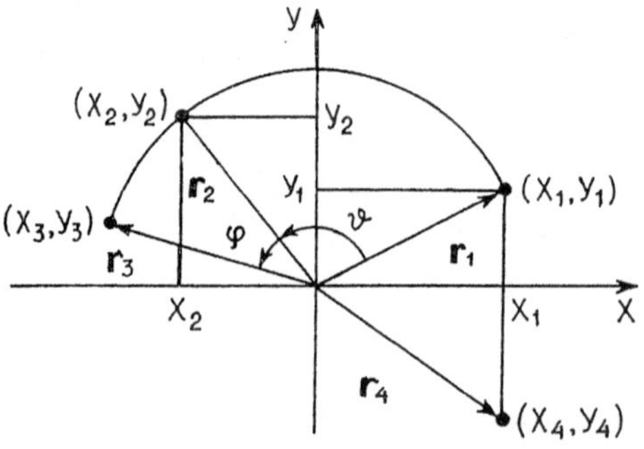


**Fig. 3.1** Transformation of vector components by rotations in the $(x, y)$ plane.

given by (3.23), we obtain a vector $\mathbf{r}_5$ with components

$$\left.\begin{aligned} x_5 &= x_1 \cos\vartheta - y_1 \sin\vartheta, \\ y_5 &= -x_1 \sin\vartheta - y_1 \cos\vartheta. \end{aligned}\right\} \tag{3.29}$$

In the matrix form

$$\begin{bmatrix} x_5 \\ y_5 \end{bmatrix} = \mathbf{D}(\vartheta)\begin{bmatrix} x_1 \\ y_1 \end{bmatrix}, \qquad \mathbf{D}(\vartheta) = \begin{bmatrix} \cos\vartheta & -\sin\vartheta \\ -\sin\vartheta & -\cos\vartheta \end{bmatrix}. \tag{3.30}$$

Here the matrix $\mathbf{D}$ corresponds to sequential application of the rotation $\hat{C}(\vartheta)$ and the reflection $\hat{\sigma}(xz)$, i.e. to the product of these operations ($\hat{C}(\vartheta)$ followed by $\hat{\sigma}(xz)$ is written as the product $\hat{\sigma}(xz)\hat{C}(\vartheta)$). On the other hand, on multiplying the associated matrices $\mathbf{D}(\vartheta)$ and $\mathbf{D}(\hat{\sigma})$ in the same order, we obtain

$$\begin{bmatrix} 1 & 0 \\ 0 & -1 \end{bmatrix}\begin{bmatrix} \cos\vartheta & -\sin\vartheta \\ \sin\vartheta & \cos\vartheta \end{bmatrix} = \begin{bmatrix} \cos\vartheta & -\sin\vartheta \\ -\sin\vartheta & -\cos\vartheta \end{bmatrix}. \tag{3.31}$$

Comparing (3.30) and (3.31), we can see that $\mathbf{D}$ in the former coincides with the matrix $\mathbf{D}(\hat{\sigma}(xz)\hat{C}(\vartheta))$ for successive rotation and reflection. The relation between the product of operations and the product of matrices is clear:

$$\mathbf{D}(\hat{\sigma}(xz)\hat{C}(\vartheta)) = \mathbf{D}(\hat{\sigma}(xz))\mathbf{D}(\hat{C}(\vartheta)). \tag{3.32}$$

The two examples given above—combination of rotations and combination of rotation and reflection—illustrate the following common feature: if the product of some geometrical operations $\hat{R}_n \cdots \hat{R}_2\hat{R}_1$ gives a transformation $\hat{Q}$ then the matrix $\mathbf{D}(\hat{Q})$ is obtained by multiplication of the matrices for the operations $\hat{R}_i$. In other words, if $\hat{R}_n \cdots \hat{R}_2\hat{R}_1 = \hat{Q}$ then

$$\mathbf{D}(\hat{R}_n)\cdots\mathbf{D}(\hat{R}_2)\mathbf{D}(\hat{R}_1) = \mathbf{D}(\hat{Q}). \tag{3.33}$$

This rule can be proved in general. Suppose that the product $\hat{R}_2\hat{R}_1$ ($\hat{R}_1$ followed by $\hat{R}_2$) is equivalent to $\hat{R}_3$ ($\hat{R}_2\hat{R}_1 = \hat{R}_3$) and that we define the matrix associated with each operator according to (3.22), then it is clear that

$$\hat{R}_2\hat{R}_1[\mathbf{i}\ \mathbf{j}] = \hat{R}_2[\mathbf{i}\ \mathbf{j}]\mathbf{D}(\hat{R}_1) = [\mathbf{i}\ \mathbf{j}]\mathbf{D}(\hat{R}_2)\mathbf{D}(\hat{R}_1). \tag{3.34}$$

The matrix associated with the operation $\hat{R}_2\hat{R}_1$ is thus

$$\mathbf{D}(\hat{R}_2\hat{R}_1) = \mathbf{D}(\hat{R}_2)\mathbf{D}(\hat{R}_1) = \mathbf{D}(\hat{R}_3). \tag{3.35}$$

In other words

$$\hat{R}_2\hat{R}_1 = \hat{R}_3 \quad \text{implies} \quad \mathbf{D}(\hat{R}_2)\mathbf{D}(\hat{R}_1) = \mathbf{D}(\hat{R}_3). \tag{3.36}$$

This relation is true for all members of a given set of operations.

Hence there is a correspondence between multiplication tables of operations and those of matrices: each geometrical operation is *represented* by a matrix, while a set of operations is *represented* by a set of matrices with the same multiplication table. It is important to note that the right-to-left order of operations is conserved in the representation.

## 3.3 GROUP REPRESENTATIONS

In Section 3.2 we have considered arbitrary geometrical operations in two-dimensional space, i.e. a plane. Let us now discuss the symmetry operations of a group, for example the group $\mathbf{C}_{2h}$, consisting of four operations: $\hat{E}$, $\hat{C}_2(z)$, $\hat{\sigma}_h$ and $\hat{I}$ (Section 2.43). We shall find the matrices corresponding to the transformations of the components of the vector $\mathbf{r}(x, y, z)$ in three-dimensional space, taking these transformations to result from the four operations of the group. The identity transformation does not change the vector components, and the corresponding matrix is a unit matrix:

$$\begin{bmatrix} 1 & 0 & 0 \\ 0 & 1 & 0 \\ 0 & 0 & 1 \end{bmatrix} \begin{bmatrix} x \\ y \\ z \end{bmatrix} = \begin{bmatrix} x \\ y \\ z \end{bmatrix}, \quad \mathbf{D}(\hat{E}) = \begin{bmatrix} 1 & 0 & 0 \\ 0 & 1 & 0 \\ 0 & 0 & 1 \end{bmatrix}. \tag{3.37}$$

The operation $\hat{C}_2$ changes the signs of $x$ and $y$, reflection in the $\sigma_h$ plane changes the sign of $z$, and the inversion $\hat{I}$ changes the signs of all components of the vector $\mathbf{r}$. Hence the set of matrices *representing* the operations of the group $\mathbf{C}_{2h}$ may be written as

$$\mathbf{D}(\hat{E}) = \begin{bmatrix} 1 & 0 & 0 \\ 0 & 1 & 0 \\ 0 & 0 & 1 \end{bmatrix}, \quad \mathbf{D}(\hat{C}_2) = \begin{bmatrix} -1 & 0 & 0 \\ 0 & -1 & 0 \\ 0 & 0 & 1 \end{bmatrix},$$

$$\mathbf{D}(\hat{\sigma}_h) = \begin{bmatrix} 1 & 0 & 0 \\ 0 & 1 & 0 \\ 0 & 0 & -1 \end{bmatrix}, \quad \mathbf{D}(\hat{I}) = \begin{bmatrix} -1 & 0 & 0 \\ 0 & -1 & 0 \\ 0 & 0 & -1 \end{bmatrix}. \tag{3.38}$$

We now construct the multiplication table of the $\mathbf{C}_{2h}$ group operations:

| $\mathbf{C}_{2h}$ | $\hat{E}$ | $\hat{C}_2$ | $\hat{\sigma}_h$ | $\hat{I}$ |
|---|---|---|---|---|
| $\hat{E}$ | $\hat{E}$ | $\hat{C}_2$ | $\hat{\sigma}_h$ | $\hat{I}$ |
| $\hat{C}_2$ | $\hat{C}_2$ | $\hat{E}$ | $\hat{I}$ | $\hat{\sigma}_h$ |
| $\hat{\sigma}_h$ | $\hat{\sigma}_h$ | $\hat{I}$ | $\hat{E}$ | $\hat{C}_2$ |
| $\hat{I}$ | $\hat{I}$ | $\hat{\sigma}_h$ | $\hat{C}_2$ | $\hat{E}$ |

Let us compare the products of operations with the products of their matrix representations. For example, $\hat{\sigma}_h \hat{C}_2 = \hat{I}$ and

$$\begin{bmatrix} 1 & 0 & 0 \\ 0 & 1 & 0 \\ 0 & 0 & -1 \end{bmatrix} \begin{bmatrix} -1 & 0 & 0 \\ 0 & -1 & 0 \\ 0 & 0 & 1 \end{bmatrix} = \begin{bmatrix} -1 & 0 & 0 \\ 0 & -1 & 0 \\ 0 & 0 & -1 \end{bmatrix}$$

$$\text{or} \quad \mathbf{D}(\hat{\sigma}_h)\mathbf{D}(\hat{C}_2) = \mathbf{D}(\hat{I}). \tag{3.39}$$

On considering all other products we find that the *multiplication table of the matrices* $\mathbf{D}(\hat{R})$ *duplicates exactly the multiplication table of the operators* $\hat{R}$ *if the latter are replaced by the corresponding* $\mathbf{D}(\hat{R})$. Each group operation has its inverse, for example, $\hat{C}_2^{-1} = \hat{C}_2$, and the same relation holds for the matrices, for example $\mathbf{D}(\hat{C}_2)\mathbf{D}(\hat{C}_2) = \mathbf{D}(\hat{E})$. Consequently, the set of matrices corresponding to the operations of the point group itself satisfies all the requirements needed to define a group (Section 1.4). The set of matrices $\mathbf{D}(\hat{R})$ is a *matrix* group and provides a *representation* of the point group.

The choice of a given group representation is not unique. The geometrical transformations in Section 3.2 were represented by 2 × 2 matrices, while when introducing the representations of $\mathbf{C}_{2h}$, we considered three-dimensional matrices. Thus the *dimension of the representation* (order of matrices) depends on the number of objects transformed under the group operations. If, for example, we choose as objects the coordinates $(x, y, z)$ of the three atoms in the $H_2O$ molecule, we obtain a nine-dimensional representation of $\mathbf{C}_{2v}$.

Let us consider in more detail the matrix structure of $\mathbf{C}_{2h}$. It is very simple: all the matrices are diagonal. This simplicity is due to the fact that the $z$ axis of the coordinate system coincides with the $C_2$ axis, and the $\mathbf{C}_{2h}$ group operations only change the signs of the coordinates. Let us now choose another coordinate system by rotating the basis vectors through 45° about the axis. The unit vector $\mathbf{j}$ remains unchanged but now $\mathbf{i}$, $\mathbf{k}$ are rotated according to the rule (3.15) and we thus obtain the basic vectors of a new coordinate system:

$$\bar{\mathbf{i}} = (\mathbf{i} + \mathbf{k})/\sqrt{2}, \bar{\mathbf{j}} = \mathbf{j}, \bar{\mathbf{k}} = (\mathbf{k} - \mathbf{i})/\sqrt{2}, \tag{3.40}$$

or, in matrix form (with $a = 1/\sqrt{2}$)

$$[\bar{\mathbf{i}} \ \bar{\mathbf{j}} \ \bar{\mathbf{k}}] = [\mathbf{i} \ \mathbf{j} \ \mathbf{k}] \begin{bmatrix} a & 0 & -a \\ 0 & 1 & 0 \\ a & 0 & a \end{bmatrix} = [\mathbf{i} \ \mathbf{j} \ \mathbf{k}]\,\mathbf{U}. \tag{3.41}$$

This may be inverted to give

$$[\mathbf{i} \ \mathbf{j} \ \mathbf{k}] = [\bar{\mathbf{i}} \ \bar{\mathbf{j}} \ \bar{\mathbf{k}}]\mathbf{U}^{-1} = [\mathbf{i} \ \mathbf{j} \ \mathbf{k}] \begin{bmatrix} a & 0 & a \\ 0 & 1 & 0 \\ -a & 0 & a \end{bmatrix}, \tag{3.42}$$

since $\mathbf{U}$ is an orthogonal matrix, for which $\mathbf{U}^{-1} = \tilde{\mathbf{U}}$, where $\tilde{\mathbf{U}}$ is the transposed matrix, i.e. the matrix resulting from "reflection" of the matrix elements across the main diagonal:

$$\mathbf{U}^{-1}\mathbf{U} = \begin{bmatrix} a & 0 & a \\ 0 & 1 & 0 \\ -a & 0 & a \end{bmatrix} \begin{bmatrix} a & 0 & -a \\ 0 & 1 & 0 \\ a & 0 & a \end{bmatrix} = \begin{bmatrix} 1 & 0 & 0 \\ 0 & 1 & 0 \\ 0 & 0 & 1 \end{bmatrix} \equiv \mathbf{1}. \quad (3.43)$$

Now let us describe the operations of $\mathbf{C}_{2h}$ in terms of the new basis. The effect of a general operation $\hat{R}$ may be referred to either basis by the alternative equations

$$\hat{R}[\mathbf{i}\ \mathbf{j}\ \mathbf{k}] = [\mathbf{i}\ \mathbf{j}\ \mathbf{k}]\mathbf{D}(\hat{R}),$$
$$\hat{R}[\bar{\mathbf{i}}\ \bar{\mathbf{j}}\ \bar{\mathbf{k}}] = [\bar{\mathbf{i}}\ \bar{\mathbf{j}}\ \bar{\mathbf{k}}]\bar{\mathbf{D}}(\hat{R}) \quad (3.44)$$

and we need to find the new representation matrix $\bar{\mathbf{D}}(\hat{R})$. This may be achieved simply by substituting for $[\mathbf{i}\mathbf{j}\mathbf{k}]$ in the first equation (3.44), using result (3.42) above, to obtain instead

$$\hat{R}[\mathbf{i}\ \mathbf{j}\ \mathbf{k}] = \hat{R}[\bar{\mathbf{i}}\ \bar{\mathbf{j}}\ \bar{\mathbf{k}}]\mathbf{U}^{-1} = [\bar{\mathbf{i}}\ \bar{\mathbf{j}}\ \bar{\mathbf{k}}]\mathbf{U}^{-1}\mathbf{D}(\hat{R}), \quad (3.45)$$

and finally multiplying both sides from the right by $\mathbf{U}$. The result is the required equation (second equation in (3.44)) when the matrix $\bar{\mathbf{D}}(\hat{R})$ is identified as

$$\bar{\mathbf{D}}(\hat{R}) = \mathbf{U}^{-1}\mathbf{D}(\hat{R})\mathbf{U}. \quad (3.46)$$

Let us now construct the set of matrices $\bar{\mathbf{D}}(\hat{R})$ for $\mathbf{C}_{2h}$ corresponding to the transformation of the new unit vectors $\bar{\mathbf{i}}$, $\bar{\mathbf{j}}$, $\bar{\mathbf{k}}$. We obtain

$$\bar{\mathbf{D}}(\hat{E}) = \begin{bmatrix} 1 & 0 & 0 \\ 0 & 1 & 0 \\ 0 & 0 & 1 \end{bmatrix}, \bar{\mathbf{D}}(\hat{C}_2) = \begin{bmatrix} 0 & 0 & 1 \\ 0 & -1 & 0 \\ 1 & 0 & 0 \end{bmatrix},$$
$$\bar{\mathbf{D}}(\hat{\sigma}_h) = \begin{bmatrix} 0 & 0 & -1 \\ 0 & 1 & 0 \\ -1 & 0 & 0 \end{bmatrix}, \bar{\mathbf{D}}(\hat{I}) = \begin{bmatrix} -1 & 0 & 0 \\ 0 & -1 & 0 \\ 0 & 0 & -1 \end{bmatrix}. \quad (3.47)$$

On multiplying the matrices $\bar{\mathbf{D}}(\hat{R})$ it can be seen that they obey the same multiplication law as the matrices (3.38). Therefore, like the matrices $\mathbf{D}(\hat{R})$ they represent the group $\mathbf{C}_{2h}$. This representation appears substantially different from the initially chosen representation $\mathbf{D}(\hat{R})$. The loss of simplicity is due to the choice of coordinate system. Nevertheless, each choice of an orthogonal coordinate system provides a new representation. Thus the single set of operations defining a particular group leads to an infinite number of representations, differing both in the matrix dimensions and structure.

## 3.4 REDUCIBLE AND IRREDUCIBLE REPRESENTATIONS

Rotation of the coordinate system (or change of basis) induces a similarity transformation of the representation matrices. The set of matrices obtained from the group representation by means of a similarity transformation also constitutes a representation. This has been verified above taking $\mathbf{C}_{2h}$ as an example; we shall now prove it in the general case. Let us check, for example, the multiplication rule. Assume $\mathbf{D}(\hat{A})\mathbf{D}(\hat{B}) = \mathbf{D}(\hat{C})$ and find the product of the similarity-transformed matrices:

$$\begin{aligned}\mathbf{D}'(\hat{A})\mathbf{D}'(\hat{B}) &= \mathbf{U}^{-1}\mathbf{D}(\hat{A})\mathbf{U}\mathbf{U}^{-1}\mathbf{D}(\hat{B})\mathbf{U} \\ &= \mathbf{U}^{-1}\mathbf{D}(\hat{A})\mathbf{D}(\hat{B})\mathbf{U} \\ &= \mathbf{U}^{-1}\mathbf{D}(\hat{C})\mathbf{U} \\ &= \mathbf{D}'(\hat{C}). \end{aligned} \tag{3.48}$$

Therefore similarity transformations result in an infinite number of representations of the same dimension. These representations are called *conjugate* or *equivalent*. In the given example the similarity transformation gives a less simple $\mathbf{C}_{2h}$ group representation, the diagonal matrices (3.38) being transformed into nondiagonal matrices (3.47). Of course, if the representation (3.47) is considered as the initial one, it is possible by means of the inverse similarity transformation

$$\mathbf{Q}^{-1}\mathbf{D}(\hat{R})\mathbf{Q} \qquad (\mathbf{Q} = \mathbf{U}^{-1})$$

to obtain the representation (3.35). It is obvious that the diagonal representation structure (3.38) is the simplest.

Let us deal now with a more complicated example. We shall write down the representations of $\mathbf{C}_{3v}$ corresponding to transformation of the coordinates $(x, y, z)$ with $(z \| C_3)$. The $\hat{C}_3$ operation gives

$$\left.\begin{aligned} x' &= x \cos \tfrac{2}{3}\pi - y \sin \tfrac{2}{3}\pi + 0 \cdot z, \\ y' &= x \sin \tfrac{2}{3}\pi + y \cos \tfrac{2}{3}\pi + 0 \cdot z, \\ z' &= 0 \cdot x + 0 \cdot y + 1 \cdot z. \end{aligned}\right\} \tag{3.49}$$

It is easy to obtain similar transformations related to the other operations of

$\mathbf{C}_{3v}$. The representation realized by the respective matrices has the form

$$\left.\begin{array}{l}
\mathbf{D}(\hat{E}) = \begin{bmatrix} 1 & 0 & 0 \\ 0 & 1 & 0 \\ 0 & 0 & 1 \end{bmatrix}, \quad \mathbf{D}(\hat{C}_3) = \left[\begin{array}{cc|c} -\frac{1}{2} & \frac{1}{2}\sqrt{3} & 0 \\ -\frac{1}{2}\sqrt{3} & -\frac{1}{2} & 0 \\ \hline 0 & 0 & 1 \end{array}\right], \\
\mathbf{D}(\hat{C}_3^2) = \left[\begin{array}{cc|c} -\frac{1}{2} & -\frac{1}{2}\sqrt{3} & 0 \\ \frac{1}{2}\sqrt{3} & -\frac{1}{2} & 0 \\ \hline 0 & 0 & 1 \end{array}\right], \quad \mathbf{D}(\hat{\sigma}_v) = \left[\begin{array}{cc|c} 1 & 0 & 0 \\ 0 & -1 & 0 \\ \hline 0 & 0 & 1 \end{array}\right], \\
\mathbf{D}(\hat{\sigma}'_v) = \left[\begin{array}{cc|c} -\frac{1}{2} & -\frac{1}{2}\sqrt{3} & 0 \\ -\frac{1}{2}\sqrt{3} & \frac{1}{2} & 0 \\ \hline 0 & 0 & 1 \end{array}\right], \quad \mathbf{D}(\hat{\sigma}''_v) = \left[\begin{array}{cc|c} -\frac{1}{2} & \frac{1}{2}\sqrt{3} & 0 \\ \frac{1}{2}\sqrt{3} & \frac{1}{2} & 0 \\ \hline 0 & 0 & 1 \end{array}\right],
\end{array}\right\} \tag{3.50}$$

where $\hat{\sigma}_v \equiv \hat{\sigma}_v(xz)$.

The matrices (3.50) have a block-diagonal structure and consist of two sets of submatrices of the second and first order. The representation (3.50) has the schematic form

$$\mathbf{D}(\hat{R}) = \left[\begin{array}{c|c} 2 \times 2 & \begin{matrix} 0 \\ 0 \end{matrix} \\ \hline 0 \;\; 0 & 1 \times 1 \end{array}\right] \equiv \left[\begin{array}{c|c} \mathbf{D}^{(1)}(\hat{R}) & 0 \\ \hline 0 & \mathbf{D}^{(2)}(\hat{R}) \end{array}\right]. \tag{3.51}$$

The different blocks are governed by the same group multiplication law as the group operations and the 3 × 3 matrices (3.51). Indeed,

$$\left.\begin{array}{l}
\hat{\sigma}_v \hat{C}_3 = \hat{\sigma}''_v, \\
\mathbf{D}^{(1)}(\hat{\sigma}_v)\mathbf{D}^{(1)}(\hat{C}_3) = \begin{bmatrix} 1 & 0 \\ 0 & -1 \end{bmatrix} \begin{bmatrix} -\frac{1}{2} & \frac{1}{2}\sqrt{3} \\ -\frac{1}{2}\sqrt{3} & -\frac{1}{2} \end{bmatrix} = \begin{bmatrix} -\frac{1}{2} & \frac{1}{2}\sqrt{3} \\ \frac{1}{2}\sqrt{3} & \frac{1}{2} \end{bmatrix} \equiv \mathbf{D}^{(1)}(\hat{\sigma}''_v), \\
\mathbf{D}^{(2)}(\hat{\sigma}_v)\mathbf{D}^{(2)}(\hat{C}_3) = [1][1] = [1] = \mathbf{D}^{(2)}(\hat{\sigma}''_v)
\end{array}\right\} \tag{3.52}$$

This means that the submatrices of the same order, occupying the same positions in the matrix **D**, themselves form a representation of the group. This example illustrates the general conclusion of Section 3.1.3 regarding the block structure of the block-diagonal matrix product.

Of course, the representation (3.50) is not the only one, since any number of equivalent representations can be obtained from it by means of similarity transformations. As we have already shown, using $\mathbf{C}_{2h}$ as an example, a similarity transformation of a representation is induced by a rotation of the coordinate system. In the case of $\mathbf{C}_{3v}$ the $z$ axis is directed along the $C_3$ axis. If we consider another coordinate system, for example (3.15), whose $z'$ axis is inclined relative to the $C_3$ axis, then the $3 \times 3$ representation no longer has a block-diagonal structure. Now let us see whether it is possible to devise a similarity transformation that will simplify the three-dimensional representation (3.50). In other words, we ask whether it is possible, using a coordinate-system transformation, to obtain a further division into blocks of the matrices in (3.50). Intuitively, it can be seen that this is impossible, since the $x$ and $y$ components of the vector are transformed into one another under the influence of $C_3$ rotations and reflections. Therefore the two-dimensional representation of $\mathbf{C}_{3v}$ is of maximum simplicity, meaning that it cannot be reduced to two one-dimensional representations. The two-dimensionality of this representation can be considered to be a property of the symmetry group, and any further simplification through similarity transformations is impossible.

Let us summarize the conclusions from the two examples considered above, assuming that the set of matrices $\mathbf{D}(\hat{R})$ is a representation of a group. Let a similarity transformation with matrix $\mathbf{U}$ reduce the representation to block-diagonal form $\mathbf{D}'(\hat{R})$:

$$\mathbf{D}' = \mathbf{U}^{-1}\mathbf{D}\mathbf{U} = \begin{bmatrix} \mathbf{D}'_1 & & & \\ & \mathbf{D}'_2 & & \\ & & \ddots & \\ & & & \mathbf{D}'_n \end{bmatrix}. \tag{3.53}$$

The set of matrices $\mathbf{D}'_i(\hat{R})$ (for all operations $\hat{R}$) that appear in any block i also form a group representation, which has a lower dimension than the initial $\mathbf{D}$. A representation $\mathbf{D}$ that can be reduced to several representations of lower dimensions by means of a similarity transformation is called a *reducible representation*. A representation whose dimension cannot be reduced by any similarity transformation is called an *irreducible representation*. The concept of reducible and irreducible representations is of great importance in group theory, and is widely used in the following chapters.

## 3.5 IRREDUCIBLE REPRESENTATIONS OF THE CUBIC GROUP

### 3.5.1 Atomic orbitals and the effect of symmetry operations

In Section 3.4 we found the matrices for transformations of vector components due to operations of the point groups $\mathbf{C}_{2h}$ and $\mathbf{C}_{3v}$. Before discussing the general features of irreducible representations, let us consider the influence of symmetry operations on more complicated objects such as the wavefunctions. Strictly speaking, a wavefunction is not a geometrical object, defined in three-dimensional space like a vector: it may for example be a function of the coordinates of all the electrons in a many-electron system and if we rotate a molecule we must define carefully what is meant by the corresponding "rotated wavefunction".

Let us start by considering the wavefunctions for the various stationary states of the electron in a hydrogen atom: these "atomic orbitals" have the general form:

$$\psi_{nlm}(\mathbf{r}) = R_{nl}(r)\,Y_{lm}(\vartheta, \varphi), \tag{3.54}$$

where $n = 0, 1, 2, \ldots$ is the principal quantum number, $l = 0, 1, 2, \ldots, n-1$ is the orbital quantum number and $m = -l, -l+1, \ldots, l-1, l$ is the magnetic quantum number. The radial part $R_{nl}(r)$ depends only on the absolute value of the radius vector of the electron, while the spherical harmonics $Y_{lm}(\vartheta, \varphi)$ depend on the angular variables. The Cartesian components $(x, y, z)$ are related to the polar ones $(r, \vartheta, \varphi)$ by

$$x = r \sin\vartheta \cos\varphi, \quad y = r \sin\vartheta \sin\varphi, \quad z = r \cos\vartheta. \tag{3.55}$$

In general, the angular components $Y_{lm}(\vartheta, \varphi)$ are complex (they are given in textbooks on quantum mechanics; see e.g. [34]. It is convenient to replace them by *real* linear combinations $A_l(\vartheta, \varphi)$. For orbitals with $l = 0, 1, 2$ (s, p, and d orbitals) such real angular functions are given in Table 3.1 The angular functions $A_l(\vartheta, \varphi)$ correspond to the values of the orbital quantum number $l$ given in the table, but magnetic moment associated with the electronic angular momentum then no longer has a definite component in the $z$ direction (it is in fact 'quenched').

How should we define the effect of rotation $\hat{R}$ (i.e. geometrical operation) applied to an orbital, which is a function of position $f(\mathbf{r})$ with a definite numerical value at any point P with coordinates $x$, $y$, $z$ in three-dimensional space? Since $f(\mathbf{r})$ refers to some physical system, such as an atom or a molecule, a "natural" definition would seem to be the following:

Definition. On applying a rotation $\hat{R}$ to a physical system, the rotated function $f'(\mathbf{r}) = \hat{R}f(\mathbf{r})$ is the function defined exactly like $f(\mathbf{r})$ but with reference to the rotated coordinate system or basis.

This definition of a "rotated function" $f(\mathbf{r})$ by symmetry operation $\hat{R}$ is illustrated in Fig. 3.2. The function $f^2(\mathbf{r})$ can in fact be interpreted as a spatial electron density function. Fig. 3.2 shows the initial wavefunction in the $p_x$-orbital which can be represented as $f(\mathbf{r}) = f(r)x$ where $f(r)$ is a fully symmetric function. When operation $\hat{\mathbf{C}}_4(z)$ is applied the $p_x$ orbital is transformed into a $p_y$-orbital ($p_y = f(r)y$) as indicated by the dotted curve.

**Table 3.1** Normalized angular functions $A_l(\vartheta, \varphi)$ of s, p and d orbitals.

| $l$ | Orbital | Linear combination $Y_{lm}$ | $A_l(\vartheta, \varphi)$ |
|---|---|---|---|
| 0 (s) | s | $Y_{00}$ | $\frac{1}{2}\pi^{1/2}$ |
|  | $p_x$ | $\sqrt{\frac{1}{2}}(Y_{11} + Y_{1-1})$ | $\left(\frac{3}{4\pi}\right)^{1/2} \sin\vartheta \cos\varphi = \left(\frac{3}{4\pi}\right)^{1/2} \frac{x}{r}$ |
| 1 (p) | $p_y$ | $\sqrt{\frac{1}{2}}i(Y_{11} - Y_{1-1})$ | $\left(\frac{3}{4\pi}\right)^{1/2} \sin\vartheta \sin\varphi = \left(\frac{3}{4\pi}\right)^{1/2} \frac{y}{r}$ |
|  | $p_z$ | $Y_{10}$ | $\left(\frac{3}{4\pi}\right)^{1/2} \cos\vartheta = \left(\frac{3}{4\pi}\right)^{1/2} \frac{z}{r}$ |
|  | $d_{yz} = d_\xi$ | $\sqrt{\frac{1}{2}}i(Y_{21} + Y_{2-1})$ | $\left(\frac{15}{4\pi}\right)^{1/2} \sin\vartheta \cos\vartheta \sin\varphi = \left(\frac{15}{4\pi}\right)^{1/2} \frac{yz}{r^2}$ |
|  | $d_{xz} = d_\eta$ | $-\sqrt{\frac{1}{2}}(Y_{21} - Y_{2-1})$ | $\left(\frac{15}{4\pi}\right)^{1/2} \sin\vartheta \cos\vartheta \cos\varphi = \left(\frac{15}{4\pi}\right)^{1/2} \frac{xz}{r^2}$ |
| 2 (d) | $d_{xy} = d_\zeta$ | $-\sqrt{\frac{1}{2}}i(Y_{22} - Y_{2-2})$ | $\left(\frac{15}{4\pi}\right)^{1/2} \sin^2\vartheta \sin 2\varphi = \left(\frac{15}{4\pi}\right)^{1/2} \frac{xy}{r^2}$ |
|  | $d_{3z^2-r^2} = d_u$ | $Y_{20}$ | $\left(\frac{5}{16\pi}\right)^{1/2} (3\cos^2\vartheta - 1) = \left(\frac{5}{16\pi}\right)^{1/2} \frac{3z^2 - r^2}{r^2}$ |
|  | $d_{x^2-y^2} = d_v$ | $\sqrt{\frac{1}{2}}(Y_{22} + Y_{2-2})$ | $\left(\frac{15}{16\pi}\right)^{1/2} \sin^2\vartheta \cos 2\varphi = \left(\frac{5}{16}\right)^{1/2} \frac{\sqrt{3}(x^2 - y^2)}{r^2}$ |

The transformation may thus be visualized as a rotation of the 'contour map' of the function. The above definition of a "rotated object" is general enough to be applied very widely, for example to general many-electron wavefunctions and to quantum-mechanical operators. For orbitals it is easy to give the

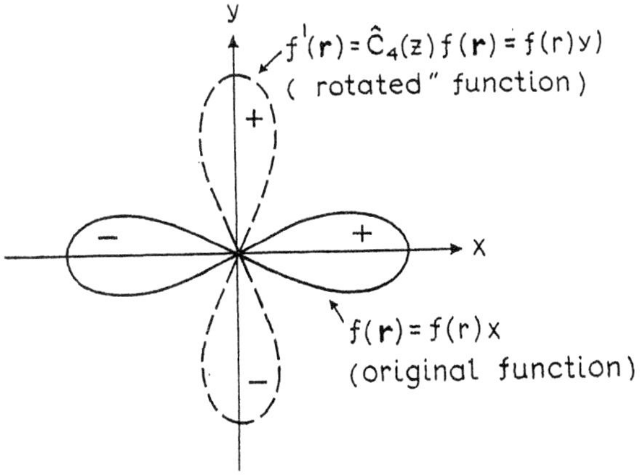


**Fig. 3.2** Application of $\hat{C}_4(z)$ operation to the function $f(\mathbf{r}) = f(r)x$.

definition of a practical form: for when the rotated function $f'(r)$ is evaluated at the rotated point P$'$ with $\mathbf{r}' = \hat{R}\mathbf{r}$, it must take the same value as $f(\mathbf{r})$ evaluated at the original point P (Fig. 3.3). In symbols, $f'(\mathbf{r}') = f'(\hat{R}\mathbf{r}) = f(\mathbf{r})$ as it is shown in Fig. 3.3 for the electronic distribution in $p$-orbitals; but since $\mathbf{r}$ is an arbitrary point we may replace it by $\hat{R}^{-1}\mathbf{r}$ to obtain

$$f'(\hat{R}\hat{R}^{-1}\mathbf{r}) = f(\hat{R}^{-1}\mathbf{r}), \tag{3.56}$$

or

$$f'(\mathbf{r}) = \hat{R}f(\mathbf{r}) = f(\hat{R}^{-1}\mathbf{r}). \tag{3.57}$$

In other words, to get the rotated function we simply take the original function and replace the coordinates $x, y, z$ by the coordinates $x', y', z'$ of a "backwards rotated" point

$$f'(x, y, z) = \hat{R}f(x, y, z) = f(x', y', z'). \tag{3.58}$$

This recipe applies at once to all the functions in Table 3.1 and assumes a particularly simple form for the orthogonal Cartesian basis; for the matrix of the rotation $\hat{R}^{-1}$ is the inverse of that associated with $\hat{R}$ and gives

$$\begin{bmatrix} x' \\ y' \\ z' \end{bmatrix} = \tilde{\mathbf{D}}(\hat{R}) \begin{bmatrix} x \\ y \\ z \end{bmatrix}, \tag{3.59}$$

since the inverse matrix in this case coincides with the transpose of the matrix $\mathbf{D}(\hat{R})$ indicated here as $\tilde{\mathbf{D}}(\hat{R})$.

### 3.5.2 Transformation of $p$ orbitals under the cubic group

As a first example let us consider the functions $p_x, p_y, p_z$ whose angular factors are shown in Table 3.1 and study the effect of the symmetry operations of the

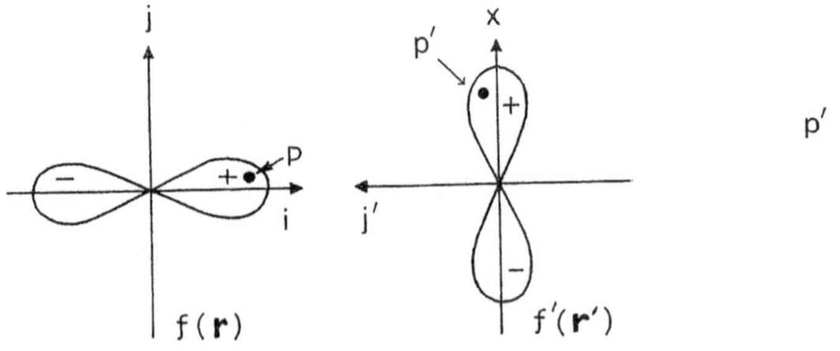


Fig. 3.3 Illustration of the statement $f(\mathbf{r}) = f'(\mathbf{r}')$.

**Table 3.2** Representation of the group **O** in the basis of p orbitals.

| $\hat{E}$ | $\hat{C}_4(x)$ | $\hat{C}_4(y)$ | $\hat{C}_4(z)$ | $\hat{C}_4^3(x)$ | $\hat{C}_4^3(y)$ |
|---|---|---|---|---|---|
| $\begin{bmatrix} 1 & 0 & 0 \\ 0 & 1 & 0 \\ 0 & 0 & 1 \end{bmatrix}$ | $\begin{bmatrix} 1 & 0 & 0 \\ 0 & 0 & -1 \\ 0 & 1 & 0 \end{bmatrix}$ | $\begin{bmatrix} 0 & 0 & 1 \\ 0 & 1 & 0 \\ -1 & 0 & 0 \end{bmatrix}$ | $\begin{bmatrix} 0 & -1 & 0 \\ 1 & 0 & 0 \\ 0 & 0 & 1 \end{bmatrix}$ | $\begin{bmatrix} 1 & 0 & 0 \\ 0 & 0 & 1 \\ 0 & -1 & 0 \end{bmatrix}$ | $\begin{bmatrix} 0 & 0 & -1 \\ 0 & 1 & 0 \\ 1 & 0 & 0 \end{bmatrix}$ |
| $\hat{C}_4^3(z)$ | $\hat{C}_4^2(x)$ | $\hat{C}_4^2(y)$ | $\hat{C}_4^2(z)$ | $\hat{C}_3(xyz)$ | $\hat{C}_3(\bar{x}yz)$ |
| $\begin{bmatrix} 0 & 1 & 0 \\ -1 & 0 & 0 \\ 0 & 0 & 1 \end{bmatrix}$ | $\begin{bmatrix} 1 & 0 & 0 \\ 0 & -1 & 0 \\ 0 & 0 & -1 \end{bmatrix}$ | $\begin{bmatrix} -1 & 0 & 0 \\ 0 & 1 & 0 \\ 0 & 0 & -1 \end{bmatrix}$ | $\begin{bmatrix} -1 & 0 & 0 \\ 0 & -1 & 0 \\ 0 & 0 & 1 \end{bmatrix}$ | $\begin{bmatrix} 0 & 0 & 1 \\ 1 & 0 & 0 \\ 0 & 1 & 0 \end{bmatrix}$ | $\begin{bmatrix} 0 & -1 & 0 \\ 0 & 0 & 1 \\ -1 & 0 & 0 \end{bmatrix}$ |
| $\hat{C}_3(x\bar{y}z)$ | $\hat{C}_3(xy\bar{z})$ | $\hat{C}_3^2(xyz)$ | $\hat{C}_3^2(\bar{x}yz)$ | $\hat{C}_3^2(x\bar{y}z)$ | $\hat{C}_3^2(xy\bar{z})$ |
| $\begin{bmatrix} 0 & -1 & 0 \\ 0 & 0 & -1 \\ 1 & 0 & 0 \end{bmatrix}$ | $\begin{bmatrix} 0 & 0 & -1 \\ 1 & 0 & 0 \\ 0 & -1 & 0 \end{bmatrix}$ | $\begin{bmatrix} 0 & 1 & 0 \\ 0 & 0 & 1 \\ 1 & 0 & 0 \end{bmatrix}$ | $\begin{bmatrix} 0 & 0 & -1 \\ -1 & 0 & 0 \\ 0 & 1 & 0 \end{bmatrix}$ | $\begin{bmatrix} 0 & 0 & 1 \\ -1 & 0 & 0 \\ 0 & -1 & 0 \end{bmatrix}$ | $\begin{bmatrix} 0 & 1 & 0 \\ 0 & 0 & -1 \\ -1 & 0 & 0 \end{bmatrix}$ |
| $\hat{C}_2(xy)$ | $\hat{C}_2(yz)$ | $\hat{C}_2(zx)$ | $\hat{C}_2(\bar{x}y)$ | $\hat{C}_2(\bar{y}z)$ | $\hat{C}_2(\bar{z}x)$ |
| $\begin{bmatrix} 0 & 1 & 0 \\ 1 & 0 & 0 \\ 0 & 0 & -1 \end{bmatrix}$ | $\begin{bmatrix} -1 & 0 & 0 \\ 0 & 0 & 1 \\ 0 & 1 & 0 \end{bmatrix}$ | $\begin{bmatrix} 0 & 0 & 1 \\ 0 & -1 & 0 \\ 1 & 0 & 0 \end{bmatrix}$ | $\begin{bmatrix} 0 & -1 & 0 \\ -1 & 0 & 0 \\ 0 & 0 & -1 \end{bmatrix}$ | $\begin{bmatrix} -1 & 0 & 0 \\ 0 & 0 & -1 \\ 0 & -1 & 0 \end{bmatrix}$ | $\begin{bmatrix} 0 & 0 & -1 \\ 0 & -1 & 0 \\ -1 & 0 & 0 \end{bmatrix}$ |

group **O**. The operation $\hat{C}_4(z)$ leads to a transformation of basis vectors $\mathbf{i} \to \mathbf{j}, \mathbf{j} \to -\mathbf{i}, \mathbf{k} \to \mathbf{k}$ according to (3.22) and hence the matrix $\mathbf{D}\,(C_4(z))$ may be written as:

$$\mathbf{D}(\hat{C}_4(z)) = \begin{bmatrix} 0 & -1 & 0 \\ 1 & 0 & 0 \\ 0 & 0 & 1 \end{bmatrix}. \tag{3.60}$$

Similarly,

$$\mathbf{D}(\hat{C}_4(x)) = \begin{bmatrix} 1 & 0 & 0 \\ 0 & 0 & -1 \\ 0 & 1 & 0 \end{bmatrix}, \quad \mathbf{D}(\hat{C}_4(y)) = \begin{bmatrix} 0 & 0 & 1 \\ 0 & 1 & 0 \\ -1 & 0 & 0 \end{bmatrix}. \tag{3.61}$$

The operation $\hat{C}_3(xyz)$ corresponds to the rotation of axes $x \to y, y \to z, z \to x$ (i.e. $\mathbf{i} \to \mathbf{i}' = \mathbf{j}$, $\mathbf{j} \to \mathbf{j}' = \mathbf{k}$, $\mathbf{k} \to \mathbf{k}' = \mathbf{i}$) with the matrix

$$\mathbf{D}(\hat{C}_3(xyz)) = \begin{bmatrix} 0 & 0 & 1 \\ 1 & 0 & 0 \\ 0 & 1 & 0 \end{bmatrix}. \tag{3.62}$$

In the same way, we can obtain all the other matrices of the rotations $\hat{C}_3$, as well as the matrices of other **O** group operations (Table 3.2). On writing the $p$-functions as $p_x(\mathbf{r}) = f(r)x, p_y(\mathbf{r}) = f(r)y$, $p_z(\mathbf{r}) = f(r)z$ (where the common radial factor $f(r)$ is not affected by the symmetry operations) and using the rule (3.58) we have

$$p'_x(\mathbf{r}) = f(r)x', p'_y(\mathbf{r}) = f(r)y', \; p'_z(\mathbf{r}) = f(r)z'.$$

If now we use (3.59) we find at once the rotated $p$-functions: thus for $\hat{C}_3(xyz)$ we obtain by transposition of $\mathbf{D}(\hat{C}_3(xyz))$, $x' = y$, $y' = z, z' = x$ and hence $p'_x = p_y$, $p'_y = p_z$, $p'_z = p_x$. This transformation is exactly like that of the Cartesian basis vectors: $\mathbf{i}' = \mathbf{j}, \mathbf{j}' = \mathbf{k}, \mathbf{k}' = \mathbf{i}$. It is easily verified that under all group operations $p_x, p_y, p_z$ transform like the basis vectors $\mathbf{i}, \mathbf{j}, \mathbf{k}$. On collecting the functions into a row matrix we may write

$$[\mathbf{i}' \;\; \mathbf{j}' \;\; \mathbf{k}'] = [\mathbf{i} \;\; \mathbf{j} \;\; \mathbf{k}]\, \mathbf{D}(\hat{R}),$$
$$[p'_x \;\; p'_y \;\; p'_z] = [p_x \;\; p_y \;\; p_z]\, \mathbf{D}(\hat{R}), \tag{3.63}$$

for all operations of the group. The wavefunctions $p_x, p_y, p_z$ therefore provide the same representation of **O** as the Cartesian unit vectors: they are basis functions, or rather, the basis of the representation.

Let us consider a set of $n$ functions $\varphi_1, \varphi_2 \cdots \varphi_n$ as a row vector $[\varphi_1, \varphi_2 \cdots \varphi_n]$ in an $n$-dimensional space. A general transformation under the action of operation $\hat{R}$ belonging to a certain symmetry group may be expressed in a matrix form as

$$[\varphi'_1 \varphi'_2 \ldots \varphi'_n] = [\varphi_1 \varphi_2 \ldots \varphi_n]\mathbf{D}(\hat{R}),$$

or

$$\hat{R}[\varphi_1 \varphi_2 \ldots \varphi_n] = [\varphi_1 \varphi_2 \ldots \varphi_n]\, \mathbf{D}(\hat{R}),$$
$$\hat{R}\varphi_j = \sum_i \varphi_i D_{ij}(\hat{R}) \tag{3.64}$$

Equation (3.64) defines an $n$-dimensional representation of a group with the basis $\varphi_1 \varphi_2 \ldots \varphi_n$.

This definition is fundamental to the theory of group representations; later on we shall use it continuously.

The set of matrices given in Table 3.2 is not block-diagonal (although some matrices have block structures). The question arises as to whether one similarity transformation can give a division of the whole three-dimensional representation considered above into blocks of smaller dimensions. The detailed mathematical theory [12] gives the criteria for a rigorous solution of this problem.

It turns out that the three-dimensional representation of **O** whose basis is the p orbitals is irreducible. Indeed, since there are four $C_3$ axes inclined to one another and three $C_4$ axes, the wavefunctions $p_x$, $p_y$ and $p_z$ are transformed into one another. They cannot be divided by a linear transformation into "closed" sets of lower dimension within which they can be transformed into one another.

It should be noted the the matrices of the six operations $\hat{C}_4$ and $\hat{C}_4^3 = \hat{C}_4^{-1}$ have identical characters:

$$\chi(\mathbf{D}(\hat{C}_4(x)) = \chi(\mathbf{D}(\hat{C}_4(y)) = \ldots = \chi(\mathbf{D}(\hat{C}_4^{-1}(z)) = 1. \tag{3.65}$$

For the eight $\hat{C}_3$ operations we find that $\chi(\mathbf{D}(\hat{C}_3)) = 0$. From Table 3.2 it can be seen that the operations belonging to a given class have the same characters. This is a common feature of the characters of irreducible representations, which follows from the equality of the characters of conjugate matrices proved in Section 3.1.4. Indeed, by definition, two operations belong to the same class when they are connected by a similarity transformation (3.12) (Section 2.2).

### 3.5.3 Transformation of $d$ wavefunctions under the group O

Let us consider the transformation laws of the angular $d$ wavefunctions ($l = 2$) $A_2(\vartheta, \varphi)$ (Table 3.1). It is easy to find the respective matrices, knowing the result of application of the **O** group operations to the functions $p_x$, $p_y$ and $p_z$. Since, for example, the $d_\xi$ wavefunction containing the factor $yz$ is proportional to the product $p_y p_z$, we obtain

$$\hat{C}_4(z)\, d_\xi = -\left(\frac{15}{4\pi}\right)^{1/2} \frac{zx}{r^2} = -d_\eta. \tag{3.66}$$

In the same way, we find

$$\left.\begin{aligned} \hat{C}_4(z)\, d_\eta &= d_\xi, \qquad \hat{C}_4(z)\, d_\zeta = -d_\zeta, \\ \hat{C}_4(z)\, d_u &= -d_u, \qquad \hat{C}_4(z)\, d_v = -d_v. \end{aligned}\right\} \tag{3.67}$$

Taking the set of five functions $A_2(\vartheta, \varphi)$ as the unit vectors of a five-dimensional basis $[d_\xi \;\; d_\eta \;\; d_\zeta \;\; d_u \;\; d_v]$, we obtain the $5 \times 5$ matrix of its

**Table 3.3** The **O** group representation in the basis of d orbitals.

| $\hat{E}$ | $\hat{C}_4(x)$ | $\hat{C}_4(y)$ | $\hat{C}_4(z)$ |
|---|---|---|---|
| $\left[\begin{array}{c\|c} \begin{matrix}1&0&0\\0&1&0\\0&0&1\end{matrix} & \mathbf{0} \\ \hline \mathbf{0} & \begin{matrix}1&0\\0&1\end{matrix}\end{array}\right]$ | $\left[\begin{array}{c\|c} \begin{matrix}-1&0&0\\0&1&0\\0&-0&1\end{matrix} & \mathbf{0} \\ \hline \mathbf{0} & \begin{matrix}-\frac{1}{2}&-\frac{1}{2}\sqrt{3}\\-\frac{1}{2}\sqrt{3}&\frac{1}{2}\end{matrix}\end{array}\right]$ | $\left[\begin{array}{c\|c} \begin{matrix}0&0&-1\\0&-1&0\\0&0&0\end{matrix} & \mathbf{0} \\ \hline \mathbf{0} & \begin{matrix}-\frac{1}{2}&\frac{1}{2}\sqrt{3}\\\frac{1}{2}\sqrt{3}&\frac{1}{2}\end{matrix}\end{array}\right]$ | $\left[\begin{array}{c\|c} \begin{matrix}0&1&0\\-1&0&0\\0&0&-1\end{matrix} & \mathbf{0} \\ \hline \mathbf{0} & \begin{matrix}1&0\\0&-1\end{matrix}\end{array}\right]$ |
| $\hat{C}_4^3(x)$ | $\hat{C}_4^3(y)$ | $\hat{C}_4^3(z)$ | $\hat{C}_4^2(x)$ |
| $\left[\begin{array}{c\|c} \begin{matrix}-1&0&0\\0&0&-1\\0&1&0\end{matrix} & \mathbf{0} \\ \hline \mathbf{0} & \begin{matrix}-\frac{1}{2}&-\frac{1}{2}\sqrt{3}\\-\frac{1}{2}\sqrt{3}&\frac{1}{2}\end{matrix}\end{array}\right]$ | $\left[\begin{array}{c\|c} \begin{matrix}0&0&1\\0&-1&0\\-1&0&0\end{matrix} & \mathbf{0} \\ \hline \mathbf{0} & \begin{matrix}-\frac{1}{2}&\frac{1}{2}\sqrt{3}\\\frac{1}{2}\sqrt{3}&\frac{1}{2}\end{matrix}\end{array}\right]$ | $\left[\begin{array}{c\|c} \begin{matrix}0&-1&0\\1&0&0\\0&0&-1\end{matrix} & \mathbf{0} \\ \hline \mathbf{0} & \begin{matrix}1&0\\0&-1\end{matrix}\end{array}\right]$ | $\left[\begin{array}{c\|c} \begin{matrix}1&0&0\\0&-1&0\\0&0&-1\end{matrix} & \mathbf{0} \\ \hline \mathbf{0} & \begin{matrix}1&0\\0&1\end{matrix}\end{array}\right]$ |
| $\hat{C}_4^2(y)$ | $\hat{C}_4^2(z)$ | $\hat{C}_3(xyz)$ | $\hat{C}_3(\bar{x}yz)$ |
| $\left[\begin{array}{c\|c} \begin{matrix}-1&0&0\\0&1&0\\0&0&-1\end{matrix} & \mathbf{0} \\ \hline \mathbf{0} & \begin{matrix}1&0\\0&1\end{matrix}\end{array}\right]$ | $\left[\begin{array}{c\|c} \begin{matrix}-1&0&0\\0&-1&0\\0&0&1\end{matrix} & \mathbf{0} \\ \hline \mathbf{0} & \begin{matrix}1&0\\0&1\end{matrix}\end{array}\right]$ | $\left[\begin{array}{c\|c} \begin{matrix}0&0&1\\1&0&0\\0&1&0\end{matrix} & \mathbf{0} \\ \hline \mathbf{0} & \begin{matrix}-\frac{1}{2}&-\frac{1}{2}\sqrt{3}\\\frac{1}{2}\sqrt{3}&-\frac{1}{2}\end{matrix}\end{array}\right]$ | $\left[\begin{array}{c\|c} \begin{matrix}0&-1&0\\0&0&1\\-1&0&0\end{matrix} & \mathbf{0} \\ \hline \mathbf{0} & \begin{matrix}-\frac{1}{2}&\frac{1}{2}\sqrt{3}\\-\frac{1}{2}\sqrt{3}&-\frac{1}{2}\end{matrix}\end{array}\right]$ |

| | | | |
|---|---|---|---|
| $\left[\begin{array}{cc} \begin{matrix} 0 & -1 & 0 \\ 0 & 0 & -1 \\ 1 & 0 & 0 \end{matrix} & \mathbf{0} \\ \hline \mathbf{0} & \begin{matrix} -\frac{1}{2} & \frac{1}{2}\sqrt{3} \\ -\frac{1}{2}\sqrt{3} & -\frac{1}{2} \end{matrix} \end{array}\right]$ | $\left[\begin{array}{cc} \begin{matrix} 0 & 0 & -1 \\ 1 & 0 & 0 \\ 0 & -1 & 0 \end{matrix} & \mathbf{0} \\ \hline \mathbf{0} & \begin{matrix} -\frac{1}{2} & -\frac{1}{2}\sqrt{3} \\ \frac{1}{2}\sqrt{3} & -\frac{1}{2} \end{matrix} \end{array}\right]$ | $\left[\begin{array}{cc} \begin{matrix} 0 & 1 & 0 \\ 0 & 0 & 1 \\ 1 & 0 & 0 \end{matrix} & \mathbf{0} \\ \hline \mathbf{0} & \begin{matrix} -\frac{1}{2} & \frac{1}{2}\sqrt{3} \\ -\frac{1}{2}\sqrt{3} & -\frac{1}{2} \end{matrix} \end{array}\right]$ | $\left[\begin{array}{cc} \begin{matrix} 0 & 0 & -1 \\ -1 & 0 & 0 \\ 0 & 1 & 0 \end{matrix} & \mathbf{0} \\ \hline \mathbf{0} & \begin{matrix} -\frac{1}{2} & -\frac{1}{2}\sqrt{3} \\ \frac{1}{2}\sqrt{3} & -\frac{1}{2} \end{matrix} \end{array}\right]$ |
| $\hat{C}_3^2(x\bar{y}z)$ | $\hat{C}_3^2(xy\bar{z})$ | $\hat{C}_2(xy)$ | $\hat{C}_2(yz)$ |
| $\left[\begin{array}{cc} \begin{matrix} 0 & 0 & 1 \\ -1 & 0 & 0 \\ 0 & -1 & 0 \end{matrix} & \mathbf{0} \\ \hline \mathbf{0} & \begin{matrix} -\frac{1}{2} & -\frac{1}{2}\sqrt{3} \\ \frac{1}{2}\sqrt{3} & -\frac{1}{2} \end{matrix} \end{array}\right]$ | $\left[\begin{array}{cc} \begin{matrix} 0 & 1 & 0 \\ 0 & 0 & -1 \\ -1 & 0 & 0 \end{matrix} & \mathbf{0} \\ \hline \mathbf{0} & \begin{matrix} -\frac{1}{2} & \frac{1}{2}\sqrt{3} \\ -\frac{1}{2}\sqrt{3} & -\frac{1}{2} \end{matrix} \end{array}\right]$ | $\left[\begin{array}{cc} \begin{matrix} 0 & -1 & 0 \\ -1 & 0 & 0 \\ 0 & 0 & 1 \end{matrix} & \mathbf{0} \\ \hline \mathbf{0} & \begin{matrix} 1 & 0 \\ 0 & 1 \end{matrix} \end{array}\right]$ | $\left[\begin{array}{cc} \begin{matrix} 1 & 0 & 0 \\ 0 & 0 & -1 \\ 0 & -1 & 0 \end{matrix} & \mathbf{0} \\ \hline \mathbf{0} & \begin{matrix} -\frac{1}{2} & -\frac{1}{2}\sqrt{3} \\ -\frac{1}{2}\sqrt{3} & \frac{1}{2} \end{matrix} \end{array}\right]$ |
| $\hat{C}_2(zx)$ | $\hat{C}_2(\bar{x}y)$ | $\hat{C}_2(yz)$ | $\hat{C}_2(\bar{z}x)$ |
| $\left[\begin{array}{cc} \begin{matrix} 0 & 0 & -1 \\ 0 & 1 & 0 \\ -1 & 0 & 0 \end{matrix} & \mathbf{0} \\ \hline \mathbf{0} & \begin{matrix} -\frac{1}{2} & \frac{1}{2}\sqrt{3} \\ \frac{1}{2}\sqrt{3} & \frac{1}{2} \end{matrix} \end{array}\right]$ | $\left[\begin{array}{cc} \begin{matrix} 0 & 1 & 0 \\ 1 & 0 & 0 \\ 0 & 0 & 1 \end{matrix} & \mathbf{0} \\ \hline \mathbf{0} & \begin{matrix} 1 & 0 \\ 0 & -1 \end{matrix} \end{array}\right]$ | $\left[\begin{array}{cc} \begin{matrix} 1 & 0 & 0 \\ 0 & 0 & 1 \\ 0 & 1 & 0 \end{matrix} & \mathbf{0} \\ \hline \mathbf{0} & \begin{matrix} -\frac{1}{2} & -\frac{1}{2}\sqrt{3} \\ -\frac{1}{2}\sqrt{3} & \frac{1}{2} \end{matrix} \end{array}\right]$ | $\left[\begin{array}{cc} \begin{matrix} 0 & 0 & 1 \\ 0 & 1 & 0 \\ 1 & 0 & 0 \end{matrix} & \mathbf{0} \\ \hline \mathbf{0} & \begin{matrix} -\frac{1}{2} & \frac{1}{2}\sqrt{3} \\ \frac{1}{2}\sqrt{3} & \frac{1}{2} \end{matrix} \end{array}\right]$ |

transformation due to the operation $\hat{C}_4(z)$:

$$\mathbf{D}(\hat{C}_4(z)) = \left[\begin{array}{ccc|cc} 0 & 1 & 0 & & \\ -1 & 0 & 0 & \multicolumn{2}{c}{0} \\ 0 & 0 & -1 & & \\ \hline & 0 & & 1 & 0 \\ & & & 0 & -1 \end{array}\right]. \tag{3.68}$$

In fact

$$[d_\eta \; d_\xi \; d_\zeta - d_u - d_v] = [d_\xi \; d_\eta \; d_\zeta \; d_u \; d_v]\left[\begin{array}{ccc|cc} 0 & 1 & 0 & & \\ -1 & 0 & 0 & \multicolumn{2}{c}{0} \\ 0 & 0 & -1 & & \\ \hline & 0 & & 1 & 0 \\ & & & 0 & -1 \end{array}\right]$$

The rotations $\hat{C}_4(y)$ result in $x \to -z$, $y \to y$, $z \to x$, so that

$$\left.\begin{array}{l} \hat{C}_4(y)\, d_\xi = d_\xi, \quad \hat{C}_4(y)\, d_\eta = -d_\eta, \quad \hat{C}_4(y)\, d_\zeta = -d_\xi, \\ \hat{C}_4(y)\, d_u = -\frac{1}{2}\, d_u + \frac{1}{2}\sqrt{3}\, d_v, \quad \hat{C}_4(y)\, d_v = \frac{1}{2}\sqrt{3}\, d_u + \frac{1}{2}\, d_v. \end{array}\right\} \tag{3.69}$$

In matrix representation the rotation $\hat{C}_4(y)$ has the form

$$\mathbf{D}(\hat{C}_4(y)) = \left[\begin{array}{ccc|cc} 0 & 0 & -1 & & \\ 0 & -1 & 0 & \multicolumn{2}{c}{0} \\ 1 & 0 & 0 & & \\ \hline & 0 & & -\frac{1}{2} & \frac{1}{2}\sqrt{3} \\ & & & \frac{1}{2}\sqrt{3} & \frac{1}{2} \end{array}\right]. \tag{3.70}$$

Similar simple calculations give the other matrices of the five-dimensional **O** group representation constructed in the basis of $d$ wavefunctions (Table 3.3).

The five-dimensional representation obtained has the block-diagonal structure

$$\left[\begin{array}{c|c} 3 \times 3 & 0 \\ \hline 0 & 2 \times 2 \end{array}\right].$$

This result follows directly from the fact that $\hat{C}_4(z)$ and $\hat{C}_4(y)$ are generators of group **O**. Full set of matrices (representation) can be obtained from (3.68) and (3.69) which are block diagonal. As it was noted in section 3.1.3, multiplication of block-diagonal matrices leads to a matrix which is block-diagonal also.

Hence this is a reducible representation, consisting of the three-dimensional and two-dimensional representations. Each of these representations is irreducible. The five $d$ wavefunctions of the hydrogen atom are thus divided into two sets: $d_{yz}$, $d_{xz}$, $d_{xy}$ and $d_{z^2}$, $d_{x^2-y^2}$. The symmetry operations of **O** mix functions only within each of the sets while functions belonging to different sets are not transformed into one another by symmetry operations. It should be emphasized that the block-diagonal matrix structure of the five-dimensional representation is related directly to the real basis $A_2(\vartheta, \varphi)$ (Table 3.1). Of course, the representation can also be constructed in another basis, using, for example, the functions $Y_{2m}$. The transformation matrices for the vector $[Y_{2-2} \quad Y_{2-1} \quad Y_{20} \quad Y_{21} \quad Y_{22}]$ do not have a block structure. In this case the reducibility of the five-dimensional representation is not obvious; in order to determine the irreducible representations, it is necessary to perform a similarity transformation corresponding to the transformation to their basis functions.

### 3.5.4 Basis functions and irreducible representations

We have found two three-dimensional irreducible representations of **O**, one with the basis set $p_x$, $p_y$, $p_z$, and the other with the basis $d_{yz}$, $d_{xz}$, $d_{xy}$; let us denote them by $\mathbf{D}^{(p)}$ and $\mathbf{D}^{(d)}$. Since both representations are three-dimensional, the question arises as to whether one of them can be obtained from the other by a similarity transformation—in other words, whether it is possible to find a $3 \times 3$ matrix $\mathbf{Q}$ such that

$$\mathbf{D}^{(p)} = \mathbf{Q}^{-1}\mathbf{D}^{(d)}\mathbf{Q}. \tag{3.71}$$

Let us choose a class, for example, $6\hat{C}_4$, and compare the characters of the representations $\mathbf{D}^{(p)}$ and $\mathbf{D}^{(d)}$ for it. From Tables 3.2 and 3.3, we find that

$$\chi^{(p)}(\hat{C}_4) = 1, \quad \chi^{(d)}(\hat{C}_4) = -1.$$

Since a similarity transformation does not change the character of a matrix, the transformations $\mathbf{D}^{(p)}$ and $\mathbf{D}^{(d)}$, having different characters, are not equivalent. In this respect they may be considered as being essentially different representations of the group **O**. Thus, from the infinite number of three-dimensional equivalent representations, we have found two inequivalent ones.

How many inequivalent irreducible representations of **O** exist? At first sight, it seems that by a careful choice of basis fuctions, it should be possible to obtain more and more irreducible representations. However, this is not the case, since a rigorous treatment leads to the following important result: *the number of inequivalent irreducible representations of a finite point group is finite.*

In Section 3.6 the properties of irreducible representations are formulated and, in particular, their number is given for each point group. Now it should be noted that there are no other three-dimensional inequivalent irreducible representations, except for the two already found in **O**. This means that even

**Table 3.4** Basis functions of the irreducible representations of the group **O**.

| $A_1$ | $A_2$ | E | $T_1$ | $T_2$ |
|---|---|---|---|---|
| $x^2+y^2+z^2$<br>$(l=0)$ | $xyz$<br>$(l=3)$ | $3z^2-r^2$<br>$\sqrt{3}(x^2-y^2)$<br>$(l=2)$ | $x, y, z$<br>$(l=1)$ | $yz, xz, xy$<br>$(l=2)$ |

by choosing as basis sets the multidimensional $A_l(\vartheta, \varphi)$ vectors with $l = 3, 4$ etc. (the $f$ and $g$ wavefunctions, etc), we shall not obtain any new irreducible three-dimensional representations. Furthermore, for a certain value of $l$, the number of inequivalent irreducible representations of all possible dimensions will be exhausted.

It has been found that **O** has five inequivalent irreducible representations: two of them are one-dimensional, one is two-dimensional and two are three-dimensional. The one-dimensional representations are denoted by $A_1$ and $A_2$, the two-dimensional by E, and three-dimensional by $T_1$ and $T_2$. The basis functions of the irreducible representations can be obtained from the angular functions $A_l(\vartheta, \varphi)$ with different values of $l$.

Table 3.4 gives the basis functions of the irreducible representations corresponding to the minimum value of $l$. The basis of the one-dimensional representation $A_1$ is a totally symmetric function that is not changed under the application of symmetry operations. Such a function may be a spherically symmetric s orbital. The representation $A_2$ is also one-dimensional, but its basis function $xyz$ is obtained from the f orbitals $A_3(\vartheta, \varphi)$. This function, in contrast to the totally symmetric one, changes its sign under the rotations $\hat{C}_4$ and $\hat{C}_2$, for example:

$$\left.\begin{aligned} \hat{C}_4(z)xyz &= y(-x)z = -xyz, \\ \hat{C}_2(xy)xyz &= yx(-z) = -xyz. \end{aligned}\right\} \tag{3.72}$$

The representation $T_1$ is realized by the $p$ wavefunctions, while the representations E and $T_2$ are realized by the $d$ wavefunctions.

## 3.6 PROPERTIES OF IRREDUCIBLE REPRESENTATIONS

In this section we formulate some properties of the irreducible representations of point groups. Their proof constitutes the basis of the detailed mathematical theory (see e.g. [12, 15, 20, 35]), and is not necessary for most practical applications. We therefore give these properties without proof.

(1) The number of inequivalent irreducible representations of a point group is equal to the number of classes in the group.

This makes it possible to determine the number of irreducible representations of any point group. It is for this reason that the enumeration of point-group operations in Section 2.4 was followed by the distribution of operations into classes. For example, the cubic group **O** contains the five irreducible respresentations obtained above. $\mathbf{O}_h$ contains ten irreducible representations. Indeed, this group is the direct product of **O** and $\mathbf{C}_i$ (the inversion group), and therefore its number of classes is twice that of **O** (Section 2.3).

(2) The sum of the squares of the dimensions of inequivalent irreducible representations is equal to the order of the group:

$$g_1^2 + g_2^2 + \cdots + g_r^2 = g, \tag{3.73}$$

where $g_1, g_2, \ldots, g_r$ are the dimensions of the irreducible representations, of which there are $r$ (equal to the number of classes), and $g$ is the order of the group.

It should be added that every group has a one-dimensional representation, whose basis is a function remaining unchanged under all symmetry operations. This function is called an *invariant function*, and the representation is a totally symmetric representation (in **O** the $s$ wavefunction of the hydrogen atom is invariant).

Let us apply the condition (3.73) to the group **O**, in which $g = 24$ and $r = 5$:

$$g_1^2 + g_2^2 + g_3^2 + g_4^2 + g_5^2 = 24. \tag{3.74}$$

This is an equation determining the dimensions of the irreducible representations. In order to find these dimensions, it is necessary to obtain the *integer* solutions of this equation. The only set of five integers satisfying (3.74) is

$$\left.\begin{aligned} &g_1 + g_2 = 1, \quad g_3 = 2, \quad g_4 = g_5 = 3, \\ &1^2 + 1^2 + 2^2 + 3^2 + 3^2 = 24. \end{aligned}\right\} \tag{3.75}$$

These representations ($A_1$, $A_2$, E, $T_1$ and $T_2$) have already been found in Section 3.5.4.

$\mathbf{C}_{2v}$ contains four operations, each forming a class. Therefore this group has four one-dimensional representations:

$$1^2 + 1^2 + 1^2 + 1^2 = 4. \tag{3.76}$$

For all Abelian groups $r = g$, and hence all $g_i = 1$ (the representations are one-dimensional). $\mathbf{C}_{3v}$ contains three classes ($\hat{E}$, $2\hat{C}_3$, $3\hat{\sigma}_v$). The dimensions of the three ($r = 3$) irreducible representations satisfy

$$g_1^2 + g_2^2 + g_3^2 = 6, \tag{3.77}$$

which provides two one-dimensional and one two-dimensional representations.

The remarkable feature of (3.73) is the *uniqueness* of its solution in terms of integers. Sorting out all the point groups, it is easy to see that for each it is possible to select only one set of $r$ integers satisfying this equation.

(3) The characters of matrices belonging to the same class are the same in any representation (either reducible or irreducible).

This property is explained by the fact that operations of the same class are interrelated by similarity transformations. The proof is given in Section 3.1.4.

(4) The sum of squared characters in each of the irreducible representations is equal to the order of the group:

$$\sum_{\hat{R}}[\chi^{(\Gamma)}(\hat{R})]^2 = g, \qquad (3.78)$$

where for brevity we have used the standard notation $\chi(\hat{R}) \equiv \chi(\mathbf{D}(\hat{R}))$. The index $\Gamma$ numbers the irreducible representations, while the summation is over all the group operations.

Let us check this *normalization equation*, for example for the $T_1$ representation of **O**:

$$[\chi^{(T_1)}(\hat{E})]^2 + 6[\chi^{(T_1)}(\hat{C}_4)]^2 + 3[\chi^{(T_1)}(\hat{C}_4^2)]^2 + 6[\chi^{(T_1)}(\hat{C}_3)]^2 + 6[\chi^{(T_1)}(\hat{C}_2)]^2 = 24. \qquad (3.79)$$

Owing to property (2), operations of the same class are combined in the sum (3.78). On substituting the characters of the matrices from Table 3.2, it is easy to see that (3.79) is satisfied.

(5) The characters of two different inequivalent irreducible representations are related by the orthogonality property†

$$\sum_{\hat{R}} \chi^{(\Gamma_1)}(\hat{R})\chi^{(\Gamma_2)}(\hat{R}) = 0 \quad (\Gamma_1 \neq \Gamma_2). \qquad (3.80)$$

The sum contains $g$ terms. The characters can be formally considered as the components of a vector in $g$-dimensional space. By analogy with the scalar product $a_x b_x + a_y b_y + a_z b_z \equiv \mathbf{a} \cdot \mathbf{b} = ab \cos \vartheta$ of two-dimensional vectors **a** and **b**, the sum in (3.80) is the scalar product of the $g$-dimensional vectors $\chi^{(\Gamma_1)}(\hat{R})$ and $\chi^{(\Gamma_2)}(\hat{R})$. Since this scalar product is zero, the equation (3.80) is called the *orthogonality relationship.*

The above properties make it possible to find the characters of irreducible representations without knowing the matrices, that is the irreducible representations themselves. Let us show this, using as an example two point groups.

**Example 3.1** The group $\mathbf{C}_{2v}$ contains four one-dimensional representations. The characters of the totally symmetric representation $A_1$ are equal to 1 for all the

† For complex representations (Section 3.7.3) in (3.80) we must use $\chi^{(\Gamma_1)}(\hat{R})^*$ (complex-conjugate value).

operations, since the basis function is invariant. They may be written as a four-dimensional vector

| | $\hat{E}$ | $\hat{C}_2$ | $\hat{\sigma}_v$ | $\hat{\sigma}'_v$ |
|---|---|---|---|---|
| $A_1$ | 1 | 1 | 1 | 1 |

The characters of the other three representations (let us call them $A_2$, $B_1$ and $B_2$), according to property (5), should be vectors orthogonal to $\chi^{(A_1)}$, and, in accordance with the normalization equation (3.78)

$$[\chi^{(\Gamma)}(\hat{E})]^2 + [\chi^{(\Gamma)}(\hat{C}_2)]^2 + [\chi^{(\Gamma)}(\hat{\sigma}_v)]^2 + [\chi^{(\Gamma)}(\sigma'_v)]^2 = 4. \tag{3.81}$$

The vectors with components $\chi^{(\Gamma)}(\hat{R}) = \pm 1$ satisfy the requirements of orthogonality and normalization, while two of the components are equal to 1 and the other two to $-1$. Therefore we obtain the *character table* of the group $\mathbf{C}_{2v}$:

| $\mathbf{C}_{2v}$ | $\hat{E}$ | $\hat{C}_2$ | $\hat{\sigma}_v$ | $\hat{\sigma}'_v$ |
|---|---|---|---|---|
| $A_1$ | 1 | 1 | 1 | 1 |
| $A_2$ | 1 | 1 | $-1$ | $-1$ |
| $B_1$ | 1 | $-1$ | 1 | $-1$ |
| $B_2$ | 1 | $-1$ | $-1$ | 1 |

**Example 3.2** The characters of the three representations of $\mathbf{C}_{3v}$ are found in the same way. For the totally symmetric representation $\chi^{(A_1)}(\hat{R}) = 1$, for all the operations,

| | $\hat{E}$ | $2\hat{C}_3$ | $3\hat{\sigma}_v$ |
|---|---|---|---|
| $A_1$ | 1 | 1 | 1 |

The group operations are divided into classes, since the operations of the same class have the same characters. The characters of the second one-dimensional representation $A_2$ satisfy the conditions of orthogonality to $\chi^{(A_1)}$ and normalization only when the vector $\chi^{(A_2)}$ has the form

| | $\hat{E}$ | $2\hat{C}_3$ | $3\hat{\sigma}_v$ |
|---|---|---|---|
| $A_2$ | 1 | 1 | $-1$ |

Calculating the characters of the two-dimensional representation (denoted by E) using the above conditions, we obtain the $\mathbf{C}_{3v}$ group character table:

| $\mathbf{C}_{3v}$ | $\hat{E}$ | $2\hat{C}_3$ | $3\hat{\sigma}_v$ |
|---|---|---|---|
| $A_1$ | 1 | 1 | 1 |
| $A_2$ | 1 | 1 | $-1$ |
| E | 2 | $-1$ | 0 |

In many applications of group theory no symmetry-operation matrices are needed, but only their characters. Therefore most reference books on group theory give only character tables.

## 3.7 CHARACTER TABLES

### 3.7.1 Structure of tables

For practical application of group theory it is not necessary to obtain characters using the methods of Section 3.6 in each particular case, since the point-group characters are listed in readily available tables. The most complete reference book [1] contains, besides the characters of the 32 crystallographic point groups, other important data, covering practically all the information necessary for practical application of group theory. The use of these data is explained in the following sections. The characters of molecular groups (other than crystallographic ones, e.g. $\mathbf{C}_5$, $\mathbf{S}_{10}$ and $\mathbf{S}_{12}$) are given in [17].

Let us illustrate the structure of the tables by way of some examples. The $\mathbf{C}_{3v}$ group characters have the form

| $\mathbf{C}_{3v}$ | $\hat{E}$ | $2\hat{C}_3$ | $3\hat{\sigma}_v$ | $f^{(\Gamma)}$ | $f^{(\Gamma)}$ |
|---|---|---|---|---|---|
| $A_1$ | 1 | 1 | 1 | $z$ | $x^2+y^2; z^2$ |
| $A_2$ | 1 | 1 | $-1$ | $R_z$ | |
| E | 2 | $-1$ | 0 | $x, y; R_x, R_y$ | $x^2-y^2, xy; xz, yz.$ |
| I | | II | | III | IV |

In the upper-left corner the Schoenflies group notation is shown. The top row of the table, above the line, contains the group operations divided into classes ($\hat{E}$, $2\hat{C}_3$, $3\hat{\sigma}_v$), so that the number of columns is equal to the number of classes. The left-hand column (I), outside the vertical line, contains the symbols of the irreducible representations $\Gamma$ (two one-dimensional representations $A_1$ and $A_2$ and one two-dimensional representation E; the symbol E for the two-dimensional representation should not be confused with the symbol $\hat{E}$ for the similarity-transformation operation). The main part of the table (II) is square, since the number of irreducible representations is equal to the number of classes. Column III contains the simplest basis functions $f^{(\Gamma)}$ of the irreducible representations $\Gamma$. The function $z$ is the basis of the totally symmetric representation, while the two functions $x$ and $y$ form the basis of the two-dimensional representation E. The functions $x$ and $y$ are said to be *transformed according to the representation* E, while $z$ is transformed over the representation $A_1$.

### 3.7.2 Polar and axial vectors

The functions $R_x$, $R_y$ and $R_z$ require special explanation. In column III $R_z$ corresponds to the representation $A_2$, while $R_x$ and $R_y$, as well as $x$ and $y$, transform over the representation E. Let us obtain the *vector product* **R** according to the familiar rule from the two vectors $\mathbf{r}_1(x_1, y_1, z_1)$ and $\mathbf{r}_2(x_2, y_2, z_2)$:

$$\mathbf{R} = \begin{vmatrix} \mathbf{i} & \mathbf{j} & \mathbf{k} \\ x_1 & y_1 & z_1 \\ x_2 & y_2 & z_2 \end{vmatrix},$$

where **i**, **j**, and **k** are unit vectors (along the $x$, $y$ and $z$ axes respectively and $|\ \ |$ denotes the determinant. Expanding the determinant, we obtain

$$\mathbf{R} = \mathbf{i}R_x + \mathbf{j}R_y + \mathbf{k}R_z,$$

$$R_x = y_1z_2 - z_1y_2, \quad R_y = z_1x_2 - x_1z_2, \quad R_z = x_1y_2 - y_1x_2. \tag{3.82}$$

Let us now find the transformation matrix of $[R_x \quad R_y \quad R_z]$ for the rotation $\hat{C}_3$, assuming that we know the matrices for $[x_1 \quad y_1 \quad z_1]$ and $[x_2 \quad y_2 \quad z_2]$, (3.49) and (3.50):

$$\begin{aligned} R'_x &= y'_1z'_2 - z'_1y'_2 = (-\tfrac{1}{2}\sqrt{3}x_1 - \tfrac{1}{2}y_1)z_2 - z_1(-\tfrac{1}{2}\sqrt{3}x_2 - \tfrac{1}{2}y_2) \\ &= -\tfrac{1}{2}(y_1z_2 - z_1y_2) + \tfrac{1}{2}\sqrt{3}(z_1x_2 - y_1z_2) = -\tfrac{1}{2}R_x + \tfrac{1}{2}\sqrt{3}R_y. \end{aligned} \tag{3.83}$$

In the same way, we find

$$R'_y = -\tfrac{1}{2}\sqrt{3}R_x - \tfrac{1}{2}R_y, \qquad R'_z = R_z. \tag{3.84}$$

On comparing these results with (3.49), we see that the rotation matrix $\mathbf{D}(\hat{C}_3)$ for the vector **R** coincides with the rotation matrix of the vector **r** (3.50). Let us now consider the reflection $\hat{\sigma}_v(xz)$:

$$\left.\begin{aligned} R'_z &= x_1(-y_2) - (-y_1)x_2 = -R_z, \\ R'_x &= -R_x, \qquad R'_y = R_y. \end{aligned}\right\} \tag{3.85}$$

Thus the reflection $\hat{\sigma}_v$ changes the sign of the $z$ components of the vector **R**, while the $z$ component of the "ordinary" vector **r** is invariant. It should also be noted that, in contrast to **r**, the vector **R** does not change its sign under inversion. A vector with such properties is called an *axial vector*, while one like **r** is called a *polar vector*. The vector product of two polar vectors is an axial vector.

In $\mathbf{C}_{3v}$ both $R_x$ and $R_y$ transform over the representation E while $R_z$ transforms over the representation $A_2$. It should be noted that the quantum mechanical angular-momentum operator **L** transforms as an axial vector, since it is the vector product of the coordinate and momentum operators:

$$\hat{p}_x = -i\hbar\frac{\partial}{\partial x}, \quad \hat{p}_y = -i\hbar\frac{\partial}{\partial y}, \quad \hat{p}_z = -i\hbar\frac{\partial}{\partial z}.$$

Therefore, in some character tables, instead of the components $R_i$, the operators $\hat{L}_x$, $\hat{L}_y$ and $\hat{L}_z$ or the spin components $\hat{S}_x$, $\hat{S}_y$ and $\hat{S}_z$, having similar properties, are given as the basis. The spin operators are discussed below.

Finally, column IV contains the basis functions of the representations Γ quadratic in $x$, $y$ and $z$. In many character tables the columns III and IV are combined.

The tables in [1] contain an additional column, which shows behaviour of the basis functions under *time inversion*. We shall not go any deeper into this here, since the problem of time inversion is discussed briefly later in connection with problems of magnetic resonance.

### 3.7.3 Complex-conjugate representations

The character tables of the rotation groups $\mathbf{C}_n$ $(n > 2)$, the rotoflection transformations $\mathbf{S}_n$ $(n > 4)$, the groups $\mathbf{C}_{nh}$ $(n > 2)$ and the group $\mathbf{T}$, as well as those of some continuous groups, contain complex numbers. Let us consider as an example the $\mathbf{C}_3$ group characters:

| $\mathbf{C}_3$ | $\hat{E}$ | $\hat{C}_3$ | $\hat{C}_3^2$ | $f^{(\Gamma)}$ |
|---|---|---|---|---|
| A | 1 | 1 | 1 | $z, R_z$ |
| $\varepsilon$ | 1 | $\omega$ | $\omega^*$ | $x + iy; R_x + iRy$ |
| $\varepsilon^*$ | 1 | $\omega^*$ | $\omega$ | $x - iy; R_x - iR_y$ |
| $\varepsilon, \varepsilon^*$; E | 2 | $-1$ | $-1$ | $x, y; R_x, R_y$ |

$\mathbf{C}_3$ is Abelian, and all of the irreducible representations are one-dimensional. The representation A is totally symmetric, with basis $z$ or $R_z$. Here $\varepsilon$ and $\varepsilon^*$ denote two one-dimensional representations, * denotes complex-conjugation, and $\omega$ denotes the complex number (cube root of unity)

$$\omega = \mathrm{e}^{2\pi \mathrm{i}/3} = \cos \tfrac{2}{3}\pi + \mathrm{i} \sin \tfrac{2}{3}\pi. \tag{3.86}$$

The basis functions $x + \mathrm{i}y$ and $x - \mathrm{i}y$ of the representations $\varepsilon$ and $\varepsilon^*$ respectively are complex conjugate. Since $\mathbf{C}_3$ contains neither the reflections $\sigma_v$ nor inversions, the complex components of the axial vector $R_x \pm \mathrm{i} R_y$ are transformed similarly to the components of the polar vector $x \pm \mathrm{i}y$. Let us write $x \pm \mathrm{i}y$ as

$$x \pm \mathrm{i}y = \rho \mathrm{e}^{\mathrm{i}\varphi}, \quad \rho = (x^2 + y^2)^{1/2}, \quad \tan \varphi = \frac{y}{x}, \tag{3.87}$$

where $\rho$ is the modulus of the complex number and $\varphi$ is the argument. The operations $\hat{C}_3$ and $\hat{C}_3^2$ result in multiplication of the basis function by $\mathrm{e}^{\pm 2\pi \mathrm{i}/3}$ (Fig. 3.4):

$$\left.\begin{aligned}\hat{C}_3(x \pm \mathrm{i}\,y) &= \hat{C}_3 \rho\, \mathrm{e}^{\pm(\varphi+2\pi/3)\mathrm{i}} = (x \pm \mathrm{i}\,y)\, \mathrm{e}^{\pm 2\pi \mathrm{i}/3},\\ \hat{C}_3^2(x \pm \mathrm{i}\,y) &= (x \pm \mathrm{i}\,y)\, \mathrm{e}^{\pm 2\pi \mathrm{i}/3},\end{aligned}\right\} \tag{3.88}$$

or

$$\left.\begin{aligned}\hat{C}_3(x + \mathrm{i}\,y) &= \omega(x + \mathrm{i}\,y), & \hat{C}_3^2(x + \mathrm{i}\,y) &= \omega^*(x + \mathrm{i}\,y),\\ \hat{C}_3(x - \mathrm{i}\,y) &= \omega^*(x - \mathrm{i}\,y), & \hat{C}_3^2(x - \mathrm{i}\,y) &= \omega(x - \mathrm{i}\,y).\end{aligned}\right\} \tag{3.89}$$

Hence $x + \mathrm{i}\,y$ and $x - \mathrm{i}\,y$ are transformed independently, i.e. over the one-dimensional representations. The characters of the irreducible representations $\varepsilon$ and $\varepsilon^*$ are complex conjugate, and such representations are called *complex-conjugate representations.*

The groups $\mathbf{C}_n$, $\mathbf{S}_n$ and $\mathbf{C}_{nh}$ with $\mathbf{C}_n$ axes also have one-dimensional complex-conjugate characters, which depend on $n$:

$$\left.\begin{aligned}\omega_n^m &= (\omega_n)^m \quad (m \leq n),\\ \omega_n &= \mathrm{e}^{2\pi \mathrm{i}/n} = \cos\frac{2\pi}{n} + \mathrm{i}\,\sin\frac{2\pi}{n},\\ \omega_n^m &= \mathrm{e}^{2\pi \mathrm{i} m/n} = \cos\frac{2\pi m}{n} + \mathrm{i}\,\sin\frac{2\pi m}{n}.\end{aligned}\right\} \tag{3.90}$$

It is easy to find the values of $\omega_n^m$: for $m = 0$ or $n$, $\omega_n^m = 1$; for $m = \frac{1}{2}n$, $\omega_n^m = -1$; for $m = \frac{1}{4}n$, $\omega_n^m = \mathrm{i}$; and finally, when $m > \frac{1}{2}n$, $\omega_n^m = (\omega_n^{n-m})^*$ (for example, $\omega_6^5 = \omega_6^*$ and $\omega_6^4 = (\omega_6^2)^*$).

Some groups have more than one pair of complex-conjugate representations. Let us consider for instance the $\mathbf{C}_6$ group characters ($\nu = \exp(\mathrm{i}\pi/3)$)

| $\mathbf{C}_6$ | $\hat{E}$ | $\hat{C}_6$ | $\hat{C}_6^2$ | $\hat{C}_6^3$ | $\hat{C}_6^4$ | $\hat{C}_6^5$ |
|---|---|---|---|---|---|---|
| $A$ | 1 | 1 | 1 | 1 | 1 | 1 |
| $B$ | 1 | $-1$ | 1 | $-1$ | 1 | $-1$ |
| $\epsilon_1$ | 1 | $\nu$ | $-\nu^*$ | $-1$ | $-\nu$ | $\nu^*$ |
| $\epsilon_1^*$ | 1 | $\nu^*$ | $-\nu$ | $-1$ | $-\nu^*$ | $\nu$ |
| $\epsilon_2$ | 1 | $-\nu^*$ | $-\nu$ | 1 | $-\nu^*$ | $-\nu$ |
| $\epsilon_2^*$ | 1 | $-\nu$ | $-\nu^*$ | 1 | $-\nu$ | $-\nu^*$ |
| $\epsilon_1, \epsilon_1^*; E_1$ | 2 | 1 | $-1$ | $-2$ | $-1$ | 1 |
| $\epsilon_2, \epsilon_2^*; E_2$ | 2 | $-1$ | $-1$ | 2 | $-1$ | $-1$ |

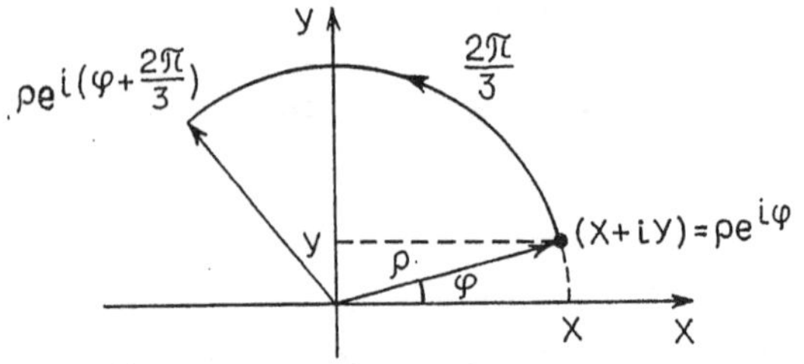


**Fig. 3.4** Action of the $\hat{C}_3$ operation on the basis function of the complex representation.

Thus $\mathbf{C}_6$ contains two pairs of complex-conjugate representations $\varepsilon_1, \varepsilon_1^*$ and $\varepsilon_2, \varepsilon_2^*$, while $\mathbf{S}_8$ and $\mathbf{S}_{12}$ have three and five pairs of such representations respectively [17].

For reasons that will be clear from the following, in physical applications the complex-conjugate representations are used as a single representation of twice the dimension. The characters of this *two-dimensional* representation E are equal to the sum of the characters $\varepsilon$ and $\varepsilon^*$; that is, they are real quantities. In the given table of $\mathbf{C}_3$ group characters this joint two-dimensional representation is located in the lower row, its basis being the real functions $x$, $y$ or $R_x$, $R_y$. In the tables in [1] the representations $\varepsilon$ and $\varepsilon^*$ are written separately, but the characters of a two-dimensional "physical" representation are easily obtained from the formula

$$\omega_n^m + \omega_n^{m*} = 2\cos\frac{2\pi m}{n}. \tag{3.91}$$

Similarly, several complex-conjugate representations can be combined in pairs. In the $C_6$ there are two two-dimensional irreducible representations $E_1$ and $E_2$ combined from pairs $\epsilon_1, \epsilon_1^*$ and $\epsilon_2, \epsilon_2^*$ respectively.

### 3.7.4 Groups with an inversion centre

Groups containing inversion can be represented as the direct product of some point group $\mathbf{G}$ with the inversion group $\mathbf{C}_i(\hat{E}, \hat{I})$ group. $\mathbf{G} \times \mathbf{C}_i$ contains twice as many classes as the original group $\mathbf{G}$ (Section 2.3), and hence twice as many irreducible representations. The basis functions are divided into even and odd ones relative to their behaviour under inversion, and consequently groups with inversion contain two types of representations—even and odd.

Let us consider, for example, the character tables of: $\mathbf{D}_3$ and $\mathbf{D}_{3d}$, which is the direct product of $\mathbf{D}_3 \times \mathbf{C}_i$:

| $\mathbf{D}_3$ | $\hat{E}$ | $2\hat{C}_3$ | $3\hat{C}_2'$ | $f^{(\Gamma)}$ |
|---|---|---|---|---|
| $A_1$ | 1 | 1 | 1 | $x^2+y^2; z^2$ |
| $A_2$ | 1 | 1 | $-1$ | $z; R_z$ |
| E | 2 | $-1$ | 0 | $x, y; R_x, R_y;$ $x^2-y^2, xy; xz, yz.$ |

| $\mathbf{D}_{3d}$ | $\hat{E}$ | $2\hat{C}_3$ | $3\hat{C}_2'$ | $\hat{I}$ | $2\hat{S}_6$ | $3\hat{\sigma}_d$ | $f^{(\Gamma)}$ |
|---|---|---|---|---|---|---|---|
| $A_{1g}$ | 1 | 1 | 1 | 1 | 1 | 1 | $x^2+y^2; z^2$ |
| $A_{2g}$ | 1 | 1 | $-1$ | 1 | 1 | $-1$ | $R_z$ |
| $E_g$ | 2 | $-1$ | 0 | 2 | $-1$ | 0 | $R_x, R_y; x^2-y^2, xy; yz, xz.$ |
| $A_{1u}$ | 1 | 1 | 1 | $-1$ | $-1$ | $-1$ | |
| $A_{2u}$ | 1 | 1 | $-1$ | $-1$ | $-1$ | 1 | $z$ |
| $E_u$ | 2 | $-1$ | 0 | $-2$ | 1 | 0 | $x, y$ |

The irreducible representations ($A_1$, $A_2$, E) of $\mathbf{D}_3$ and $\mathbf{D}_{3d}$ are denoted by identical symbols. However, $\mathbf{D}_{3d}$ contains twice as many representations, differing in the subscripts g and u, which indicate respectively the even (German, "gerade") and odd ("ungerade") basis functions relative to inversion. The components of a polar vector **r** are transformed over the odd representations $A_{2u}(z)$ and $E_u(x, y)$, while the components of the axial vector **R** are transformed over the even representations $A_{2g}(R_z)$ and $E_g(R_x, R_y)$. All the functions quadratic (and of other even powers) of the coordinates are transformed over the even representations. In $\mathbf{D}_3$ the respective components of the polar and axial vectors are transformed over one and the same representation, for example $A_1(z; R_z)$, $E(x, y; R_xR_y)$. Therefore the characters of $\mathbf{D}_{3d}$ are obtained from the characters of the original group $\mathbf{D}_3$ according to a simple rule. The even representations of $\mathbf{D}_{3d}$ coincide with the $\mathbf{D}_3$ group representations, the odd ones also coincide for operations not including inversion, while the odd representations for operations including inversion have the opposite sign. Schematically, this is written as

| $\mathbf{D}_{3d}$ | $\hat{E}$ $2\hat{C}_3$ $3\hat{C}_2'$ | $\hat{I}$ $2\hat{S}_6$ $3\hat{\sigma}_d$ |
|---|---|---|
| $A_{1g}$ | | |
| $A_{2g}$ | $\chi$ | $\chi$ |
| $E_g$ | | |
| $A_{1u}$ | | |
| $A_{2u}$ | $\chi$ | $-\chi$ |
| $E_u$ | | |

Here $\chi$ is the character table of $\mathbf{D}_3$. In many publications the character tables of groups of the type $\mathbf{G} \times \mathbf{C}_i$ are not given at all, since they are easily obtained from the characters of **G**.

### 3.7.5 Systems of notation

There are several systems of notation for the irreducible representations of the point groups. The most widely used is the following Mulliken system, which is generally accepted in chemistry and spectroscopy:

(1) All one-dimensional representations are denoted by A or B, the two-dimensional by E and the three-dimensional by T (or sometimes by F).

(2) One-dimensional representations whose basis functions are symmetric relative to rotations about the principal $C_n$ axis are denoted by A, and their characters for the operations $\hat{C}_n^m$ are equal to 1. Antisymmetric representations, for which $\chi(\hat{C}_n) = -1$, are denoted by B;

(3) Subscripts 1 and 2 on the symbols A and B denote respectively symmetric and antisymmetric behaviour of the basis function relative to the $C_2'$ axes (perpendicular to $C_n$) or to the vertical $\sigma_v$ and $\sigma_d$ planes;

(4) Representations that are symmetric or antisymmetric with respect to the horizontal plane $\sigma_h$ are indicated by a prime or two primes respectively;

(5) The point groups $\mathbf{D}_2$ and $\mathbf{D}_{2h}$, each having three perpendicular $C_2$ axes, have representations $B_1$, $B_2$ and $B_3$ that are symmetric with respect to rotations about these axes. The $B_1$ basis is antisymmetric with respect to the operations $\hat{C}_2(x)$ and $\hat{C}_2(y)$, the $B_2$ basis is antisymmetric with respect to $\hat{C}_2(x)$ and $\hat{C}_2(z)$, and the $B_3$ basis is antisymmetric with respect to $\hat{C}_2(y)$ and $C_2(z)$;

(6) An index g indicates representations that are even under inversion, while u indicates odd representations;

(7) The three-dimensional representations of the cubic groups $\mathbf{T}_d$, **O** and $\mathbf{O}_h$ are divided into $T_1$ and $T_2$ depending on their behaviour under the operations $\hat{S}_4$ and $\hat{\sigma}_d$ of $\mathbf{T}_d$ and the operations $\hat{C}_2'$ and $\hat{C}_4$ of **O**;

(8) The icosahedral groups have representations of dimensions 4 and 5 denoted by U and V respectively.

The Bethe system of notation is also widely used. In this the irreducible representations are denoted by the $\Gamma_1$, $\Gamma_2$ etc. Even representations are denoted by $\Gamma_i^+$ and the odd ones by $\Gamma_i^-$. In **O**, for example, the Mulliken notation $A_1$, $A_2$, E, $T_1$, $T_2$ corresponds to the Bethe notation $\Gamma_1$, $\Gamma_2$, $\Gamma_3$, $\Gamma_4$, $\Gamma_5$. This system is less suggestive, but it is used in the tables in [1].

## PROBLEMS

**3.1** Find the set of matrices realizing the representation of the vector components in $\mathbf{D}_3$. Show that there is a mutually single-valued correspondence between the multiplication tables of these matrices and the $\mathbf{D}_3$ group operations.

**3.2** Find the irreducible representations realized by the *d* functions in $\mathbf{D}_{2d}$.

**3.3** Using (3.53), find the dimensions of the irreducible representations of $\mathbf{C}_{3v}$, $\mathbf{D}_{4h}$ and $\mathbf{T}_d$.

**3.4** Using the character tables of $\mathbf{D}_4$, write down the character table of $\mathbf{D}_{4h} = \mathbf{D}_4 \times \mathbf{C}_i$.

# 4 Crystal Field Theory for One-Electron Ions

## 4.1 QUALITATIVE DISCUSSION

In 1929 Bethe published the remarkable paper "Term splitting in crystals" [36], which dealt with atomic levels in crystals and complexes and laid the foundations of contemporary spectro- and magnetochemistry. It is assumed that an atom or ion with an unfilled electron shell is surrounded by point charges. The electric field of these charges, called a *crystal field*, influences the electron shell of the metal ion, and partially or completely removes the degeneracy of the atomic levels. A point-ion model of the crystal field does not take into account the real electronic structure of ligands, the effects of covalent chemical bonding and many other factors. In spite of subsequent progress in the theory, Bethe's original treatment has not lost its importance for physics and chemistry after over 60 years. The main component of this work is a group theoretical analysis of ion states in the crystal field, based only on symmetry concepts, and is not dependent on the approximate character of the point-ion model of the crystal field. Therefore the results of the group theoretical part of Bethe's paper are fully applicable to the more accurate models of complexes that take into account the covalence and the many-electron nature of the ligands.

Let us explain the qualitative behaviour of an atomic level in a crystal field by way of a simple example. Consider an atom containing one p electron outside a filled shell in the field of six point ligands situated at the vertices of an octahedron. The electron density in the three p states, $p_x$, $p_y$ and $p_z$, has the form of "dumb-bells" directed along the $x$, $y$ and $z$ axes (Fig. 4.1). The total electronic energy comprises the energy of a free hydrogen-like atom and the mean energy of repulsion of the electron from the negatively charged point ligands. Since the p state is degenerate, the energies of the three atomic p orbitals are identical. From Fig. 4.1 it is clear that the crystal field affects

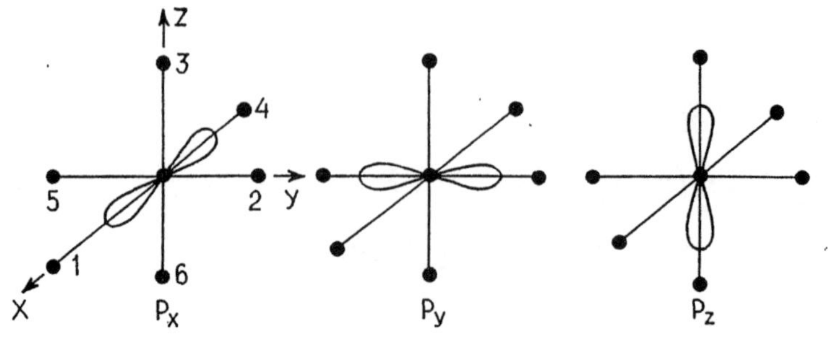


**Fig. 4.1** *p* orbitals of a hydrogen like ion in octahedral ($\mathbf{O}_h$) surroundings.

equally the electron in the $p_x$, $p_y$ and $p_z$ states because of the equivalence of the three diagonals of the octahedron. Hence both the p level in the cubic field and the atomic p level are threefold-degenerate. It should be noted that the wavefunctions $p_x$, $p_y$ and $p_z$ (or $x$, $y$ and $z$) form the basis of the irreducible representation $T_{1u}$ of $\mathbf{O}_h$.

Let us now consider how the electron energy is changed in the crystal field if its degree of symmetry is decreased. Assume, for example, that for some reason or other (ion substitution, deformation etc.) one of the diagonals of the octahedron has become elongated (Fig. 4.2a), so that the point group of the deformed octahedron is $\mathbf{D}_{4h}$. Since the negative ligands 3 and 6 are now more remote, this leads to a decrease in crystal field energy for the $p_z$ orbital. The $p_x$ and $p_y$ states have higher and equal energies, since the $x$ and $y$ axes are equivalent. Reducing the cubic symmetry to tetragonal symmetry leads to a partial splitting of the threefold-degenerate energy level (Fig. 4.2b). From the character tables of the $\mathbf{D}_{4h}$, we see that the $p_z$ wavefunction is a basis of the one-dimensional representation $A_{2u}$ of this group, while $p_x$ and $p_y$ transform according to the two-dimensional representation $E_u$.

Summarizing this qualitative analysis, we come to the following important conclusions.

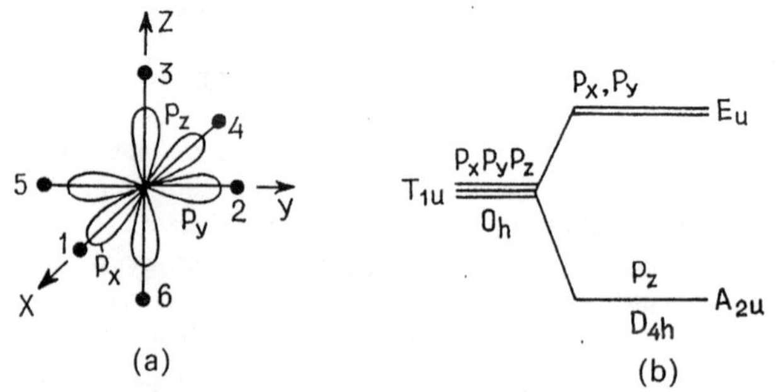


**Fig. 4.2** p-level splitting: (a) p orbitals in tetragonal ($\mathbf{D}_{4h}$) ligand surroundings; (b) splitting of p-level energy arising from the symmetry decrease $\mathbf{O}_h \rightarrow \mathbf{D}_{4h}$.

(1) Some atomic levels maintain their degeneracy in the crystal field; the wavefunctions of atomic states in this case provide the basis of an irreducible representation of the point group of the surrounding crystal.

(2) Lowering the crystal field symmetry results in a splitting of the energy level. The wavefunctions of each energy level then transform according to the irreducible representations of the lower-symmetry point group.

## 4.2 SCHRÖDINGER EQUATION AND IRREDUCIBLE REPRESENTATIONS

The qualitative discussion of the energy levels in the crystal field has revealed the relationship between the wavefunctions, the energy levels of the quantum system and the irreducible representations of its point group. Let us now consider a more rigorous formulation of these concepts. The stationary state wavefunctions $\psi$ of a quantum system are the solutions of the Schrödinger equation

$$\hat{H}\psi_n = E_n\psi_n, \tag{4.1}$$

where $\hat{H}$ is the Hamiltonian, $E_n$ is the energy of the $n$th level (i.e. its eigenvalue) and $\psi_n$ is the corresponding eigenfunction. We assume that the atomic nuclei have certain fixed positions $\boldsymbol{R}_k$ in space. The Hamiltonian $\hat{H}$ contains the kinetic energy operators of all of the valence electrons, the Coulomb energy of their repulsion and the energy of electron interaction with the crystal environment. In the particular case of point ligands the interaction with the crystal field is expressed as

$$\hat{V}^{\mathrm{c}} = \sum_{i=1}^{N}\sum_{k=1}^{N'} \frac{eq_k}{|\boldsymbol{r}_i - \boldsymbol{R}_k|}, \tag{4.2}$$

where the indices $i$ and $k$ run over the $N$ electrons and $N'$ nuclei respectively, and $|\boldsymbol{r}_i - \boldsymbol{R}_k|$ is the distance between the $i$th electron and $k$th nucleus with charge $q_k$.

Here and elsewhere we use mixed Gaussian (C.G.S.) units: with SI units a factor $4\pi\epsilon_0$ appears in the denominator. Let us subject the Hamiltonian $\hat{H}$ to the point symmetry operation $\hat{R}$ changing neither the electron kinetic energy nor the interaction between electrons, nor even the potential field in which they move, since a symmetry operation only permutes identical nuclei. Point group operations do not change the electron-nuclear potential energy – all nuclei including those in (4.2) going into the indistinguished positions. Therefore the Hamiltonian is invariant under point-group transformations. We can say that the invariance of the Hamiltonian under a particular group of transformations means that the quantum system "belongs to" this symmetry group.

Let us now formulate the principal theorem, relating group theory to quantum mechanics:

> If the Hamiltonian is invariant under a particular symmetry group then the eigenfunctions corresponding to a given energy level normally[†] form the basis of an irreducible representation of this group.

Let us consider a case with energy levels. Suppose that both sides of the Schrödinger equation (4.1) are subjected to the group operation $\hat{R}$:

$$\hat{R}(\hat{H}\psi_n) = \hat{R}E_n\psi_n. \tag{4.3}$$

The energy $E_n$ can be taken outside the $\hat{R}$ operation; since the operator is invariant under the group, the same can be done with $H$ :

$$\hat{H}(\hat{R}\psi_n) = E_n(\hat{R}\psi_n). \tag{4.4}$$

From this equation it is clear that the function $\hat{R}\psi_n$ is also an eigenfunction corresponding to the same eigenvalue $E_n$, with $\hat{R}\psi_n = \pm\psi_n$. Hence the wavefunction $\psi_n$ is transformed according to a one-dimensional representation, which is, of course, irreducible.

Suppose that $E_n$ is a *degenerate* energy level, with corresponding wavefunctions $\psi_{n\alpha}$, where the index $\alpha$ takes values 1, 2, 3, ..., $g$ (with $g$ the degree of degeneracy). Each of these functions satisfies the Schrödinger equation:

$$\left.\begin{aligned} \hat{H}\psi_{n1} &= E_n\psi_{n1}, \\ \hat{H}\psi_{n2} &= E_n\psi_{n2}, \\ &\vdots \\ \hat{H}\psi_{ng} &= E_n\psi_{ng}. \end{aligned}\right\} \tag{4.5}$$

and different solutions may, without loss of generality, be assumed orthogonal and normalized. It should be noted that any linear combination $\phi_\beta = \Sigma_{\alpha=1}^{g} c_{\alpha\beta}\psi_{n\alpha}$ of the wavefunctions of a degenerate state is also a wavefunction corresponding to the same energy level:

$$\begin{aligned} \hat{H}\phi_\beta &= \sum_\alpha c_{\alpha\beta}\hat{H}\psi_{n\alpha} = \sum_\alpha c_{\alpha\beta}E_n\psi_{n\alpha} \\ &= E_n\sum_\alpha c_{\alpha\beta}\psi_{n\alpha} = E_n\phi_\beta. \end{aligned} \tag{4.6}$$

[†] Rare cases of "accidental" degeneracy (e.g. between the ns and np orbitals of an isolated hydrogen atom) cannot be considered as exceptions to the theorem. When an accidental degeneracy occurs the Hamiltonian of the system often belongs to a more general symmetry group. Chapter 14 provides examples "accidental" degeneracy in the systems of exchange coupled ions.

The functions $\phi_\beta$ may also be normalized:

$$\int \phi_\beta^* \phi_\beta \, \mathrm{d}\tau = 1, \tag{4.7}$$

where $\tau$ indicates the set of electron coordinates, and the coefficients $c_{\alpha\beta}$ then satisfy

$$\sum_\alpha |c_{\alpha\beta}|^2 = 1. \tag{4.8}$$

Let us apply the group operation $\hat{R}$ to one of the equations (4.5) and take into account the invariance of the Hamiltonian ($\hat{H}' = \hat{H}$):

$$\hat{H}\hat{R}\psi_{n\alpha} = E_n \hat{R}\psi_{n\alpha}. \tag{4.9}$$

Since $\hat{R}\psi_{n\alpha}$ is a wavefunction corresponding to the same level $E_n$, it is a linear combination of the functions $\psi_{n\alpha}$:

$$\hat{R}\psi_{n\alpha} = \sum_\beta r_{\beta\alpha}\psi_{n\beta} \equiv \sum_\beta \psi_{n\beta} D(\hat{R})_{\beta\alpha}, \tag{4.10}$$

which must hold for all $\alpha$. Eq. (4.10) is also often written in the matrix form

$$\hat{R}[\psi_{n1}\psi_{n2}\ldots\psi_{ng}] = [\psi_{n1}\psi_{n2}\ldots\psi_{ng}]D(\hat{R})$$

Comparing (4.10) and (4.11) with the equation (3.64) that defines the representation property one can be convinced that the set of wavefunctions $\psi_{n1}\psi_{n2}\ldots\psi_{ng}$ form the basis of a representation of the symmetry point group. The linear combination in (4.10) includes only the wavefunctions of the degenerate level which provide the basis of an irreducible representation.

The statement proved above is called the *Wigner theorem on the classification of quantum states*: each energy level is related to a certain irreducible representation of the symmetry point group corresponding wavefunctions forming the basis of an *irreducible representation*. Since, in general, the number of levels is not limited, while the number of irreducible representations of the point group is finite, a particular irreducible representation may occur more than once.

The complex-conjugate representations (section 3.7.3) require special discussion. Suppose some wavefunction $\psi_n$ satisfies the Schrödinger equation (4.1). Since the Hamiltonian $\hat{H}$ is a real operator, the complex-conjugate function $\psi_n^*$ satisfies the same Schrödinger equation:

$$\hat{H}\psi_n^* = E_n\psi_n^*. \tag{4.11}$$

If the energy level $E_n$ is degenerate then both wavefunctions $\psi_n$ and $\psi_n^*$

correspond to it, forming the basis of the complex-conjugate representation. Using the Wigner theorem, it is necessary to account for this. Because the states $\psi_n$ and $\psi_n^*$ are degenerate, the complex-conjugate representations are combined into one "physical" two-dimensional representation (Section 3.7.3). It should be noted here that the twofold degeneracy occurs only in the absence of a magnetic field—in a nonzero magnetic field the complex-conjugate representations are associated with different energy levels.

## 4.3 SPLITTING OF ONE-ELECTRON LEVELS IN CRYSTAL FIELDS

### 4.3.1 Formula for reduction of representations

Let us again consider the examples analysed qualitatively in Section 4.1. The Wigner theorem shows that the degeneracy of the p level in a cubic field ($\mathbf{O}_h$) results from the fact that the three p functions form the basis of the irreducible representation $T_{1u}$. In a tetragonal crystal field ($\mathbf{D}_{4h}$) the three $p$ functions transform over the two irreducible representations $A_{2u}(p_z)$ and $E_u(p_x, p_y)$, which is indicative of the existence of the two levels. In other words, the triplet $T_{1u}$ splits into a singlet (a nondegenerate level $A_{2u}$) and a doublet $E_u$ (a twofold-degenerate level).

The Wigner theorem provides a *method of classification* of the atomic states in a crystal field. In order to find the atomic state a ligand environment of point symmetry **G**, it is necessary to subject the angular parts of the wavefunctions $Y_{lm}$ to symmetry operations. The matrices constructed in this way form the representation $D^{(l)}$ of the group **G**. If this representation is irreducible, the atomic level is not split. If this representation is reducible, the level is split. From the Wigner theorem, the number of levels in this case is equal to the number of irreducible representations according to which the $Y_{lm}$ functions transform in **G**. It was in this way that it was shown in Section 3.5 that the $p$ functions form the basis of one irreducible representation in **O**, while the $d$ functions are transformed according to two representations, namely the three-dimensional representation $T_2$ and the two-dimensional representation E. This latter fact implies that the fivefold-degenerate d level is split by the octahedral environment into a triplet ($T_2$) and a doublet (E).

In practice, however, this method is difficult to apply. First, construction of the representation $D^{(l)}$ is rather tedious, especially for atomic states with $l > 2$. Secondly, the reducibility of representations is not always obvious, because of an inappropriate choice of basis (see Section 3.4).

It turns out that, to solve the problem of the energy-level classification, it is necessary to know only the *characters* of the representations and *not* the representations themselves (i.e. the transformation matrices of the wavefunctions).

Suppose that in some group $\mathbf{G}_0$ the wavefunctions corresponding to the energy level $E$ are transformed according to the representation $\Gamma$. When the symmetry decreases, the degenerate level $E$ is split into levels $E_1, E_2, \ldots, E_k$. These levels have corresponding representations $\Gamma_1, \Gamma_2, \ldots \Gamma_k$ of a lower symmetry group $\mathbf{G}$ which is a subgroup of $\mathbf{G}_0$ (Table 2.3). The representation $\Gamma$ in $\mathbf{G}$ appears to be reducible. Determination of the irreducible representations $\Gamma_i$ ($i = 1, 2, \ldots, k$) is known as the *decomposition of the reducible representation into irreducible ones*, and is performed using the *reduction formula*

$$a(\Gamma_i) = \frac{1}{g}\sum_{\hat{R}} \chi^{(\Gamma)}(\hat{R})\chi^{(\Gamma_i)}(\hat{R}), \tag{4.12}$$

where $a(\Gamma_i)$ is the number of times $\Gamma_i$ occurs in the reduction.

This formula follows from the properties of representations [15]. In order for us to use it in practice, we do not require its proof—it is only necessary that we understand clearly the meaning of the quantities involved.

(1) $\hat{R}$ is an operation of the lower symmetry group $\mathbf{G}$; since $\mathbf{G}$ is a subgroup of $\mathbf{G}_0$, $\hat{R}$ is simultaneously an operation of the latter; the summation is over such operations, and $g$ is the order of $\mathbf{G}$.

(2) $\chi^{(\Gamma_i)}(\hat{R})$ is the character of the $i$th irreducible representation of $\mathbf{G}$ for the operation $\hat{R}$.

(3) $\chi^{(\Gamma)}(\hat{R})$ is the character of $\hat{R}$ in an irreducible representation $\Gamma$ of $\mathbf{G}_0$. The representation $\Gamma$ is irreducible in $\mathbf{G}_0$, while it is reducible in $\mathbf{G}$.

(4) $a_i \equiv a(\Gamma_i)$ is the number of times that the irreducible representation $\Gamma_i$ of $\mathbf{G}$ is contained in the reducible representation $\Gamma$ (in the same group). If some representation $\Gamma_i$ is absent then $a_i = 0$. Equation (4.12) always gives $a_i$ as an integer.

All the quantities on the right-hand side of (4.12) are available in the character tables for the given representation ($\Gamma$) and the irreducible representations ($\Gamma_i$).

### 4.3.2 Splitting of the *p* level in tetragonal, trigonal and rhombic fields

We shall illustrate the application of the methods of decomposition of representations by some simple examples. The reduction formula contains only the characters of the point group $\mathbf{G}_0$ and its subgroup $\mathbf{G}$, and is not connected with any particular molecular system. In order to clarify this, we consider below an octahedral complex $ML_6$ and its distorted low-symmetry conformations.

**Example 4.1** Let us consider the splitting of the p level arising from a decrease in the cubic symmetry **O** (the group $\mathbf{G}_0$) down to the tetragonal $\mathbf{D}_4$ (the group **G**). In **O** the $p_x$, $p_y$ and $p_z$ functions are transformed according to the irreducible representation $\Gamma = \mathrm{T}_1$. The characters $\chi^{(\mathrm{T}_1)}(\hat{R})$ of this representation, selected from the general table of the **O** group characters, are as follows:

| O | $\hat{E}$ | $6\hat{C}_4$ | $3\hat{C}_4^2$ | $8\hat{C}_3$ | $6\hat{C}_2$ |
|---|---|---|---|---|---|
| . | . | . | . | . | . |
| $\mathrm{T}_1$ | 3 | 1 | −1 | 0 | −1 |
| . | . | . | . | . | . |

To use the reduction formula, it is necessary to have the characters $\chi^{(\Gamma_i)}(\hat{R})$ of all the irreducible representations of $\mathbf{D}_4$:

| $\mathrm{D}_4$ | $\hat{E}$ | $2\hat{C}_4$ | $\hat{C}_4^2$ | $2\hat{C}_2$ | $2\hat{C}_2'$ |
|---|---|---|---|---|---|
| $\mathrm{A}_1$ | 1 | 1 | 1 | 1 | 1 |
| $\mathrm{A}_2$ | 1 | 1 | 1 | −1 | −1 |
| $\mathrm{B}_1$ | 1 | −1 | 1 | −1 | 1 |
| $\mathrm{B}_2$ | 1 | −1 | 1 | 1 | −1 |
| E | 2 | 0 | −2 | 0 | 0 |

Stretching (or contraction) of an octahedron† along one of its axes (e.g. the $z$ axis, Fig. 4.2a) or substitution of ligands in the *trans* position while forming the $\mathrm{ML_4L_2'}$ complex decreases the symmetry: the $C_3$ axes disappear; the two axes $C_4(x)$ and $C_4(y)$ of fourth-order are reduced to $C_2$ axes; and out of the six $C_2$ axes only two, $C_2(xy)$ and $C_2(\bar{x}y)$, are left. In the $\mathbf{D}_4$ ($g = 8$) subgroup the latter are denoted by $C_2'$, while the $C_2(x)$, $C_2(y)$ class is denoted by $2C_2$. After comparing the operations $\hat{R}$ of **O** and $\mathbf{D}_4$, we proceed to the reduction formula:

$$a(\mathrm{A}_1) = \tfrac{1}{8}[3\cdot 1 + 2\cdot 1\cdot 1 + (-1)\cdot 1 + 2(-1)\cdot 1 + 2\cdot(-1)\cdot 1] = 0,$$

$$a(\mathrm{A}_2) = \tfrac{1}{8}[3\cdot 1 + 2\cdot 1\cdot 1 + (-1)\cdot 1 + 2\cdot(-1)\cdot(-1) + 2\cdot(-1)\cdot(-1)] = 1,$$

$$a(\mathrm{B}_1) = \tfrac{1}{8}[3\cdot 1 + 2\cdot 1\cdot(-1) + (-1)\cdot 1 + 2\cdot(-1)\cdot(-1) + 2(-1)\cdot 1] = 0,$$

$$a(\mathrm{B}_2) = \tfrac{1}{8}[3\cdot 1 + 2\cdot 1\cdot(-1) + (-1)\cdot 1 + 2\cdot(-1)\cdot 1 + 2\cdot(-1)\cdot(-1)] = 0,$$

$$a(\mathrm{E}) = \tfrac{1}{8}[3\cdot 2 + 2\cdot 1\cdot 0 + (-1)(-2) + 2\cdot(-1)\cdot 0 + 2\cdot(-1)\cdot 0] = 1.$$

Hence $a(\mathrm{A}_2) = 1$ and $a(\mathrm{E}) = 1$; in other words, the representation $\mathrm{T}_1$ of **O** is decomposed into two irreducible representations $\mathrm{A}_2$ and E of $\mathbf{D}_4$, which agrees with the qualitative analysis given in Section 4.1.

†Here we omit the inversion symmetry (see Section 3.7.4), replacing $\mathrm{O_h}$ by O.

The decomposition of a reducible representation into irreducible ones is denoted symbolically in the form (for the above example)

$$D^{(T_1)} \doteq D^{(A_2)} + D^{(E)} \quad \text{or} \quad T_1 \doteq A_2 + E.$$

Reduction of a representation means that, when the symmetry decreases, the degenerate level is split, and the dimensions of the representations $\Gamma_i$ give the degree of degeneracy of energy levels in the lower-symmetry crystal field. In the above example the triple degeneracy is partially removed—one component turns out to be twofold-degenerate.

**Example 4.2** Let us analyse the behaviour of the p($T_1$) level when the **O** cubic symmetry is reduced to the $\mathbf{D}_3$ trigonal symmetry (Table 2.5). For definiteness, assume that we have a six-coordinate complex $MX_6$, whose axes are the symmetry elements of **O**. In the symmetric octahedral complex the ligands L occupy the centres of the six faces of the cube (the filled circles in Fig. 4.3a). A trigonal distortion is obtained by displacement of ligand triads along the cube face diagonals towards the opposite vertices of the cube (open circles in Fig. 4.3a). When such a deformation occurs, the $C_4$ axes are lost, out of four $C_3$ axes only one is left ([III]), and out of six $C_2$ axes three are left. $\mathbf{D}_3$ is a subgroup of **O**; its character table is given in Section 3.7.4. Using the reduction formula, we obtain

$$a(A_1) = \tfrac{1}{6}[3\cdot 1 + 0\cdot 2\cdot 1 + (-1)\cdot 3\cdot 1] = 0,$$

$$a(A_2) = \tfrac{1}{6}[3\cdot 1 + 0\cdot 2\cdot 1 + (-3)\cdot 3\cdot (-1)] = 1,$$

$$a(E) = \tfrac{1}{6}[3\cdot 2 + 0\cdot 2\cdot (-1) + (-1)\cdot 3\cdot 0] = 1.$$

Hence $a(A_2) = a(E) = 1$, so that a triply degenerate $T_1$ level is split into a singlet $A_2$ and a doublet E: $T_1 \doteq A_2 + E$.

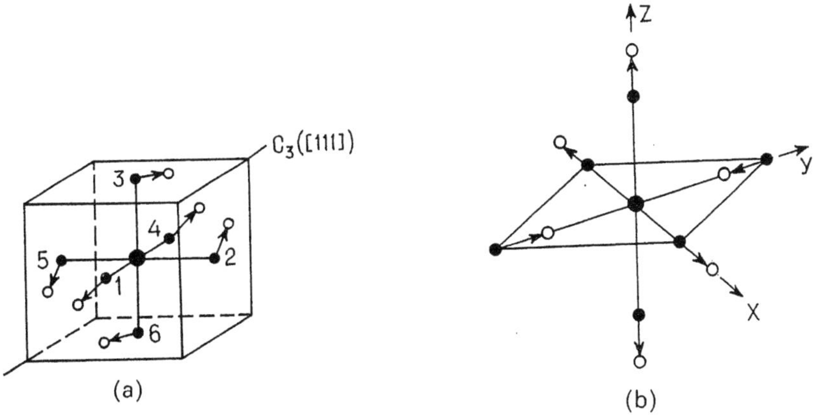


**Fig. 4.3** Trigonal (a) and rhombic (b) distortions of an octahedral complex $ML_6$.

**Example 4.3** Let us examine a rhombic distortion of the complex $ML_6$. This distortion displaces the ligands along the $x$, $y$ and $z$ axes so that all the diagonals of the octahedron become different (Fig. 4.3(b)). $C_3$ axes are absent, while the $C_4$ axes of the octahedron become $C_2$ axes of the distorted complex. The symmetry group $\mathbf{D}_2$ ($g = 4$) thus obtained belongs to the rhombic system (Table 2.3). The character table has the form

| $\mathbf{D}_2$ | $\hat{E}$ | $\hat{C}_2(z)$ | $\hat{C}_2(y)$ | $\hat{C}_2(x)$ |
|---|---|---|---|---|
| A | 1 | 1 | 1 | 1 |
| $B_1$ | 1 | 1 | $-1$ | $-1$ |
| $B_2$ | 1 | $-1$ | 1 | $-1$ |
| $B_3$ | 1 | $-1$ | $-1$ | 1 |

The reduction formula gives

$$a(\mathrm{A}) = \tfrac{1}{4}[3 \cdot 1 + (-1) \cdot 1 + (-1) \cdot 1 + (-1) \cdot 1] = 0,$$

$$a(\mathrm{B}_1) = \tfrac{1}{4}[3 \cdot 1 + (-1) \cdot 1 + (-1) \cdot (-1) + (-1) \cdot (-1)] = 1,$$

$$a(\mathrm{B}_2) = \tfrac{1}{4}[3 \cdot 1 + (-1) \cdot (-1) + (-1) \cdot 1 + (-1)(-1)] = 1,$$

$$a(\mathrm{B}_3) = \tfrac{1}{4}[3 \cdot 1 + (-1) \cdot (-1) + (-1) \cdot (-1) + (-1) \cdot 1] = 1.$$

The result shows that the decrease of symmetry $\mathbf{O} \to \mathbf{D}_2$ leads to a full splitting of the $T_1$ level into three nondegenerate levels: $T_1 \doteq B_1 + B_2 + B_3$. The corresponding wavefunctions transform according to the representations $B_1$, $B_2$ and $B_3$ of $\mathbf{D}_2$.

$\mathbf{D}_2$ is also a subgroup of $\mathbf{D}_4$, with the $C_2(x)$, $C_2(y)$ and $C_2(z)$ axes coinciding with $C_4(z)$ and $2C_2$ of $\mathbf{D}_2$. Decomposing the representations $A_2$ and E of $\mathbf{D}_4$ (obtained through the reduction of $T_1$), according to the reduction formula, into irreducible representations in $\mathbf{D}_2$, we obtain

$$A_2 \doteq B_1, \qquad E \doteq B_2 + B_3.$$

Thus the reduction chain $T_1 \doteq A_2 + E \doteq B_1 + B_2 + B_3$ ($\mathbf{O} \to \mathbf{D}_4 \to \mathbf{D}_2$) gives the same result as that obtained when the symmetry decreases according to the scheme $\mathbf{O} \to \mathbf{D}_2$. Figure 4.4 shows schematically the splitting of the $T_1$ level by the sequential decrease in symmetry $\mathbf{O} \to \mathbf{D}_4 \to \mathbf{D}_2$. As stated above, in tetragonal surroundings the $p_z$ state ($A_2$) has a lower energy, while the degeneracy of the E level is related to the energy equivalence of the $p_x$ and $p_y$ orbitals. The rhombic distortion, lengthening and shortening the $x$ and $y$ diagonals respectively, decreases the energy of electron repulsion from ligands in the $p_x$ state and increases it in the $p_y$ state. This explains the splitting of the tetragonal E level in the rhombic field and the level sequence shown in Figs. 4.4(c,d). In the case of a weak rhombic distortion the $p_z$ state ($B_1$) becomes the ground state, while in a strong rhombic field the $B_3(p_x)$ level becomes the ground one.

The above examples illustrate the efficiency of the group-theoretical method in solving quantum mechanical problems. Indeed, *without solving the Schrödinger equation*, we have managed to find the number of levels of a complex and their degree of degeneracy in different conformations. On the other hand, it should be clearly understood that the group-theoretical approach is appropriate only for problems whose solution is based on symmetry principles. In the examples discussed above these problems involve classification of levels and wavefunctions according to irreducible representations of a point group. The absolute values of splitting of the energy levels and their relative positions can be obtained only by solving the Schrödinger equation through available approximate methods. Let us consider from this viewpoint the levels $E(p_x, p_y)$ and $A(p_z)$ (Figs. 4.2 and 4.4) of the $\mathbf{D}_4$ complex. The singlet level $A_2$ appears to be lower than the doublet in the "elongated" octahedron; it is obvious that in the "compressed" octahedron the order of levels is reversed, since the $p_z$ electron is repelled from the ligands more strongly than the $p_x$ and $p_y$ electrons are. Meanwhile, both the elongated and compressed octahedra belong to the same symmetry group, irrespective of the amount and sign of distortion. Therefore, both the degree of splitting, as well as the order of levels, must be obtained by solving the Schrödinger equation,

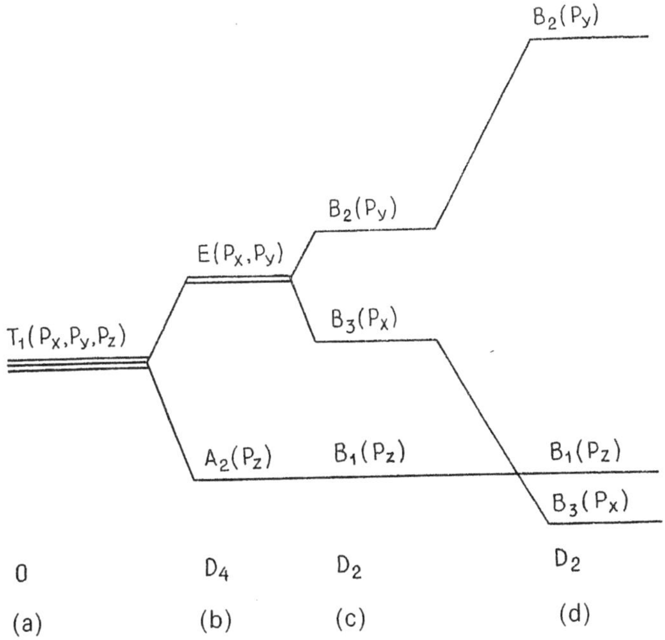


**Fig. 4.4** Splitting of a cubic (**O**) $T_1$ level in low-symmetry ligand surroundings: (a) octahedral field; (b) tetragonal ($\mathbf{D}_4$) distortion; (c) weak and (d) strong rhombic ($\mathbf{D}_2$) distortions.

although group theory greatly simplifies the solution in this case (Sections 5.6 and 5.7).

### 4.3.3 Characters of rotation groups

So far, we have considered reduction of representations associated with decreases in symmetry of finite point groups with known characters. Let us now consider a more general problem, namely the calculation of atomic and ion states in a crystal or a complex with a certain point symmetry group. A free atom belongs to the continuous *rotation group* **K**. Hence its symmetry is higher than that of a system belonging to any point group. The decrease in symmetry when the atom is introduced into a crystal or when a complex is formed leads, according to the Wigner theorem, to splitting of the atomic energy levels. In one particular case the problem of classification of atomic levels in the crystal field has already been solved. In Section 3.5.3 it was shown that five d orbitals are transformed according to the reducible five-dimensional representation of the group **O**, which is split into two irreducible representations: the three-dimensional representation $T_2$ and the two-dimensional representation E. From the viewpoint of the Wigner theorem, this means that the fivefold-degenerate atomic level in cubic surroundings is split into a triplet and a doublet. Qualitative evaluations based on examination of electron density maps help in understanding the nature of the splitting and the cause of the degeneracy. Figure 4.5 shows the $d_\xi$, $d_\eta$ and $d_\zeta$ orbitals which are of equal energy owing to the equivalence of the $x$, $y$ and $z$ axes. The origin of the degeneracy of the E level is less obvious; however, from Fig. 4.5 it can be seen that the ground state is the triplet $T_2$.

It is not so simple to obtain irreducible representations of point groups in the basis of atomic functions with angular factors $Y_{lm}$. In the case of d orbitals, discussed in Section 3.5.3, the basis functions of the irreducible representations were essentially obtained by guesswork.

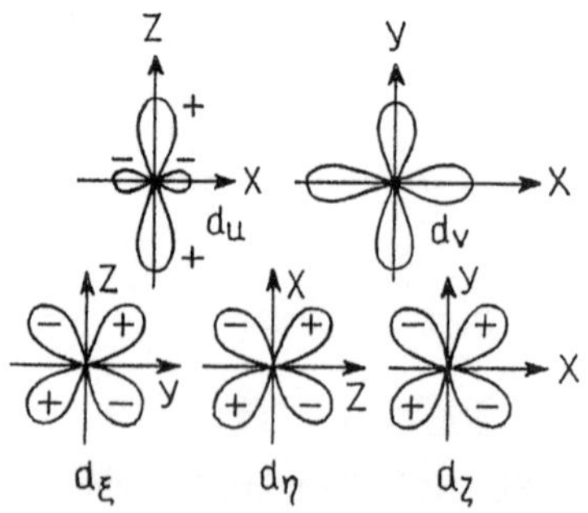


**Fig. 4.5** d orbitals in octahedral ligand surroundings.

The classification of atomic states in the crystal field is based on the reduction formula (4.12). Here, for a high symmetry group $\mathbf{G}_0$, we take the rotation group **K**, with **G** being the group of ligand surroundings or the ion point group in a crystal. The irreducible representations of the group **K** are enumerated by the orbital angular momentum quantum number $l$. The spherical functions $Y_{lm}$ with given $l$ form the basis of the irreducible representation of **K** with dimension $2l+1$. They can be represented as the product [14, 34]:

$$Y_{lm}(\vartheta,\varphi)=\frac{1}{(2\pi)^{1/2}}\Theta_{lm}(\vartheta)\mathrm{e}^{im\varphi}. \tag{4.13}$$

The operation of rotation through an arbitrary angle $\alpha$ about the $z$ axis of angular momentum quantization belongs to the group **K** and leads to multiplication of the basis function $Y_{lm}(\vartheta,\varphi)$ by $\mathrm{e}^{im\alpha}$:

$$\begin{aligned}\hat{R}_\alpha Y_{lm}&=\frac{1}{(2\pi)^{1/2}}\Theta_{lm}(\vartheta)\hat{R}_\alpha\mathrm{e}^{\mathrm{i}m\varphi}=\frac{1}{(2\pi)^{1/2}}\Theta_{lm}(\vartheta)\mathrm{e}^{\mathrm{i}m(\varphi+\alpha)}\\&=\mathrm{e}^{\mathrm{i}m\alpha}Y_{lm}.\end{aligned} \tag{4.14}$$

Hence the $(2l+1)$-dimensional matrix of such a rotation has the form

$$\mathbf{D}^{(l)}(\hat{R}_\alpha)=\begin{bmatrix}\mathrm{e}^{\mathrm{i}l\alpha} & 0 & \cdot & \cdot & \cdot & \cdot & 0\\ 0 & \mathrm{e}^{\mathrm{i}(l-1)\alpha} & \cdot & \cdot & \cdot & \cdot & 0\\ \cdot & \cdot & \cdot & \cdot & \cdot & \cdot & 0\\ \cdot & \cdot & \cdot & \cdot & \cdot & \mathrm{e}^{-\mathrm{i}(l-1)\alpha} & 0\\ \cdot & \cdot & \cdot & \cdot & \cdot & \cdot & \mathrm{e}^{-\mathrm{i}l\alpha}\end{bmatrix}. \tag{4.15}$$

The character of this operation can be found easily, allowing for the fact that the diagonal elements of $\mathbf{D}^{(l)}(\hat{R}_\alpha)$ form a geometrical progression:

$$\chi^{(l)}(\alpha)=\frac{\sin(l+\frac{1}{2})\alpha}{\sin\frac{1}{2}\alpha}. \tag{4.16}$$

After substituting the values of the angles $\alpha$ corresponding to the $\hat{C}_n^m$ rotations of finite groups, we obtain [14]

$$\alpha=\pi,\qquad \chi^{(l)}(\hat{C}_2)=(-1)^l,$$

$$\alpha=\tfrac{1}{2}\pi,\qquad \chi^{(l)}(\hat{C}_4)=\begin{cases}(-1)^{l/2} & (\text{even } l),\\ (-1)^{(l-1)/2} & (\text{odd } l),\end{cases}$$

$$\alpha=\tfrac{2}{3}\pi,\qquad \chi^{(l)}(\hat{C}_3)=\begin{cases}1 & (l=3k),\\ 0 & (l=3k+1),\\ -1 & (l=3k+2),\end{cases}$$

with $k$ an integer. Finally, for $\alpha = 0$ we obtain the character of the identity transformation:

$$\chi^{(l)}(\hat{E}) = 2l + 1$$

The characters $\chi^{(l)}$ of the rotation group thus obtained may then be substituted into the reduction formula (4.12) in place of the quantities $\chi^{(\Gamma)}(\hat{R})$.

### 4.3.4 Classification of one-electron states in crystal fields

Given two rotations, $\hat{R}_\alpha$ and $\hat{R}'_\alpha$, through equal angles but around different axes, it is possible to find an operation that brings these axes into coincidence: they thus belong to the same class. Hence the character formula derived for the rotations $R_\alpha$ around the $z$ axis of quantization of the electron angular momentum holds for rotations through an angle $\alpha$ around *any* axis (although the rotation matrix depends significantly upon the selection of the axis). The characters of the rotation group for point group **O** operations in the bases of the s, p, d and f orbitals are given in Table 4.1.

**Table 4.1** Characters of **K** for the bases of states with $l = 0, 1, 2$ and 3.

| $l$ | $\chi^{(l)}(\hat{E})$ | $\chi^{(l)}(\hat{C}_2)$ | $\chi^{(l)}(\hat{C}_3)$ | $\chi^{(l)}(\hat{C}_4)$ |
|---|---|---|---|---|
| 0(s) | 1 | 1 | 1 | 1 |
| 1(p) | 3 | −1 | 0 | 1 |
| 2(d) | 5 | 1 | −1 | −1 |
| 3(f) | 7 | −1 | 1 | −1 |

Let us now use the reduction formula for expansion of the $D^{(l)}$ representation into irreducible parts in **O**. The group characters are found from the following table:

| O | $\hat{E}$ | $6\hat{C}_4$ | $3\hat{C}_4^2$ | $8\hat{C}_3$ | $6\hat{C}_2$ |
|---|---|---|---|---|---|
| $A_1$ | 1 | 1 | 1 | 1 | 1 |
| $A_2$ | 1 | −1 | 1 | 1 | −1 |
| E | 2 | 0 | 2 | −1 | 0 |
| $T_1$ | 3 | 1 | −1 | 0 | −1 |
| $T_2$ | 3 | −1 | −1 | 0 | 1 |

On substituting $\chi^{(2)}(\hat{R})$ and $\chi^{(\Gamma_i)}(\hat{R})$ ($\Gamma_i = \mathrm{A_1, A_2, E, T_1, T_2}$) into the reduction formula, we obtain

$$a(\mathrm{A_1}) = \tfrac{1}{24}[5\cdot 1 + 6\cdot(-1)\cdot 1 + 3\cdot 1\cdot 1 + 8\cdot(-1)\cdot 1 + 6\cdot 1\cdot 1] = 0,$$
$$a(\mathrm{A_2}) = \tfrac{1}{24}[5\cdot 1 + 6\cdot(-1)\cdot(-1) + 3\cdot 1\cdot 1 + 8\cdot(-1)\cdot 1 + 6\cdot 1\cdot(-1)] = 0,$$
$$a(\mathrm{E}) = \tfrac{1}{24}[5\cdot 2 + 6\cdot(-1)\cdot 0 + 3\cdot 1\cdot 2 + 8\cdot(-1)\cdot(-1) + 6\cdot 1\cdot 0] = 1,$$
$$a(\mathrm{T_1}) = \tfrac{1}{24}[5\cdot 3 + 6\cdot(-1)\cdot 1 + 3\cdot 1\cdot(-1) + 8\cdot(-1)\cdot 0 + 6\cdot 1\cdot(-1)] = 0,$$
$$a(\mathrm{T_2}) = \tfrac{1}{24}[5\cdot 3 + 6\cdot(-1)\cdot(-1) + 3\cdot 1\cdot(-1) + 8\cdot(-1)\cdot 0 + 6\cdot 1\cdot 1] = 1.$$

Therefore the representation $\mathrm{D}^{(2)}$, being reducible in **O**, is reduced according to the scheme

$$\mathrm{D}^{(2)} \doteq \mathrm{D}^{(\mathrm{E})} + \mathrm{D}^{(\mathrm{T_2})}.$$

It should be emphasized that this result has been obtained using only characters, without direct calculation of the representation $\mathrm{D}^{(2)}$. In a similar way it can be shown that the atomic f level is split in the cubic **O** field into two triplets $\mathrm{T_1}$ and $\mathrm{T_2}$ and one singlet $\mathrm{A_2}$:

$$\mathrm{D}^{(3)} \doteq \mathrm{D}^{(\mathrm{A_2})} + \mathrm{D}^{(T_1)} + \mathrm{D}^{(\mathrm{T_2})}.$$

The results for the levels with $l = 0$–$6$ are given in Table 4.2.

It is shown below that the results of reduction of the **K** group representations are also valid for many-electron atoms. Therefore Table 4.2 is valid both for one-electron atoms with orbital angular momentum $l$ and for many-electron atoms with the total angular momentum quantum number $L$ (the corresponding spectroscopic notations are S, P etc.).

The classification of one-electron states in crystal fields of different symmetry can be performed in the same way. The results for some point groups are given in Table 4.3 [13]. The irreducible representations of point groups in this table are denoted by lower-case letters ($\mathrm{a_1, e, t_2}$ etc.) instead of the corresponding capital letters ($\mathrm{A_1, E, T_2}$ etc.) used previously. This system of notation

**Table 4.2** Splitting of levels in a cubic field.

| $l$ or | $L$ | Splitting in the cubic **O** field |
|---|---|---|
| 0 | s(S) | $\mathrm{A_1}$ |
| 1 | p(P) | $\mathrm{T_1}$ |
| 2 | d(D) | $\mathrm{E + T_2}$ |
| 3 | f(F) | $\mathrm{A_2 + T_1 + T_2}$ |
| 4 | g(G) | $\mathrm{A_1 + E + T_1 + T_2}$ |
| 5 | h(H) | $\mathrm{E + 2T_1 + T_2}$ |
| 6 | i(I) | $\mathrm{A_1 + A_2 + E + T_1 + 2T_2}$ |

**Table 4.3** Splitting of one-electron levels in crystal fields of different symmetry [13].

| Level type | $O_h$ | $T_d$ | $D_{4h}$ | $D_3$ | $D_{2h}$ |
|---|---|---|---|---|---|
| s | $a_{1g}$ | $a_1$ | $a_{1g}$ | $a_1$ | $a_1$ |
| p | $t_{1u}$ | $t_2$ | $a_{2u} + e_u$ | $a_2 + e$ | $b_2 + e$ |
| d | $e_g + t_{2g}$ | $e + t_2$ | $a_{1g} + b_{1g} + b_{2g} + e_g$ | $a_1 + 2e$ | $a_1 + b_1 + b_2 + e$ |
| f | $a_{2u} + t_{1u} + t_{2u}$ | $a_2 + t_1 + t_2$ | $a_{2u} + b_{1u} + b_{2u} + 2e_u$ | $a_1 + 2a_2 + 2e$ | $a_1 + a_2 + b_2 + 2e$ |
| g | $a_{1g} + e_g + t_{1g} + t_{2g}$ | $a_1 + e + t_1 + t_2$ | $2a_{1g} + a_{2g} + b_{1g}$ $+b_{2g} + 2e_g$ | $2a_1 + a_2 + 3e$ | $2a_1 + a_2 + b_1 + b_2 + 2e$ |
| h | $e_u + 2t_{1u} + t_{2u}$ | $e + t_1 + 2t_2$ | $a_{1u} + 2a_{2u} + b_{1u}$ $+b_{2u} + 3e_u$ | $a_1 + 2a_2 + 4e$ | $a_1 + a_2 + b_1$ $+2b_2 + 3e$ |
| i | $a_{1g} + a_{2g} + e_g$ $+t_{1g} + 2t_{2g}$ | $a_1 + a_2 + e + t_1 + 2t_2$ | $2a_{1g} + a_{2g} + 2b_{1g}$ $+2b_{2g} + 3e_g$ | $3a_1 + 2a_2 + 4e$ | $2a_1 + a_2 + 2b_1$ $+2b_2 + 3e$ |

emphasizes that the irreducible representations belong to one-electron atoms (or ions containing one electron outside the inner closed shell). In many-electron ions this notation is used for one-electron wavefunctions, while capital letters are retained as notations for irreducible representations in a many-electron basis (Chapter 5).

A special explanation should be given of the *parity rule*. Applied to the functions $Y_{lm}(\vartheta, \varphi)$ the operation of inversion results in the factor +1 or −1 for even and odd functions respectively for all values of $m$. Hence the characters of the operations $\hat{I}$ are

$$\chi^{(l)}(\hat{I}) = \pm(2l+1). \tag{4.17}$$

Characters of the mirror reflection $\hat{\sigma}$ in the plane of symmetry and of the improper rotation $\hat{S}(\vartheta)$ can be obtained using the relations derived in Chapter 1:

$$\hat{\sigma} = \hat{I}\hat{C}_2, \hat{S}(\varphi) = \hat{I}\hat{C}(\pi + \varphi). \tag{4.18}$$

The representations of groups containing inversion are characterized by a certain parity (in Table 4.3 these are $\mathbf{O}_h$ and $\mathbf{D}_{4h}$). Parity, as usual, is denoted by a subscript u or g (see Section 3.7.4), groups without inversion do not require such notation. It follows from Table 4.3 that the s, d, g and i states result in the even (g) irreducible representations of symmetry groups with inversion, while the p, f and h states lead to odd irreducible representations (u). This shows the symmetry behaviour of one-electron wavefunctions under inversion: states with even values of the quantum number $l$ are even, while those with odd $l$ are odd: the parity is determined by the factor $(-1)^l$.

In non-centrisymmetrical groups the wavefunctions of different parity can form the basis of one representation. Therefore it is possible to obtain the reduction schemes for groups without inversion centres from Table 4.3. For example, in order to obtain the splitting diagram in the $\mathbf{O}$ and $\mathbf{D}_4$ groups, it is necessary to eliminate symbols u or g in the notations for the $\mathbf{O}_h$ and $\mathbf{D}_{4h}$ groups.

### 4.3.5 Splitting of the *d* level in cubic fields

In the preceding sections we have obtained a qualitative pattern of atomic level splitting in crystal fields. Let us now analyse the quantitative results for a d electron in a cubic ligand field. In order to determine the energy levels in a cubic field, it is necessary to solve the Schrödinger equation for a one-electron atom with potential energy including the repulsion of the electron from the ligands. When taking into account only the nearest ligands in (4.2), $N' = 4$, 6 and 8 for tetrahedral ($\mathbf{T}_d$), octahedral ($\mathbf{O}_h$) and cubic ($\mathbf{O}_h$) complexes respectively, and $\mathbf{R}_k$ are taken to be the ligand coordinates. If the energy of the crystal field is less than the interatomic interaction energy, the splittings can be calculated using perturbation theory. In the case under discussion, for the five fold-degenerate d level, it is necessary to solve the fifth order secular equation:

$$|V^{c}_{mm'} - \epsilon\delta_{mm'}| = 0, \tag{4.19}$$

where $|\ldots|$ denotes the determinant and $\hat{V}^{c}_{mm'}$ are the matrix elements of the crystal field operator $(m, m' = -2, -1, 0, 1, 2)$ calculated using the d-electron wavefunctions (3.54):

$$V^{c}_{mm'} = \sum_{k=1}^{N'} \int \varphi^{*}_{n2m}(\mathbf{r}) \frac{eq_k}{|\mathbf{r} - \mathbf{R}_k|} \varphi_{n2m'}(\mathbf{r})\, d\tau, \tag{4.20}$$

or in the more convenient Dirac notation for $\varphi_{nlm} \equiv |nlm\rangle$.

$$V^{c}_{mm'} = \langle n2m|\hat{V}^{c}(\mathbf{r})|n2m'\rangle. \tag{4.21}$$

In (4.19), $\delta_{mm'} = 1\ (m = m')$, $\delta_{mm'} = 0\ (m \neq m')$ is the Kronecker symbol. The secular equation (4.19) determines the corrections to the energy $\epsilon_i$, describing the level splitting.

In the second stage of the calculation it is necessary to determine from perturbation theory the appropriate functions in the zeroth-order approximation, that is the linear combinations of the initial *d* functions corresponding to the energy levels $\epsilon_i$:

$$\psi_{n2i}(r) = \sum_{m=-2}^{2} c^{(i)}_{m} \varphi_{n2m}(r). \tag{4.22}$$

The coefficients $c^{(i)}_{m}$ for each level $\epsilon_i$ satisfy the system of linear homogeneous equations, the so-called "secular equations",

$$\sum_{m'} (V^{c}_{mm'} - \epsilon_i) c^{(i)}_{m'} = 0, \tag{4.23}$$

or, more explicitly,

$$\left.\begin{array}{l}
(V^{c}_{-2,-2} - \epsilon_i)c^{(i)}_{-2} + V^{c}_{-2,-1}c^{(i)}_{-1} + \ldots + V^{c}_{2,2}c^{(i)}_{2} = 0,\\
V^{c}_{-1,-2}c^{(i)}_{-2} + (V^{c}_{-1,-1} - \epsilon_i)c^{(i)}_{-1} + \ldots + V^{c}_{12}c^{(i)}_{2} = 0,\\
\qquad\vdots \qquad\qquad\qquad\qquad \vdots\\
V^{c}_{2,-2}c^{(i)}_{-2} \qquad + \ldots + \qquad (V^{c}_{22} - \epsilon_i)c^{(i)}_{2} = 0.
\end{array}\right\} \tag{4.24}$$

Using the electron states classification, we can anticipate that the secular equation (4.17) has two roots, one being threefold-degenerate (the representation $T_{2g}$ in $\mathbf{O}_h$ and $T_2$ in $\mathbf{T}_d$), and the other twofold-degenerate (the representations $E_g$ and E in $\mathbf{O}_h$ and $\mathbf{T}_d$ respectively). According to the Wigner theorem, the energy levels of the system correspond to the irreducible representations. Therefore the linear combinations of *d* functions transforming according to the representations $E_g$ and $T_{2g}$ (E and $T_2$) will be eigenfunctions in the zeroth-order approximation. The angular components $A_2(\vartheta, \varphi)$ of these functions are given in Table 3.1. Using these components and adding the radial factors (from (3.54)), we obtain for the triplets $T_{2g}$ and $T_2$

$$\left.\begin{aligned} \varphi_{T_2\xi} &= \sqrt{\tfrac{1}{2}}\,i(\varphi_{n21} + \varphi_{n2-1}), \\ \varphi_{T_2\eta} &= -\sqrt{\tfrac{1}{2}}(\varphi_{n21} - \varphi_{n2-1}), \\ \varphi_{T_2\zeta} &= -\sqrt{\tfrac{1}{2}}\,i\,(\varphi_{n22} - \varphi_{n2-2}). \end{aligned}\right\} \tag{4.25}$$

The wavefunctions of the orbital doublets have the forms

$$\varphi_{Eu} = \varphi_{n20}, \qquad \varphi_{Ev} = \sqrt{\tfrac{1}{2}}\,(\varphi_{n22} + \varphi_{n2-2}). \tag{4.26}$$

This result shows the possibility of an alternative direct approach to the problem of solving secular equations. In contrast to the direct approach, we have managed *to construct the wavefunctions of a perturbed system without solving the secular equation (4.19) and the system of equations (4.24) for the coefficients* $c_m^{(i)}$. The matrix of the operator $\hat{V}^c(\mathrm{r})$ in the basis of the wavefunctions $\varphi_{T_2\gamma}(\gamma = \xi, \eta, \zeta)$ and $\varphi_{E\gamma}(\gamma = u, v)$ is diagonal with its elements being the corrections to the atomic $n$d-level energy in the crystal field:

$$\left.\begin{aligned} \epsilon(\mathrm{E}) &= \langle \varphi_{E\gamma} | \hat{V}^c | \varphi_{E\gamma} \rangle \quad (\gamma = u, v), \\ \epsilon(\mathrm{T}_2) &= \langle \varphi_{T_2\gamma} | \hat{V}^c | \varphi_{T_2\gamma} \rangle \quad (\gamma = \xi, \eta, \zeta). \end{aligned}\right\} \tag{4.27}$$

The calculation of splitting in a crystal field is reduced to the calculation of two matrix elements (4.27). Without going into details (see e.g. [14, 16, 46, 47]), we shall present the final results. The effect of cubic fields ($\mathbf{O}_h, \mathbf{T}_d$) can be expressed in the form of two terms. The first increases (destabilizes) the atomic energy level without splitting it; that is, it acts as a spherically symmetric field. The second splits the $n$d level into a doublet and a triplet. This splitting (without taking destabilization into account) is shown diagrammatically in Fig. 4.6. In the case of an octahedrally coordinated ion the ground state turns out to be the $T_{2g}$ triplet, while for a tetrahedral or cubic complex it is the doublet E or $E_g$. In all the above cases the absolute value of the splitting is $10Dq$. For octahedral surroundings $\epsilon(t_{2g}) = -4Dq$ and $\epsilon(e_g) = +6Dq$. The quantity $\Delta = 10Dq$ is called the *cubic crystal field parameter*; for a octahedral complex it is given by

$$Dq = \frac{eq\langle r^4 \rangle_{nd}}{6a^5}, \tag{4.28}$$

where $q$ is the ligand charge, $a$ is the metal–ligand distance and $\langle r^4 \rangle$ is the mean

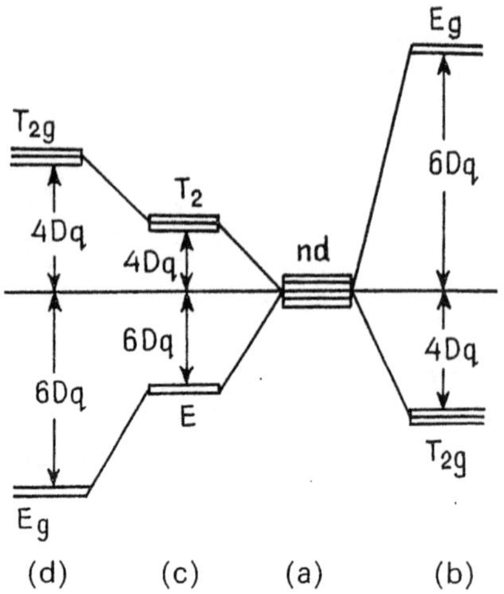


**Fig. 4.6** d-level splitting in a crystal field: (a) d level of a free ion; and splitting in the (b) octahedral, (c) tetrahedral and (d) cubic ligand surroundings.

value of $r^4$ for an $n$d-shell electron:

$$\langle r^4 \rangle_{nd} = \int_0^\infty R_{n2}^2(r) r^4 r^2 \, dr. \tag{4.29}$$

The parameter $(Dq)_{\text{tet}}$ for tetrahedral coordination is equal to $\frac{4}{9}(Dq)_{\text{oct}}$, and the level order is reversed, the ground state being a doublet (Fig. 4.6). For cubic coordination the level order is the same as for tetrahedral coordination, but the parameter $Dq$ is twice as large. Hence

$$(Dq)_{\text{tet}} = -\tfrac{4}{9}(Dq)_{\text{oct}} = -\tfrac{1}{2}(Dq)_{\text{cub}}$$

Let us summarize the results that follow from group theory:

(1) the determination of the nature of the atomic $nl$-level splitting in the crystal field, and of sublevel numbers and degrees of degeneracy;

(2) the classification of these sublevels according to irreducible representations of point groups;

(3) the determination of the wavefunctions diagonalizing the electron interaction with the crystal field without solving the secular equations directly.

It should be noted that classification of states and the determination of the nature of the splitting by means of the Wigner theorem can also be performed for more general cases of many-electron ions and noncubic fields. On the other hand, it is not always possible to obtain the wavefunctions in explicit form as in this section. Group-theoretical methods, as shown below, allow considerable simplification of this problem.

Other results such as the expression for the parameter $Dq$ and the level sequence in a crystal field are not directly connected with symmetry concepts but rather with the model employed. Thus both $Dq$ and the level sequence depend on the coordination number of the complex ion, the oxidation state, the electron distribution (which influences $\langle r^4 \rangle_{nd}$) and the ligand charge. These factors determine the strength and symmetry of the crystal field.

### 4.3.6 Splitting of the *d* level in low-symmetry fields

The classification of atomic states in fields of different symmetry given in Table 4.3 is not based on any assumptions regarding the structure of the complex, in a particular coordination compound. In this section we consider the splitting of the d level in low-symmetry crystal fields, resulting from distortions of the cubic surroundings (Section 4.3.5), as well as some other coordination polyhedra.

Let us consider the case of a *tetragonally* distorted octahedral complex. The metal–ligand M–L distance in the equatorial plane $(x, y)$ is denoted by $a$, while the corresponding distance for axial ligands is denoted by $b$. This distortion reduces the $\mathbf{O}_h$ symmetry to $\mathbf{D}_{4h}$. The d level undergoes further splitting (Fig. 4.7a) according to reduction of the representations $T_{2g}$ and $E_g$:

$$T_{2g} \doteq E_g + B_{2g}, \qquad E_g \doteq A_{1g} + B_{1g}.$$

Of course, the same representations of $\mathbf{D}_{4h}$ are obtained by direct reduction of the representation $D^{(2)}$ (Table 4.3). From the $\mathbf{D}_{4h}$ group character table we see that the basis of the irreducible representation $E_g$ comprises the functions $xz$ and $yz$, and the basis of the one-dimensional representations $B_{2g}$, $A_{1g}$ and $B_{1g}$ are $xy$, $3z^2 - r^2$ and $x^2 - y^2$, respectively. This information determines the wavefunctions of the split states.

Calculation of the splittings in the point-ion model gives the following expressions [14, 16, 37]:

$$\left.\begin{aligned} \epsilon(B_{1g}) &= 6Dq + 2Ds - Dt, \\ \epsilon(A_{1g}) &= 6Dq - 2Ds - 6Dt, \\ \epsilon(B_{2g}) &= -4Dq + 2Ds - Dt, \\ \epsilon(E_g) &= -4Dq - Ds + 4Dt. \end{aligned}\right\} \qquad (4.30)$$

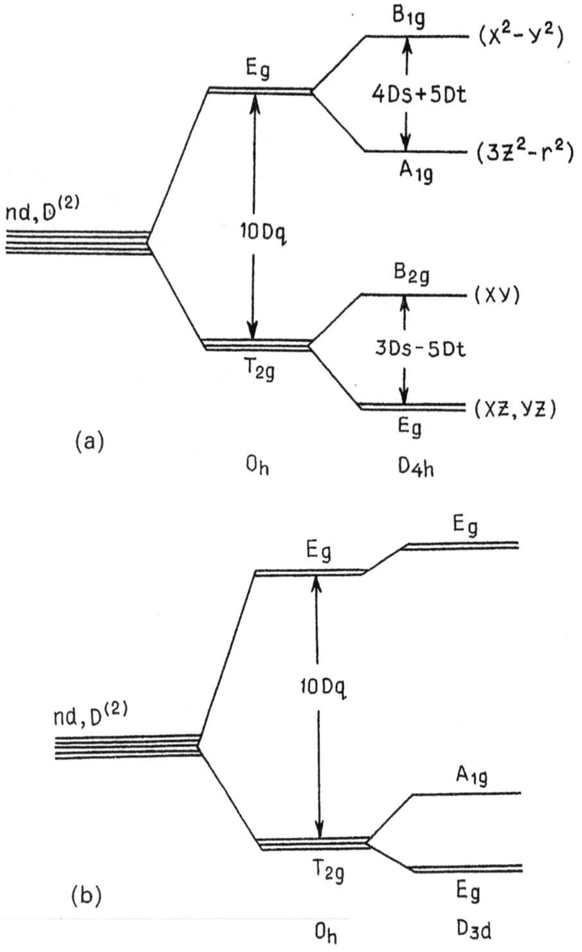


**Fig. 4.7** d-level splitting by octahedron distortions: (a) tetragonal field $\mathbf{D}_{4h}$; (b) trigonal field $\mathbf{D}_{3d}$.

Here $Ds$ and $Dt$ are two parameters describing the tetragonal field:

$$\left.\begin{aligned} Ds &= \frac{2eq}{7}\langle r^2\rangle_{nd}\left(\frac{1}{a^3}-\frac{1}{b^3}\right), \\ Dt &= \frac{2eq}{21}\langle r^4\rangle_{nd}\left(\frac{1}{a^5}-\frac{1}{b^5}\right). \end{aligned}\right\} \tag{4.31}$$

They are positive for an elongated octahedron ($a < b$) and negative for a compressed octahedron ($a < b$). This shows that these quantities, like $Dq$, are determined not by symmetry but by the crystal field strength. When $b \to \infty$ we obtain the crystal field parameters for a square-planar complex. It should be noted that $Ds$ and $Dt$ depend upon $\langle r^2 \rangle_{nd}$ and $\langle r^4 \rangle_{nd}$, respectively. Usually, we do not know the electron density distribution for an ion in a crystal exactly. Therefore, when calculating these quantities, we meet with certain difficulties and they can be determined only approximately. There is a generally used phenomenological approach to the description of complex ions according to which the crystal field parameters are obtained from comparison of the calculated energy expressions with spectroscopic data. In this approach the parameters $Ds$ and $Dt$ (and $Dq$) should be considered as *independent*, since they are not connected by universal relations and their relative values are different in different complexes having the same symmetry and coordination. Of course, the group-theoretical approach does not provide a method of calculation for $Ds$ and $Dq$. However, as shown below, it does allow determination of the *number of independent parameters* (with the meaning discussed above) determining the energy levels of the ion in the crystal field.

In the case of trigonal distortion of an octahedron (Fig. 4.3a) the symmetry decreases $\mathbf{O}_h \to \mathbf{D}_{3d}$ and reduction of representations takes place according to

$$\mathrm{T}_{2g} \doteq \mathrm{A}_{1g} + \mathrm{E}_g, \qquad \mathrm{E}_g \doteq \mathrm{E}_g.$$

This scheme corresponds to the expansion $\mathrm{D}^{(2)} \doteq \mathrm{A}_{1g} + 2\mathrm{E}_g$ (Table 2.3). Since the trigonal component of the crystal field does not remove the degeneracy, of the cubic level $\mathrm{E}_g$, the d-ion energy spectrum consists in this case of three levels (Fig. 4.7b). Their relative position depends on the two parameters $D\sigma$ and $D\tau$, which are similar to $Ds$ and $Dt$; however, their explicit expressions are more complex. Attention should be paid to the two doublet levels present, whose wavefunctions form two type-$\mathrm{E}_g$ basis sets of $\mathbf{D}_{3d}$. Similar representations formed by different bases are called *repeating representations*. In the case under discussion, of a small trigonal distortion, one of the repeating trigonal representations $\mathrm{E}_g$ is formed by cubic $\mathrm{T}_{2g}$ functions, while the other is formed by cubic $\mathrm{E}_g$ functions. The repeating representations complicate the calculation in the framework of the crystal field theory discussed below. The qualitative splitting diagrams of the d level in five-coordinate ligand surroundings of the trigonal bipyramidal ($\mathbf{D}_{3h}$) and square pyramidal ($\mathbf{C}_{4v}$) types are shown in Fig. 4.8. A large number of different types of coordination polyhedra are described in [37].

### 4.3.7 Representation-reduction tables. External fields

The above methods of representation reduction upon decrease in symmetry may be applied to any point group. The tables in [1] are convenient for practical use. Each table section, containing data concerning the point group,

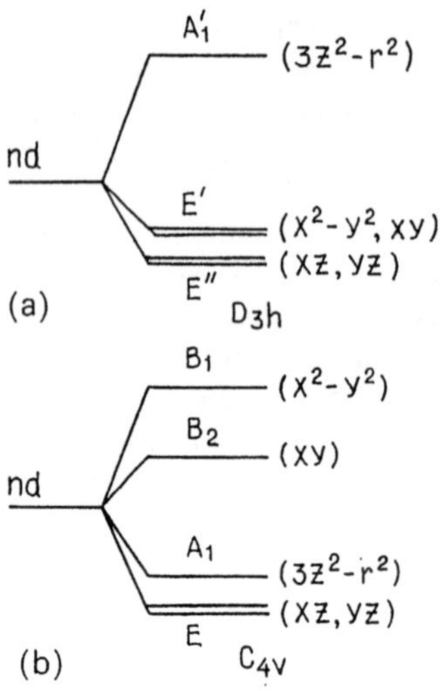


**Fig. 4.8** Diagrams of d-level splittings in five-coordinate surrounding of (a) trigonal-bipyramidal and (b) square-pyramidal types.

also includes *branching diagrams* (*compatibility tables*) of this group's irreducible representations in all the subgroups (Table 2.5). Let us consider the rules for using the branching diagrams (sometimes called *correlation or reduction tables*) by means of an example.

**Example 4.4** The compatibility table for the $\mathbf{O}_h$ group representations is shown in Table 4.4 [1]. The Bethe notation used in [1] is retained here—it is easy to pass to the Mulliken notation by comparing the representation characters. In $\mathbf{O}_h$ the symbols $\Gamma_1^+$, $\Gamma_2^+$, $\Gamma_3^+$, $\Gamma_4^+$ and $\Gamma_5^+$ correspond to the even representations $A_{1g}$, $A_{2g}$, $E_g$, $T_{1g}$ and $T_{2g}$, while $\Gamma_i^-$ correspond to the odd representations $A_{1u}$, $A_{2u}$, $E_u$, $T_{1u}$ and $T_{2u}$. The first column shows the subgroups of $\mathbf{O}_h$, and the rows show the corresponding irreducible representation branches of $\mathbf{O}_h$. Passing to the inversionless cubic groups $\mathbf{O}$ and $\mathbf{T}_d$ does not lead to splitting of the representations $\Gamma_i^{\pm}$, the even and odd basis functions $\varphi(\Gamma_i^{\pm})$ being transformed according to the same representation $\Gamma_i$. In the groups of lower symmetry the representations $\Gamma_i^{\pm}$ are, generally speaking, split into representations of lower dimension. The reduction tables also contain important information on the decrease in symmetry due to external electric and magnetic fields. The point symmetry of the system in the presence of an external field depends on the field direction relative to the symmetry axes of $\mathbf{O}_h$.

In [1] and Table 4.4 the [100] directions (the $C_4(z)$ axis), [110] direction ($C_2(xy)$ axis) and [111] direction ($C_3(xyz)$ axis) of the octahedron and cube are denoted by $z$, $v$ and $w$ respectively, $E(z)$, $E(v)$ and $E(w)$ are the electric fields directed along these axes, and $H(z)$, $H(v)$ and $H(w)$ are the magnetic fields. The influence of the electric field is obvious from Fig. 4.9, which shows the displacements of the positive metal ion and negatively charged ligand. The fields $E(z)$, $E(v)$ and $E(w)$ reduce the symmetry of the octahedron to $\mathbf{C}_{4v}$, $\mathbf{C}_{2v}$ and $\mathbf{C}_{3v}$, respectively. From Table 4.4 we find that in the electric field $E(w)$

**Table 4.4** Compatibility table for $\mathbf{O}_h$ group representations [1].

| $\mathbf{O}_h$ | $\Gamma_1^+$ | $\Gamma_2^+$ | $\Gamma_3^+$ | $\Gamma_4^+$ | $\Gamma_5^+$ | $\Gamma_1^-$ | $\Gamma_2^-$ | $\Gamma_3^-$ | $\Gamma_4^-$ | $\Gamma_5^-$ |
|---|---|---|---|---|---|---|---|---|---|---|
| $\mathbf{O}$ | $\Gamma_1$ | $\Gamma_2$ | $\Gamma_3$ | $\Gamma_4$ | $\Gamma_5$ | $\Gamma_1$ | $\Gamma_2$ | $\Gamma_3$ | $\Gamma_4$ | $\Gamma_5$ |
| $\mathbf{T}_d$ | $\Gamma_1$ | $\Gamma_2$ | $\Gamma_3$ | $\Gamma_4$ | $\Gamma_5$ | $\Gamma_2$ | $\Gamma_1$ | $\Gamma_3$ | $\Gamma_5$ | $\Gamma_4$ |
| $\mathbf{T}_h$ | $\Gamma_1^+$ | $\Gamma_1^+$ | $\Gamma_2^+ + \Gamma_3^+$ | $\Gamma_4^+$ | $\Gamma_4^+$ | $\Gamma_1^-$ | $\Gamma_1^-$ | $\Gamma_2^- + \Gamma_3^-$ | $\Gamma_4^-$ | $\Gamma_4^-$ |
| $\mathbf{D}_{4h}$ | $\Gamma_1^+$ | $\Gamma_3^+$ | $\Gamma_1^+ + \Gamma_3^+$ | $\Gamma_2^+ + \Gamma_5^+$ | $\Gamma_4^+ + \Gamma_5^+$ | $\Gamma_1^-$ | $\Gamma_3^-$ | $\Gamma_1^- + \Gamma_3^-$ | $\Gamma_2^- + \Gamma_5^-$ | $\Gamma_4^- + \Gamma_5^-$ |
| $\mathbf{D}_{3d}$ | $\Gamma_1^+$ | $\Gamma_2^+$ | $\Gamma_3^+$ | $\Gamma_2^+ + \Gamma_3^+$ | $\Gamma_1^+ + \Gamma_3^+$ | $\Gamma_1^-$ | $\Gamma_2^-$ | $\Gamma_3^-$ | $\Gamma_2^- + \Gamma_3^-$ | $\Gamma_1^- + \Gamma_3^-$ |
| $\mathbf{C}_{4h}$ : H($z$) | $\Gamma_1^+$ | $\Gamma_2^+$ | $\Gamma_1^+ + \Gamma_2^+$ | $\Gamma_1^+ + \Gamma_3^+ + \Gamma_4^+$ | $\Gamma_2^+ + \Gamma_3^+ + \Gamma_4^+$ | $\Gamma_1^-$ | $\Gamma_2^-$ | $\Gamma_1^- + \Gamma_2^-$ | $\Gamma_1^- + \Gamma_3^- + \Gamma_4^-$ | $\Gamma_2^- + \Gamma_3^- + \Gamma_4^-$ |
| $\mathbf{C}_{2h}$ : H($v$) | $\Gamma_1^+$ | $\Gamma_2^+$ | $\Gamma_1^+ + \Gamma_2^+$ | $2\Gamma_2^+ + \Gamma_1^+$ | $2\Gamma_1^+ + \Gamma_2^+$ | $\Gamma_1^-$ | $\Gamma_2^-$ | $\Gamma_1^- + \Gamma_2^-$ | $2\Gamma_2^- + \Gamma_1^-$ | $2\Gamma_1^- + \Gamma_2^-$ |
| $\mathbf{C}_{3i}$ : H($w$) | $\Gamma_1^+$ | $\Gamma_1^+$ | $\Gamma_2^+ + \Gamma_3^+$ | $\Gamma_1^+ + \Gamma_2^+ + \Gamma_3^+$ | $\Gamma_1^+ + \Gamma_2^+ + \Gamma_3^+$ | $\Gamma_1^-$ | $\Gamma_1^-$ | $\Gamma_2^- + \Gamma_3^-$ | $\Gamma_1^- + \Gamma_2^- + \Gamma_3^-$ | $\Gamma_1^- + \Gamma_2^- + \Gamma_3^-$ |
| $\mathbf{C}_{2v}$ : E($v$) | $\Gamma_1$ | $\Gamma_2$ | $\Gamma_1 + \Gamma_2$ | $\Gamma_2 + \Gamma_3 + \Gamma_4$ | $\Gamma_1 + \Gamma_3 + \Gamma_4$ | $\Gamma_3$ | $\Gamma_4$ | $\Gamma_3 + \Gamma_4$ | $\Gamma_1 + \Gamma_2 + \Gamma_4$ | $\Gamma_1 + \Gamma_2 + \Gamma_3$ |
| $\mathbf{C}_{3v}$ : E($w$) | $\Gamma_1$ | $\Gamma_2$ | $\Gamma_3$ | $\Gamma_2 + \Gamma_3$ | $\Gamma_1 + \Gamma_3$ | $\Gamma_2$ | $\Gamma_1$ | $\Gamma_3$ | $\Gamma_1 + \Gamma_3$ | $\Gamma_2 + \Gamma_3$ |
| $\mathbf{C}_{4v}$ : E($z$) | $\Gamma_1$ | $\Gamma_3$ | $\Gamma_1 + \Gamma_3$ | $\Gamma_2 + \Gamma_5$ | $\Gamma_4 + \Gamma_5$ | $\Gamma_2$ | $\Gamma_4$ | $\Gamma_2 + \Gamma_4$ | $\Gamma_1 + \Gamma_5$ | $\Gamma_3 + \Gamma_5$ |

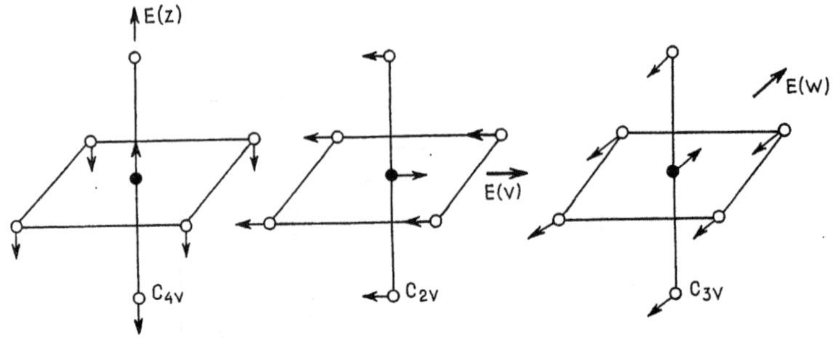


**Fig. 4.9** Lowering of the symmetry an octahedral complex in an electric field.

the $\Gamma_4^+(T_{1g})$ level is split into two levels: $\Gamma_2$ and $\Gamma_3$ ($A_2$ and E), while the $\Gamma_3$ ($A_2$ and E), while the $\Gamma_4^-$ ($T_{1u}$) level is split into the levels $\Gamma_1$ and $\Gamma_3$ ($A_1$ and E). The situation in the presence of a magnetic field is more complicated, for $H(z)$, $H(v)$ and $H(w)$ the $\mathbf{O}_h$ symmetry is reduced to $\mathbf{C}_{4h}$, $\mathbf{C}_{2h}$ and $\mathbf{C}_{3i}$ respectively; the corresponding level splitting is obtained from Table 4.4. The electric field removes the inversion centre, while the magnetic field preserves it. In the latter case the representation branching retains its parity.

If there is no inversion centre in the initial group, the electric and magnetic fields can decrease the symmetry in the same way. For example, both $H(z)$ and $E(z)$ cause the symmetry decrease $\mathbf{O} \rightarrow \mathbf{C}_4$ [1].

## PROBLEMS

**4.1.** Derive the results shown in Table 4.2.

**4.2** Determine the irreducible representations according to which the atomic s, p, d and f states are transformed in complexes with symmetry $\mathbf{T}_d$, $\mathbf{D}_{4h}$, $\mathbf{D}_3$ and $\mathbf{D}_{2h}$.

**4.3.** Obtain the splitting diagrams of the $\mathbf{O}_h$ cubic group terms $\Gamma_3^{\pm}, \Gamma_4^{\pm}$ and $\Gamma_5^{\pm}$ in electric and magnetic fields, directed along the $C_4$, $C_3$ and $C_2$ axes (Table 4.4).

# 5 Many-electron Ions in Crystal Fields

## 5.1 QUANTUM STATES OF A FREE ATOM

Let us consider an atom or an ion containing several electrons outside the filled shells. These shells have a spherically symmetric electron density distribution shielding the nucleus, which can therefore be taken to have an effective charge less than its actual charge. It can be assumed as an approximation that electrons in the unfilled shells are attracted to the nucleus by this effective charge, and are repelled by each other. Each electron state in the field of the nucleus and the other electrons is characterized by a set of quantum numbers $n$, $l$, $m$, $s$ and $m_s$, where $s = \frac{1}{2}$ is the electron spin and $m_s = \pm\frac{1}{2}$ is its projection. With interelectron interaction included, the atomic energy levels have the corresponding quantum numbers of the *total orbital angular momentum* $L$ and *its projection* $M_L$, as well as those of the *total spin* $S$ and *its projection* $M_S$. In contrast with one-electron states, the corresponding *many-electron states* (Table 4.1) are denoted by capital letters S, P, D, F, G, H, I, .... The values of the total orbital angular momentum $L$ are obtained by summing the total orbital angular momenta $l_i$ of the individual electrons according to quantum mechanical rules. For two electrons with quantum numbers $l_1$ and $l_2$ the *summation rule* has the form [34, 38]

$$L = l_1 + l_2,\ l_1 + l_2 - 1,\ \ldots,\ |l_1 - l_2|. \tag{5.1}$$

For electrons with identical orbital angular momentum quantum numbers $l_1 = l_2 = l$, the total orbital angular momentum quantum number $L$ takes $2l + 1$ integral values from $2l$ to 0. The summation rule for angular momenta can be visualized through the "vector model", as shown in Fig. 5.1 for two d electrons ($l = 2$). This indicates the "couplings" of two orbital angular momentum vectors leading to $L = 4, 3, 2, 1, 0$. The summation rule is applicable to angular momenta of all physical types, in particular to electron spin. Therefore the total

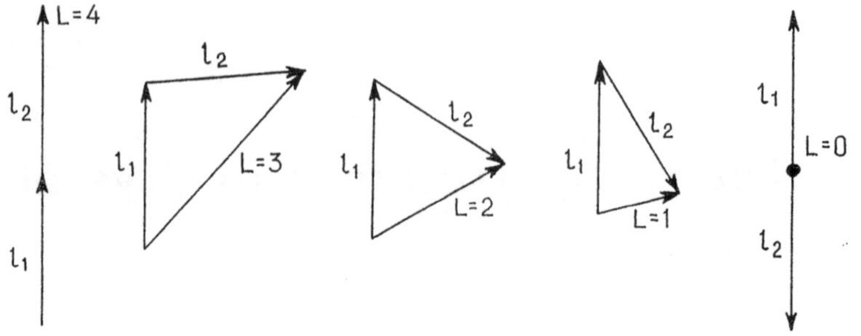


**Fig. 5.1** Vector summation model for the angular momenta of two d electrons ($l_1 = l_2 = 2$).

spin of two electrons ($s_1 = s_2 = s = \frac{1}{2}$) takes the values $S = 0$ and 1, while for three spins $S = \frac{1}{2}$, 1 and $\frac{3}{2}$.

The states numbered by quantum numbers $L$ and $S$ are called *atomic terms.* Each term is $(2L+1)$-fold degenerate over the orbital angular momentum projection $M_L$ and $(2S+1)$-fold degenerate over the full spin projection $M_S$; hence each $LS$ multiplet is $(2L+1)(2S+1)$-fold degenerate. The quantum numbers $L$ and $S$ are sufficient to describe the atomic electron states when their mutual interaction is stronger than that between each electron's orbital angular momentum and its spin (spin–orbit interaction). In this case we speak of $LS$ or Russell–Saunders coupling. The values $L$ and $S$ of a Russell–Saunders term are indicated by a capital letter (S, P etc.) and by a superscript indicating the spin multiplicity $2S+1$. For example, $^4$F denotes the state with $L = 3$ and $S = \frac{3}{2}$.

The $LS$-coupling approximation is adequate for transition metal ions with unfilled 3d shells. In rare-earth ions the spin–orbital interaction is much stronger; therefore their states are enumerated by the quantum numbers of the total momentum $J$. In accordance with the vector diagram, the total angular momentum $j$ of one electron may take two values $j = l + s = l + \frac{1}{2}$ and $j = l - s = l - \frac{1}{2}$. In a many-electron system the values of the total angular momentum $J$ may be found from the orbital and spin angular momenta by the vector coupling scheme

$$J = L + S,\ \ L + S - 1,\ \ \ldots,\ \ |L - S|.$$

The total angular momentum is denoted by a subscript. For example, for the $^3$P term, states with $J = 2, 1, 0$ are possible, i.e. the states $^3P_2$, $^3P_1$ and $^3P_0$. Every $J$-multiplet is $(2J+1)$-fold degenerate, according to the possible values of the quantum number of the total angular momentum projection $M_J = -J, -J+1, \ldots, +J$.

The Pauli exclusion principle imposes certain restrictions on the possible terms of a many-electron ion with an unfilled shell. The simplest way of illustrating this is by means of the example of two equivalent electrons in an s

**Table 5.1** Russell–Saunders terms of equivalent electron $d^n$-configurations.

| Configurations | Terms |
|---|---|
| $d^1, d^9$ | $^2D$ |
| $d^2, d^8$ | $^1S$, $^1D$, $^1G$, $^3P$, $^3F$ |
| $d^3, d^7$ | $2^2D$, $^2P$, $^2F$, $^2G$, $^2H$, $^4P$, $^4F$ |
| $d^1, d^6$ | $2^1S$, $2^1D$, $2^1G$, $^1F$, $^1I$, $^3P$, $^3D$, $^3F$, $^3G$, $^3H$, $^5D$ |
| $d^5$ | $3^2D$, $^2P$, $2^2F$, $2^2G$, $^2H$, $^2S$, $^2I$, $^4P$, $^4F$, $^4D$, $^1G$, $^6S$ |
| $d^{10}$ | $^1S$ |

shell. For the electron $ns^2$ configuration the term $^1S(m_1 = m_2 = 0,\ m_{s1} = \frac{1}{2},\ m_{s2} = -\frac{1}{2})$ is allowed, while the $^3S$ term is forbidden, since all the quantum numbers of the two electrons are identical. If the $s^2$ configuration contains two *inequivalent* electrons, $n_1$s and $n_2$s, the Pauli principle does not restrict the possible $LS$ states; hence the terms $^3S$ and $^1S$ are allowed. For d electrons there are more possibilities of producing different terms. The allowed terms of the $d^n$-configurations of equivalent electrons are listed in Table 5.1.

The brief sketch given above, based on an independent particle model, is sufficient for an understanding of the group-theoretical classication of the states of atoms or ions in crystal fields.

## 5.2 CLASSIFICATION OF LEVELS IN CRYSTAL FIELDS

### 5.2.1 Classification method for the *LS* scheme

The method of classification of one-electron states in crystal fields (Section 4.3) is easily generalized to the case of many-electron ions. Two factors that allow this generalization in the Russell–Saunders scheme should be noted:

(1) a crystal field, being electrostatic, does not influence the spin directly;

(2) many-electron wavefunctions with a given quantum number $L$ constitute the basis of the $(2L + 1)$-dimensional representation of the rotation group.

The first statement is of an approximate character, since the crystal field does influence the electronic orbital motion, which is coupled with the spin through the spin–orbit interaction. The second statement follows from the spherical symmetry of the atom or ion, irrespective of the number of electrons. Therefore, in contrast with the first statement, the second is not restricted by the Russell–Saunders coupling scheme.

Taking the above into account, the representation reduction formula (4.12) may be used for many-electron ions in the form

$$a(\Gamma_i) = \frac{1}{g}\sum_{\hat{R}} \chi^{(L)}(\hat{R})\chi^{(\Gamma_i)}(\hat{R}). \tag{5.2}$$

The rotation group characters $\chi^{(L)}(\hat{R})$ of many-electron states are calculated from (4.16), in which $L$ should be substituted for $l$. Therefore, as shown in Section 4.3.4, the results of Table 4.2 are applicable to one-electron and many-electron ions. Using this table, it is possible to determine how an atomic $LS$ state is split in a cubic field **O**. In the atomic $LS$-coupling scheme the crystal components (*crystal terms*) are characterized by the irreducible representations $\Gamma_i$ and the spin $S$ of the initial atomic level. By analogy with the notation for atomic terms, a notation is introduced for spin multiplicity. For example, $^4T_2$ is the symbol for the crystal term in the **O** field whose orbital wavefunctions transform† according to the representation $T_2$ of the group **O**, while the total spin is $S = \frac{3}{2}$.

## 5.2.2 Parity rule

The classification of many-electron states in centrosymmetric groups requires the use of the *parity rule* for atomic states. This rule can be easily established by means of simple examples. Suppose that the state of a two-electron atom arises from 3d and 4p electrons. The wavefunction for the two electrons is written in the form of products $\varphi_{3\mathrm{d}m_1}(\mathbf{r}_1)\varphi_{4\mathrm{p}m_2}(\mathbf{r}_2)$ (spin variables are omitted). All of these products reverse sign under inversion, since the $d$ functions are even ($l_1 = 2$), while the $p$ functions are odd ($l_2 = 1$). The quantum number $L$ takes the values $1, 2, 3$ and the total spin $S = 0, 1$; and the Pauli principle permits the terms $^1P$, $^1D$, $^1F$, $^3P$, $^3D$, $^3F$ (or, briefly, $^{1,3}P$, $^{1,3}D$, $^{1,3}F$). The wavefunctions of these terms are odd, because they are composed of the products $\varphi_{3\mathrm{d}}\varphi_{4\mathrm{p}}$. Therefore reduction of the representation $D^{(L)}$ in a centrosymmetric crystal field gives only odd representations in the case under consideration. The results for a number of point groups are given in Table 5.2. In $\mathbf{O}_h$ and $\mathbf{D}_{4h}$ the states are odd, while in $\mathbf{T}_d$, which does not contain the inversion operation, the parity symbol is absent. Attention should be paid to notation: in contrast with Table 4.3 for one-electron states, the irreducible representations describing many-electron terms are denoted by capital letters. Two electrons in the d shell give only even states, since the product $\varphi_{n_1\mathrm{d}m_1}\varphi_{n_2\mathrm{d}m_2}$ is even. We can formulate the following *parity rule for many-electron LS states*:

† The expression is used to denote transformation of the wavefunction when the symmetry operations are applied only to the 'orbital' (i.e. *spatial*) variables.

**Table 5.2** Terms of the two-electron 3d 4p configuration in crystal fields.

| Terms of a free ion | Splitting in crystal fields | | |
|---|---|---|---|
| | $O_h$ | $T_d$ | $D_{4d}$ |
| $^{1,3}P$ | $^{1,3}T_{1u}$ | $^{1,3}T_2$ | $^{1,3}A_{2u} + {}^{1,3}E_u$ |
| $^{1,3}D$ | $^{1,3}E_u + {}^{1,3}T_{2u}$ | $^{1,3}E + {}^{1,3}T_1$ | $^{1,3}A_{2u} + {}^{1,3}B_{1u} + {}^{1,3}B_{2u} + {}^{1,3}E_u$ |
| $^{1,3}F$ | $^{1,3}A_{2u} + {}^{1,3}T_{1u} + {}^{1,3}T_{2u}$ | $^{1,3}A_1 + {}^{1,3}T_1 + {}^{1,3}T_2$ | $^{1,3}A_{2u} + {}^{1,3}B_{1u} + {}^{1,3}B_{2u} + 2{}^{1,3}E$ |

> The parity of many-electron LS states is determined by the factor $(-1)^{l_1+l_2+\ldots+l_k}$, where $k$ is the total number of electrons in unfilled shells and $l_1, l_2, \ldots, l_k$ are their orbital angular momentum quantum numbers.

Thus, parity is determined by the arithmetic (not vectorial!) sum of the one-electron momenta and is by no means equal to $(-1)^L$. Atomic states with a given $L$ may be either even or odd, depending on the numbers of electrons in different shells. The corresponding irreducible representations describing the splitting of the *LS* terms in crystal fields have the same parity as the initial atomic state. Therefore, in particular, all the terms originating from $d^n$ configurations are even (Table 5.1). Using the reduction formula, it is easy to find the point group irreducible representations for any representation $D^{(L)}$. The results for the $d^3(d^7)$ configuration are listed in Table 5.3. All terms in centrosymmetric crystal fields are even.

### 5.2.3 Reduction tables for representations of the full rotation group

In order to determine the crystal terms without using (5.2), it is possible to use the tables in [1] giving the results of the decomposition of the representations $D^{(L)}$ for $L = 0$–6. Since for a given $L$ the atomic state may be either even or odd, the tables in [1] contain branching schemes of the representations of the full rotation group $\mathbf{K}_h$, i.e. the spherical group involving inversion.

**Example 5.1** Table 5.4 shows results for the group $\mathbf{T}_d$ using the notations of [1]. The value of $L$ is shown as the subscript on the representation D, and the + and − signs indicate the parity. Thus $D_2^-$ is a five-dimensional odd representation of the full spherical group whose basis comprises functions with $L = 2$. In $\mathbf{T}_d$ this representation is reduced to the two-dimensional $\Gamma_3(E)$ and three-dimensional $\Gamma_4(T_1)$ representations. The representation $D_2^+$ gives $D_2^+ = \Gamma_3 + \Gamma_5(E + T_2)$. It should be noted that the decompositions of the representations $D_L^+$ and $D_L^-$ are, generally speaking, different, although $\mathbf{T}_d$ contains no inversion. In the group $\mathbf{O}$ the representations $D_L^\pm$ are reduced in a similar way.

**Table 5.3** Splitting of terms of the $d^3(d^7)$ configuration in crystal fields.

| Terms of a free ion | Splittings in crystal fields | | |
|---|---|---|---|
| | $O_h$ | $T_d$ | $D_{4h}$ |
| $2^2D$ | $2(^2E_g + {}^2T_{2g})$ | $2(^2E + {}^2T_2)$ | $2(^2A_{1g} + {}^2B_{1g} + {}^2B_{2g} + {}^2E_g)$ |
| $^2P$ | $^2T_{1g}$ | $^2T_1$ | $^2A_{2g} + {}^2E_g$ |
| $^2F$ | $^2A_{2g} + {}^2T_{1g} + {}^2T_{2g}$ | $^2A_2 + {}^2T_1 + {}^2T_2$ | $^2A_{2g} + {}^2B_{1g} + {}^2B_{2g} + 2^2E_g$ |
| $^2G$ | $^2A_{1g} + {}^2E_g + {}^2T_{1g} + {}^2T_{2g}$ | $^2A_1 + {}^2E + {}^2T_1 + {}^2T_2$ | $2^2A_{1g} + {}^2A_{2g} + {}^2B_{1g} + {}^2B_{2g} + 2^2E_g$ |
| $^2H$ | $^2E_g + 2^2T_{1g} + {}^2T_{1u}$ | $^2E + 2^2T_1 + {}^2T_1$ | $^2A_{1g} + 2^2A_{2g} + {}^2B_{1g} + {}^2B_{2g} + 3^2E_g$ |
| $^4P$ | $^4T_{1g}$ | $^4T_1$ | $^4A_{2g} + {}^4E_g$ |
| $^4F$ | $^4A_{2g} + {}^4T_{1g} + {}^4T_{2g}$ | $^4A_2 + {}^4T_1 + {}^4T_2$ | $^4A_{2g} + {}^4B_{1g} + {}^4B_{2g} + 2^4E_g$ |

**Table 5.4** Reduction of the representations $D_L^{\pm}$ in the group $\mathbf{T}_d$ of the full spherical group $\mathbf{K}_h$.

| | | | |
|---|---|---|---|
| $D_0^+$ | $\Gamma_1$ | $D_0^-$ | $\Gamma_2$ |
| $D_1^+$ | $\Gamma_4$ | $D_1^-$ | $\Gamma_2$ |
| $D_2^+$ | $\Gamma_3+\Gamma_5$ | $D_2^-$ | $\Gamma_3+\Gamma_4$ |
| $D_3^+$ | $\Gamma_2+\Gamma_4+\Gamma_5$ | $D_3^-$ | $\Gamma_1+\Gamma_4+\Gamma_5$ |
| $D_4^+$ | $\Gamma_1+\Gamma_3+\Gamma_4+\Gamma_5$ | $D_4^-$ | $\Gamma_2+\Gamma_3+\Gamma_4+\Gamma_5$ |
| $D_5^+$ | $\Gamma_3+2\Gamma_4+\Gamma_5$ | $D_5^-$ | $\Gamma_3+\Gamma_4+2\Gamma_5$ |
| $D_6^+$ | $\Gamma_1+\Gamma_2+\Gamma_3+\Gamma_4+2\Gamma_5$ | $D_6^-$ | $\Gamma_1+\Gamma_2+\Gamma_3+2\Gamma_4+\Gamma_5$ |

## 5.3 STRONG-CRYSTAL-FIELD SCHEME

The classification of crystal terms given in Section 5.2 starts from the atomic states. The crystal field is considered as a perturbation influencing the electronic states of a free atom or ion. In the extreme case of a *strong crystal field*, first its action on each electron is considered and then the electron interaction is taken into account as a perturbation [14, 16].

Let us assume that in an octahedral crystal field there are two equivalent d electrons; that is, we have an ion with the $(n\mathrm{d})^2 \equiv \mathrm{d}^2$ configuration. We write the Hamiltonian of the two-electron ion as

$$\hat{H} = \hat{H}_0 + \hat{V}^{\mathrm{c}}(\mathbf{r}_1) + \hat{V}^{\mathrm{c}}(\mathbf{r}_2) + \hat{V}_{12}, \tag{5.3}$$

where $H_0$ is the interaction of the d electron with the atomic core, $\hat{V}^{\mathrm{c}}(\mathbf{r}_1)$ and $\hat{V}^{\mathrm{c}}(\mathbf{r}_2)$ are the Hamiltonians of the d electrons in the octahedral field, and $\hat{V}_{12} = e^2/r_{12}$ is their interaction ($r_{12} = |\mathbf{r}_1 - \mathbf{r}_2|$ and $e$ is the electron charge). According to the strong-field scheme, in the first stage in calculating the energy and wavefunctions we shall omit the interelectron interaction $\hat{V}_{12}$. The energy of the zeroth-order approximation (without considering the spherically symmetric contribution of the crystal field) may then be written as

$$\varepsilon_{0kl} = \varepsilon_k + \varepsilon_l. \tag{5.4}$$

This energy is composed of two energies $\varepsilon_k$ and $\varepsilon_l$ of noninteracting electrons in the octahedral field, $k$ and $l$ denoting the irreducible representations $\mathrm{T}_{2g}$ and $\mathrm{E}_g$ of $\mathbf{O}_h$. It is obvious that (5.4) describes three energy levels. In the lowest level both electrons are in the $\mathrm{t}_2$ state; is the first excited level one electron is in the $\mathrm{t}_{2g}$ state, while the other is in the $\mathrm{e}_g$ state; and in the highest energy level both electrons are in the $\mathrm{e}_g$ state. By analogy with the terminology used in atomic spectroscopy, the one-electron states $\mathrm{t}_{2g}$ and $\mathrm{e}_g$ are called *shells*, while the two-electron states $\mathrm{t}_{2g}^2$, $\mathrm{t}_{2g}\mathrm{e}_g$ and $\mathrm{e}_g^2$ are called *strong-field configurations*. The energies of these configurations are calculated simply, as follows:

$$\left.\begin{aligned} \varepsilon(t_{2g}^2) &= \varepsilon(t_{2g}) + \varepsilon(t_{2g}) = -4Dq - 4Dq = -8Dq, \\ \varepsilon(t_{2g}e_g) &= \varepsilon(t_{2g}) + \varepsilon(e_g) = -4Dq + 6Dq = 2Dq, \\ \varepsilon(e_g^2) &= \varepsilon(e_g) + \varepsilon(e_g) = +6Dq + 6Dq = 12Dq. \end{aligned}\right\} \tag{5.5}$$

Thus the three equidistant levels are separated by intervals $10Dq$ corresponding to electron excitations from the $t_{2g}$ to the $e_g$ shell (Fig. 5.2a).

The wavefunctions of a two-electron ion are constructed from products

$$\Psi^{(0)}_{km_1,lm_2}(\mathbf{r}_1,\mathbf{r}_2) = \varphi_k(\mathbf{r}_1)\chi(\tfrac{1}{2}m_1)\varphi_l(\mathbf{r}_2)\chi(\tfrac{1}{2}m_2). \tag{5.6}$$

Here $\varphi_k$ and $\varphi_l$ are the one-electron $d$ functions (4.25) and (4.26), $k$ and $l$ enumerate the three functions of the representation $t_{2g}$ $(\xi,\eta,\zeta)$ and the two functions of the representation $e_g$ ($u$ and $v$). It should be remembered that the notation $t_{2g}$ and $e_g$ has been adopted for one-electron states. $\chi(sm)$ is the electron spin function with $s=\frac{1}{2}$ and $m=\pm\frac{1}{2}$. Furthermore, $\chi(\frac{1}{2}\,\frac{1}{2})$ and

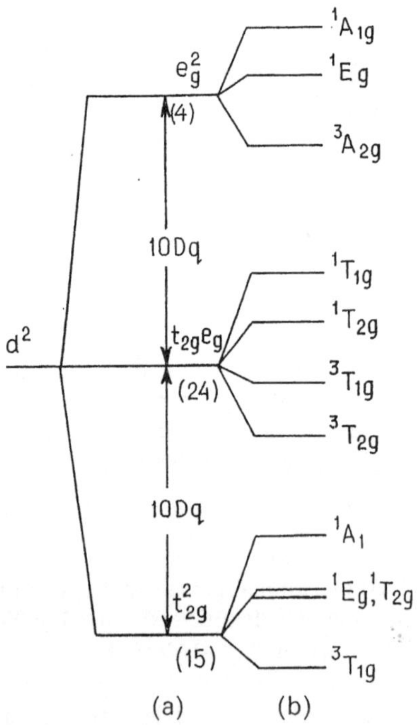


**Fig. 5.2** Energy levels of a $d^2$ ion in a strong octahedral field: (a) splitting in a crystal field (the degrees of degeneracy are given in parentheses); (b) taking account of interelectron interaction.

$\chi(\frac{1}{2}\ -\frac{1}{2})$ are denoted, as usual, by $\alpha$ and $\beta$. The wavefunctions of many-electron systems should satisfy the Pauli principle; that is, they should be antisymmetric under permutation of any two electrons. This requirement is easily satisfied if the products in (5.6) are replaced by determinants such as:

$$\sqrt{\frac{1}{2}}\begin{vmatrix} \varphi_k(\mathbf{r}_1)\chi(\frac{1}{2}m_1) & \varphi_l(\mathbf{r}_1)\chi(\frac{1}{2}m_1) \\ \varphi_k(\mathbf{r}_2)\chi(\frac{1}{2}m_2) & \varphi_l(\mathbf{r}_2)\chi(\frac{1}{2}m_2) \end{vmatrix} \equiv |\varphi_k\chi(\tfrac{1}{2}m_1)\varphi_l\chi(\tfrac{1}{2}m_2)|, \tag{5.7}$$

the factor $\sqrt{1/2}$ being introduced for normalization. Since the determinant reverses sign under permutation of rows or columns, the determinantal functions are antisymmetric. As a result, the two electrons cannot have identical sets of quantum numbers; if $k = l$ and $m_1 = m_2$, the determinant vanishes, because it possesses identical columns.

Let us consider in more detail the $t_{2g}^2$ configuration. The one-electron $t_{2g}$ level is sixfold-degenerate: threefold orbitally and twofold according to spin projection. It can be imagined that the electron is situated in one of the six cells of the following table:

| | orbital states $\xi$ | $\eta$ | $\zeta$ |
|---|---|---|---|
| spin $\alpha$ | | | |
| states $\beta$ | | | |

According to the Pauli principle, the quantum numbers of two electrons should be different; hence the electrons may be put in any two *different* squares. The total number of *permitted states* is equal to the number of combinations of 6 taken 2 at a time:

$$C_2^6 = \frac{6 \times 5}{2} = 15.$$

Thus the energy level is 15-fold-degenerate. Each of the 15 wavefunctions is a determinant of the form (5.7):

$$\left.\begin{matrix} |\xi\ \eta|, & |\bar{\xi}\ \bar{\eta}|, & |\eta\ \zeta|, & |\bar{\eta}\ \bar{\zeta}|, & |\xi\ \zeta|, & |\bar{\xi}\ \bar{\zeta}|, \\ |\xi\ \bar{\eta}|, & |\bar{\xi}\ \eta|, & |\eta\ \bar{\zeta}|, & |\bar{\eta}\ \zeta|, & |\zeta\ \bar{\xi}|, & |\bar{\xi}\ \zeta|, \\ |\zeta\ \bar{\zeta}|, & |\eta\ \bar{\eta}|, & |\xi\ \bar{\xi}|, & & & \end{matrix}\right\} \tag{5.8}$$

with the following notation:

$$|\xi\ \eta| = |\varphi_\xi\alpha\ \varphi_\eta\alpha|, \qquad |\bar{\xi}\ \eta| = |\varphi_\xi\beta\ \varphi_\eta\alpha|, \qquad \text{etc.} \tag{5.9}$$

The states (5.8) are expressed graphically by schemes such as that shown in Fig. 5.3.

In a similar way it is possible to obtain the number of permitted states of the $e_g^2$ configuration by arranging the two electrons in the squares of

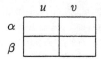

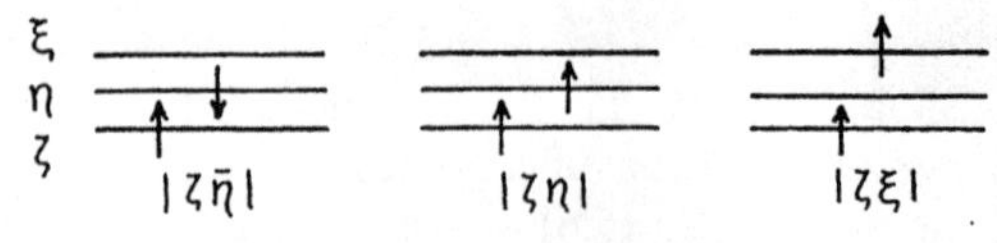


**Fig. 5.3** Diagrammatic representation of determinantal states of $t_2^2$ configuration.

the table

| | $u$ | $v$ |
|---|---|---|
| $\alpha$ | | |
| $\beta$ | | |

Their number is $C_2^4 = (4 \times 3)/2 = 6$. Finally, in the $t_{2g}e_g$ shell the electron orbital quantum numbers are different; hence the Pauli principle does not set a limit on the possible spin states. For a given orbital state, for example $\varphi_\xi(\mathbf{r}_1)\varphi_u(\mathbf{r}_2)$, the two electron spins may be parallel or antiparallel: $\alpha\alpha$, $\beta\beta$, $\alpha\beta$, $\beta\alpha$. The total degree of degeneracy of the level $\varepsilon(t_{2g}e_g)$ is $3 \times 2 \times 4 = 24$.

Let us consider these conclusions from the viewpoint of the Wigner theorem. The maximum dimension of the irreducible representations of **O** is three (excluding the so-called double-valued representations, discussed later). Therefore in crystal fields of cubic symmetry there may exist energy levels with degrees of orbital degeneracy no more than three. The high degeneracy of the $t_{2g}^2$, $e_g^2$ and $t_{2g}e_g$ configurations is related to the approximate nature of the analysis, which does not take account of the interelectron interaction $\hat{V}_{12}$. This interaction should split the levels $\varepsilon(t_{2g}^2)$, $\varepsilon(e_g^2)$ and $\varepsilon(t_{2g}e_g)$. In order to calculate, for example, the $\varepsilon(t_{2g}^2)$ level splitting, it is necessary to write down $15 \times 15 = 225$ matrix elements of the operator $\hat{V}_{12}$ with the determinantal functions (5.7). The matrix elements obtained (e.g. $\langle|\xi\ \bar{\eta}|\,|\hat{V}_{12}|\,|\eta\ \zeta|\rangle$ etc.) determine a 15th-order secular equation. Its eigenvalues are the energy levels in first order of perturbation theory, while its eigenfunctions are expressed as linear combinations of the determinants:

$$\sum_{k,l}\sum_{m_1,m_2} c(klm_1m_2)|\varphi_k(1)\chi(\tfrac{1}{2}m_1)\ \varphi_l(2)\chi(\tfrac{1}{2}m_2)|. \qquad (5.10)$$

This procedure is cumbersome owing to the high degree of degeneracy of the initial levels. Group-theoretical methods allow the construction of the correct zeroth-order functions without solving high order secular equations. The procedure is to construct linear combinations (5.10) that form the basis functions of the irreducible representations of $\mathbf{O}_h$ and correspond to definite values of the total spin of two electrons $S = 0$ and 1. In order to solve this problem, it is necessary to introduce a number of new group-theoretical concepts discussed in the next section.

## 5.4 THE DIRECT PRODUCT OF REPRESENTATIONS

### 5.4.1 Definition of the direct product

Let us consider two sets of functions, each being the basis of an irreducible representation of a certain point group. We shall denote these irreducible representations by $\Gamma_1$ and $\Gamma_2$, and the indices enumerating the basis functions by $\gamma_1$ and $\gamma_2$. Suppose that $\varphi(\Gamma_1\gamma_1)$ and $\varphi(\Gamma_2\gamma_2)$ are the basis functions of these representations. A *direct* or *Kronecker product of the representations* $\Gamma_1$ and $\Gamma_2$ is a representation whose basis consists of products $\varphi(\Gamma_1\gamma_1)\varphi(\Gamma_2\gamma_2)$. Let us denote the dimension of representation $\Gamma$ by $[\Gamma]$; then the dimension of the direct product of the representations $\Gamma_1$ and $\Gamma_2$ is $[\Gamma_1][\Gamma_2]$. The direct product of the representations itself is denoted by $\Gamma_1 \times \Gamma_2$.

Suppose, for example, that $\Gamma_1$ and $\Gamma_2$ are two $T_{1u}$ representations of $\mathbf{O}_h$, while the basis functions are the components of two vectors $\mathbf{r}_1(x_1, y_1, z_1)$ and $\mathbf{r}_2(x_2, y_2, z_2)$. The vectors $\mathbf{r}_1$ and $\mathbf{r}_2$ may be the coordinates of two electrons or the angular components of the $p$ functions. The basis of the direct product $T_{1u} \times T_{1u}$ comprises the nine functions $x_1x_2$, $x_1y_2$, $x_1z_2$, $y_1x_2$, $y_1y_2$, $y_1z_2$, $z_1x_2$, $z_1y_2$ and $z_1z_2$. This nine-dimensional representation is reducible, since there are no nine-dimensional irreducible representations in $\mathbf{O}_h$ (or, in general, in any of the discrete point groups).

### 5.4.2 Characters of the direct product

In the above example it is possible to write down linear combinations of the direct product basis functions forming bases of the irreducible representations of $\mathbf{O}_h$. One of these combinations is the scalar product of the vectors $\mathbf{r}_1$ and $\mathbf{r}_2$:

$$\sqrt{1/3}\,\mathbf{r}_1 \cdot \mathbf{r}_2 = \sqrt{1/3}\,(x_1x_2 + y_1y_2 + z_1z_2). \tag{5.11}$$

This is invariant under all the group operations, and is transformed according to the totally symmetric representation $A_{1g}$ ($\sqrt{1/3}$ is a normalization factor). The three components $R_x$, $R_y$ and $R_z$ of the vector product $\mathbf{R} = \sqrt{1/2}\,\mathbf{r}_1 \times \mathbf{r}_2$, (3.82) and (3.83), form the basis of the representation $T_{1g}$. The $\mathbf{O}_h$ group operations transform the three linear combinations $\sqrt{1/2}\,(y_1z_2 + z_1y_2)$, $\sqrt{1/2}\,(z_1x_2 + x_1z_2)$ and $\sqrt{1/2}\,(x_1y_2 + y_1x_2)$ as $d$ functions $d_\xi$, $d_\eta$ and $d_\zeta$; that is, they are the basis of the three-dimensional representation $T_{2g}$. Finally, the functions $\sqrt{1/6}\,(3z_1z_2 - \mathbf{r}_1 \cdot \mathbf{r}_2)$ and $\sqrt{1/2}\,(x_1x_2 - y_1y_2)$ together with $d_u$ and $d_v$ form a basis of the representation $E_g$. Thus the nine-dimensional representation $T_{1u} \times T_{1u}$ is reduced to a one-dimensional representation $A_{1g}$, a two-dimensional representation $E_g$ and two three-dimensional representations $T_{1g}$ and $T_{2g}$. The direct product reduction is denoted as follows:

$$T_{1u} \times T_{1u} \doteq A_{1g} + E_g + T_{1g} + T_{2g}.$$

**Table 5.5** Direct products of irreducible representations of $\mathbf{C}_{4v}$.

| $C_{4v}$ | $\hat{E}$ | $2\hat{C}_4$ | $\hat{C}_2$ | $2\hat{\sigma}_v$ | $2\hat{\sigma}_d$ | $f^{(\Gamma)}$ |
|---|---|---|---|---|---|---|
| $A_1$ | 1 | 1 | 1 | 1 | 1 | $z; x^2+y^2$ |
| $A_2$ | 1 | 1 | 1 | −1 | −1 | $R_z$ |
| $B_1$ | 1 | −1 | 1 | 1 | −1 | $x^2-y^2$ |
| $B_2$ | 1 | −1 | 1 | −1 | 1 | $xy$ |
| E | 2 | 0 | −2 | 0 | 0 | $x, y; R_x R_y; yz, xz$ |
| $A_1 \times A_1$ | 1 | 1 | 1 | 1 | 1 | |
| $A_1 \times A_2$ | 1 | 1 | 1 | −1 | 1 | |
| $A_1 \times B_1$ | 1 | −1 | 1 | 1 | −1 | |
| $A_1 \times B_2$ | 1 | −1 | 1 | −1 | 1 | |
| $A_1 \times E$ | 2 | 0 | −2 | 0 | 0 | |
| $A_2 \times A_2$ | 1 | 1 | 1 | 1 | 1 | |
| $A_2 \times B_1$ | 1 | −1 | 1 | −1 | 1 | |
| $A_2 \times B_2$ | 1 | −1 | 1 | 1 | −1 | |
| $A_2 \times E$ | 2 | 0 | −2 | 0 | 0 | |
| $B_1 \times B_1$ | 1 | 1 | 1 | 1 | 1 | |
| $B_1 \times B_2$ | 1 | 1 | 1 | −1 | −1 | |
| $B_1 \times E$ | 2 | 0 | −2 | 0 | 0 | |
| $B_2 \times B_2$ | 1 | 1 | 1 | 1 | 1 | |
| $B_2 \times E$ | 2 | 0 | −2 | 0 | 0 | |
| $E \times E$ | 4 | 0 | 4 | 0 | 0 | |

It is said that the representations $A_{1g}$, $E_g$, $T_{1g}$ and $T_{2g}$ are *contained* in the direct product $T_{1u} \times T_{1u}$. The basis functions and the corresponding irreducible representations in the example discussed here have been constructed by analogy with other known cases [6].

In the general case it is necessary to have a systematic procedure for reduction of the direct product $\Gamma_1 \times \Gamma_2$. The problem is divided into two parts:

(1) classification of the irreducible representations $\Gamma_i$ contained in the product $\Gamma_1 \times \Gamma_2$ ($\Gamma_1 \times \Gamma_2 \doteq \Sigma_i \Gamma_i$);

(2) construction of the basis functions $\psi(\Gamma_i \gamma_i)$ of these irreducible representations as linear combinations of the direct product functions $\varphi(\Gamma_1 \gamma_1)\varphi(\Gamma_2 \gamma_2)$.

In order to find the irreducible representations $\Gamma_i$, it is necessary to know the characters of the reducible representation of dimension $[\Gamma_1][\Gamma_2]$, which is conveniently denoted by $\chi^{(\Gamma_1 \times \Gamma_2)}$. We state without proof the following important theorem:

the character $\chi^{(\Gamma_1 \times \Gamma_2)}(\hat{R})$ of a representation whose basis is the direct product of two representations $\Gamma_1$ and $\Gamma_2$ is equal to the product of the characters $\chi^{(\Gamma_1)}(\hat{R})$ and $\chi^{(\Gamma_2)}(\hat{R})$:

$$\chi^{(\Gamma_1 \times \Gamma_2)}(\hat{R}) = \chi^{(\Gamma_1)}(\hat{R})\chi^{(\Gamma_2)}(\hat{R}). \tag{5.12}$$

On substituting the point group irreducible representations into (5.12), we can obtain the character table for any direct product.

Let us illustrate this, using as an example $\mathbf{C}_{4v}$, by adding to the table of characters of the irreducible representations all their direct products (Table 5.5).

Knowing the direct product characters, it is possible to decompose them into irreducible representations.

### 5.4.3 Decomposition of a direct product into irreducible parts

Let us use the general formula (4.12) for reduction of irreducible representations after writing it as

$$a(\Gamma_i) = \frac{1}{g}\sum_{\hat{R}} \chi^{(\Gamma_1 \times \Gamma_2)}(\hat{R})\chi^{(\Gamma_i)}(\hat{R}), \tag{5.13}$$

where $\Gamma_i$ are irreducible representations of the group, and $a(\Gamma_i)$ are nonzero if $\Gamma_i$ is contained in the direct product $\Gamma_1 \times \Gamma_2$.

We shall use (5.13) to decompose the direct products of the irreducible representations of $\mathbf{C}_{4v}$. From the table we obtain, for example, for the case of $\mathrm{E} \times \mathrm{E}$,

$$a(\mathrm{A}_1) = \tfrac{1}{8}(4 \cdot 1 + 2 \cdot 0 \cdot 1 + 4 \cdot 1 + 2 \cdot 0 \cdot 1 + 2 \cdot 0 \cdot 1) = 1, \quad \text{etc.}$$

The result is

$$\mathrm{E} \times \mathrm{E} \doteq \mathrm{A}_1 + \mathrm{A}_2 + \mathrm{B}_1 + \mathrm{B}_2.$$

Using the same procedure, it is possible to compose a *multiplication table for irreducible representations* of $\mathbf{C}_{4v}$ (Table 5.6). The multiplication table is symmetric ($\Gamma_1 \times \Gamma_2 = \Gamma_2 \times \Gamma_1$); hence the full information is contained in just one half of it. *Reference* [1] *gives multiplication tables for all the irreducible representations of the point groups.*

In the cubic groups products of two- and three-dimensional representations are encountered; hence the multiplication table is richer. That for $\mathbf{O}$ is shown as Table 5.7.

For $\mathbf{O}_h$ and other centrosymmetrical groups the *parity rule* holds: this states that taking products of representations of identical parity gives even

**Table 5.6** Multiplication table for irreducible representations of $\mathbf{C}_{4v}$.

| $\mathbf{C}_{4v}$ | $\mathrm{A}_1$ | $\mathrm{A}_2$ | $\mathrm{B}_1$ | $\mathrm{B}_2$ | E |
|---|---|---|---|---|---|
| $\mathrm{A}_1$ | $\mathrm{A}_1$ | $\mathrm{A}_2$ | $\mathrm{B}_1$ | $\mathrm{B}_2$ | E |
| $\mathrm{A}_2$ | $\mathrm{A}_2$ | $\mathrm{A}_1$ | $\mathrm{B}_2$ | $\mathrm{B}_1$ | E |
| $\mathrm{B}_1$ | $\mathrm{B}_1$ | $\mathrm{B}_2$ | $\mathrm{A}_1$ | $\mathrm{A}_2$ | E |
| $\mathrm{B}_2$ | $\mathrm{B}_2$ | $\mathrm{B}_1$ | $\mathrm{A}_2$ | $\mathrm{A}_1$ | E |
| E | E | E | E | E | $\mathrm{A}_1 + \mathrm{A}_2 + \mathrm{B}_1 + \mathrm{B}_2$ |

**Table 5.7** Multiplication table for irreducible representations of **O**.

| O | $A_1$ | $A_2$ | E | $T_1$ | $T_2$ |
|---|---|---|---|---|---|
| $A_1$ | $A_1$ | $A_2$ | E | $T_1$ | $T_2$ |
| $A_2$ | $A_2$ | $A_1$ | E | $T_2$ | $T_1$ |
| E | E | E | $A_1+A_2+E$ | $T_1+T_2$ | $T_1+T_2$ |
| $T_1$ | $T_1$ | $T_2$ | $T_1+T_2$ | $A_1+E+T_1+T_2$ | $A_2+E+T_1+T_2$ |
| $T_2$ | $T_2$ | $T_1$ | $T_1+T_2$ | $A_2+E+T_1+T_2$ | $A_1+E+T_1+T_2$ |

representations, while taking the products of representations of different parity gives odd representations:

$$g \times g = g, \quad g \times u = u, \quad u \times u = g. \tag{5.14}$$

Therefore in $\mathbf{O}_h$

$$E_g \times T_{2g} \doteq T_{1g} + T_{2g}, \qquad E_u \times T_{2u} \doteq T_{1g} + T_{2g},$$
$$E_g \times T_{2u} = E_u \times T_{2g} \doteq T_{1u} + T_{2u}$$

Similarly, we can obtain the decomposition found above for $T_{1u} \times T_{1u}$:

$$T_{1u} \times T_{1u} \doteq A_{1g} + E_g + T_{1g} + T_{2g}.$$

It is possible to formulate the following multiplication rules [13]:

$$A \times A = A, \qquad B \times B = A, \qquad A \times B = B, \qquad A \times T = T,$$

where the indices $A$ and $B$ are combined as $1 \times 1 = 1$, $2 \times 2 = 1$ and $1 \times 2 = 2$, excluding the groups $\mathbf{D}_2$ and $\mathbf{D}_{2h}$, for which $1 \times 2 = 3$ and $2 \times 3 = 1$. Of course, these rules follow from the multiplication tables [1]. Multiplying the totally symmetric representation $A_1$ by any representation $\Gamma$ we obtain $\Gamma$ ($A_1 \times \Gamma \doteq \Gamma$). It should be noted that *the product of two identical representations $\Gamma$ contains a totally symmetric representation*: $\Gamma \times \Gamma \doteq A_1 + \ldots$. The opposite statement is also true: the *direct product of different representations contains no totally symmetric representation.*

### 5.4.4 Clebsch–Gordan coefficients

Let us return to the problem of constructing basis functions of irreducible representations $\Gamma$, contained in the direct product $\Gamma_1 \times \Gamma_2$.

Basis functions $\psi(\Gamma\gamma)$ with $\Gamma$ contained in the direct product $\Gamma_1 \times \Gamma_2$ are expressed as linear combinations of the form

$$\psi(\Gamma\gamma) = \sum_{\gamma_1, \gamma_2} \varphi(\Gamma_1\gamma_1)\varphi(\Gamma_2\gamma_2)\langle\Gamma_1\gamma_1\Gamma_2\gamma_2|\Gamma\gamma\rangle. \tag{5.15}$$

The quantities $\langle\Gamma_1\gamma_1\Gamma_2\gamma_2|\Gamma\gamma\rangle$ are called the *Clebsch–Gordan coefficients* of the point group. They are nonzero if $\Gamma$ is contained in $\Gamma_1 \times \Gamma_2$.

Clebsch–Gordan coefficients form a square matrix $\mathbf{U}$ with elements $\mathbf{U}_{\gamma_1\gamma_2,\Gamma\gamma} = \langle \Gamma_1\gamma_1\Gamma_2\gamma_2|\Gamma\gamma\rangle$, whose rows and columns are labelled by the index *pairs*, $\gamma_1\gamma_2$ and $\Gamma\gamma$, respectively. The matrix $\mathbf{U}$ is the unitary matrix, hence $\langle \Gamma_1\gamma_1\Gamma_2\gamma_2|\Gamma\gamma\rangle = \langle \Gamma\gamma|\Gamma_1\gamma_1\Gamma_2\gamma_2\rangle^*$. The order of this matrix is $[\Gamma_1][\Gamma_2]$, just the number of values taken by the index $\gamma_1\gamma_2$ and $\Gamma\gamma$. In fact, $\mathbf{U}$ describes a similarity transformation (3.53), in which $D \equiv D^{(\Gamma_1\times\Gamma_2)}$ and $D'_1, D'_2, \ldots$ are the irreducible representations $\Gamma_i$. In the case considered in Section 5.4.1, $T_{1u}\times T_{1u}$, this transformation has the form

$$\mathbf{U}^{-1}\mathbf{D}^{(T_{1u}\times T_{1u})}(\hat{R})\mathbf{U} = \begin{bmatrix} \mathbf{D}^{(A_{1g})}(\hat{R}) & & & \\ & \mathbf{D}^{(E_g)}(\hat{R}) & & \\ & & \mathbf{D}^{(T_{1g})}(\hat{R}) & \\ & & & \mathbf{D}^{(T_{2g})}(\hat{R}) \end{bmatrix}, \quad (5.16)$$

where $\mathbf{U}^{-1}$ is the inverse matrix, with elements $\langle \Gamma\gamma|\Gamma_1\gamma_1\Gamma_2\gamma_2\rangle$.

In order to obtain the Clebsch–Gordan coefficients, the point-group representations are needed [14]. For practical purposes it is necessary to use the tables of these coefficients given for all point groups in [1] (in which, and in other works, they are called *coupling coefficients*). For each point group, tables are given corresponding to the possible products $\Gamma_1 \times \Gamma_2$.

Let us consider as an example the **O** group product $E \times T_2$. The Clebsch–Gordan coefficients are shown in Table 5.8 [14]. To the left of the vertical line are shown the six basis functions of the direct product $E\times T_2$ ($\gamma_1 = u, v$; $\gamma_2 = \xi, \eta, \zeta$; $\gamma_1\gamma_2 = u\xi, u\eta, u\zeta, v\xi, v\eta, v\zeta$). The column headings give the bases of the irreducible representations $\Gamma = T_1(\alpha, \beta, \gamma)$ and $T_2(\xi, \eta, \zeta)$ ($E\times T_2 \doteq T_1 + T_2$). The body of the table shows the Clebsch–Gordan coefficients $\langle EuT_2\xi|T_1\alpha\rangle = -\frac{1}{2}\sqrt{3}$, $\langle EuT_2\zeta|T_1\beta\rangle = 0$ etc. From (Table 5.8) we obtain all the

**Table 5.8** Clebsch–Gordan coefficients for the **O** group product $E\times T_2$

| $E\times T_2$ | | $\Gamma$ | $T_1$ | | | $T_2$ | | |
|---|---|---|---|---|---|---|---|---|
| $\gamma_1$ | $\gamma_2$ | $\gamma$ | $\alpha$ | $\beta$ | $\gamma$ | $\xi$ | $\eta$ | $\zeta$ |
| $u$ | $\xi$ | | $-\frac{1}{2}\sqrt{3}$ | 0 | 0 | $-\frac{1}{2}$ | 0 | 0 |
| | $\eta$ | | 0 | $\frac{1}{2}\sqrt{3}$ | 0 | 0 | $-\frac{1}{2}$ | 0 |
| | $\zeta$ | | 0 | 0 | 0 | 0 | 0 | 1 |
| $v$ | $\xi$ | | $-\frac{1}{2}$ | 0 | 0 | $\frac{1}{2}\sqrt{3}$ | 0 | 0 |
| | $\eta$ | | 0 | $-\frac{1}{2}$ | 0 | 0 | $-\frac{1}{2}\sqrt{3}$ | 0 |
| | $\zeta$ | | 0 | 0 | 1 | 0 | 0 | 0 |

basis functions of $E \times T_2$. The coefficients corresponding to the function $\psi(\Gamma\gamma)$ are located in the column headed by $\Gamma\gamma$ (for example those for $\psi(T_1\beta)$ are outlined by the vertical dashed box): thus

$$\left.\begin{aligned} \psi(T_1\alpha) &= -\tfrac{1}{2}\sqrt{3}\,\varphi(Eu)\varphi(T_2\xi) - \tfrac{1}{2}\varphi(Ev)\varphi(T_2\xi), \\ \psi(T_1\beta) &= \tfrac{1}{2}\sqrt{3}\,\varphi(Eu)\varphi(T_2\eta) - \tfrac{1}{2}\varphi(Ev)\varphi(T_2\eta), \\ \psi(T_2\gamma) &= \varphi(Ev)\varphi(T_2\zeta), \qquad \text{etc.} \end{aligned}\right\} \tag{5.17}$$

It should be noted that the functions $\psi(\Gamma\gamma)$ are normalized if the original products $\varphi(\Gamma_1\gamma_1)\varphi(\Gamma_2\gamma_2)$ are normalized and orthogonal. Tables of the Clebsch–Gordan coefficients for **O** may be also used for $\mathbf{O}_h$ if the parity rules are taken into account. Table 5.8 is valid for the three products $E_u \times T_{2u} \doteq T_{1g} + T_{2g}$, $E_u \times T_{2g} = E_g \times T_{2u} \doteq T_{1u} + T_{2u}$ and $E_g \times T_{2g} \doteq T_{1g} + T_{2g}$. For example, the basis function $\psi(T_{1u}x)$ of the set $E_u \times T_{2g}$ has the form

$$\psi(T_{1u}x) = -\tfrac{1}{2}\sqrt{3}\,\varphi(E_u u)\varphi(T_{2g}\xi) - \tfrac{1}{2}\varphi(E_u v)\varphi(T_{2g}\xi).$$

The Clebsch–Gordan coefficients satisfy the *orthogonality relations*

$$\sum_{\gamma_1,\gamma_2}\langle\Gamma\gamma|\Gamma_1\gamma_1\Gamma_2\gamma_2\rangle\langle\Gamma_1\gamma_1\Gamma_2\gamma_2|\Gamma'\gamma'\rangle = \delta(\Gamma\Gamma')\delta(\gamma\gamma'), \tag{5.18}$$

$$\sum_{\Gamma,\gamma}\langle\Gamma_1\gamma_1\Gamma_2\gamma_2|\Gamma\gamma\rangle\langle\Gamma\gamma|\Gamma_1\gamma_1'\Gamma_2\gamma_2'\rangle = \delta(\gamma_1\gamma_1')\delta(\gamma_2\gamma_2'), \tag{5.19}$$

where $\delta(\ldots)$ is the Kronecker symbol ($\delta(\Gamma\Gamma') = 1$ if $\Gamma = \Gamma'$ and $\delta(\Gamma\Gamma') = 0$ if $\Gamma \neq \Gamma'$).

These relations make it possible to derive an *inverse transformation*; that is, to express the basis $\varphi(\Gamma_1\gamma_1)\varphi(\Gamma_2\gamma_2)$ in terms of the functions $\psi(\Gamma\gamma)$:

$$\varphi(\Gamma_1\gamma_1)\varphi(\Gamma_2\gamma_2) = \sum_{\Gamma,\gamma}\psi(\Gamma\gamma)\langle\Gamma\gamma|\Gamma_1\gamma_1\Gamma_2\gamma_2\rangle. \tag{5.20}$$

Table 5.8 allows direct use of (5.20): the functions $\varphi(\Gamma_1\gamma_1)\varphi(\Gamma_2\gamma_2)$ can be read along the rows. For example, the product $\varphi(Ev)\psi(T_2\eta)$ (outlined by the horizontal dashed box) has the form

$$\varphi(Ev)\varphi(T_2\eta) = -\tfrac{1}{2}\psi(T_1\beta) - \tfrac{1}{2}\sqrt{3}\,\psi(T_2\eta). \tag{5.21}$$

The transformation (5.20) turns out to be useful for many applications.

### 5.4.5 Wigner coefficients

Just as in the case of finite point groups, we may for the rotation group introduce the concept of the direct product of representations. The basis of the irreducible representations of the rotation groups and **K** and $\mathbf{K}_h$ comprises the eigenfunctions of the squared angular momentum operator $\hat{j}^2$ and its projection $\hat{j}_z$. For electron orbital angular momentum $j = l$, and the basis functions are the spherical functions $Y_{lm}$:

$$\hat{\mathbf{l}}^2 Y_{lm}(\vartheta, \varphi) = l(l+1) Y_{lm}(\vartheta, \varphi), \qquad \hat{l}_z Y_{lm}(\vartheta, \varphi) = m Y_{lm}(\vartheta, \varphi). \tag{5.22}$$

In many-electron atoms $j = L$ is the total orbital angular momentum quantum number:

$$\hat{L}^2 \psi_{LM} = L(L+1)\psi_{LM}, \qquad \hat{L}_z \psi_{LM} = M\psi_{LM}, \tag{5.23}$$

where $\psi_{LM}$ are the orbital wavefunctions. The spin functions $\chi(Sm)$ are the eigenfunctions of the operators $\hat{S}^2$ and $\hat{S}_z$:

$$\hat{S}^2 \chi(Sm) = S(S+1)\chi(Sm), \qquad \hat{S}_z \chi(Sm) = m\chi(Sm). \tag{5.24}$$

Finally, $j$ may be the total angular momentum quantum number arising from the orbital and spin angular momenta ($\hat{\mathbf{j}} = \hat{\mathbf{l}} + \hat{\mathbf{s}}$ or $\hat{\mathbf{J}} = \hat{\mathbf{L}} + \hat{\mathbf{S}}$). In all of these cases the angular momentum quantum number enumerates the irreducible representations $\mathrm{D}^{(j)}$ of the rotation group.

The direct product of two irreducible representations $\mathrm{D}^{(l_1)}$ and $\mathrm{D}^{(l_2)}$ (for definiteness, we shall consider the orbital angular momenta of two electrons) is formed by the $(2l_1 + 1)(2l_2 + 1)$-dimensional basis $Y_{l_1 m_1} Y_{l_2 m_2}$. This direct product is denoted by $\mathrm{D}^{(l_1)} \times \mathrm{D}^{(l_2)}$ and is reducible. As in the point groups (Section 5.4.3), it is necessary, first, to classify the irreducible representations $\mathrm{D}^{(L)}$ contained in the direct product $\mathrm{D}^{(l_1)} \times \mathrm{D}^{(l_2)}$ and, secondly, to construct the basis functions of the representations $\mathrm{D}^{(L)}$.

The first problem is solved using the summation rule for angular momentum, (5.1):

$$L = l_1 + l_2,\ l_1 + l_2 - 1,\ \ldots,\ l_1 - l_2 \quad (l_1 > l_2). \tag{5.25}$$

The formula for reduction of the direct product has the form

$$\mathrm{D}^{(l_1)} \times \mathrm{D}^{(l_2)} \doteq \mathrm{D}^{(l_1+l_2)} + \mathrm{D}^{(l_1+l_2-1)} + \ldots + \mathrm{D}^{(l_1-l_2)} \quad (l_1 > l_2). \tag{5.26}$$

The basis functions $\psi_{LM}$ are expressed as linear combinations:

$$\psi_{LM} = \sum_{m_1 m_2} \langle l_1 m_1 l_2 m_2 | LM \rangle Y_{l_1 m_1} Y_{l_2 m_2}, \tag{5.27}$$

where the quantities $\langle l_1 m_1 l_2 m_2 | LM \rangle$ used as Clebsch–Gordan coefficients of the rotation group are called the *Wigner coefficients*.†

† Very often these coefficients are also called Clebsch–Gordan coefficients or vector-coupling coefficients.

It should be noted that (5.26) and (5.27) are applicable for the summation of angular momenta of any type.

The *Racah V coefficients and Wigner 3j symbols*:

$$\left.\begin{aligned} \langle l_1 m_1 l_2 m_2 | LM\rangle &= (-1)^{-l_1+l_2-M}(2L+1)^{\frac{1}{2}}\begin{pmatrix} l_1 & l_2 & L \\ m_1 & m_2 & -M \end{pmatrix}, \\ \langle l_1 m_1 l_2 m_2 | LM\rangle &= (-1)^{L+M}(2L+1)^{\frac{1}{2}}V(l_1 l_2 L;\, m_1 m_2 - M). \end{aligned}\right\} \qquad (5.28)$$

The Wigner coefficients satisfy the orthogonality conditions

$$\sum_{L,M}\langle l_1 m_1 l_2 m_2 | LM\rangle\langle LM | l_1 m_1' l_2 m_2'\rangle = \delta(m_1 m_1')\delta(m_2 m_2'), \qquad (5.29)$$

$$\sum_{m_1,m_2}\langle LM | l_1 m_1 l_2 m_2\rangle\langle l_1 m_1 l_2 m_2 | L'M'\rangle = \delta(LL')\delta(MM'), \qquad (5.30)$$

and are nonzero if $|l_1 - l_2| \leq L \leq l_1 + l_2$ and $M = m_1 + m_2$.

Methods for their calculation are described in books on group theory (see e.g. [12, 20, 39–43]). For certain quantum numbers $l_1$, $l_2$ and $L$ there exist simple analytical expressions for the Wigner coefficients [2]; for $l_2 = \frac{1}{2}$ and $l_2 = 1$ they are given in Tables 5.9 and 5.10 [39].

When summing the spin momenta of two electrons, $j_1 = s_1 = \frac{1}{2}$, $j_2 = s_2 = \frac{1}{2}$, $j = s$, and $S = 0$ and 1. The results obtained in this particular case from Table 5.10 are given in Table 5.11.

On substituting $S = 0$, $M = 0$ and $S = 1$, $M = -1, 0, 1$, we obtain the known spin functions $\chi(SM)$ of the helium atom or the hydrogen molecule:

$$\chi(00) = \sqrt{1/2}[\alpha(1)\beta(2) - \alpha(2)\beta(1)], \qquad (5.31)$$

$$\left.\begin{aligned} \chi(11) &= \alpha(1)\alpha(2), \\ \chi(10) &= \sqrt{1/2}[\alpha(1)\beta(2) + \alpha(2)\beta(1)], \\ \chi(1-1) &= \beta(1)\beta(2). \end{aligned}\right\} \qquad (5.32)$$

A detailed description of the quantum theory of angular momentum is given

**Table 5.9** Wigner coefficients $\langle l_1 m_1 \frac{1}{2} m_2 | LM\rangle$ ($l_2 = \frac{1}{2}$, $M = m_1 + m_2$).

| $L$ | $m_2 = \frac{1}{2}$ | $m_2 = -\frac{1}{2}$ |
|---|---|---|
| $l_1 + \frac{1}{2}$ | $\left(\frac{l_1 + M + \frac{1}{2}}{2l_1 + 1}\right)^{1/2}$ | $\left(\frac{l_1 - 1 + \frac{1}{2}}{2l_1 + 1}\right)^{1/2}$ |
| $l_1 - \frac{1}{2}$ | $-\left(\frac{l_1 - M - \frac{1}{2}}{2l_1 + 1}\right)^{1/2}$ | $\left(\frac{l_1 + M + \frac{1}{2}}{2l_1 + 1}\right)^{1/2}$ |

**Table 5.10** Wigner coefficients $\langle l_1 m_1 1 m_2 | LM \rangle$ $(l_2 = 1, M = m_1 + m_2)$.

| $L$ | $m_2 = 1$ | $m_2 = 0$ | $m_2 = -1$ |
|---|---|---|---|
| $l_1 + 1$ | $\left[\frac{(l_1+M)(l_1+M+1)}{(2l_1+1)(2l_2+2)}\right]^{1/2}$ | $\left[\frac{(l_1-M+1)(l_1+m+1)}{(2l_1+1)(l_1+1)}\right]^{1/2}$ | $\left[\frac{(l_1-M)(l_1-M+1)}{(2l_1+1)(2l_2+2)}\right]^{1/2}$ |
| $l_1$ | $-\left[\frac{(l_1+M)(l_1-M+1)}{2l_1(l_1+1)}\right]^{1/2}$ | $\left[\frac{M}{l_1(l_1+1)}\right]^{1/2}$ | $\left[\frac{(l_1-M)(l_1+M+1)}{2l_1(l_1+1)}\right]^{1/2}$ |
| $l_1 - 1$ | $\left[\frac{(l_1-M)(l_1-M+1)}{2l_1(2l_1+1)}\right]^{1/2}$ | $-\left[\frac{(l_1-M)(l_1+M)}{l_1(2l_1+1)}\right]^{1/2}$ | $\left[\frac{(l_1+M+1)(l_1+M)}{2l_1(2l_1+1)}\right]^{1/2}$ |

**Table 5.11** Wigner coefficients $\langle \frac{1}{2}m_1 \frac{1}{2}m_2|SM\rangle$ $(M = m_1 + m_2)$.

| $S$ | $m_2 = \frac{1}{2}$ | $m_2 = -\frac{1}{2}$ |
|---|---|---|
| 1 | $[\frac{1}{2}(1+M)]^{1/2}$ | $[\frac{1}{2}(1-M)]^{1/2}$ |
| 0 | $-[\frac{1}{2}(1-M)]^{1/2}$ | $[\frac{1}{2}(1+M)]^{1/2}$ |

**Table 5.12** Some Wigner coefficients $C^{2m}_{5/2m_1\,3/2m_2}$.

| $j_1$ | $m_1$ | $j_2$ | $m_2$ | | $C^{2m}_{5/2m_1\,3/2m_2}$ |
|---|---|---|---|---|---|
| $\frac{5}{2}$ | $\frac{5}{2}$ | $\frac{3}{2}$ | $-\frac{1}{2}$ | $\sqrt{2\times 5}/\sqrt{3\times 7}$ | 0.690 066 |
| $\frac{5}{2}$ | $\frac{5}{2}$ | $\frac{3}{2}$ | $-\frac{3}{2}$ | $\sqrt{5}/\sqrt{2\times 7}$ | 0.597 616 |
| $\frac{5}{2}$ | $\frac{3}{2}$ | $\frac{3}{2}$ | $\frac{1}{2}$ | $-2\sqrt{2}/\sqrt{3\times 7}$ | −0.617 213 |
| $\frac{5}{2}$ | $\frac{3}{2}$ | $\frac{3}{2}$ | $-\frac{1}{2}$ | $1/\sqrt{2\times 3\times 7}$ | 0.154 303 |
| $\frac{5}{2}$ | $\frac{3}{2}$ | $\frac{3}{2}$ | $-\frac{3}{2}$ | $\sqrt{3}/\sqrt{7}$ | 0.654 654 |

in [2, 39, 40]. The most complete information on the Wigner coefficients† is provided in [2, 39–43]. The most convenient tables for practical use are those in [2], where the commonly accepted notation

$$C^{jm}_{j_1m_1j_2m_2} = \langle j_1m_1j_2m_2|jm\rangle = (-1)^{j_1+j_2-j}\langle j_2m_2j_1m_1|jm\rangle \tag{5.33}$$

is adopted and the values of $C^{\cdot\cdot}_{\cdot\cdot\cdot}$ are given for $j_1, j_2, j \leq 3$. As an example, consider Table 5.12, which is part of one of the tables from [2]. From this we obtain

$$\langle \tfrac{5}{2}\tfrac{5}{2}\tfrac{3}{2} - \tfrac{1}{2}|22\rangle = \sqrt{2\times 5}/\sqrt{3\times 7} = 0.690\,066, \ \text{etc.}$$

The following relation should be noted

$$C^{jm}_{j_1m_1j_2m_2} = (-1)^{j_1+j_2-j}\,C^{j-m}_{j_1-m_1j_2-m_2} = C^{j-m}_{j_2-m_2j_1-m_1}$$

There are also numerous algebraic relations and formulae for sums, containing the products of two, three and four Wigner coefficients [2], that are very useful for applications. Many of the latter are based on the so-called graphical methods of spin algebra [43], which considerably simplify calculations in atomic and nuclear many-particle problems.

## 5.5 TWO-ELECTRON TERMS IN A STRONG CUBIC FIELD

The concept of the direct product of irreducible representation discussed in Section 5.4 makes it possible to determine the crystal terms of two electrons and

† Clebsch–Gordan coefficients and Wigner $3j$-symbols are both used, the last are the more symmetric relative to the permutations of symbols.

to construct the wavefunctions corresponding to these terms. Let us consider the $t_{2g}^2$ configuration, whose wavefunctions consist of the determinants

$$|\varphi_{t_{2g}\gamma_1}(1)\chi(\tfrac{1}{2}m_1)\ \varphi_{t_{2g}\gamma_2}(2)\chi(\tfrac{1}{2}m_2)|, \tag{5.34}$$

denoted below for brevity as $|\varphi(t_2\gamma_1 m_1)\ \varphi(t_2\gamma_2 m_2)|$. The determinants (5.34) form a basis for the $\mathbf{O}_h$ group direct product $T_{2g} \times T_{2g}$ in the electron coordinate space as well as the basis of the direct product $D^{(1/2)} \times D^{(1/2)}$ in the rotation group in spin space. Let us decompose these direct products into irreducible parts:

$$\left.\begin{aligned} T_{2g} \times T_{2g} &\doteq A_{1g} + E_g + T_{1g} + T_{2g}, \\ D^{(1/2)} \times D^{(1/2)} &\doteq D^{(0)} + D^{(1)}. \end{aligned}\right\} \tag{5.35}$$

The first equation determines the orbital states of the two electrons, while the second determines the total spins $S = 0$ and 1. Equations (5.35) indicate the following terms of the $t_{2g}^2$ configuration:

$$\begin{aligned} &{}^1A_{1g}(1), \quad {}^1E_g(2), \quad {}^1T_{1g}(3), \quad {}^1T_{2g}(3), \\ &{}^3A_{1g}(3), \quad {}^3E_g(6), \quad {}^3T_{1g}(9), \quad {}^3T_{2g}(9), \end{aligned}$$

where the degrees of degeneracy are shown in parentheses, the spin being taken into account.

The classification of terms thus obtained does not take into account the Pauli principle, which prohibits certain states. Therefore the total degree of degeneracy (36) exceeds the number of allowed states (15) determined in Section 5.3 for the $t_{2g}^2$ configuration. We shall postpone for a while the discussion of the problem of classification of allowed terms, and pass to the construction of the strong-crystal-field wavefunctions.

Let us multiply the function (5.34) by the Clebsch–Gordan coefficient $\langle T_{2g}\gamma_1 T_{2g}\gamma_2|\Gamma\gamma\rangle$ and sum over $\gamma_1, \gamma_2$ ($\gamma_1, \gamma_2 = \xi, \eta, \zeta$):

$$\sum_{\gamma_1,\gamma_2} |\varphi(t_{2g}\gamma_1 m_1)\varphi(t_{2g}\gamma_2 m_2)|\langle T_{2g}\gamma_1 T_{2g}\gamma_2|\Gamma\gamma\rangle. \tag{5.36}$$

The wavefunction obtained, using the definition (5.15) of the Clebsch–Gordan coefficient, is transformed according to the representation $\Gamma$ (which is contained in $T_{2g} \times T_{2g}$). Multiplying (5.36) by the Wigner coefficient $\langle \tfrac{1}{2}m_1 \tfrac{1}{2}m_2|SM\rangle$ and summing over $m_1$ and $m_2$ ($m_1, m_2 = \pm\tfrac{1}{2}$), we obtain [14]

$$\Psi(t_{2g}^2 S\Gamma M\gamma) = \sum_{m_1,m_2}\sum_{\gamma_1,\gamma_2} |\varphi(t_2\gamma_1 m_1)\varphi(t_2\gamma_2 m_2)|\langle T_{2g}\gamma_1 T_{2g}\gamma_2|\Gamma\gamma\rangle\langle \tfrac{1}{2}m_1\tfrac{1}{2}m_2|SM\rangle. \tag{5.37}$$

The wavefunctions $\Psi(t_2^2 S\Gamma M\gamma)$ satisfy all the necessary requirements:

(1) they form the basis of an irreducible representation $\Gamma$ from $T_{2g} \times T_{2g}$ and transform like the basis vector with index $\gamma$ (or $\gamma$-th basis function of the irreducible representation $\gamma$, see equation (3.64)).

(2) they correspond to total spin $S = 0$ or 1 and to its projection $M$;

(3) they are antisymmetric under electron permutation (i.e. the Pauli principle is taken into account), which means that, on constructing the wavefunctions according to (5.37), we obtain the allowed terms.

Let us write, for example, the wavefunction of the $^1A_{1g}$ term:

$$\Psi(t_2^2\,{}^1A_{1g}) = \sum_{m_1,m_2}\sum_{\gamma_1,\gamma_2} |\varphi(t_2\gamma_1 m_1)\varphi(t_2\gamma_2 m_2)|\langle T_{2g}\gamma_1 T_{2g}\gamma_2|A_{1g}e_1\rangle\langle\tfrac{1}{2}m_1\tfrac{1}{2}m_2|00\rangle, \tag{5.38}$$

where $e_1$ is the basis of $A_{1g}$. From the tables in [14] we find that only the following Clebsch–Gordan coefficients are nonzero:

$$\left.\begin{aligned}\langle T_{2g}\xi T_{2g}\xi|A_{1g}e_1\rangle &= \langle T_{2g}\eta T_{2g}\eta|A_{1g}e_1\rangle\\ &= \langle T_{2g}\zeta T_{2g}\zeta|A_{1g}e_1\rangle = \sqrt{1/3},\\ \Psi(t_2^2 A_{1g}) &= \sqrt{1/3}\sum_{m_1,m_2}\sum_{\gamma=\xi,\eta,\zeta}|\varphi(t_2\gamma m_1)\varphi(t_2\gamma m_2)|\langle\tfrac{1}{2}m_1\tfrac{1}{2}m_2|00\rangle.\end{aligned}\right\} \tag{5.39}$$

Since for $M = 0$, $m_1 = -m_2 = \frac{1}{2}$ or $m_1 = m_2 = -\frac{1}{2}$, the sum (5.39) contains only the determinants

$$|\xi\ \bar{\xi}| = -|\bar{\xi}\ \xi|, \qquad |\eta\ \bar{\eta}| = -|\bar{\eta}\ \eta|, \qquad |\zeta\ \bar{\zeta}| = -|\bar{\zeta}\ \zeta|,$$

By substituting the Wigner coefficients from Table 5.11, we obtain

$$\Psi(t_2^2\,{}^1A_{1g}) = \sqrt{1/3}(|\xi\ \bar{\xi}| + |\eta\ \bar{\eta}| + |\zeta\ \bar{\zeta}|). \tag{5.40}$$

From (5.39) it can be seen that $\Psi(t_2^2\,{}^3A_{1g}) = 0$. Indeed, the wavefunction of the $^3A_{1g}$ term with $M = 1$ consists of the determinants $|\xi\ \xi|$, $|\eta\ \eta|$ and $|\zeta\ \zeta|$, each of which is zero. Therefore the Pauli principle prohibits the term $^3A_{1g}$. Similarly, we can show that the wavefunctions of the terms $^1T_{1g}$, $^3E_g$ and $^3T_{2g}$ are zero. Only the terms $^1A_{1g}$, $^1E_g$, $^3T_{1g}$ and $^1T_{2g}$ are allowed for the $t_{2g}^2$ shell. Their total degree of degeneracy is 15; that is, it coincides with the number of determinantal functions (5.8).

The application of (5.37) leads to the following expressions for the remaining wavefunctions of the allowed terms of the $d^2$ shell:

$$\left.\begin{aligned}\Psi(t_2^2\,{}^1E_g u) &= \sqrt{1/6}(|\xi\ \bar{\xi}| - |\eta\ \bar{\eta}| + 2|\zeta\ \bar{\zeta}|,\\ \Psi(t_2^2\,{}^1E_g v) &= \sqrt{1/2}(|\xi\ \bar{\xi}| - |\eta\ \bar{\eta}|),\\ \Psi(t_2^2\,{}^3T_{1g}, M = 1, \gamma) &= |\xi\ \eta|,\\ \Psi(t_2^2\,{}^1T_2, \zeta) &= \sqrt{1/2}(|\xi\ \bar{\eta}| + |\eta\ \bar{\zeta}|),\end{aligned}\right\} \tag{5.41}$$

where for the term $^3T_{1g}$ there is given one function corresponding to $\gamma$ row and $M = 1$, while for the term $^1T_{2g}$ only one function is given of the $\zeta$ type.

Formulae of the type (5.37) can be written for any two-electron configuration of the $d^2$ shell:

$$\Psi(t_2^k e^{2-k} S\Gamma M\gamma) = \sum_{m_1,m_2} \sum_{\gamma_1,\gamma_2} |\varphi(\Gamma_1\gamma_1 m_1)\varphi(\Gamma_2\gamma_2 m_2)| \times \langle\Gamma_1\gamma_1\Gamma_2\gamma_2|\Gamma\gamma\rangle\langle\tfrac{1}{2}m_1\tfrac{1}{2}m_2|SM\rangle, \qquad (5.42)$$

where $k = 2\,(t_2^2)$, $1\,(t_2e)$ or $0\,(e^2)$. The wavefunctions are calculated as for the $t_2^2$ configuration. The nonzero functions give the permitted terms of the $t_{2g}^2$, $t_{2g}e_g$ and $e_g^2$ configurations of the two equivalent d electrons shown in Table 5.13. The bottom row of this table gives the terms of the $d^2$ ion.

The classification of allowed states can be obtained by direct calculation of the wavefunctions. This procedure is inconvenient for many-electron ions, since the calculation becomes very cumbersome. In [44, 45] (see also [46]) a group-theoretical classification of permitted terms has been developed. This procedure is based on permutation-group theory [18, 22], and makes it possible to determine permitted terms without the tiresome procedure of constructing all the many-particle wavefunctions. In particular cases it is possible to make use of Bethe's method of symmetry lowering (see [37]). Let us consider the states with $S = 1$ and $M = 1$ of the $t_{2g}^2$ configuration. They can include only three determinants: $|\xi\ \eta|$, $|\xi\ \zeta|$ and $|\eta\ \zeta|$. The rotations $\hat{C}_3$ and $\hat{C}_4$ transform these into each another; hence the spin triplet is at the same time an orbital triplet $^3T_{1g}$ or $^3T_{2g}$. If the symmetry decreases to $\mathbf{D}_{4h}$ then $|\xi\ \eta|$ is the basis of $^3A_{2g}$, while $|\eta\ \zeta|$ and $|\xi\ \zeta|$ comprise the basis of $^3E_g$. From the reduction tables (see Table 4.4) we obtain

$$T_{1g}(\mathbf{O}_h) \doteq A_{2g} + E_g(\mathbf{D}_{4h}), \qquad T_{2g}(\mathbf{O}_h) \doteq B_{2g} + E_g(\mathbf{D}_{4h}).$$

Thus the spin triplet is $^3T_{1g}$, and for the $t_{2g}^2$ configuration we find the terms $^1A_{1g}$, $^1E_g$, $^1T_{2g}$ and $^3T_{1g}$, which coincides with the first line of Table 5.13.

Let us now consider the results obtained assuming a weak crystal field. The atomic $LS$ terms of the two equivalent electrons (Table 5.1) are split in the $\mathbf{O}_h$ field according to the following scheme (see the reduction in Table 4.3):

$$^1S \to {}^1A_{1g}, \quad {}^1D \to {}^1E_g + {}^1T_{2g}, \quad {}^1G \to {}^1A_{1g} + {}^1E_g + {}^1T_{1g} + {}^1T_{2g},$$
$$^3P \to {}^3T_{1g}, \quad {}^3F \to {}^3A_{2g} + {}^3T_{1g} + {}^3T_{2g}.$$

**Table 5.13** Allowed terms of the $d^2$ shell.

| | Terms $S\Gamma$ |
|---|---|
| $t_{2g}^2$ | $^1A_{1g}$, $^1E_g$, $^1T_{2g}$, $^3T_{1g}$ |
| $t_{2g}e_g$ | $^1T_{1g}$, $^1T_{2g}$, $^3T_{1g}$, $^3T_{2g}$ |
| $e_g^2$ | $^1A_{1g}$, $^1E_g$, $^3A_{2g}$ |
| $d^2$ | $2\,^1A_{1g}$, $2\,^1E_g$, $^1T_{2g}$, $2\,^3T_{1g}$, $2\,^1T_{2g}$, $^3A_{2g}$ $^3T_{2g}$. |

The set of crystal terms of the $d^2$ shell in a weak field coincides with that in a strong field, given in the last line of Table 5.13. Therefore the atomic terms in a weak field are correlated with the strong-field terms:

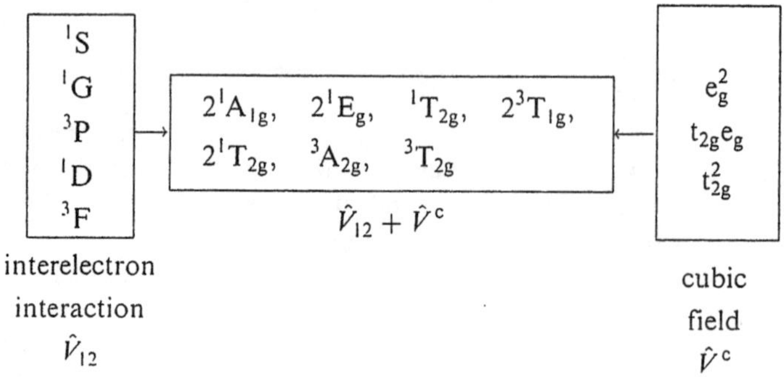


We shall return to this problem when discussing energy levels.

The construction of wavefunctions given above is a general method. The formula (5.42) is also applicable for constructing wavefunctions of two inequivalent electrons. If, for example, one 3d electron is excited to the 4p state then, in (5.42), $\Gamma_1 = E_g$ and $T_{2g}$, while $\Gamma_2 = T_{1u}$. In this case the atomic configuration 3d 4p gives terms determined by the direct products $E_g \times T_{1u} = T_{1u} + T_{2u}$ and $T_{2g} \times T_{1u} = A_{2u} + E_u + T_{1u} + T_{2u}$. All of the terms $2\,^{1,3}T_{1u}$, $2\,^{1,3}T_{2u}$, $^{1,3}A_{2u}$ and $^{1,3}E_u$ are allowed, since in the case of inequivalent electrons the Pauli principle does not lead to prohibition of states. The same result is obtained from the atomic terms $^{1,3}P$, $^{1,3}D$ and $^{1,3}F(3d\,4p)$ in a weak cubic field (Table 5.2).

## 5.6 ENERGY LEVELS OF A TWO-ELECTRON d ION IN THE STRONG-CRYSTAL-FIELD SCHEME

### 5.6.1 Nonrepeating representations

The wavefunctions $\Psi(\alpha S\Gamma M\gamma)$ $(\alpha = t_2^2 e^{2-k})$ allow the computation of energy levels. Let us consider the $t_2^2$ configuration. The functions $\Psi(t_2^2 S\Gamma M\gamma)$ take account of the cubic field exactly and are appropriate functions in the zeroth-order approximation relative to the interelectron interaction $\hat{V}_{12}$, which has been omitted. Corrections to the energy are calculated as diagonal matrix elements:

$$\Delta\varepsilon(t_2^2 S\Gamma) = \langle t_2^2 S\Gamma M\gamma|\hat{V}_{12}|t_2^2 S\Gamma M\gamma\rangle. \tag{5.43}$$

To the $t_2^2$-configuration terms, to first order of perturbation theory, there correspond the following energies:

$$\varepsilon(t_2^2 S\Gamma) = -8Dq + \Delta\varepsilon(t_2^2 S\Gamma). \tag{5.44}$$

**Table 5.14** First-order perturbation-theory corrections for the $d^2$-ion energies in a strong field [14, 46, 47].

| $t_2^2 S\Gamma$ | $\Delta\varepsilon(t_2^2 S\Gamma)$ | $t_2 e S\Gamma$ | $\Delta\varepsilon(t_2 e S\Gamma)$ | $e^2 S\Gamma$ | $\Delta\varepsilon(e^2 S\Gamma)$ |
|---|---|---|---|---|---|
| $^3T_{1g}$ | A − 5B | $^3T_{2g}$ | A − 8B | $^3A_{2g}$ | A − 8B |
| $^1T_{2g}$ | A + B + 2C | $^3T_{1g}$ | A + 4B | $^1E_g$ | A + 2C |
| $^1E_g$ | A + B + 2C | $^1T_{2g}$ | A + 2C | $^1A_{1g}$ | A + 8B 4C |
| $^1A_{1g}$ | A + 10B + 5C | $^1T_{1g}$ | A + 4B + 2C | | |

Similarly, it is possible to represent the term energies of the remaining configurations:

$$\varepsilon(t_2 e S\Gamma) = +2Dq + \Delta\varepsilon(t_2 e S\Gamma), \tag{5.45}$$

$$\varepsilon(e^2 S\Gamma) = +12Dq + \Delta\varepsilon(e^2 S\Gamma). \tag{5.46}$$

The remainder of the calculation reduces to the computation of interelectron interaction-matrix elements using the determinantal wavefunctions. Appropriate methods are given in [14, 16], and are not directly related to group theory. Therefore we shall present only the results. Energy corrections are expressed in the two-electron Coulomb and exchange integrals or the so-called *Slater integrals*—the *Slater–Condon parameters* $F_0$, $F_2$ and $F_4$. It is often more convenient to use the *Racah parameters* $A$, $B$ and $C$ [14] generally accepted in spectroscopy. The quantities $\Delta\varepsilon(t_2^{2-k}e^k S\Gamma)$ are listed in Table 5.14 in increasing order of magnitude.

Thus, to first order of perturbation theory, the interelectron interaction splits the strong-field configurations (Fig. 5.2b). According to the Wigner theorem, to the energy levels there correspond irreducible representations of $\mathbf{O}_h$. The only exceptions are the terms $^1T_{2g}(t_2^2)$ and $^1E_g(t_2^2)$, which have equal energies. This additional degeneracy should be referred to as *accidental* (see Secton 4.2). It does not imply the violation of the Wigner theorem: accidental degeneracy results simply from the fact that the point-ion model of the crystal field is approximate. In a more exact model taking into account covalence, for example, the degeneracy is removed.

The levels calculated at first order in perturbation theory describe the spectrum well in the case of a very strong field $10Dq \gg A, B, C$.

In this section we have considered the case of *nonrepeating terms*—in each strong-field configuration the $S\Gamma$ term is encountered only once. A more accurate calculation of the energy is achieved when so-called *repeating terms* are taken into account.

### 5.6.2 Configuration mixing

Before passing to a more accurate energy calculation, we shall make some important points.

(1) The interelectron interaction $\hat{V}_{12} = e^2/r_{12} = \hat{V}_{A_{1g}}$ is invariant under all the transformations of any point group, either infinite and finite, in particular those of $\mathbf{O}_h$. Indeed, the distance between electrons $r_{12}$ remains the same under any transformation; hence the *operator* $\hat{V}_{12}$ *is spherically symmetric* and belongs to the totally symmetric representation. It is called a *scalar in the group*, which is reflected in the notation $\hat{V}_{A_{1g}}$.

(2) For any scalar operator all matrix elements calculated using the basis functions of *different* irreducible representations $\Gamma_1$ and $\Gamma_2$ are zero:

$$\langle \Gamma_1\gamma_1|\hat{V}_{A_{1g}}|\Gamma_2\gamma_2\rangle = 0 \quad \text{if } \Gamma_1 \neq \Gamma_2. \tag{5.47}$$

(3) The matrix elements of a scalar operator are nonzero only if the irreducible representations $\Gamma_1$ and $\Gamma_2$ are the same:

$$\langle \Gamma_1\gamma_1|\hat{V}_{1g}|\Gamma_2\gamma_2\rangle \neq 0 \quad \text{only if } \Gamma_1 = \Gamma_2. \tag{5.48}$$

Here the indices 1 and 2 indicate different basis sets of one representation. As an example we may take two basis sets of the representation $T_{2g}$ formed by wavefunctions of the configurations $t_2^2$ and $t_2e$. This also applies to the case of a single basis set. In particular, the diagonal matrix elements (5.43) are corrections to the energy.

(4) The matrix elements of the scalar operator (5.48) are nonzero and equal only for the basis functions which transform in the same way, $\gamma_1 = \gamma_2$. Suppose, for example, that $E_1$ and $E_2$ are two E representations of $\mathbf{O}$. This means that

$$\left.\begin{aligned} \langle E_1u|\hat{V}_{A_1}|E_2u\rangle &= \langle E_1v|\hat{V}_{A_1}|E_2v\rangle, \\ \langle E_1u|\hat{V}_{A_1}|E_2v\rangle &= \langle E_1v|\hat{V}_{A_1}|E_2u\rangle = 0. \end{aligned}\right\} \tag{5.49}$$

The above properties of the matrix elements will be proved later using the Wigner–Eckart theorem (Section 9.2). Now it is clear that in the matrix of the operator $\hat{V}_{12}$ the only nonzero elements are those that relate the repeating representations from different electronic configurations. It follows from Table 5.14 that for a $d^2$ ion only twice-repeating representations exist:

$$^3T_{1g}(t_2^2, t_2e), \quad ^1T_{2g}(t_2^2, t_2e), \quad ^1E_g(t_2^2, e^2), \quad ^1A_{1g}(t_2^2, e^2).$$

In the case of orbital singlets $^1A_1$ the nonzero matrix element of the so-called *configuration mixing element* $\langle t_2^2\,{}^1A_1|\hat{V}_{12}|e^2\,{}^1A_1\rangle$, is given in terms of the Racah parameters [14] as

$$\langle t_2^2\,{}^1A_1|\hat{V}_{12}|e^2\,{}^1A_1\rangle = \sqrt{6}(2B + C).$$

When configuration mixing is taken into account, the energy is determined from the second-order secular equation:

$$\begin{matrix} t_2^2 \\ e^2 \end{matrix}\begin{vmatrix} -8Dq + \Delta\varepsilon(t_2^2\,{}^1A_1) - \varepsilon & \langle t_2^2\,{}^1A_1|\hat{V}_{12}|e^2\,{}^1A_1\rangle \\ \langle e_2^2\,{}^1A_1|\hat{V}_{12}|t_2^2\,{}^1A_1\rangle & 12Dq + \Delta\varepsilon(e^2\,{}^1A_1) - \varepsilon \end{vmatrix} = 0. \quad (5.50)$$

Inserting $\Delta\varepsilon$ (Table 5.14) into this equation and solving, we obtain the new energies $\varepsilon_a$ and $\varepsilon_b$ of the two ${}^1A_{1g}$ terms. The wavefunctions taking account of configuration mixing are of the form

$$\left.\begin{aligned} \Psi(a\,{}^1A_1) &= \cos\vartheta\,\Psi(t_2^2\,{}^1A_1) - \sin\vartheta\,\Psi(e^2\,{}^1A_1), \\ \Psi(b\,{}^1A_1) &= \sin\vartheta\,\Psi(t_2^2\,{}^1A_1) + \cos\vartheta\,\Psi(e^2\,{}^1A_1), \end{aligned}\right\} \quad (5.51)$$

where the angle $\vartheta$ is given by the equation

$$\tan 2\vartheta = \frac{2\langle t_2^2\,{}^1A_1|\hat{V}_{12}|e^2\,{}^1A_1\rangle}{20Dq + \Delta\varepsilon(e^2\,{}^1A_1) - \Delta\varepsilon(t_2^2\,{}^1A_1)}. \quad (5.52)$$

The symbols $a$ and $b$ enumerate the repeating representations ${}^1A_1$ that, on taking mixing into account, are not related to a certain configuration ($t_2^2$ or $e^2$) in the strong field. The energy corrections imply "*term repulsion*" as shown schematically in Fig. 5.4. It follows from statement (3) above that repulsion takes place only between terms of identical symmetry, that is, between those corresponding to repeating irreducible representations.

Let us pass now to *degenerate repeating terms* and consider the simplest case of the doublets ${}^1E_g(t_{2g}^2)$ and ${}^1E_g(e_g^2)$. The fourth order secular equation of the operator $\hat{V}_{12}$ has the form

$$\begin{array}{c|cc:cc|} & t_{2g}^2\,{}^1E_g u & t_{2g}^2\,{}^1E_g v & e_g^2\,{}^1E_g u & e_g^2\,{}^1E_g v \\ t_{2g}^2\,{}^1E_g u & \Delta_1 - \varepsilon & 0 & Q & 0 \\ t_{2g}^2\,{}^1E_g v & 0 & \Delta_1 - \varepsilon & 0 & Q \\ \hdashline e_g^2\,{}^1E_g u & Q & 0 & \Delta_2 - \varepsilon & 0 \\ e_g^2\,{}^1E_g v & 0 & Q & 0 & \Delta_2 - \epsilon \end{array} = 0. \quad (5.53)$$

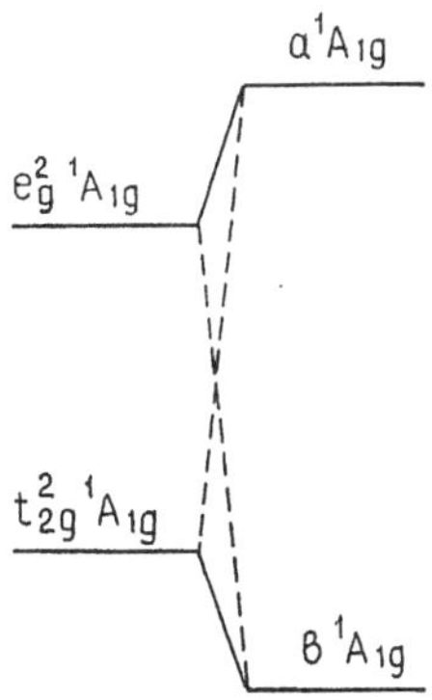


**Fig. 5.4** "Repulsion" of terms of equal symmetry; the dashed lines show state mixing.

where the notations

$$\Delta_1 = -8Dq + \Delta\epsilon(t_{2g}^2\,{}^1E_g),$$
$$\Delta_2 = 12Dq + \Delta\epsilon(e_g^2\,{}^1E_g)$$

are used. The matrix elements $\langle t_{2g}^2\,{}^1E_g u|\hat{V}_{12}|e_g^2\,{}^1E_g u\rangle$ and $\langle t_{2g}^2\,{}^1E_g v|\hat{V}_{12}|e_g^2\,{}^1E_g v\rangle$ are denoted by $Q$; they are equal according to statement (4) above, from which it also follows that all the other matrix elements vanish. The configuration mixing matrix of degenerate terms has a characteristic structure: its two nondiagonal blocks are multiples of 2 × 2 unit matrices. By changing the basis function sequence in the matrix (5.53) we obtain the block structure of the secular equation:

$$\begin{array}{c|cc|cc|} & t_{2g}^2\,{}^1E_g u & e_g^2\,{}^1E_g u & t_{2g}^2\,{}^1E_g v & e_g^2\,{}^1E_g v \\ t_{2g}^2\,{}^1E_g u & \Delta_1-\varepsilon & Q & 0 & 0 \\ e_g^2\,{}^1E_g u & Q & \Delta_2-\varepsilon & 0 & 0 \\ \hline t_{2g}^2\,{}^1E_g v & 0 & 0 & \Delta_1-\varepsilon & Q \\ e_g^2\,{}^1E_g v & 0 & 0 & Q & \Delta_2-\varepsilon \end{array} = 0. \qquad (5.54)$$

Both 2 × 2 blocks are identical and their roots coincide in pairs. In practice, this means that E$u$ states are subject to "repulsion" similarly to E$v$ states. In other words, configuration mixing does not remove the degeneracy of the repeating terms. This conclusion is not unexpected. Indeed, reduction of the representation occurs only when the symmetry is reduced (Section 4.3.1), while the perturbation symmetry $\hat{V}_{12}$ (spherical symmetry) is higher than the $\mathbf{O}_h$ symmetry of the unperturbed Hamiltonian.

Similarly, we can show that, for the three repeating doublets $E_1$, $E_2$ and $E_3$, with energies $\varepsilon_1$, $\varepsilon_2$ and $\varepsilon_3$, the secular equation of configuration mixing is

$$\begin{array}{c|cccccc|} u & \varepsilon_1-\varepsilon & 0 & Q_1 & 0 & Q_2 & 0 \\ v & 0 & \varepsilon_1-\varepsilon & 0 & Q_1 & 0 & Q_2 \\ u & Q_1 & 0 & \varepsilon_2-\varepsilon & 0 & Q_3 & 0 \\ v & 0 & Q_1 & 0 & \varepsilon_2-\varepsilon & 0 & Q_3 \\ u & Q_2 & 0 & 0 & Q_3 & \varepsilon_3-\varepsilon & 0 \\ v & 0 & Q_2 & Q_3 & 0 & 0 & \varepsilon_3-\varepsilon \end{array} = 0, \qquad (5.55)$$

where $Q_1$, $Q_2$ and $Q_3$ are the *independent* mixing matrix elements. Transition to a new basis splits the matrix $V_{12}$ into two identical 3 × 3 blocks:

$$\begin{array}{c|ccc|} E_1\gamma & \varepsilon_1-\varepsilon & Q_1 & Q_2 \\ E_2\gamma & Q_1 & \varepsilon_2-\varepsilon & Q_3 \\ E_3\gamma & Q_2 & Q_3 & \varepsilon_3-\varepsilon \end{array} = 0 \quad (\gamma = u \text{ or } v). \qquad (5.56)$$

We thus have a general law, stating that *the order of a secular equation is equal to the number of repeating representations*, but by no means to their total degree of degeneracy. It should be noted that further simplification of the configuration mixing matrix is not possible. Equations of the type (5.56) can be solved easily by numerical methods.

Up to now we have been considering spin singlets ($S = 0$). The existence of spin does not introduce any additional difficulties, since *the operator* $\hat{V}_{12}$ *connects only states with equal spin* ($S$) *projections*. Consequently, for example, the $18 \times 18$ matrix of mixing terms $^3T_{1g}(t_2^2)$ and $^3T_{1g}(t_2e)$ reduces to nine identical two-dimensional blocks (according to the number of repeating representations).

## 5.7 MANY-ELECTRON TERMS IN A STRONG CUBIC FIELD

### 5.7.1 Classification of three-electron terms

The three-electron wavefunctions are constructed using products of irreducible representations (Section 5.5). Consider a two-electron configuration $t_2^2$ and introduce into the system another electron in the $t_2$ shell. Its wavefunction is transformed according to the representation $T_2$ (the parity index is omitted for brevity). We shall consider three cases:

(1) the three electrons are inequivalent and their states differ, say, in their principal quantum numbers (3d, 4d, 5d);

(2) the $t_2^2$ configuration consists of equivalent electrons ($3d^2$), while the $t_2'$ electron originates from another atomic shell (4d);

(3) the three electrons are equivalent ($t_2' = t_2$).

Let us start with the case of three inequivalent electrons. The wavefunctions of the $t_2t_2't_2''$ configuration are constructed from the products of the three one-electron $t_2$ orbitals $\varphi(t_2\gamma_1m_1)\varphi(t_2'\gamma_2m_2)\varphi(t_2''\gamma_3m_3)$. Therefore they form the basis of the direct product of the three irreducible $T_2$ representations: $T_2 \times T_2 \times T_2$. Using the multiplication table 5.7, this product may be reduced in two steps:

$$T_2 \times T_2 \times T_2 \doteq \underbrace{(A_1 + E + T_1 + T_2)}_{\Gamma_1} \times T_2$$

$$\doteq \underbrace{T_2}_{A_1 \times T_2} + \underbrace{T_1 + T_2}_{E \times T_2} + \underbrace{A_2 + E + T_1 + T_2}_{T_1 \times T_2} + \underbrace{A_1 + E + T_1 + T_2}_{T_2 \times T_2}$$

$$\doteq A_1 + A_2 + 2E + 3T_1 + 4T_2. \qquad (5.57)$$

Here $\Gamma_1$ are the irreducible representations obtained after the first multiplication step. The values of the total spin of the three electrons are determined

**Table 5.15** Terms of three inequivalent $t_2$ electrons.

| $t_2t_2'(S_1\Gamma_1)$ | $t_2t_2'(S_1\Gamma_1)t_2''S\Gamma$ |
|---|---|
| $^1A_1$ | $^2T_2$ |
| $^1E$ | $^2T_1$, $^2T_2$ |
| $^1T_1$ | $^2A_2$, $^2E$, $^2T_1$, $^2T_2$ |
| $^1T_2$ | $^2A_1$, $^2E$, $^2T_1$, $^2T_2$ |
| $^3A_1$ | $^2T_2$, $^4T_2$ |
| $^3E$ | $^2T_1$, $^2T_2$, $^4T_1$, $^4T_2$ |
| $^3T_1$ | $^2A_2$, $^2E$, $^2T_1$, $^2T_2$, $^4A_2$, $^4E$, $^4T_1$, $^4T_2$ |
| $^3T_2$ | $^2A_1$, $^2E$, $^2T_1$, $^2T_2$, $^4A_1$, $^4E$, $^4T_1$, $^4T_2$ |

by decomposing the direct product of three representations $D^{(1/2)}$ of the rotation groups (using the rule for summation of angular momenta):

$$D^{(1/2)} \times D^{(1/2)} \times D^{(1/2)} = \underbrace{(D^{(0)} + D^{(1)})}_{D^{(S_1)}} \times D^{(1/2)}$$

$$= \underbrace{D^{(1/2)}}_{D^{(0)} \times D^{(1/2)}} + \underbrace{D^{(1/2)} + D^{(3/2)}}_{D^{(1)} \times D^{(1/2)}} = 2D^{(1/2)} + D^{(3/2)} \quad (5.58)$$

where $S_1 = 0$ and 1 are the *intermediate spins* in the coupling scheme of the three momenta. The resulting spin takes two values: $S = \frac{1}{2}$ (twice) and $S = \frac{3}{2}$. The results (5.57) and (5.58) provide the term classification of a three-electron ion in the case of inequivalent electrons (Table 5.15). In contrast with the two-electron states, it is now necessary to indicate the *intermediate quantum numbers* $S_1\Gamma_1$ or the *term genealogy*.

If the third $t_2'$ electron is added to two equivalent ($t_2^2$) electrons, certain intermediate states $S_1\Gamma_1$ are prohibited by the Pauli principle (Section 5.5). The allowed $S\Gamma$ terms of the $t_2^2t_2'$ configuration are determined by all the direct products $\Gamma_1(t_2^2) \times T_2$ and $D^{(S_1)} \times D^{(1/2)}$, where the intermediate quantum numbers $S_1\Gamma_1(t_2^2)$ are given in Table 5.13. The allowed terms $t_2^2(S_1\Gamma_1)t_2'$ are given in Table 5.16.

**Table 5.16** Allowed terms $t_2^2(S_1\Gamma_1)t_2'S\Gamma$.

| $t_2^2(S_1\Gamma_1)$ | $t_2^2(S_1\Gamma_1)t_2'S\Gamma$ |
|---|---|
| $^1A_1$ | $^2T_2$ |
| $^1E$ | $^2T_1$, $^2T_2$ |
| $^1T_2$ | $^2A_1$, $^2E$, $^2T_1$, $^2T_2$ |
| $^3T_1$ | $^2A_2$, $^2E$, $^2T_1$, $^2T_2$, $^2A_2$, $^4E$, $^4T_1$, $^4T_2$ |

## 5.7.2 Wavefunctions of three electrons

The wavefunctions of the $t_2^2t_2'$ configuration are linear combinations of third-order normalized determinants $|\varphi(t_2\gamma_1 m_1)\ \varphi(t_2\gamma_2 m_2)\ \varphi(t_2\gamma_3 m_3)|$:

$$\Psi(t_2^2(S_1\Gamma_1)t_2' S\Gamma M\gamma) = \sum_{M_1\bar{\gamma}_1}\sum_{\gamma_1\gamma_2}\sum_{m_1 m_2} |\varphi(t_2\gamma_1 m_1)\varphi(t_2\gamma_2 m_2)\varphi(t_2'\gamma_3 m_3)|$$
$$\times \langle T_2\gamma_1 T_2\gamma_2|\Gamma_1\bar{\gamma}_1\rangle\langle \tfrac{1}{2}m_1\tfrac{1}{2}m_2|S_1 M_1\rangle$$
$$\times \langle T_2\gamma_3\Gamma_1\bar{\gamma}_1|\Gamma\gamma\rangle\langle \tfrac{1}{2}m_3 S_1 M_1|SM\rangle. \tag{5.59}$$

In (5.59) the summation over $\gamma_1, \gamma_2$ and $m_1, m_2$, together with the Clebsch–Gordan coefficients $\langle T_2\gamma_1 T_2\gamma_2|\Gamma_1\bar{\gamma}_1\rangle$ and Wigner coefficients $\langle \frac{1}{2}\,m_1\,\frac{1}{2}\,m_2|S_1 M_1\rangle$, gives, according to (5.37), a determinantal function of the form $|\Psi(t_2^2 S_1\Gamma_1 M_1\bar{\gamma}_1)\Phi(t_2'\gamma_3 m_3)|$. This function transforms as the simple product $\Psi(t_2^2 S_1\Gamma_1 M_1\bar{\gamma}_1)\Phi(t_2'\gamma_3 M_3)$ under $\mathbf{O}_h$ group operations, and at the same time it is *antisymmetric* with respect to permutations not only of the two $t_2$ electrons but also of all three $t_2^2$ and $t_2'$ electrons. The summations over $M_1$ and $\bar{\gamma}_1$, with the corresponding Clebsch–Gordan and Wigner coefficients, give the $S\Gamma$ basis.

The complicated nature of (5.59) is only apparent—the calculational procedure is fairly simple. Let us determine the wavefunctions for certain states of the configuration $t_2^3$.

**Example 5.2** The wavefunctions of the $t_2^2(^1A_1)t_2^2T_2$ term are given by

$$\Psi(t_2^2(^1A_1)t_2\,^2T_2 M\gamma) = \sum_{\gamma_3 m_3} |\Psi(t_2^2\,^1A_1)\varphi(t_2\gamma_3 m_3)|\langle A_1 e_1 T_2\gamma_3|T_2\gamma_3\rangle\langle 00\tfrac{1}{2}m_3|\tfrac{1}{2}M\rangle. \tag{5.60}$$

The coefficient $\langle A_1 e_1|T_2\gamma_3 T_2\gamma\rangle$ is nonzero and equal to $\sqrt{1/3}$ when $\gamma_3 = \gamma$, and $\langle 00\frac{1}{2}m_3|\frac{1}{2}M\rangle = 1$ when $m_3 = M$. Therefore in the sums over $\gamma_3$ and $m_3$ only one member remains. The function with $\gamma = \xi$ and $m_3 = M = \frac{1}{2}$ takes the form

$$\Psi(t_2^2(^1A_1)t_2\,^2T_2 M = \tfrac{1}{2}, \gamma = \xi) = |\Psi(t_2^2\,^1A_1)\xi\alpha|. \tag{5.61}$$

On substituting $\Psi(t_2^2\,^1A_1)$ into this, (5.40), we obtain

$$\Psi(t_2^2\,(^1A_1)^2T_2 M = \tfrac{1}{2}, \gamma = \zeta) = \sqrt{1/3}(|\xi\ \bar{\xi}\ \xi| + |\eta\ \bar{\eta}\ \xi| + |\zeta\ \bar{\zeta}\ \xi|). \tag{5.62}$$

Bearing in mind that $|\xi\ \bar{\xi}\ \xi| = 0$ (the first and third columns coincide), after the final normalization of the function we obtain

$$\Psi(t_2^2(^1A_1)t_2\,^2T_2, M = \tfrac{1}{2}, \gamma = \xi) = \sqrt{1/2}(|\eta\ \bar{\eta}\ \xi| + |\zeta\ \bar{\zeta}\ \xi|). \tag{5.63}$$

**Example 5.3** Let us construct the $^4E(t_2^3)$-term wavefunction, proceeding from the intermediate state $^3T_1(t_2^2)$ $(S_1 = 1, \Gamma_1 = T_1)$:

$$\Psi(t_2^2(^3T_1)t_2\,^4E, M = \tfrac{3}{2}, \gamma = u) = \sum_{M_1=0,\pm 1}\ \sum_{\bar{\gamma}_1=\alpha,\beta,\gamma}\ \sum_{\gamma_3=\xi,\eta,\zeta}\ \sum_{m_3=\pm\frac{1}{2}} |\Psi(t_2^2(^3T_1)M_1\bar{\gamma}_1)\Phi(t_2\gamma_3 m_3)|$$
$$\times \langle T_1\bar{\gamma}_1 T_2\gamma_3|Eu\rangle\langle 1M_1\tfrac{1}{2}m_3|\tfrac{3}{2}\tfrac{3}{2}\rangle. \tag{5.64}$$

In the Wigner coefficient, $M_1 + m_3 = \frac{3}{2}$; that is, $M_1 = 1$, $m_3 = \frac{1}{2}$ and $\langle 11\frac{1}{2}\frac{1}{2}|\frac{3}{2}\frac{3}{2}\rangle = 1$.

From the tables in [14] we obtain the nonzero Clebsch–Gordan coefficients:

$$\langle T_1\alpha T_2\xi|Eu\rangle = -\langle T_1\beta T_2\eta|Eu\rangle = -\sqrt{1/2}.$$

Therefore in the sums over $\bar{\gamma}_1$ and $\gamma_3$ only two members are left: $\bar{\gamma}_1 = \alpha$, $\gamma_3 = \xi$ and $\bar{\gamma}_1 = \beta$, $\gamma_3 = \eta$. The components $\alpha$, $\beta$ and $\gamma$ of the function $\Psi(t_2^2\,{}^3T_1, M = 1, \gamma)$ are $|\eta\zeta|$, $|\zeta\;\xi|$ and $|\xi\;\eta|$ respectively (see (5.41)). Substituting these into (5.64), we obtain

$$\Psi(t_2^2\,({}^3T_1)t_2\,{}^4E, M = \tfrac{3}{2}, \gamma = u) = |\eta\;\zeta\;\xi|\sqrt{1/2} + |\zeta\;\xi\;\eta|(-\sqrt{1/2}) = 0. \qquad (5.65)$$

Since the wavefunction is zero, the term $t_2^2({}^3T_1)t_2\,{}^4E$ is forbidden in the $t_2^3$ configuration of the three equivalent electrons. In the same way, we obtain the wavefunctions of all the terms. The nonzero functions provide a classification of the allowed terms of the $t_2^3$ configuration:

$${}^4A_2(t_2^3),\quad {}^2E(t_2^3),\quad {}^2T_1(t_2^3),\quad {}^2T_2(t_2^3).$$

By excitation of one electron in the e shell, the configuration $t_2^2e$ is formed. In this case the Pauli principle does not introduce any restrictions; hence all terms are allowed with $S = \frac{1}{2}$ and $\frac{3}{2}$ and $\Gamma \doteq \Gamma_1(t_2^2) \times E$. Taking into consideration that, in **O**, $A_1 \times E \doteq E$, $E \times E \doteq A_1 + A_2 + E$ and $E \times T_1 = E \times T_2 \doteq T_1 + T_2$ (Table 5.7), we obtain the terms listed in Table 5.17.

The wavefunctions are constructed according to the equation

$$\Psi(t_2^2(S_1\Gamma_1)eS\Gamma M\gamma) = \sum_{M_1\bar{\gamma}_1}\sum_{m_3\gamma_3} |\Psi(t_2^2S_1\Gamma_1M_1\bar{\gamma}_1)\Phi(em_3\gamma_3)| \times \langle S_1M_1\tfrac{1}{2}m_3|SM\rangle\langle\Gamma_1\bar{\gamma}_1E\gamma_3|\Gamma\gamma\rangle, \qquad (5.66)$$

**Example 5.4** Let us determine the wavefunctions of the term $t_2^2({}^3T_1)e\,{}^4T_2$. Taking into account that $M = \frac{3}{2}$ when $M_1 = 1$ and $m_3 = \frac{1}{2}$, and that $\langle 1 1 \frac{1}{2} \frac{1}{2} | \frac{3}{2} \frac{3}{2}\rangle = 1$, we obtain

$$\Psi(t_2^2({}^3T_1)e\,{}^4T_2, M = \tfrac{3}{2}, \gamma = \xi) = \sum_{\bar{\gamma}_1=\alpha,\beta,\gamma}\;\sum_{\gamma_3=\xi,\eta,\zeta} |\Psi(t_2^2({}^3T_1), 1, \bar{\gamma}_1)\Phi(E\gamma_3\tfrac{1}{2})|\langle T_1\bar{\gamma}_1E\gamma_3|T_2\xi\rangle. \qquad (5.67)$$

From the tables in [14] we obtain the nonzero Clebsch–Gordan coefficients:

$$\langle EuT_1\alpha|T_2\xi\rangle = \tfrac{1}{2}\sqrt{3}, \qquad \langle EvT_1\alpha|T_2\xi\rangle = \tfrac{1}{2}.$$

The final result is

$$\Psi(t_2^2({}^3T_1)e\,{}^4T_2, M = \tfrac{3}{2}, \gamma = \xi) = \sqrt{1/2}(\sqrt{3}|\eta\;\zeta\;u| + |\eta\;\zeta\;v|), \qquad (5.68)$$

All the other allowed states are obtained in the same way; they are given in [14].

**Table 5.17** Allowed terms of the $t_2^2e$ configuration.

| $t_2^2(S_1\Gamma_1)$ | $t_2^2(S_1\Gamma_1)eS\Gamma$ |
|---|---|
| ${}^1A_1$ | ${}^2E$ |
| ${}^1E$ | ${}^2A_1$, ${}^2A_2$, ${}^2E$ |
| ${}^1T_2$ | ${}^2T_1$, ${}^2T_2$ |
| ${}^3T_1$ | ${}^2T_1$, ${}^2T_2$, ${}^4T_1$, ${}^4T_2$ |

### 5.7.3 Many-electron wavefunctions

A similar programme for calculation of wavefunctions can be realized for many-electron ions in crystal fields using the methods for multiplying representations described above. Let us consider, for example, the $t_2^2e^2$ configuration of a $d^4$ ion. The classification of allowed terms involves two stages:

(1) multiplication of the irreducible representations $\Gamma_1(t_2^2)$ and $\Gamma_2(e^2)$, and reduction of the direct product $\Gamma_1 \times \Gamma_2 \doteq \Sigma\Gamma$;

(2) determining the spin states $D^{(S_1)} \times D^{(S_2)} = \Sigma D^{(S)}$.

Since the intermediate configurations $t_2^2$ and $e^2$ contain inequivalent (relative to one another) electrons, all $S\Gamma$ terms are allowed. The intermediate states $S_1\Gamma_1(t_2^2)$ and $S_2\Gamma_2(e^2)$ are shown in Table 5.13. The following scheme gives the terms $S\Gamma(t_2^2e^2)$ as the result of the multiplication of $\Gamma_1 \times \Gamma_2$ and $D^{(S_1)} \times D^{(S_2)}$:

$$S_1\Gamma_1:\quad \underbrace{{}^1A_1, {}^1E, {}^1T_2, {}^3T_1}_{{}^1A_1}\quad \underbrace{{}^1A_1, {}^1E, {}^1T_2, {}^3T_1}_{{}^1E}\quad \underbrace{{}^1A_1, {}^1E, {}^1T_2, {}^3T_1}_{{}^3A_2}$$

$$S_2\Gamma_2:\quad {}^1A_1, {}^1E, {}^1T_2, {}^3T \quad {}^1E, {}^1A_1, {}^1A_2, {}^1E, {}^1T_1, {}^1T_2, {}^3T_1, {}^3T_2 \quad {}^3A_2, {}^3E, {}^3T_1, {}^1T_2, {}^3T_2$$

The wavefunctions are represented as linear combinations of fourth-order determinants:

$$\Psi(t_2^2(S_1\Gamma_1)e^2(S_2\Gamma_2)S\Gamma M\gamma) = \sum_{M_1,M_2}\sum_{\gamma_1\gamma_2} |\varphi(t_2^2 S_1\Gamma_1 M_1\gamma_1)\varphi(e^2 S_2\Gamma_2 M_2\gamma_2)| \times \langle \Gamma_1\gamma_1\Gamma_2\gamma_2|\Gamma\gamma\rangle\langle S_1M_1S_2M_2|SM\rangle. \qquad (5.69)$$

This equation generalizes (5.42) for the case of two interacting subsystems, each containing several electrons in given configurations.

### 5.7.4 Energy levels

The energy levels of many-electron (in particular, three-electron) ions are calculated according to the same scheme as for two-electron ions. In the first stage the first-order corrections $\Delta\varepsilon$ to the energies of the strong-field $t_2^{m}e^{m'}$ configurations are calculated; in the second stage the problem of configuration mixing is solved. For the case of three equivalent electrons, for example, for the $t_2^3$ configuration the correction to the energy $(-12Dq)$ is

$$\Delta\varepsilon(t_2^2(S_1\Gamma_1)t_2S\Gamma) = \langle t_2^2(S_1\Gamma_1)t_2S\Gamma M\gamma|\hat{V}_{12}|t_2^2(S_1\Gamma_1)t_2S\Gamma M\gamma\rangle. \qquad (5.70)$$

Calculations are performed by using determinantal wavefunctions of the type (5.68), while the energy corrections are expressed through the Racah

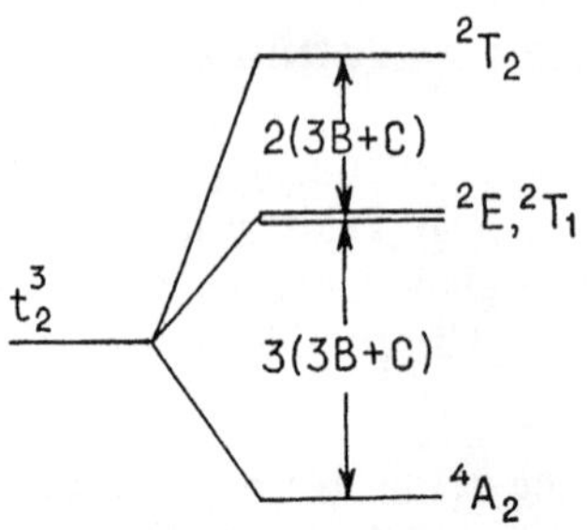


**Fig. 5.5** Terms of $t_2^3$ configuration in a first approximation.

parameters (Fig. 5.5). Matrix elements for configuration mixing of repeating terms are calculated in the same way. For all transition ions $d^1, \ldots, d^9$ the configuration mixing matrices are given in [14]. Let us illustrate the results, taking as an example the $^2E$ term of a $d^3$ ion (Table 5.18).

The $^2E$ term originates from the configurations $t_2^3$, $t_2^2e(S_1\Gamma_1 = {}^1A_1, {}^1E)$ and $e^3$, which lead to a fourth-order matrix. Of course, in the extreme case of a weak field the four atomic states are mixed: $a\,^2D$, $b\,^2D$, $^2G$, $^2H$ (where $a$ and $b$ indicate the repeating terms $^2D$); that is, the energy matrix is of the same order. Finally, the matrix is symmetric, which is why only half is shown in Table 5.18.

**Table 5.18** Configuration mixing matrix for the $^2E$ term of a $d^3$ ion.

| $t_2^3$ | $t_2^2(^1A_1)e$ | $t_2^2(^1E)e$ | $e^3$ |
|---|---|---|---|
| $-6B+3C$ | $-6\sqrt{2}B$ | $-3\sqrt{2}B$ | 0 |
| | $8B+6C$ | $10B$ | $\sqrt{3}(2B+C)$ |
| | | $-B+3C$ | $2\sqrt{3}B$ |
| | | | $-8B+4C$ |

## 5.7.5 Correlation diagrams

In practical uses of the crystal field theory results it is often necessary to know not only the symmetry but also the intensity of the field. The properties of the intensity are not related to group-theoretical concepts. Nevertheless, an approach based on symmetry makes it possible to follow the qualitative dependence of the energy levels of a coordinated ion on the crystal field intensity. This dependence is useful for interpreting optical spectra and other properties of ions in crystal fields.

Since the terms in strong and weak crystal fields are correlated with each other, they may be expressed as functions of the field parameter $Dq$. The relative positions of the energy levels, the *correlation diagram*, is determined by the relative values of the field strength $Dq$ and by the interelectron interaction $\hat{V}_{ee}$. Figure 5.6 shows the correlation diagram of a $d^2$ ion in an octahedral field

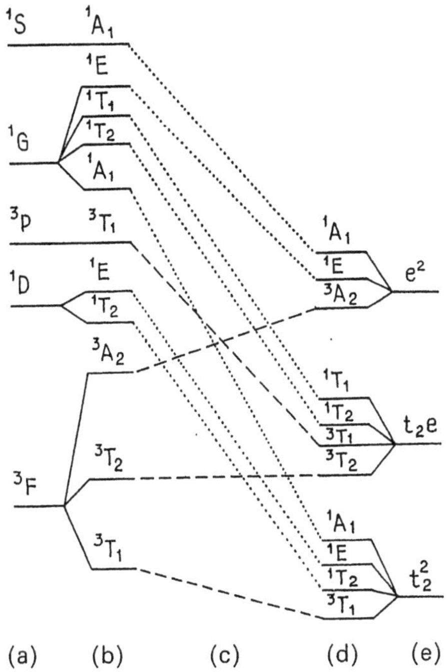


**Fig. 5.6** Correlation diagram of $d^2$-ion terms in an octahedral field: (a) atomic states; (b) a weak field; (c) an intermediate field; (d) a strong field and weak interelectronic repulsion; (e) a strong field (– – –, spin-triplets; . . . ., spin-singlets).

[37]. The left-hand side corresponds to the atomic levels, and the right-hand side to a field of maximum strength. The dashed lines connect high-spin terms ($S = 1$) while the dotted lines connect low-spin terms ($S = 0$). Since terms of identical symmetry "repel each other" (Section 5.6.2), the corresponding lines connecting their strong and weak-field levels do not intersect. Therefore, for example, the strong-field lower level ${}^3\mathrm{T}_1(\mathrm{t}_2^2)$ is correlated with the weak-field lower level ${}^3\mathrm{T}_1({}^3\mathrm{F})$, while the upper level ${}^3\mathrm{T}_1(\mathrm{t}_2^2)$ is correlated with the upper level ${}^3\mathrm{T}_1({}^3\mathrm{P})$.

## 5.7.6 Tanabe–Sugano diagrams

Tanabe and Sugano (see [14]) have developed and applied the above group-theoretical methods of calculating wavefunctions to calculate the energy levels of the $S\Gamma(\mathrm{t}_2^k\mathrm{e}^{n-k})$ terms of all $\mathrm{d}^n$ ions in a cubic field. In the zeroth-order approximation (i.e. when there is no interelectron interaction) the energy levels are determined from

$$\varepsilon_0(\mathrm{t}_2^k\mathrm{e}^{n-k}) = -4Dqk + 6Dq(n-k). \tag{5.71}$$

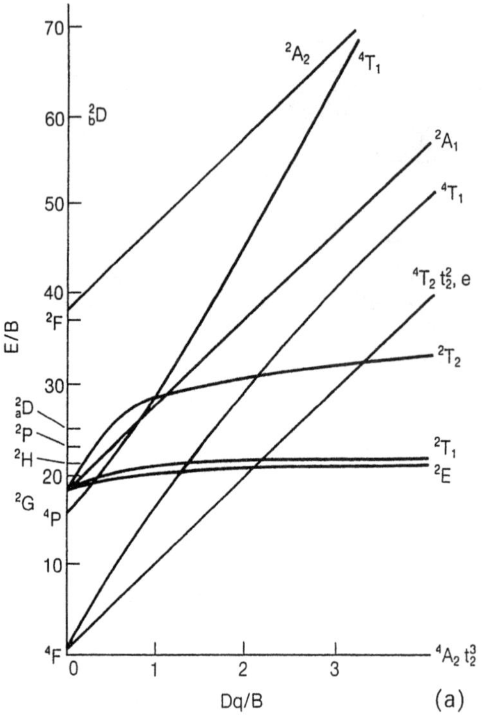


**Fig. 5.7** Tanabe–Sugano diagrams for (a) $d^3$ and (b) $d^4$ ions.

After numerically solving the secular equations corresponding to the configuration mixing of repeating terms, Tanabe and Sugano have constructed the reduced energies $E(S\Gamma)/B$ as functions of the dimensionless parameter $Dq/B$ of the cubic field. The plots thus obtained are known as *Tanabe–Sugano diagrams*. Diagrams for all transition metal ions are given, for example, in [14, 37]. Figure 5.7 shows diagrams for $d^3$ and $d^4$ ions. The ground state is denoted by the horizontal axis, the excited state energies being determined relative to it. From Fig. 5.7(b) it can be seen that for certain values of the crystal field the ground state $^5E$ of the $d^4$ ion is changed into $^3T_1(t_2^4)$; that is, there is a tran-sition from the high-spin to the low-spin state. Using Tanabe–Sugano diagrams, it is possible to relate optical absorption bands to definite vertical sections of the diagrams and to determine the value of the crystal field parameter. This lies beyond the scope of the group-theoretical approach, and has been discussed in a number of papers on crystal spectroscopy (see e.g. [46, 47]).

A generalization of the Tanabe–Sugano results, taking into account trigonal and tetragonal crystal fields, as well as spin–orbit interaction, has been obtained by König and Kremer [48], who describe many-electron calculation methods and give energy level diagrams for all $d^n$ ions, taking into account

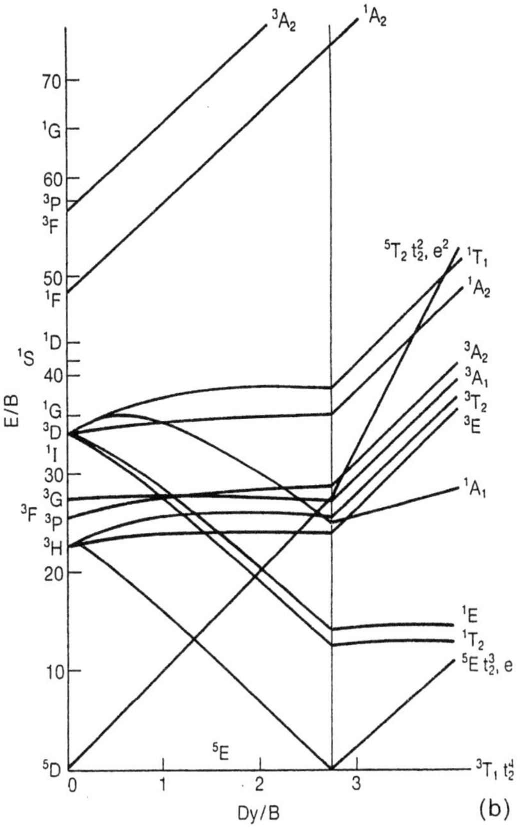


**Fig. 5.7** Continued

configuration mixing in cubic and low-symmetry fields as well as spin–orbit interaction (for $d^2$ and $d^8$ ions).

## PROBLEMS

**5.1** Classify the terms of 3d 4p configuration in $\mathbf{D}_{4h}$, $\mathbf{D}_{2d}$ and $\mathbf{C}_{2v}$ crystal fields.

**5.2** Find the reduction of $D^{(L)}$ in $\mathbf{T}_d$ (Table 5.4).

**5.3** Compose the multiplication tables for the irreducible representations of $\mathbf{O}_h$, $\mathbf{D}_{4h}$, $\mathbf{D}_{3h}$ and $\mathbf{C}_{3v}$.

**5.4** Find the wavefunctions of the $t_2^2$, $t_2e$ and $e^2$ configurations (**O** symmetry) and determine the allowed states.

# 6 Semiempirical Crystal Field Theory

## 6.1 CRYSTAL FIELD HAMILTONIAN

In the point-ion model of the crystal field the electron–ligand interaction operator is given by (4.2). In order to compute the matrix elements of this operator, it is convenient to represent it as an expansion in spherical harmonics $Y_{km}(\vartheta_i\varphi_i)$ (where $r_i, \vartheta_i, \varphi_i$ are the coordinates of the $i$th electron [14]):

$$\hat{V}^c(\mathbf{r}) = \sum_{k=0}^{\infty} \sum_{m=-k}^{k} A_m^k \sum_i r_i^k Y_{km}(\vartheta_i\varphi_i). \tag{6.1}$$

The coefficients $A_m^k$ depend on the ligand coordinates, i.e. on the geometry of the complex:

$$A_m^k = -\frac{2(4\pi)^{1/2}e}{2k+1} \sum_\alpha \frac{q_\alpha}{r_\alpha^{k+1}} \Theta_{km}(\vartheta_\alpha) e^{-im\varphi_\alpha}, \tag{6.2}$$

where $(r_\alpha, \vartheta_\alpha, \varphi_\alpha)$ are the spherical coordinates at the $\alpha$th ligand, $q_\alpha$ is its charge and $\Theta_{km}(\vartheta_\alpha)$ are associated Legendre polynomials. The summation in (6.2) is over all the atoms of the complex or crystal, although in practice it is confined to a few nearest neighbours of the ion with an unfilled shell.

Let us consider the action of the crystal field operator (6.1) on the atomic states $\kappa SL$ (Section 5.1), where $\kappa$ enumerates repeating terms, like, for instance, the twice-repeated terms $^2D(d^3)$, and thrice-repeated terms $^2D(d^5)$ (Table 5.1). In order to determine the crystal splittings, we must compute matrix elements of the form $\langle \kappa SLM_SM_L|\hat{V}^c|\kappa' S'L'M_{S'}M_{L'}\rangle$ and solve the corresponding secular equation. These matrix elements depend on the average values $\langle r^k \rangle$ of different powers of $r$ for the unfilled shell ion. Since the crystal field and the interelectronic interaction strongly influence the radial distribution of electron density, $\langle r^k \rangle$ cannot be computed ab initio with a sufficient degree of accuracy. On the other hand, the "geometrical parameters" $A_m^k$ depend on both the crystal symmetry and interatomic distances. The crystal symmetry, in particular, imposes restrictions on possible $k$ and $m$ values in the sum (6.1). Therefore in complex and impurity ion spectroscopy a *semiempirical approach* is widely used. Here the quantities $\langle r^k \rangle$ and $A^k_m$ are not computed, but rather are determined by comparison of experimental spectra with those predicted theoretically. In this case the theoretical description of the spectrum makes maximum use of the symmetries of the Hamiltonian and group-theoretical procedures for constructing it. These are based on several concepts in group theory that we shall introduce below.

## 6.2 WIGNER–ECKART THEOREM FOR SPHERICAL TENSORS

### 6.2.1 Spherical tensors

A *spherical tensor* $\hat{\mathbf{T}}^k_m$ is a set of $2k+1$ values $(m = -k, -k+1, \ldots, k)$ that transform under rotation of the coordinate system in the same way as the spherical functions $Y_{lm}(\vartheta, \varphi)$ [2, 40].† Spherical tensors are also called *irreducible tensors* or *irreducible tensor operators of rank k of the rotation group*. This terminology corresponds to the definition of the basis functions of irreducible representations of **K**; the spherical harmonics $Y_{lm}$ act as these functions. In fact the rank $k$ of an irreducible tensor is the angular momentum quantum number, while the index $m$ labelling the tensor components is the angular momentum projection quantum number. A spherical tensor of rank 1 has three components $(\hat{\mathbf{T}}^1_1, \hat{\mathbf{T}}^1_0, \hat{\mathbf{T}}^1_{-1})$, while a rank-2 tensor has five components $(\hat{\mathbf{T}}^2_2, \hat{\mathbf{T}}^2_1, \hat{\mathbf{T}}^2_0, \hat{\mathbf{T}}^2_{-1}, \hat{\mathbf{T}}^2_{-2})$.

The simplest example of an irreducible tensor is a function of the form $f(r) Y_{em}(\vartheta, \varphi)$, where $f(r)$ is a function of only the radius $r$, i.e. a function with spherical symmetry. The functions $\hat{\mathbf{T}}^k_m(r) = r^k Y_{km}(\vartheta, \varphi)$ are spherical tensors of rank $k$ in the coordinate space of the $i$th electron. Hence the sum

$$\hat{\mathbf{T}}^k_m = \sum_i r_i^k Y_{km}(\vartheta_i, \varphi_i) \equiv \sum_i \hat{\mathbf{T}}^k_m(i) \tag{6.3}$$

appearing in (6.1) is also a component of an irreducible tensor operator of rank $k$.

### 6.2.2 Matrix elements of tensor operators

Let $\psi_{\kappa LM} \equiv |\kappa LM\rangle$ be spherical functions corresponding to angular momentum $L$ and its projection $M$, the index $\kappa$ indicating the number of repeating representations $\mathrm{D}^{(L)}$ of the rotation group. These functions are eigenfunctions

† More rigorous determination; see in [2].

of the operators $\hat{L}^2$ and $\hat{L}_z$, which means that they satisfy (5.23). Let us consider matrix elements of the form $\langle \kappa LM|\hat{\mathbf{T}}^k_m|\kappa' L' M'\rangle$, where the spherical tensor $\hat{\mathbf{T}}^k_m$ depends on the coordinates of the same electrons as the wavefunctions $|\kappa LM\rangle$ and $|\kappa' L' M'\rangle$. This notation unites the set of $(2L+1)(2L'+1)(2k+1)$ matrix elements. It is intuitively clear that there are relations among them resulting from the symmetry properties under simultaneous transformations of wavefunctions and operators $\hat{\mathbf{T}}^k_m$ when the coordinate system is rotated. These relations are established by the *Wigner–Eckart theorem*, which is of fundamental importance in group theory and its applications. The mathematical expression of this theorem is as follows [2]:

$$\langle \kappa LM|\hat{\mathbf{T}}^k_m|\kappa' L' M'\rangle = \frac{(-1)^{2k}}{(2L+1)^{1/2}}\langle \kappa L||\hat{\mathbf{T}}^k||\kappa' L'\rangle\langle LM|L'M'km\rangle. \tag{6.4}$$

The matrix element of the operator $\hat{\mathbf{T}}^k_m$ includes two factors: a Wigner (Clebsch–Gordan) coefficient and the quantity $\langle \kappa L||\hat{\mathbf{T}}^k||\kappa' L'\rangle$, called a *reduced matrix element*. It should be noted that the reduced matrix element does not depend on $M$, $M'$ or $m$; in other words, it is the same for the whole set of matrix elements of the operator $\hat{\mathbf{T}}^k_m$. The dependence on $M$, $M'$ and $m$ is contained entirely in the Wigner coefficients. Hence the matrix of any component of an irreducible tensor contains the Wigner coefficients $\langle LM|L'M'km\rangle = \langle L'M'km|LM\rangle$ and a common factor, outside the matrix. This characteristic, which expresses the essence of the Wigner–Eckart theorem, is particularly evident in matrix form:

$$\hat{\mathbf{T}}^k_m = \frac{(-1)^{2k}}{(2L+1)^{1/2}}\langle \kappa L||\hat{\mathbf{T}}^k||\kappa' L'\rangle \mathbf{O}^k_m, \tag{6.5}$$

where $\mathbf{O}^k_m$ is a matrix composed from the Wigner elements:

$$(\mathbf{O}^k_m)_{MM'} = \langle LM|L'M'km\rangle. \tag{6.6}$$

It follows from the properties of the Wigner coefficients that a matrix element of the operator $\hat{\mathbf{T}}^k_m$ is nonzero if the *triangle condition* is satisfied, i.e. if $|L-L'| \le k \le L+L'$ and $m = M - M'$. Such conditions are called *selection rules*, since they select the nonvanishing matrix elements.

**Example 6.1** Let $L = L' = 1$, so that the operator $\hat{\mathbf{T}}^k_m$ can be represented by a square matrix of order 3 ($M, M' = 0, \pm 1$):

$$\mathbf{O}^k_m = \begin{array}{c|ccc} M \backslash M' & -1 & 0 & +1 \\ \hline -1 & \langle 1\,-1|1\,-1km\rangle & \langle 1\,-1|10km\rangle & \langle 1\,-1|11km\rangle \\ 0 & \langle 10|1\,-1km\rangle & \langle 10|10km\rangle & \langle 10|11km\rangle \\ +1 & \langle 11|1\,-1km\rangle & \langle 11|10km\rangle & \langle 11|11km\rangle \end{array}. \tag{6.7}$$

Inserting the Wigner coefficients from Table 5.10 (for $j_1 = j = 1$) into this, we obtain the matrices $\mathbf{O}^1_m$ for the rank-1 tensor:

$$\mathbf{O}^1_{+1} = \frac{1}{\sqrt{2}}\begin{bmatrix} 0 & 0 & 0 \\ 1 & 0 & 0 \\ 0 & 1 & 0 \end{bmatrix}, \quad \mathbf{O}^1_0 = \frac{1}{\sqrt{2}}\begin{bmatrix} -1 & 0 & 0 \\ 0 & 0 & 0 \\ 0 & 0 & 1 \end{bmatrix}, \quad \mathbf{O}^1_{-1} = \frac{1}{\sqrt{2}}\begin{bmatrix} 0 & 1 & 0 \\ 0 & 0 & 1 \\ 0 & 0 & 0 \end{bmatrix}. \quad (6.8)$$

To obtain all the matrices of any operator $\hat{\mathbf{T}}^k_m$ in a given basis (i.e. for fixed $\kappa L$ and $\kappa' L'$), only the reduced matrix element $\langle \kappa L \| \hat{\mathbf{T}}^k \| \kappa' L' \rangle$ need be known. It can be found by computing just one matrix element $\langle \kappa LM | \hat{\mathbf{T}}^k_m | \kappa' L' M' \rangle$ and using (6.4):

$$\langle \kappa L \| \hat{\mathbf{T}}^k \| \kappa' L' \rangle = (-1)^{2k} \frac{(2L+1)^{1/2} \langle \kappa LM | \hat{\mathbf{T}}^k_m | \kappa' L' M' \rangle}{\langle LM | L' M' km \rangle}. \quad (6.9)$$

Thus between the matrix elements of an irreducible tensor there exist relations resulting from symmetry. These relations are valid for wavefunctions $\psi_{\kappa LM}$ and operators $\hat{\mathbf{T}}^k_m$ of any physical nature. On the other hand, the reduced matrix elements depend on the physical structure of an atom (e.g. the number of electrons, the radial parameters of the wavefunctions and the electron distribution).

Bearing (6.3) in mind, let us represent the crystal field operator (6.1) as a linear combination of irreducible tensors:

$$\hat{V}^{\mathrm{c}}(\mathbf{r}) = \sum_{k=0}^{\infty} \sum_{m=-k}^{k} A^k_m \hat{\mathbf{T}}^k_m(\mathbf{r}), \quad (6.10)$$

where the vector $\mathbf{r}$ denotes the set of coordinates $\mathbf{r}_1, \mathbf{r}_2, \ldots, \mathbf{r}_n$ of all the electrons of the atom or ion. When calculating the matrix elements of the operator $\hat{V}^{\mathrm{c}}$, it should be remembered that the atomic wavefunctions $|\kappa SLM_S M_L\rangle$ in the Russell–Saunders coupling scheme also depend on the quantum numbers $S$ and $M_S$. It can be shown that the operator $V^{\mathrm{c}}(\mathbf{r})$, being dependent only on the spatial coordinates of the electrons, does not connect atomic states with different quantum numbers $SM_S$. Hence in this case the Wigner–Eckart theorem can be written as

$$\langle \kappa SLM_S M_L | \hat{\mathbf{T}}^k_m(\mathbf{r}) | \kappa' S' L' M_{S'} M_{L'} \rangle = \frac{(-1)^{2k}}{(2L+1)^{1/2}} \langle \kappa SL \| \hat{\mathbf{T}}^k \| \kappa' S' L' \rangle \times \langle LM | L' M' km \rangle \delta_{SS'} \delta_{M_S M_{S'}}, \quad (6.11)$$

where $\delta_{ij}$ is the Kronecker symbol (Section 4.3.5). Thus the matrix elements of the operator $V^{\mathrm{c}}$ are nonzero only for states with identical spin $S$ and spin projection $M_S$. In the basis of these states the matrix $V^{\mathrm{c}}$, as in the example above, is proportional to the universal matrix $\mathbf{O}^k_m$. It is of fundamental importance, however, that the reduced matrix element in (6.11) depends on $S$; therefore it is determined by the $LS$ term.

## 6.3 PROJECTION OPERATORS

### 6.3.1 Spherical tensors in point groups

This section deals with the elucidation of the possible structure of the crystal field Hamiltonian, taking account of symmetry properties. Let us assume that an ion with an incomplete $d^n$ shell having the ground *LS* state is situated in the crystal field of a point group **G**. Determining the structure of the $\hat{V}^c$ Hamiltonian involves

(1) determining the rank $k$ of the tensor operators $\hat{\mathbf{T}}^k_m$ appearing in the expansion (6.10);

(2) finding the linear combination of tensor operators and determining the independent parameters $A^k_m$.

The Hamiltonian of the crystal field is invariant under all the transformations of the symmetry point group **G** (Section 4.2). It is therefore transformed according to the totally symmetric representation of the group (which we shall denote by $A_1$). The invariance of the Hamiltonian determines the rank $k$ of the spherical tensors $\hat{\mathbf{T}}^k_m$ that may be involved in the crystal field expansion. Now it is obvious that this expansion can comprise only those harmonics $\hat{\mathbf{T}}^k_m$ that can form the basis of the totally symmetric representation of **G**. The ranks can be determined by expanding the irreducible representations $D^{(k)}$ of the group of spherical symmetry in terms of irreducible representations $\Gamma_i$ of the point group **G**:

$$D^{(k)} \doteq \sum_i \Gamma_i. \qquad (6.12)$$

If this expansion contains the totally symmetric representation $A_1$ (that is, $\Gamma_i = A_1$) then the spherical tensor $\hat{\mathbf{T}}^k_m$ appears within the crystal field operator. Otherwise, the components $\hat{\mathbf{T}}^k_m$ cannot form a linear combination invariant under operations of the group **G**.

The reduction of the representations $D^{(k)}$ into irreducible components has been described in Section 4.3.4. Some examples are given in Tables 4.2 and 4.3.

**Example 6.2** Consider octahedral coordination $\mathbf{O}_h$. From Table 4.3 we can see that the totally symmetric representation is obtained by the reduction of $D^{(0)}$, $D^{(4)}$, $D^{(6)}$ (the results for $l > 6$ are not given). The zeroth-rank operator $\hat{\mathbf{T}}^0_0$ should be omitted, since it does not depend on the electron coordinates and therefore leads to an additive shift of all energy levels. This means that a crystal field of $\mathbf{O}_h$ symmetry can be composed from the operators $\hat{\mathbf{T}}^4_m$ and $\hat{\mathbf{T}}^6_m$.

**Example 6.3** Consider a tetragonal crystal field of $\mathbf{D}_{4h}$ symmetry. It follows from Table 4.3 that the operator $\hat{V}^c(\mathbf{r})$ includes spherical tensors of types $\hat{\mathbf{T}}^2_m$, $\hat{\mathbf{T}}^4_m$ and $\hat{\mathbf{T}}^6_m$. It should be noted that the expansion of the cubic field begins with 4th-rank harmonics

**Table 6.1** Representation reduction of the full spherical group for $\mathbf{C}_{4v}$.

| | |
|---|---|
| $D_0^+$ $\textcircled{\Gamma_1}$ | $D_0^-$ $\Gamma_2$ |
| $D_1^+$ $\Gamma_2+\Gamma_3$ | $D_1^-$ $\textcircled{\Gamma_1}+\Gamma_5$ |
| $D_2^+$ $\textcircled{\Gamma_1}+\Gamma_3+\Gamma_4+\Gamma_5$ | $D_2^-$ $\Gamma_2+\Gamma_3+\Gamma_4+\Gamma_5$ |
| $D_3^+$ $\Gamma_2+\Gamma_3+\Gamma_4+\Gamma_5$ | $D_3^-$ $\textcircled{\Gamma_1}+\Gamma_3+\Gamma_4+\Gamma_5$ |
| $D_4^+$ $\textcircled{2\Gamma_1}+\Gamma_2+\Gamma_3+2\Gamma_5$ | $D_4^-$ $\textcircled{\Gamma_1}+2\Gamma_2+\Gamma_3+\Gamma_4+2\Gamma_5$ |
| $D_5^+$ $\textcircled{\Gamma_1}+2\Gamma_2+\Gamma_3+\Gamma_4+3\Gamma_5$ | $D_5^-$ $\textcircled{2\Gamma_1}+\Gamma_2+\Gamma_3+\Gamma_4+3\Gamma_5$ |
| $D_6^+$ $\textcircled{2\Gamma_1}+\Gamma_2+2\Gamma_3+2\Gamma_4+3\Gamma_5$ | $D_6^-$ $\textcircled{\Gamma_1}+2\Gamma_2+2\Gamma_3+2\Gamma_4+3\Gamma_5$ |

($k = 4$), while the tetragonal field contains operators with $k = 2$. Note also that, when the representation $D^{(4)}$ is reduced in $\mathbf{D}_{4h}$, the totally symmetric representation $A_{1g}$ occurs twice. Therefore *two* invariant linear combinations can be constructed from the spherical tensors $\hat{\mathbf{T}}_m^4$ within $\mathbf{D}_{4h}$. In groups of lower symmetry (see Table 2.3) the number of repeating representations increases.

The symmetry groups considered above contain inversion operations. Therefore the crystal field contains only spherical tensor operators of even rank. Such operators, according to the general rules (Section 4.3.4), are even, since for them $(-1)^k = 1$.† For operators of odd rank in the expansion (6.12) there can exist only odd representations of $\Gamma_u$ type and therefore the totally symmetric $A_{1g}$ does not occur. In noncentrosymmetric groups the crystal field can contain tensor operators of odd rank. The corresponding contributions to the operator $\hat{V}^c(\mathbf{r})$ change sign under inversion; however, since the group does not contain the inversion operation, the odd tensors $\hat{\mathbf{T}}_m^k$ can form the basis of the totally symmetric representation $A_1$. It will be shown below that the odd component of the crystal field determines the optical line intensities and polarization dependence.

The results for any crystallographic point group can be obtained using tables of full spherical group reductions in [1] (Section 5.2.3).

**Example 6.4** Consider the group $\mathbf{C}_{4v}$. Table 6.1 shows the reduction scheme for $D_n^\pm$ (with the notation adopted in [1]; see Section 5.2.3). Since there is no inversion in $\mathbf{C}_{4v}$, in the notation for the representations $\Gamma_i$ there is no parity symbol u or g. The parity of the tensor operators $\hat{\mathbf{T}}_m^k$ is determined by the factor $(-1)^k$; therefore in the left-hand half of Table 6.1 the rows corresponding to the representations $D_1^+$, $D_3^+$ and $D_5^+$ should be excluded, as should the rows $D_0^-$, $D_2^-$ and $D_4^-$ in the right-hand half. The totally symmetric representations $\Gamma_i$ are encircled. It can be seen that the even part of the $\mathbf{C}_{4v}$ symmetry crystal field includes the tensor operators $\hat{\mathbf{T}}_{m}^2$,. $\hat{\mathbf{T}}_m^4$ and $\hat{\mathbf{T}}_m^6$; for each rank $k = 4$ and 6 there are two invariant linear combinations. It is obvious from the right-hand half of the table that the odd part of the operator $\hat{V}^c(\mathbf{r})$ includes the $D_1^-$, $D_3^-$ and $D_5^-$ (two combinations) operators.

† This rule does not relate to the so-called pseudo-tensors (see [2]).

### 6.3.2 Projection operator method

To construct the Hamiltonian of the crystal field, it is necessary to find the linear combinations of the spherical tensors that form the basis of the totally symmetric representation of the group. This problem can be solved using the *projection operator* method, which allows construction of the basis functions of the irreducible representations of symmetry groups.

We shall confine ourselves to the basic facts needed for practical use of the method, omitting all proofs. A *projection operator* is an operator $\hat{P}_\gamma^\Gamma$ of the form [15]

$$\hat{P}_\gamma^\Gamma = \frac{g(\Gamma)}{g} \sum_{\hat{R}} D_{\gamma\gamma}^{(\Gamma)}(\hat{R})^* \hat{R}, \tag{6.13}$$

where, as before, $g$ is the order of the group, $g(\Gamma)$ is the dimension of the irreducible representation $\Gamma$ and $\hat{R}$ is a group operation. Finally, $\mathbf{D}^{(\Gamma)}(\hat{R})$ is the matrix of the operation $\hat{R}$ in the irreducible representation $\Gamma$, $D_{\gamma\gamma}^{(\Gamma)}(\hat{R})$ is a diagonal element of this matrix and the asterisk denotes complex conjugation. The matrix representation of group operations was introduced in Section 3.3; in conformity with the definition of the matrix of the operation $\hat{R}$ in the representation $\Gamma$, the following general relation is valid:

$$\hat{R}\varphi_{\Gamma\gamma} = \sum_{\gamma'} D_{\gamma'\gamma}^{(\Gamma)}(\hat{R})\varphi_{\Gamma\gamma'}, \tag{6.14}$$

where $\varphi_{\Gamma\gamma}$ are certain ($\gamma$-th) basis functions of the irreducible representation $\Gamma$.

The projection operator method is based on the following property: *the operator $\hat{P}_\gamma^\Gamma$ leaves unchanged the functions $\Psi_{\Gamma\gamma}$ that constitute the basis of the irreducible representation $\Gamma$ and destroys functions belonging to other representations $\Gamma' \neq \Gamma$*. The proof of this can be found for instance in [15]. Let us assume now that the projection operator $\hat{P}_\gamma^\Gamma$ is applied to a certain collection of functions $\Psi_i$ $(i = 1, 2, \ldots, f)$ constituting the basis of any reducible representation $\bar{\Gamma}$:

$$\hat{P}_\gamma^\Gamma \Psi_1, \quad \hat{P}_\gamma^\Gamma \Psi_2, \quad \ldots, \quad \hat{P}_\gamma^\Gamma \Psi_f.$$

We assume also that the irreducible representation $\Gamma$ is contained in the reducible $\bar{\Gamma}$. From the property stated above of the operator $\hat{P}_\gamma^\Gamma$, it can be seen that application of the projection operator gives the $\gamma$th basis function $\Psi_{\Gamma\gamma}$ of the irreducible representation $\Gamma$. To find the other functions of the representation $\Gamma$, we apply the operator

$$\hat{P}_{\gamma\gamma'}^\Gamma = \frac{g(\Gamma)}{g} \sum_{\hat{R}} D_{\gamma\gamma'}^{(\Gamma)}(\hat{R})^* \hat{R}, \tag{6.15}$$

to the function $\Psi_{\Gamma\gamma}$:

$$\hat{P}_{\gamma'\gamma}^\Gamma \Psi_{\Gamma\gamma} = \Psi_{\Gamma\gamma'}. \tag{6.16}$$

By repeated application of $\hat{P}^{\Gamma}_{\gamma'\gamma}$ to $\Psi_{\Gamma\gamma}$, we can obtain all $g(\Gamma)$ basis functions.† It may turn out that they are not normalized, and therefore normalization must be performed separately. It should be mentioned that, by using the "nondiagonal" operator (6.13), we can construct *all* of the basis functions $\Psi_{\Gamma\gamma}$, by fixing the index $\gamma$ and acting on the arbitrary function $\Psi$ of the reducible representation.

**Example 6.5** The inversion group $\mathbf{C}_i$ (Section 2.4.2) consists of two operations, $\hat{E}$ and $\hat{I}$ $(g = 2)$. In this group there are two one-dimensional $(g(\Gamma) = 1)$ irreducible representations $A_g$ and $A_u$, with even and odd basis functions respectively. The former can be, for instance, the components of the axial vector $\mathbf{R}$ while the latter can be the components of the polar vector $\mathbf{r}$. The character table is very simple (Table 6.2). The matrices $D^{(\Gamma)}$ are one-dimensional, i.e. they are numbers. We apply the projection operators $\hat{P}^{A_g}$ and $P^{A_u}$ to the arbitrary function $\Psi(x)$:

$$\left.\begin{aligned} \hat{P}^{A_g}\Psi(x) &= \tfrac{1}{2}[D^{(A_g)}(\hat{E})^*\hat{E}\Psi(x) + D^{(A_g)}(\hat{I})^*\hat{I}\Psi(x)], \\ \hat{P}^{A_u}\Psi(x) &= \tfrac{1}{2}[D^{(A_u)}(\hat{E})^*\hat{E}\Psi(x) + D^{(A_u)}(\hat{I})^*\hat{I}\Psi(x)]. \end{aligned}\right\} \tag{6.17}$$

Since $\hat{E}\Psi(x) = \Psi(x)$ and $\hat{I}\Psi(x) = \Psi(-x)$ (inversion changes the sign of $x$; see Section 2.1), we obtain

$$\left.\begin{aligned} \Psi_{A_g}(x) &= \tfrac{1}{2}[\Psi(x) + \Psi(-x)], \\ \Psi_{A_u}(x) &= \tfrac{1}{2}[\Psi(x) - \Psi(-x)]. \end{aligned}\right\} \tag{6.18}$$

The function $\Psi_{A_g}$ is even, while $\Psi_{A_u}$ is odd, which means that they are really bases for the above-mentioned representations of $\mathbf{C}_i$. In this simple example the matrices $D^{(\Gamma)}$ are one-dimensional; therefore they are contained in the character table.

**Table 6.2** Characters of $\mathbf{C}_i$.

| $\mathbf{C}_i$ | $\hat{E}$ | $\hat{I}$ | |
|---|---|---|---|
| $A_g$ | 1 | 1 | $R_x$; $R_y$; $R_z$ |
| $A_u$ | 1 | $-1$ | $x$; $y$; $z$ |

The projection operator method is used below for constructing basis functions of irreducible representations of point groups from basis functions of irreducible representations of the full rotation group; that is, from tensor operators $\mathbf{T}^k_m$ or spherical harmonics $Y_{lm}$. It should be emphasized here that in

† Equation (6.16) is true only for so-called unitary representations. More generally, the operators are [15]

$$\hat{P}_{\gamma'\gamma} = \frac{g(\Gamma)}{g}\sum_{\hat{R}} \tilde{D}^{(\Gamma)}(\hat{R})_{\gamma'\gamma}\hat{R}$$

where $\tilde{D}^{(\Gamma)}(\hat{R}) \equiv \tilde{D}(R^{-1})$. When $D^{(\Gamma)}$ is unitary $\tilde{D}^{(\Gamma)}_{\gamma'\gamma}(R) = D^{(\Gamma)}_{\gamma\gamma'}(R)^*$ and we get (6.15) and (6.16). Full and deep consideration of the projection operators is given in [15].

order to solve classification problems—to determine the rank $k$ of crystal harmonics (Section 6.3.1) or to find the terms in a crystal field (Section 4.3)—only the characters of the rotation group and the point group are necessary. The application of the projection operator requires knowledge of the matrices of the irreducible representations of the point group, as well as an analytical expression for the action of $\hat{R}$ upon $Y_{lm}$. As we shall see below, this requires the matrix forms of representations of the rotation group.

### 6.3.3 Euler angles, and irreducible representations of the rotation group

Any transformation of the coordinate system due to a rotation group operation can be performed by three successive rotations through the Euler angles $\alpha$, $\beta$ and $\gamma$ [2, 5]:

(1) rotation about the $z$ axis through an angle $\gamma$ ($0 \leq \gamma < 2\pi$) (Fig. 6.1a);

(2) rotation about the $y$ axis through an angle $\beta$ ($0 \leq \beta \leq \pi$) (Fig. 6.1b);

(3) rotation about the $z$ axis through an angle $\alpha$ ($0 \leq \alpha < 2\pi$) (Fig. 6.1c).

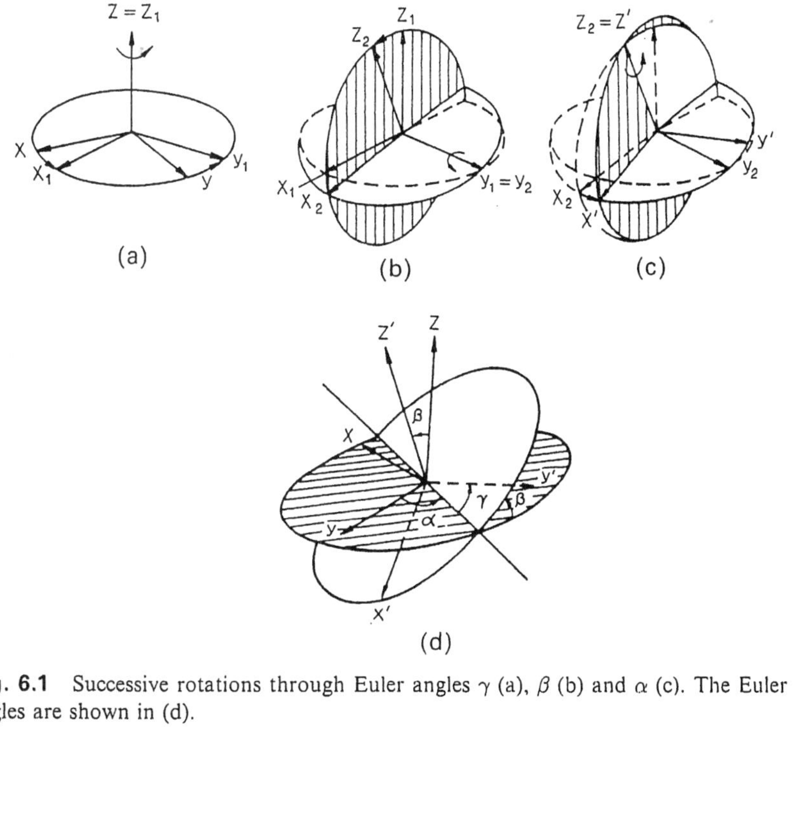


**Fig. 6.1** Successive rotations through Euler angles $\gamma$ (a), $\beta$ (b) and $\alpha$ (c). The Euler angles are shown in (d).

Hence each rotation group operation $\hat{R}$ is defined by the Euler angles: $\hat{R} = \hat{R}(\alpha, \beta, \gamma)$. Therefore the matrix of an operation $\hat{R}$ in a representation with number $l$ (angular momentum quantum number) can be denoted by $\mathbf{D}^{(l)}(\alpha\beta\gamma)$. Since the irreducible representations of the rotation group are defined in the basis of spherical functions $Y_{lm}$, the matrix, of dimension $2l+1$, has elements $D^{(l)}_{m'm}(\alpha\beta\gamma) = \int Y^*_{lm'} \hat{R} Y_{lm} \, d\tau$. In accordance with the general definition of the representation, the action of the operation $\hat{R}$ on spherical functions is expressed by

$$\hat{R}(\alpha\beta\gamma) Y_{lm} = \sum_{m'=-l}^{l} D^{(l)}_{m'm}(\alpha\beta\gamma) Y_{lm'}. \tag{6.19}$$

The elements of $\mathbf{D}^{(l)}$, in terms of the Euler angles, are [2]

$$D^{(l)}_{m'm}(\alpha\beta\gamma) = \sum_{\kappa} (-1)^{\kappa} \frac{[(l+m)!(l-m)!(l+m')!(l-m')!]^{1/2}}{(l-m'-\kappa)!(l+m-\kappa)!\kappa!(\kappa+m'-l)!} \times e^{\mathrm{i}m'\alpha} (\cos\tfrac{1}{2}\beta)^{2l-m-m'-2\kappa} (\sin\tfrac{1}{2}\beta)^{2\kappa+m'-m} e^{\mathrm{i}m\gamma}. \tag{6.20}$$

The summation is over integer $\kappa$ from the greater of 0 and $m - m'$ up to the smaller of $l - m'$ and $l + m$. Substituting the Euler angles into (6.20) and using (6.19), one can obtain the result of the action of any transformation on the spherical functions $Y_{lm}$ or tensors $\mathbf{T}^k_m$.

### 6.3.4 Matrices of irreducible representations of point groups

To each operation $\hat{R}$ of a point group there corresponds a certain set of Euler angles. For example, for the simplest case of the rotations $\hat{C}_3(z)$ and $\hat{C}_4(z)$, $\{\alpha\beta\gamma\} = \{00\tfrac{2}{3}\pi\}$ and $\{00\tfrac{1}{4}\pi\}$ respectively. To compute the matrices of the irreducible representations $\Gamma$, of the point group, one can choose as a basis the spherical harmonics $Y_{lm}$ giving the representation $\Gamma$ and possessing the minimum values of $l$ in the reduction scheme $\mathrm{D}^{(l)} \doteq \Sigma_i \Gamma_i$. Acting on these simplest basis functions with the operations $\hat{R}(\alpha\beta\gamma)$ of the point group, we can obtain the matrices of the irreducible representations. The matrices of all the point groups are collected in [15] (the matrices of some point groups are given in [49–55]). Full results, including double-valued representations, are given in the tables in [7], which are divided into seven parts, in accordance with the seven crystal systems (see Section 2.5), and contain the following data:

(1) the operations of the point group and the corresponding Euler angles, $\hat{R}(\alpha, \beta, \gamma)$;

(2) the bases of all the irreducible representations of each point group;

(3) the matrices of the irreducible representations.

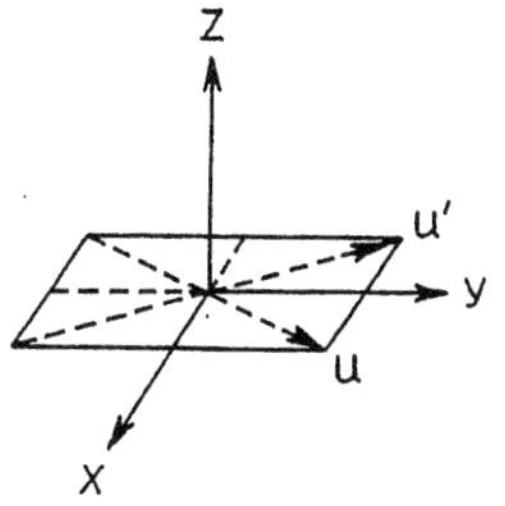


**Fig. 6.2** Symmetry axes of $\mathbf{D}_4$.

Since matrices of irreducible representations are important for many physical and chemical applications, we shall take as an example of the structure of the tables in [7] a limited part related to the group $\mathbf{D}_4$ of a tetragonal crystal system.†

The *symmetry operations* (Fig. 6.2) are

$$\hat{E}(0\ 0\ 0),\quad \hat{C}_{z4}^2(0\ 0\ \pi),\quad \hat{C}_{z4}(0\ 0\ \tfrac{1}{2}\pi),\quad \hat{C}_{z4}^3(0\ 0\ \tfrac{3}{2}\pi),$$
$$\hat{C}_{x2}(0\ \pi\ \pi),\quad \hat{C}_{y2}(0\ \pi\ 0),\quad \hat{C}_{u2}(0\ \pi\ \tfrac{1}{2}\pi),\quad \hat{C}_{u'2}(0\ \pi\ \tfrac{3}{2}\pi).$$

The *bases* are

$$\mathrm{A}_2, \Gamma_1: \quad Y_{20},\quad -(3z^2-r^2),\quad -\sqrt{1/6}\,[3\hat{S}_z^2 - S(S+1)];$$
$$\mathrm{A}_2, \Gamma_2: \quad Y_{10},\quad \mathrm{i}z,\quad \mathrm{i}\hat{S}_z;$$
$$\mathrm{B}_1, \Gamma_3: \quad \sqrt{1/2}\,(Y_{22}+Y_{2-2}),\quad -(x^2-y^2),\quad -\tfrac{1}{2}\sqrt{1/2}\,(\hat{S}_+^2+\hat{S}_-^2);$$
$$\mathrm{B}_2, \Gamma_4: \quad \frac{i}{\sqrt{2}}(Y_{22}-Y_{2-2}),\quad xy,\quad -\tfrac{1}{2}\sqrt{1/2}\,\mathrm{i}\,(\hat{S}_+^2-\hat{S}_-^2);$$
$$\mathrm{E}, \Gamma_5: \quad \begin{cases} (1) & Y_{11}, \quad -\mathrm{i}x_+, \quad -\sqrt{1/2}\,\mathrm{i}\,\hat{S}_+; \\ (2) & Y_{1-1}, \quad \mathrm{i}x_-, \quad \sqrt{1/2}\,\mathrm{i}\,S_-. \end{cases}$$

The *matrices of operations* are

| $\mathbf{D}_4$ | $\hat{E}$ | $\hat{C}_4^2(z)$ | $\hat{C}_4^3(z)$ | $\hat{C}_4(z)$ | $\hat{C}_2(x)$ | $\hat{C}_2(y)$ | $\hat{C}_2(u')$ | $\hat{C}_2(u)$ |
|---|---|---|---|---|---|---|---|---|
| $\mathrm{A}_1, \Gamma_1$ | 1 | 1 | 1 | 1 | 1 | 1 | 1 | 1 |
| $\mathrm{A}_2, \Gamma_2$ | 1 | 1 | 1 | 1 | −1 | −1 | −1 | −1 |
| $\mathrm{B}_1, \Gamma_3$ | 1 | 1 | −1 | −1 | 1 | 1 | −1 | −1 |
| $\mathrm{B}_2, \Gamma_4$ | 1 | 1 | −1 | −1 | −1 | −1 | 1 | 1 |
| $\mathrm{E}, \Gamma_5$ | $\begin{bmatrix}1 & 0\\1 & 0\end{bmatrix}$ | $\begin{bmatrix}-1 & 0\\0 & -1\end{bmatrix}$ | $\begin{bmatrix}\mathrm{i} & 0\\0 & -\mathrm{i}\end{bmatrix}$ | $\begin{bmatrix}-\mathrm{i} & 0\\0 & \mathrm{i}\end{bmatrix}$ | $\begin{bmatrix}0 & -1\\-1 & 0\end{bmatrix}$ | $\begin{bmatrix}0 & 1\\1 & 0\end{bmatrix}$ | $\begin{bmatrix}0 & -\mathrm{i}\\\mathrm{i} & 0\end{bmatrix}$ | $\begin{bmatrix}0 & \mathrm{i}\\-\mathrm{i} & 0\end{bmatrix}$ |

As the basis for the irreducible representations three types of object have been chosen: spherical functions $Y_{lm}$, coordinates $z$ and $x_\pm = x \pm \mathrm{i}y$, and spin

† The matrices for some selected point groups are given in the Appendix I. For the full set of matrices see [15].

operators $\hat{S}_z$ and $\hat{S}_\pm = \hat{S}_x \pm \mathrm{i}\hat{S}_y$. We remark here that the spin operators are transformed as the components of an axial vector (Section 3.7.2) under transformations of the coordinate system. A single-element basis corresponds to a one-dimensional representation; two elements associated by braces are connected with the two-dimensional ($\Gamma_5$) representation.

Using the definition (6.12) of the matrix of a group operation, it is possible to obtain the result of the action of any operation on an element of a basis.

**Example 6.6** Consider the rotation $\hat{C}_2(u)$ (Fig. 6.2) on the spherical function $Y_{11}$. From the list of basis elements, we find that $Y_{11}$ is the first element of the two-dimensional representation $\Gamma_5$. Therefore

$$\begin{aligned}\hat{C}_2(u)Y_{11} &= D_{11}^{(\Gamma_5)}(\hat{C}_2(u))^* Y_{11} + D^{(\Gamma_5)}(\hat{C}_2(u))^* Y_{1-1} \\ &= 0 \cdot Y_{11} + (-\mathrm{i})^* Y_{1-1} = \mathrm{i}\, Y_{1-1}.\end{aligned}$$

For one-dimensional representations the results are trivial; for instance,

$$\hat{C}_4(z)Y_{10} = 1 \cdot Y_{10}, \qquad \hat{C}_2(y)Y_{10} = (-1)Y_{10} \qquad \text{etc.}$$

We now have all that is necessary to construct the bases of irreducible representations of point groups from spherical tensors $\mathbf{T}_m^k$.

### 6.3.5 Basis functions of irreducible representations of point groups

The functions that transform according to the irreducible representations of point the groups are constructed by applying the projection operator to the spherical tensor components or to the spherical functions $Y_{lm}$. The computational procedure comprises the following steps:

(1) expanding the initial representation of the rotation group $\mathrm{D}^{(l)}$ in terms of irreducible (for the given point group) representations $\Gamma_i$, which are obtained using (4.16) and (4.12) or the reduction tables in [1] (Section 5.2.3);

(2) expressing the operations $\hat{R}$ of the point group through Euler transformations;

(3) obtaining the results of the action of the operations $\hat{R}(\alpha\beta\gamma)$ on the basis elements $Y_{lm}$ or $\mathbf{T}_m^k$ of the reducible (in the point group) representation, (6.19), (6.20);

(4) computing the functions $\Psi_{\Gamma_i\gamma_i}$, transformed according to the irreducible representations $\Gamma_i$ as linear combinations of spherical harmonics, using the projection operator method, (6.13), (6.15) and the matrices of the point group representations (see the tables in [7]).

**Example 6.7** Let us construct the basis functions of the group $\mathbf{D}_3$ from the spherical harmonics $Y_{2m}$ $(l = 2)$.

(1) We reduce the five-dimensional representation $\mathrm{D}^{(2)}$ in $\mathbf{D}_3$, thus obtaining one one-dimensional representation $(\Gamma_1)$ and two two-dimensional representations $(\Gamma_3)$ [1]:

$$\mathrm{D}^{(2)} \doteq \Gamma_1 + 2\Gamma_3.$$

(2) We find the Euler angles corresponding to the operations of $\mathbf{D}_3$ (Fig. 6.3):

$$\hat{E}(0\ 0\ 0),\quad \hat{C}_3(0\ 0\ \tfrac{2}{3}\pi),\quad \hat{C}_3^2(0\ 0\ \tfrac{4}{3}\pi),\quad \hat{C}_2(\tfrac{4}{3}\pi\ 0),\quad \hat{C}_2'(0\ \pi\ 0),\quad \hat{C}_2''(\tfrac{2}{3}\pi\pi\ 0).$$

(3) From the general expression for the matrix $D^{(l)}_{m'm}(\alpha\beta\gamma)$ we determine the nonvanishing elements; for the elements $D^{(2)}_{m0}$ we have

$$D^{(2)}_{00}(0\ 0\ 0) = D^{(2)}_{00}(0\ 0\ \tfrac{2}{3}\pi) = D^{(2)}_{00}(0\ 0\ \tfrac{4}{3}\pi) = D^{(2)}_{00}(\tfrac{4}{3}\pi\pi\ 0)$$
$$= D^{(2)}_{00}(0\ \pi\ 0) = D^{(2)}_{00}(\tfrac{2}{3}\pi\ \pi\ 0) = 1,$$

so that

$$\hat{C}_3 Y_{20} = 1 \cdot Y_{20}, \qquad \hat{C}_2' Y_{20} = 1 \cdot Y_{20}, \qquad \text{etc.}$$

(4) From the tables in [7], we find the matrices of the representations of $\mathbf{D}_3$, for example

$$\mathbf{D}^{(\Gamma_3)}(\hat{C}_3) = \begin{bmatrix} \omega & 0 \\ 0 & \omega^* \end{bmatrix}, \qquad \mathbf{D}^{(\Gamma_3)}(\hat{C}_2) = \begin{bmatrix} 0 & \omega^2 \\ \omega^{*2} & 0 \end{bmatrix},$$

where $\omega = e^{2\pi i/3}$. The basis of the two-dimensional representation comprises $Y_{11}$ and $Y_{1-1}$, and, applying the projection operator, we obtain

$$\Psi_{\Gamma_1} = Y_{20}, \qquad \Psi^{(1)}_{\Gamma_3} = \begin{cases} Y_{22}, \\ Y_{2-2}, \end{cases} \qquad \Psi^{(2)}_{\Gamma_3} = \begin{cases} Y_{21}, \\ Y_{2-1}, \end{cases}$$

where the superscripts enumerate the components of the representation $\Gamma_3$.

The projection operator method has been used by Leushin [5] to find the complete tables of functions forming the bases of the irreducible representations of all 32 point groups. Results have been obtained for angular

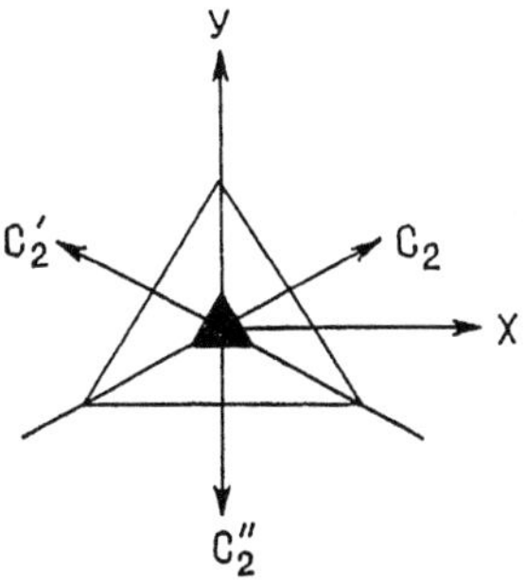


**Fig. 6.3** Symmetry axes of $\mathbf{D}_3$.

**Table 6.3** Coefficients determining the basis functions of **O** ($l = 4$).

| $D^{(4)} \doteq A_1 + E + T_1 + T_2$ | | |
|---|---|---|
| $A_1, \Gamma_1$: | (1) | $c_4 = c_{-4} = \frac{5}{24}$, $c_0 = \frac{7}{12}$ |
| $E, \Gamma_3$: | (1) | $c_4 = c_{-4} = \frac{7}{24}$, $c_0 = {}^{*}\frac{5}{12}$ |
| | (2) | $c_2 = c_{-2} = \frac{1}{2}$. |
| $T_1, \Gamma_4$: | (1) | $c_1 = \frac{7}{8}$, $c_{-3} = \frac{1}{8}$ |
| | (2) | $c_4 = -c_{-4} = {}^{*}\frac{1}{2}$ |
| | (3) | $c_3 = {}^{*}\frac{1}{8}$, $c_1 = {}^{*}\frac{7}{8}$ |
| $T_2, \Gamma_5$: | (1) | $c_1 = {}^{*}\frac{1}{8}$, $c_{-3} = \frac{7}{8}$ |
| | (2) | $c_2 = -c_{-2} = \frac{1}{8}$ |
| | (3) | $c_3 = \frac{7}{8}$, $c_{-1} = {}^{*}\frac{1}{8}$ |

momentum values $l$ up to and including $\frac{17}{2}$. Similar tables for different particular cases are given in [11, 51–55]. For the full results for all space groups see [8–10].

The tables in [5] contain the coefficients $c_m$ that determine the linear combinations of the spherical harmonics $Y_{lm}$ and constitute the bases of irreducible representations of the point groups.†

**Example 6.8** Let us construct the basis functions of the irreducible representations of the group **O** from the components of the spherical tensor with $l = 4$. For a cubic crystal system the coefficients $c_m$ are shown in Table 6.3 (for consistency with the notation adopted here they have been slightly modified from the original [5]). The table gives the reduction of the representation $D^{(4)}$ in **O** (see Table 4.2), providing a classification of the basis functions $\Psi_{\Gamma\gamma}$, as well as the coefficients $c_m$. To construct the basis, the square root of $c_m$ should be taken; an asterisk indicates that the root is to be taken with the minus sign (e.g. if $c_3 = \frac{1}{8}$ then $c_3^{1/2} = \frac{1}{2}\sqrt{1/2}$, while if $c_3 = {}^{*}\frac{1}{8}$ then $c_3^{1/2} = -\frac{1}{2}\sqrt{1/2}$). The numbers in parentheses following braces enumerate the components of the irreducible representation. The asterisk at the top left of a number indicating a basis function of an irreducible representation shows that the corresponding numbers $c_m$ are purely imaginary. In these cases the positive value $\sqrt{c_m}$ must be multiplied by the imaging unit. They do not give any information about the basis functions, and therefore it should be kept in mind that the enumeration adopted in [5] corresponds to the order of the simplest (Section 6.3.4) basis elements presented in the tables in [5] (an example can be found in Section 6.3.4). Taking these remarks into account, we obtain at once from

† Some results are collected in the Appendix II.

Table 6.3

$$\Psi_{A_1} = \sqrt{5/24}(Y_{44} + Y_{4-4}) + \sqrt{7/12}\,Y_{40}, \tag{6.21}$$

$$\left.\begin{aligned}\Psi_{E1} &= \sqrt{7/24}(Y_{44} + Y_{4-4}) - \sqrt{5/12}\,Y_{40},\\ \Psi_{E2} &= \sqrt{1/2}(Y_{42} + Y_{4-2}).\end{aligned}\right\} \tag{6.22}$$

The other functions ($\Psi_{T_1\gamma}$ and $\Psi_{T_2\gamma}$) are obtained in a similar way. The simplest basis elements of the irreducible representation E are obtained from the components of the second-rank tensor ($D^{(2)} \doteq E + T_2$); from the tables in [5] we find that the basis functions $\Psi_{E1}$ and $\Psi_{E2}$, (6.22), are transformed either as the pairs $Y_{20}$, $\sqrt{1/2}(Y_{22} + Y_{2-2})$ or $-(3z^2 - r^2)$, $-\sqrt{3}(x^2 - y^2)$, or as combinations of spin-operator components $-\sqrt{1/6}[3(\hat{S}_z^2 - S(S+1)]$, $-\frac{1}{2}\sqrt{1/2}(\hat{S}_+^2 + \hat{S}_-^2)$. In other words, the standard matrices of irreducible representations apply equally well both to the simplest basis elements and to complicated basis functions, constructed from tensor operators of higher rank.

The projection operator method makes it possible to avoid guesswork in finding the basis functions, as was used in Section 3.5.3 when illustrating the concept of irreducible representations.

## 6.4 CRYSTAL FIELD EFFECTIVE HAMILTONIAN

### 6.4.1 Rules for construction of invariants

Let us return to the issue raised in Section 6.3.1 of finding linear combinations of tensor operators that are invariant under the operations of the ion point group in a crystal field. This problem was substantially solved in the previous section—the linear combinations obtained there being transformed according to the totally symmetric representation.

Let us consider the invariant (6.21) of the group $\mathbf{O}_h$, constructed from irreducible rank-4 tensors:

$$\tfrac{1}{2}\sqrt{5/6}[\hat{\mathbf{T}}_4^4(\mathbf{r}) - \hat{\mathbf{T}}_{-4}^4(\mathbf{r})] + \tfrac{1}{2}\sqrt{7/3}\,\hat{\mathbf{T}}_0^4(\mathbf{r}).$$

Comparing this expression with the crystal field operator (6.10), we see that the presence of symmetry prohibits the existence of some constants $A_m^k$ ($A_1^4 = A_{-1}^4 = A_3^4 = A_{-3}^4 = 0$), and establishes relations between others:

$$A_4^4 = A_{-4}^4, \qquad A_4^4 : A_0^4 = \tfrac{1}{2}\sqrt{5/6} : \tfrac{1}{2}\sqrt{7/3}. \tag{6.23}$$

Therefore the contribution of rank-4 tensors to the crystal field operator for a cubically coordinated ion can be expressed as follows:

$$\tfrac{1}{2}\sqrt{7/3}\,A_4^4[\hat{\mathbf{T}}_0^4 + \sqrt{5/14}(\hat{\mathbf{T}}_4^4 + \hat{\mathbf{T}}_{-4}^4)]. \tag{6.24}$$

Similarly, using the tables in [3], we can compose an invariant from a spherical tensor of rank 6. The corresponding contribution to the operator $\hat{V}^{c}$ has the form

$$-\tfrac{1}{2}\sqrt{1/2}A_4^6[\hat{\mathbf{T}}_0^6 - \tfrac{1}{2}\sqrt{7/3}(\hat{\mathbf{T}}_4^6 + \hat{\mathbf{T}}_{-4}^6)]. \tag{6.25}$$

Adopting for convenience the notation $\frac{1}{2}\sqrt{7/3}A_4^4 = D_4$ and $-\frac{1}{2}\sqrt{1/2}A_4^6 = D_6$, we can write the crystal field operator in the form

$$\hat{V}^{c}(\mathbf{r}) = D_4[\hat{\mathbf{T}}_0^4 + \sqrt{5/14}(\hat{\mathbf{T}}_4^4 + \hat{\mathbf{T}}_{-4}^4)] + D_6[\hat{\mathbf{T}}_0^6 - \tfrac{1}{2}\sqrt{7/3}(\hat{\mathbf{T}}_4^6 + \hat{\mathbf{T}}_{-4}^6)]. \tag{6.26}$$

Before describing the construction procedures for the crystal field Hamiltonian, let us consider some examples of low-symmetry groups.

**Example 6.9** Let us take the centrosymmetric group $\mathbf{D}_{4h}$ of the tetragonal crystal group. The totally symmetric representation $A_{1g}$ is contained in the expansion $D^{(2)}$ and $D^{(4)}$ (twice). If we take spherical tensors of only second and fourth orders, we thus obtain three invariant operators. Using the tables in [5], we finally have

$$\hat{\mathbf{T}}_0^2,\quad \tfrac{1}{2}\sqrt{7/3}\hat{\mathbf{T}}_0^4 + \tfrac{1}{2}\sqrt{5/6}(\hat{\mathbf{T}}_4^4 + \hat{\mathbf{T}}_{-4}^4),\quad -\tfrac{1}{2}\sqrt{5/3}\hat{\mathbf{T}}_0^4 + \tfrac{1}{2}\sqrt{7/6}(\hat{\mathbf{T}}_4^4 + \hat{\mathbf{T}}_{-4}^4).$$

Multiplying each invariant by its "structural" constant $D$, we obtain the operator of the $\mathbf{D}_{4h}$ symmetry crystal field:

$$\hat{V}_{\text{tetr}}^{c}(\mathbf{r}) = D_2\hat{\mathbf{T}}_0^2 + D_4[\hat{\mathbf{T}}_0^4 + \sqrt{5/14}(\hat{\mathbf{T}}_4^4 + \hat{\mathbf{T}}_{-4}^4)] + D_4'[\hat{\mathbf{T}}_0^4 - \sqrt{7/10}(\hat{\mathbf{T}}_4^4 + \hat{\mathbf{T}}_{-4}^4)]. \tag{6.27}$$

One of the invariants (that including $D_4$) coincides with the cubic field operator (6.24). This is not accidental, since $\mathbf{D}_4$ is a subgroup of $\mathbf{O}_h$, whose totally symmetric representation, when reduced a subgroup, remains totally symmetric.

**Example 6.10** Consider the noncentrosymmetric group $\mathbf{C}_{4v}$, belonging to the tetragonal crystal group. Without restating the reasoning of Section 6.3.1, we mention here that the operator $\hat{V}^{c}(\mathbf{r})$ contains both an even component and an odd one. The former is constructed according to Table 6.1 from the spherical tensors $\hat{\mathbf{T}}_m^2$ and $\hat{\mathbf{T}}_m^4$; the latter is constructed from the tensors $\hat{\mathbf{T}}_m^1$ and $\hat{\mathbf{T}}_m^3$ ($k \le 4$). From the tables in [5] it is easy to see that the even part of the crystal field of $\mathbf{C}_{4v}$ coincides with the field of $\mathbf{D}_{4h}$ symmetry (see (6.27)). The odd component $\hat{V}_u^{c}(\mathbf{r})$ contains two invariants:

$$\hat{V}_u^{c}(\mathbf{r}) = A_1\hat{\mathbf{T}}_0^1 + A_3(\hat{\mathbf{T}}_2^3 - \hat{\mathbf{T}}_{-2}^3), \tag{6.28}$$

where $A_1$ and $A_3$ are "structural" parameters of the form (6.2).

**Example 6.11** Consider the noncentrosymmetric group $\mathbf{C}_{3v}$ of trigonal symmetry. The even component of the crystal field ($k \le 4$) has the form

$$V_{\text{trig}}^{c} = D_2\hat{\mathbf{T}}_0^2 + D_4\hat{\mathbf{T}}_0^4 + D_4'(\hat{\mathbf{T}}_3^4 - \hat{\mathbf{T}}_{-3}^4), \tag{6.29}$$

the $z$ axis coinciding with the trigonal axis $C_3$. For the odd field we obtain

$$\hat{V}_u^{c} = A_1\hat{\mathbf{T}}_0^1 + A_3\hat{\mathbf{T}}_0^3 + A_3'(\hat{\mathbf{T}}_3^3 - \hat{\mathbf{T}}_{-3}^3). \tag{6.30}$$

The crystal field of $\mathbf{C}_{3v}$ thus contains six parameters.

The structural parameters $D_k$ and $A_k$ can be computed using (6.2) if the structure of the complex is known; that is, if the coordinates $(R_\alpha, \vartheta_\alpha, \varphi_\alpha)$ of all the ions are known. However, in this case too the crystal field parameters cannot be computed with sufficient accuracy, since the effective charges $q_\alpha$ can differ considerably from the pure ion charges, and the average values $\langle r^k \rangle$ are far from the atomic values. That is why a *semiempirical* or *phenomenological* approach is often used. In such an approach the parameters $D_i$ and $A_i$ are not computed; rather, they are determined by comparing computed diagrams of ion levels with spectral line frequencies and intensities obtained experimentally. Semiempirical crystal field theory uses only a "model" Hamiltonian, constructed by means of the group-theoretical methods described above and does not involve approximate computation of parameters—it is based only on symmetry considerations and is absolutely exact from this point of view. A Hamiltonian constructed on the basis of invariance considerations and containing phenomenological parameters is called an *effective Hamiltonian.* Of course, group theory does not provide any relations between the parameters of the effective Hamiltonian; it only excludes those that are zero owing to symmetry considerations. Having said this, let us now describe the procedure for constructing the effective crystal field Hamiltonian:

(1) determine the ion point group **G** in the crystal (Section 2.6);

(2) construct the reduction scheme for the irreducible representations of the rotation group $\mathrm{D}^{(k)}$ $(k = 1, 2, 3, \ldots)$ in **G** (Section 6.3.1);

(3) select the spherical tensors $\hat{\mathbf{T}}^k_m$ from which the invariants in **G** can be constructed (Section 6.3.1);

(4) using either the projection operator method or the tables in [5], construct the invariants $\Psi^{(k)}_{\mathrm{A}_1}$ from the components of the tensors $\hat{\mathbf{T}}^k_m$ (see this section);

(5) on multiplying each invariant by its semiempirical parameter and then summing the products, we obtain the effective crystal field Hamiltonian.

### 6.4.2 Energy levels and wavefunctions

Let us consider the splitting of an atomic $LS$ term by a crystal field of given symmetry. To first order of perturbation theory it is necessary to set up a secular equation of the form

$$|\langle \kappa SLM_S M_L|\hat{V}^{\mathrm{c}}|\kappa SLM'_S M'_L\rangle - \varepsilon \delta_{M_S M'_S}\delta_{M_L M'_L}| = 0, \tag{6.31}$$

where $\varepsilon$ is the correction to the energy $E(\kappa SL)$ of the atomic level due to the crystal field. As mentioned above (see (6.11)), the crystal field does not mix states with different $M_S$. Therefore (6.31) can be written as

$$|\langle \kappa SLM_S M_L|\hat{V}^{\mathrm{c}}|\kappa SLM_S M'_L\rangle - \varepsilon \delta_{M_L M'_L}| = 0. \tag{6.32}$$

As in Section 4.3.5, we shall use an approach based on the Wigner theorem. To each crystal term there corresponds a basis function of the irreducible representation. This makes it possible to construct the correct zeroth-order wavefunctions, corresponding to the levels in the crystal field, without solving the secular equation (6.32). For this it is enough to construct the linear combinations of the atomic functions forming the basis of the irreducible representations $\Gamma$ of the point group:

$$\Psi(\kappa SLM_S\Gamma\gamma) = \sum_{M_L=-L}^{L} c(LM_L)|\kappa SLM_SM_L\rangle, \qquad (6.33)$$

the representations $\Gamma$ being contained in the expansion of $D^{(L)}$ for the point group. Thus we shall again find it necessary to use the projection operator method (Section 6.3). The wavefunctions (6.33) are constructed by applying the projection operator to the atomic $LS$ functions:

$$\Psi(\kappa SLM_S\Gamma\gamma) = \hat{P}_\gamma^\Gamma|\kappa SLM_SM_L\rangle. \qquad (6.34)$$

For the practical application of this method it is sufficient to use the tables in [5] of functions forming the bases of the irreducible representations of point groups.

**Example 6.12** Let us construct the wavefunctions of the $^4F$ term of a $d^3$ ion in octahedral coordination. In the $\mathbf{O}_h$ point group the $^4F(d^3)$ term is split into three levels: $^4A_{2g}$, $^4T_{1g}$ and $^4T_{2g}$ (see Table 5.3). Denoting $|\tfrac{3}{2}3M_SM_L\rangle$ by $\Psi_{M_L}$ for simplicity, we obtain the basis functions of the irreducible representations of $\mathbf{O}_h$:

$$\Psi(^4A_{2g}) = \sqrt{(1/2)}\,i(\Psi_2 - \Psi_{-2}), \qquad (6.35)$$

$$\left.\begin{aligned} \Psi_1(^4T_{2g}) &= \tfrac{1}{2}\sqrt{3/2}\,\Psi_1 + \tfrac{1}{2}\sqrt{5/2}\,\Psi_{-3}, \\ \Psi_2(^4T_{1g}) &= i\Psi_0, \\ \Psi_3(^4T_{1g}) &= \tfrac{1}{2}\sqrt{5/2}\,\Psi_3 + \tfrac{1}{2}\sqrt{3/2}\,\Psi_{-1}, \end{aligned}\right\} \qquad (6.36)$$

$$\left.\begin{aligned} \Psi_1(^4T_{2g}) &= -\tfrac{1}{2}\sqrt{5/2}\,\Psi_1 + \tfrac{1}{2}\sqrt{3/2}\,\Psi_{-3}, \\ \Psi_2(^4T_{2g}) &= \sqrt{(1/2)}\,i(\Psi_2 + \Psi_{-2}), \\ \Psi_3(^4T_{2g}) &= -\tfrac{1}{2}\sqrt{3/2}\,\Psi_3 + \tfrac{1}{2}\sqrt{5/2}\,\Psi_{-1}. \end{aligned}\right\} \qquad (6.37)$$

The enumeration of the functions of the three-dimensional representations ($\Psi_1, \Psi_2, \Psi_3$) corresponds to that of the simplest basis elements in the tables of matrices in [7]. Thus for the triplet $^4T_{2g}$ the indices 1, 2 and 3 correspond to the simplest basis $\Psi_{lm}$ with $l = 2$: $\Psi_{21}$, $\sqrt{(1/2)}\,i(\Psi_{22} - \Psi_{2-2})$ and $\Psi_{2-1}$.

**Example 6.13** Let us consider the $^4F(d^3)$ term in the tetragonal crystal field $\mathbf{D}_{4h}$. According to the reduction scheme

$$D^{(3)} = A_2 + B_1 + B_2 + 2E,$$

we find that the $^4F$ level is split into three singlets, $^4A_{2g}$, $^4B_{1g}$ and $^4B_{2g}$, and two doublets,

$2\,^4E_g$ (see Table 5.3). $\mathbf{D}_{4h}$ is a subgroup of $\mathbf{O}_h$; therefore its representations can be obtained as the result of branching of the $\mathbf{O}_h$ group representations (see Table 4.4). In the case considered here the branching diagram has the form

$$\begin{array}{ccccccccc} & & & D^{(3)}(^4F) & & & & & \mathbf{K}_h \\ & & \swarrow & \downarrow & \searrow & & & & \downarrow \\ & ^4A_{2g} & & ^4T_{1g} & & ^4T_{2g} & & & \mathbf{O}_h \\ \swarrow & & \swarrow & \downarrow & & \downarrow & \searrow & & \downarrow \\ ^4B_{1g} & & ^4A_{2g} & ^4E_g^{(1)} & & ^4B_{2g} & & ^4E_g^{(2)} & \mathbf{D}_{4h} \end{array}$$

Here the repeating representations $E_g$ are indicated by superscripts. According to this diagram, the basis functions of the subgroup are obtained from the basis functions (6.35)–(6.37) of $\mathbf{O}_h$:

$$\begin{array}{ll} \mathbf{D}_{4h} \qquad \mathbf{O}_h & \mathbf{D}_{4h} \qquad \mathbf{O}_h \\ \Psi(^4B_{1g}) = \Psi(^4A_{2g}), & \Psi_1(E_g^{(2)}) = -\Psi_1(^4T_{2g}) \\ \Psi(^4A_{2g}) = i\Psi_2(^4T_{1g}), & \Psi_2(E^{(2)}) = \Psi_2(^4T_{2g}), \\ \Psi_1(^4Eg^{(1)}) = \Psi_1(^4T_{1g}), & \Psi(^4B_{2g}) = \Psi_2(^4T_{2g}). \\ \Psi_2(^4E_g^{(1)}) = \Psi_3(^4T_{1g}), & \end{array} \tag{6.38}$$

As in the case of a strong field (Section 5.6), two cases must be distinguished—nonrepeating and repeating crystal terms $S\Gamma$. In the first case the wavefunctions $\Psi(\kappa SLM_S\Gamma\gamma)$ diagonalize the interaction of the electrons with the crystal field, and the energy can be computed directly:

$$E = E(\kappa SL) + \langle \kappa SLM_S\Gamma\gamma|\hat{V}^c|\kappa SLM_S\Gamma\gamma\rangle, \tag{6.39}$$

where $E(\kappa SL)$ is the energy of the atomic level. If there are repeating terms, we must take configuration mixing into account, since matrix elements of the form $\langle \kappa SLM_S\Gamma\gamma\|\kappa' SL'M_S\Gamma\gamma\rangle$ are nonzero. In particular, the operator $\hat{V}^c$ connects the repeating representations resulting from one atomic $LS$ term. Therefore for repeating terms (6.39) gives the energy only to first order of perturbation theory, which is nearly exact for weak crystal fields. In this connection it should be emphasized that for nonrepeating $S\Gamma$ terms (6.39) gives the exact result for the energies of atomic states of a $d^n$ electron configuration.

We shall now find the sublevels of the $^4F(d^3)$ term in a cubic crystal field. The states $^4A_{2g}(^4F)$ and $^4T_{2g}(^4F)$ are nonrepeating, while the state $^4T_{1g}$ occurs twice, since it results from the atomic $^4F$ and $^4P$ terms (see Table 5.3). For transition ions ($l = 2$) we can confine attention to the spherical tensors with $k \leq 4$ in the crystal field operator (6.26). The correction to the energy is the diagonal matrix element calculated using the functions $\Gamma(^4A_{2g})$:

$$\begin{aligned} \varepsilon(^4A_{2g}) &= D_4 < \tfrac{3}{2}3\tfrac{3}{2}A_{2g}|\hat{T}_0^4 + \sqrt{5/14}(\hat{T}_4^4 + \hat{T}_{-4}^4)|\tfrac{3}{2}3\tfrac{3}{2}A_{2g}\rangle \\ &= \tfrac{1}{2}D_4\langle \Psi_2 - \Psi_{-2}|\hat{T}_0^4 + \sqrt{5/14}(\hat{T}_4^4 + \hat{T}_{-4}^4)|\Psi_2 - \Psi_{-2}\rangle. \end{aligned} \tag{6.40}$$

Let us now apply the Wigner–Eckart theorem in the form (6.11):

$$\langle \tfrac{3}{2}3M_SM_L|\hat{T}_m^4|\tfrac{3}{2}\,3M_SM_L'\rangle = \sqrt{1/7}\langle \tfrac{3}{2}\,3\|\hat{T}^4\|\tfrac{3}{2}3\rangle\langle 3M_L|3M_L'4M\rangle. \tag{6.41}$$

The reduced matrix element is the common factor; according to the selection rules, the matrix elements of the operator $\hat{T}_0^4$ are nonzero when $M_L = M'_L$; for the operators $\hat{T}_4^4$ and $\hat{T}_{-4}^4$, $M_L - M'_L = 4$ and $-4$ respectively. Taking this into account, we write (6.40) as

$$\varepsilon(^4A_{2g}) = B_4[\langle 3\ 2|3\ 2\ 4\ 0\rangle + \langle 3\ -2|3\ -2 4 0\rangle - \sqrt{5/14}(\langle 3\ 2|3\ -2\ 4\ 4\rangle + \langle 3\ -2|3\ 2\ 4\ -4\rangle)], \qquad (6.42)$$

where $B_4 = \frac{1}{2}\sqrt{1/7}\,D_4\langle \frac{3}{2}\ 3\|\hat{T}^4|\frac{3}{2}\ 3\rangle$ is a new phenomenological parameter. In a similar way $\varepsilon(^4T_{1g})$ and $\varepsilon(^4T_{2g})$ may be expressed through the single parameter $B_4$ and the Wigner coefficients. The computation of the reduced matrix element $\langle d^3\,\frac{3}{2}3\|\hat{T}^4\|d^3\,\frac{3}{2}3\rangle$ shows that this element is expressed through the one-electron matrix elements within d states, i.e. through $\langle r^4\rangle$. Therefore the energies $\varepsilon(S\Gamma)$ can be expressed through the parameter of the cubic field $Dq$:

$$\varepsilon(^4A_{2g}) = -12Dq, \quad \varepsilon(^4T_{2g}) = -2Dq, \quad \varepsilon(^4T_{1g}) = +6Dq.$$

If we take configuration mixing $^4T_{1g}(^4F)$–$^4T_{1g}(^4P)$ into account, the expression for $\varepsilon(^4T_{1g})$ becomes more complicated.

### 6.4.3 Low-symmetry conformations of octahedral complexes

When the semiempirical approach is used, the crystal field parameters are considered as independent, or, in other words, without the interrelations resulting from group theory. In coordination chemistry there exist high-symmetry complexes with relatively slight structural distortions. Numerous examples of such "nearly octahedral" systems include isomeric forms of six-coordinate metal complexes, whose structures are represented schematically in Figs. 2.41–2.43. The symmetry of these complexes is lower than octahedral, although the departures from high symmetry are relatively slight. It is obvious that the assertion of strong or weak departure from high symmetry is in itself without great significance. Thus, for instance, the *trans* form of the $MA_4B_2$ complex (Fig. 2.41) belongs to the point group $\mathbf{D}_{4h}$, irrespective of the nature of the ligand or the MA and MB interatomic distances. The degree of lowering of symmetry can be expressed quantitatively through the ratio between the octahedral field parameter $D_4$ and the tetragonal field parameters $D'_4$ and $D_2$ in the Hamiltonian (6.27). If the A and B ligand charges are close, and the MA and MB bond lengths are short, we may take $D_2$ and $D'_4$ to be small ($|D_2|, |D'_4| \ll |D_4|$). In this sense we can talk about slight departures from the cubic ($\mathbf{O}_h$) symmetry of the complex. Such considerations are useful, since they enable us to determine the *hierarchy* of phenomenological parameters, independent in the above-mentioned sense. For slight structural distortions, the correct wavefunctions that diagonalize the weak low-symmetry field operator can be obtained in the form of linear combinations of the cubic term functions. It is not necessary to use the projection operator—it is enough to

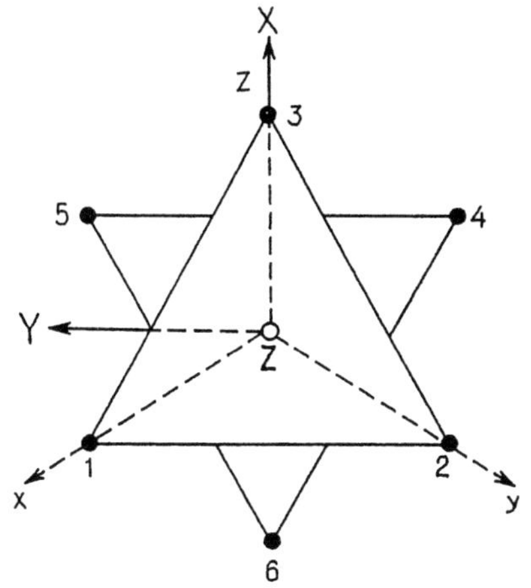


**Fig. 6.4** Trigonal coordinate system of an octahedral complex.

choose a coordinate system in which one of the axes coincides with the direction of distortion; that is, with the main axis of the low-symmetry group.

**Example 6.14** Consider trigonal distortion of an octahedral complex. When the degree of symmetry is lowered, $\mathbf{O} \to \mathbf{D}_3$, the representations are reduced according to the scheme $A_1 \doteq A_1$, $A_2 \doteq A_2$, $E \doteq E$, $T_1 \doteq A_2 + E$, $T_2 \doteq A_1 + E$. It is convenient to change to a *trigonal coordinate system* $(X, Y, Z)$ (Fig. 6.4), whose axis $Z$ is parallel to $C_3$. The "initial" (tetragonal) system is related to the "new" one as follows [14]:

$$\begin{bmatrix} x \\ y \\ z \end{bmatrix} = \begin{bmatrix} -\sqrt{1/6} & \sqrt{1/2} & \sqrt{1/3} \\ -\sqrt{1/6} & -\sqrt{1/2} & \sqrt{1/3} \\ \sqrt{2/3} & 0 & \sqrt{1/3} \end{bmatrix} \begin{bmatrix} X \\ Y \\ Z \end{bmatrix}. \tag{6.43}$$

Transforming the basis functions of E, $T_1$ and $T_2$ in accordance with (6.43), we obtain

**Table 6.4** Trigonal basis of **O** group degenerate representations [14].

| Irreducible representations | Components | Basic functions |
|---|---|---|
| E | $u_+$ | $-\sqrt{1/3}(Y_{22} - \sqrt{2}Y_{2-1})$ |
| | $u_-$ | $\sqrt{1/2}(Y_{22} + \sqrt{2}Y_{2-1})$ |
| $T_1$ | $a_+$ | $Y_{11}$ |
| | $a_-$ | $Y_{1-1}$ |
| | $a_0$ | $Y_{10}$ |
| $T_2$ | $x_+$ | $-\sqrt{1/3}(\sqrt{2}Y_{2-2} + Y_{21})$ |
| | $x_-$ | $\sqrt{1/3}(Y_{22} - Y_{21})$ |
| | $x_0$ | $Y_{20}$ |

the so-called *trigonal basis* (Table 6.4). An important property of the new basis is that the functions $(u_+, u_-)$, $(x_+, x_-)$ and $(a_+, a_-)$ are the basis functions of $\mathbf{D}_3$ group E representations, and $x_0$ and $a_0$ are the bases of $A_1$ and $A_2$. Therefore the corresponding wavefunctions are correct for the trigonal crystal field, i.e. they diagonalize the trigonal field Hamiltonian.

A large number of tables for spherical harmonics, quantized along the octahedron axes $C_4$, $C_3$ and $C_2$, are given in [37]. This can be used for field investigation in low-symmetry conformations of octahedral complexes. The full theory of the crystal field is described in [14, 15, 34, 37, 46, 47]. The results quoted here merely exemplify the basic concepts and methods of group theory, introduced in Chapters 4–6: reduction of irreducible representations of a rotation group, direct multiplication of representations, Wigner (3j–) and Clebsch–Gordan coefficients for point groups, the Wigner–Eckart theorem, and the projection operator method.

## PROBLEMS

**6.1** Using the Wigner–Eckart theorem, construct the component matrix of the second-rank tensor $\hat{\mathbf{T}}^k_m$ $(k = 2)$ by using the basis functions $Y_{LM}$ with $L = 2$.

**6.2** Determine the types (the values $k$) of those spherical tensors $\hat{\mathbf{T}}^k_m$ from which it is possible to form crystal fields possessing $\mathbf{D}_{2h}$, $\mathbf{D}_{2d}$, $\mathbf{C}_{3v}$ and $\mathbf{T}_d$ symmetry.

**6.3** Using the tables in [5, 6, 51] construct from spherical harmonics with $l = 6$ and 8 the basis functions $\varphi_{\Gamma\gamma}$ of $\mathbf{O}$. Solve the same problem for $\mathbf{D}_{3h}$ and $\mathbf{C}_{4v}$.

**6.4** Derive expressions for crystal fields possessing $\mathbf{C}_{3v}$, $\mathbf{C}_{4h}$, $\mathbf{D}_{3h}$, $\mathbf{D}_{2h}$ and $\mathbf{D}_2$ symmetry.

**6.5** Find expressions for the wavefunctions of a d electron in a cubic crystal field with weak trigonal distortion.

# 7 Theory of Directed Valence

## 7.1 DIRECTED VALENCE

A qualitative and somewhat oversimplified explanation of chemical bonding is that based on the concept of overlap of atomic orbitals in molecules. This concept, developed in the 1930s by Hückel, Pauling and Slater, is of importance in the discussion of stereochemistry.

In the simplest case of two hydrogen atoms in the spherically symmetric 1s groundstate the degree of overlap depends only on the interatomic distance (Fig. 7.1). The oxygen atom has two p valence electrons. The maximum overlap with hydrogen orbitals takes place along two mutually perpendicular p dumbbells (Fig. 4.1) and the H–O–H bond angle in the water molecule is expected (Fig. 7.2) to differ only slightly from 90° (its increase up to 105° is due to the repulsion between the hydrogen atoms). In the hydrogen peroxide molecule the H–O bonds are the result of sp overlap, while the O–O bonding is due to pp overlap [56, 57] (Fig. 7.3).

In the above examples (ss, sp and pp bonds) the electron cloud is symmetric under rotation about the bond line. Such bonds are called *$\sigma$ bonds*. A more general feature can be observed: *the electron distribution for $\sigma$ bonds has no nodal plane containing the bond axis*. In reality the electron density does not usually possess full axial symmetry under rotations around the bond, owing to the influence of the surrounding atoms.

Another type of bonding can occur as a result of overlap of the dumbbell-like p orbitals extended along parallel axes (Fig. 7.4). In this case the electron density becomes zero within a plane perpendicular to the dumbbell and containing the bond axis. Such bonds are called *$\pi$ bonds* and are characterized by the presence of a *nodal plane*, where the atomic orbitals change sign. $\pi$ bonds are always double, and the dumbbells of the atomic $\pi$ orbitals are mutually perpendicular, as well as their nodal planes. As an example we can take the $\pi_x$ and $\pi_y$ bonds in the nitrogen molecule, whose $\sigma$ bond lies along the molecular $z$ axis.

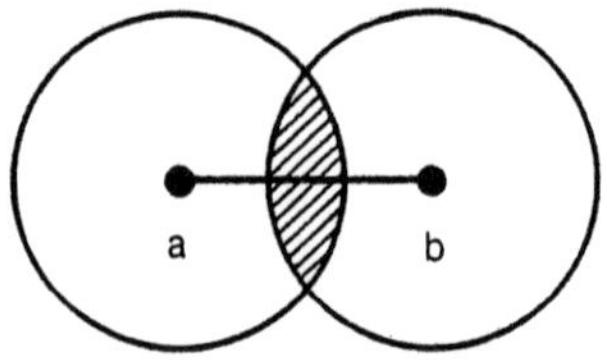


Fig. 7.1 Overlapping of s orbitals in the $H_2$ molecule.

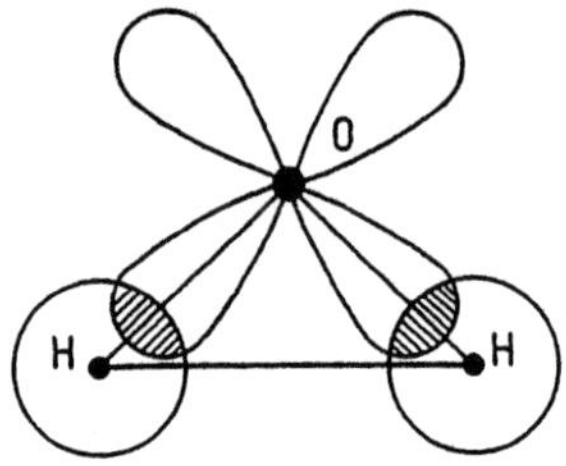


Fig. 7.2 Directed bonds in a water molecule.

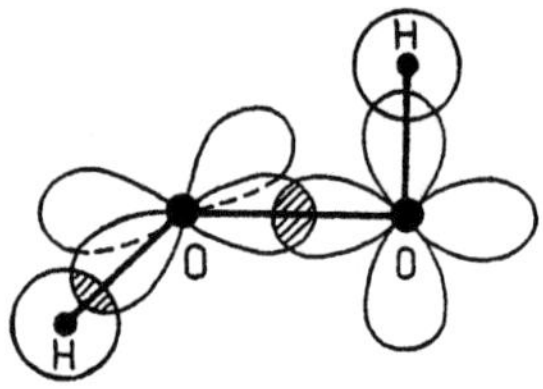


Fig. 7.3 Directed bonds in a hydrogen peroxide molecule.

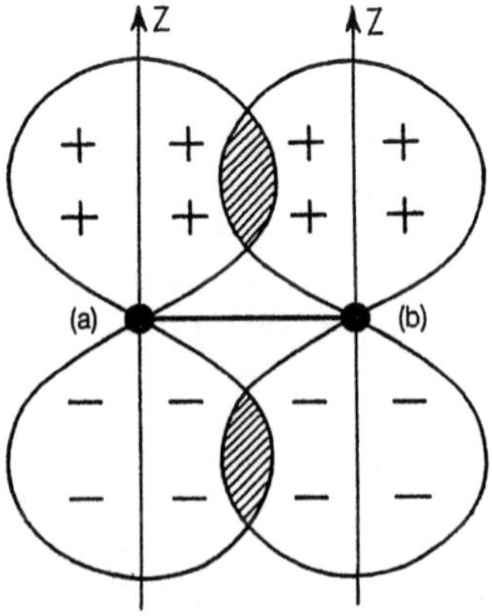


Fig. 7.4 Overlapping of electron orbitals for the case of a $\pi$ bond.

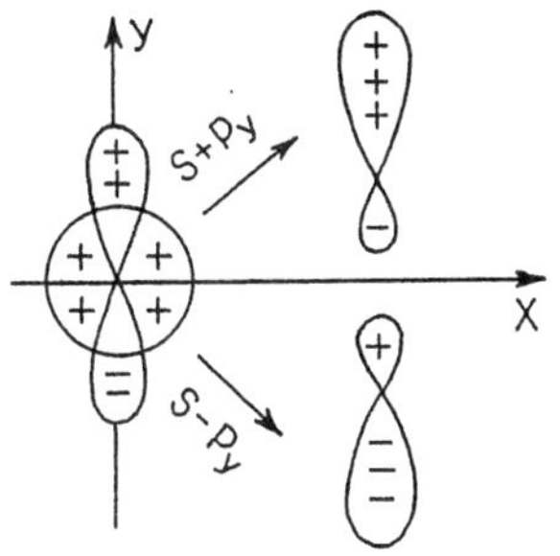


**Fig. 7.5** Hybridization of s and $p_y$ orbitals.

A simple pattern of bonds involving one atomic orbital each cannot always explain molecular stereochemistry. A good example is provided by the methane molecule. The tetrahedral symmetry of $CH_4$ with a central carbon atom cannot be explained by the $sp^3$ configuration of the carbon atom. The equivalence of the four tetrahedral bonds is the result of *hybridization* of atomic orbitals. From a quantum mechanical aspect, in the minimum-energy stationary state of the $CH_4$ molecule (i.e. the state with the strongest chemical bonds) the wavefunction is the *superposition* of the carbon s and p orbitals and the hydrogen s orbitals. In this *hybrid* state the electron density of the carbon orbitals is oriented along the tetrahedral trigonal axes, thus providing maximum overlap with the hydrogen s orbitals, and leading to tetrahedral $\sigma$ bonds. A simple example, illustrating the essential features of hybridization, is shown in Fig. 7.5, where the deformation of the electron cloud when s and $p_y$ orbitals are hybridized can be seen. A similar illustration is given by the pattern of hybrid $\pi$ bonds.

The problems posed when using the group-theoretical approach to hybrid bonding are the determination of orbital types for the atoms involved in bond formation and the classification of wavefunctions according to the irreducible representations of the point groups of the hybrid orbitals. These problems are solved using reduction of representations and the projection operator method.

## 7.2 CLASSIFICATION OF DIRECTED $\sigma$-BONDS

### 7.2.1 Hybrid tetrahedral bonds

Let us consider a hybrid tetrahedral molecule of $AB_4$ type and represent the four $\sigma$ bonds by the vectors $\boldsymbol{\sigma}_1$, $\boldsymbol{\sigma}_2$, $\boldsymbol{\sigma}_3$ and $\boldsymbol{\sigma}_4$ (Fig. 7.6). $\mathbf{T}_d$ group symmetry operations transform the vectors $\boldsymbol{\sigma}_i$ among themselves; hence the four vectors $\boldsymbol{\sigma}_i$ constitute the basis of a four-dimensional representation of the group. This representation is reducible since $\mathbf{T}_d$ has no irreducible representations of dimension greater than three. The transformations among the $\boldsymbol{\sigma}$ vectors

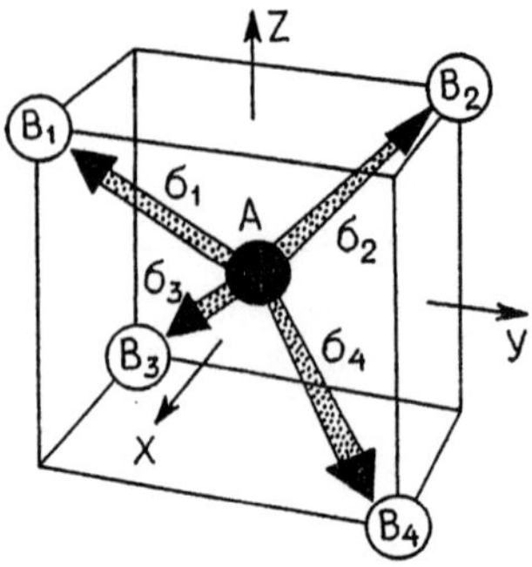


**Fig. 7.6** Tetrahedral $\sigma$ bonds in an $AB_4$ molecule.

resulting from $\mathbf{T}_d$ group operations (see Fig. 2.33a) are

$$\left.\begin{aligned}\hat{E}\boldsymbol{\sigma}_1 &= 1\cdot\boldsymbol{\sigma}_1 + 0\cdot\boldsymbol{\sigma}_2 + 0\cdot\boldsymbol{\sigma}_3 + 0\cdot\boldsymbol{\sigma}_4,\\ \hat{E}\boldsymbol{\sigma}_2 &= 0\cdot\boldsymbol{\sigma}_1 + 1\cdot\boldsymbol{\sigma}_2 + 0\cdot\boldsymbol{\sigma}_3 + 0\cdot\boldsymbol{\sigma}_4, \quad \text{etc.}\end{aligned}\right\} \tag{7.1}$$

The identity representation matrix is

$$\mathbf{D}(\hat{E}) = \begin{bmatrix} 1 & 0 & 0 & 0\\ 0 & 1 & 0 & 0\\ 0 & 0 & 1 & 0\\ 0 & 0 & 0 & 1\end{bmatrix}.$$

This matrix has character $\chi(\hat{E}) = 4$. A rotation $\hat{C}_3$ about the $C_3$ axis (parallel to $\boldsymbol{\sigma}_1$) is performed as follows:

$$\left.\begin{aligned}\hat{C}_3\boldsymbol{\sigma}_1 &= 1\cdot\boldsymbol{\sigma}_1 + 0\cdot\boldsymbol{\sigma}_2 + 0\cdot\boldsymbol{\sigma}_3 + 0\cdot\boldsymbol{\sigma}_4,\\ \hat{C}_3\boldsymbol{\sigma}_2 &= 0\cdot\boldsymbol{\sigma}_1 + 0\cdot\boldsymbol{\sigma}_2 + 1\cdot\boldsymbol{\sigma}_3 + 0\cdot\boldsymbol{\sigma}_4, \quad \text{etc.}\end{aligned}\right\} \tag{7.2}$$

The corresponding matrix is

$$\mathbf{D}(\hat{C}_3) = \begin{bmatrix} 1 & 0 & 0 & 0\\ 0 & 0 & 0 & 1\\ 0 & 1 & 0 & 0\\ 0 & 0 & 1 & 0\end{bmatrix},$$

so that

$$\hat{C}_3[\boldsymbol{\sigma}_1\ \boldsymbol{\sigma}_2\ \boldsymbol{\sigma}_3\ \boldsymbol{\sigma}_4] = [\boldsymbol{\sigma}_1 \cdots \boldsymbol{\sigma}_4]\mathrm{D}(\hat{C}_3)$$

and has character $\chi(\hat{C}_3) = 1$. Computing in a similar way the characters of the operations $\hat{C}_2$, $\hat{S}_4$ and $\hat{\sigma}_d$, and taking into account the fact that the characters within a class of operations are equal (Section 3.7), we obtain the following:

| $\mathbf{T}_d$ | $\hat{E}$ | $8\hat{C}_3$ | $3\hat{C}_2$ | $6\hat{S}_4$ | $6\hat{\sigma}_d$ |
|---|---|---|---|---|---|
| $\chi_\sigma$ | 4 | 1 | 0 | 0 | 2 |

Here $\chi_\sigma$ are the characters of the reducible representation of $\mathbf{T_d}$ whose basis consists of four vectors $\boldsymbol{\sigma}_i$ representing the $\sigma$ bonds. Using the reduction formula (4.12), the representation $\Gamma_\sigma$ can be expanded into irreducible representations. From the character table [1] of $\mathbf{T_d}$ we obtain

$$\Gamma_\sigma \doteq \mathrm{A_1} + \mathrm{T_2}.$$

This result provides a classification of the atomic orbitals involved in the formation of four hybrid tetrahedral bonds. Each hybrid orbital is the superposition of a nondegenerate ($\mathrm{A_1}$) atomic wavefunction and three degenerate ($\mathrm{T_2}$) functions of the central atom A. The values of the orbital momenta can easily be found for these atomic states using the results of the reduction of the representations $\mathrm{D}^{(l)}$ of the rotation group in the point group $\mathbf{T_d}$. According to Table 5.4, we obtain for $l = 0, 1, 2$

$$\mathrm{D}^{(0)} \doteq \mathrm{A_1}, \qquad \mathrm{D}^{(1)} \doteq \mathrm{T_2}, \qquad \mathrm{D}^{(2)} \doteq \mathrm{E} + \mathrm{T_2}.$$

Since hybrid wavefunctions are formed from one-electron atomic functions, the representations $\mathrm{D}^{(l)}$ are characterized by parity $(-1)^l$ (Section 4.3.4). Therefore from Table 5.4 the even representations $\mathrm{D}^{(0)}$ and $\mathrm{D}^{(2)}$ and the odd representations $\mathrm{D}^{(1)}$ have been selected. Thus the hybrid orbitals are formed from the atomic orbitals s ($l = 0$), p ($l = 1$) and d ($l = 2$). The wavefunctions of s and p type form the bases of the representations $\mathrm{A_1}$ and $\mathrm{T_2}$. To construct hybrid orbitals from five *d* functions, construction of the $\mathrm{T_2}$ basis is necessary. The problem of constructing the basis of the irreducible representation $\Gamma_\sigma$ is solved by applying projection operators (Section 6.3). In the case under consideration basis functions from the character tables in [1] can be used; it follows that the $\mathrm{T_2}$ basis (as also in other cubic groups) comprises the functions $d_{yz}$, $d_{xz}$ and $d_{xy}$ (Section 3.5.3). We have thus determined the atomic basis of the hybrid orbitals:

orbital $\mathrm{A_1}$ : $s$;

orbitals $\mathrm{T_2}$ : $\mathrm{p}_x, \mathrm{p}_y, \mathrm{p}_z$; $\mathrm{d}_{yz}, \mathrm{d}_{xz}, \mathrm{d}_{xy}$.

From this basis we can construct two sets of four hybrid wavefunctions; conventionally they are denoted by $\mathrm{sp}^3$ and $\mathrm{sd}^3$. Of course, atomic states having higher $l$ values also provide $\mathrm{A_1}$ and $\mathrm{T_2}$ representations suitable for hybridization; however, the energies of such atomic levels lie much higher, and they contribute only slightly to the ground state of the molecule. The group-theoretical approach does not provide an answer to the question of the choice of an energetically appropriate atomic basis, but it does enable the determination, by specifying the values of $l$, of the possibility of participation of an atomic $l$ state in hybrid bonding from the point of view of symmetry.

The example considered above leads us to a general rule for determining the characters of the irreducible representation $\Gamma_\sigma$: *the character $\chi_\sigma(\hat{R})$ of the representation $\Gamma_\sigma$ for a point-group operation $\hat{R}$ is equal to the number of hybrid bonds (or of the corresponding vectors $\boldsymbol{\sigma}_i$) left unchanged after the operation has been applied.*

### 7.2.2 Inequivalent hybrid bonds

In molecules of tetrahedral symmetry all four bonds are equivalent, i.e. they are transformed into one another by the operations of the group. An example of another kind is provided by molecules of bipyramidal structure $AB_3C_2$, belonging to the point group $\mathbf{D}_{3h}$ (e.g. $PF_5$, $PCl_5$ [13, 58]). In such systems (Fig. 2.2) the equatorial atoms are equivalent (Section 2.1) and form equivalent bonds with the central atom. However, the equatorial and axial $\sigma$ bonds are inequivalent. Using the method described above, from the transformation matrices of the five vectors $\boldsymbol{\sigma}_i$ we obtain the characters of the reducible representations $\Gamma_\sigma$:

| $\mathbf{D}_{3h}$ | $\hat{E}$ | $2\hat{C}_3$ | $3\hat{C}_2$ | $\hat{\sigma}_h$ | $2\hat{S}_3$ | $3\hat{\sigma}_v$ |
|---|---|---|---|---|---|---|
| $\chi_\sigma$ | 5 | 2 | 1 | 3 | 0 | 3 |

Reducing this representation according to (4.12), we obtain

$$\Gamma_\sigma \doteq 2A_1' + A_2'' + E'.$$

From the tables of the functions forming the bases of the irreducible representations of $\mathbf{D}_{3h}$, we obtain the types of orbitals with $l = 0, 1, 2$ involved in hybrid bonds:

orbitals $A_1'$ : s; $d_{z^2}$;

orbital $A_2''$ : $p_z$;

orbitals $E'$ : $p_x, p_y; d_{xy}, d_{x^2-y^2}$.

The hybrid wavefunctions involving all the orbital types can be constructed in six ways:

(1) $s, p_z, p_x, p_y$ ($sp^3$);

(2) $s, p_z, d_{xy}, d_{x^2-y^2}$ ($spd^2$);

(3) $s, d_z^2, p_z, p_x, p_y$ ($dsp^3$);

(4) $s, d_z^2, p_z, d_{xy}, d_{x^2-y^2}$ ($d^3sp$);

(5) $d_{z^2}, p_z, p_x, p_y$ ($p^3d$);

(6) $d_z^2, p_z, d_{xy}, d_{x^2-y^2}$ ($d^3p$).

The energetically favoured hybridization diagrams result from energy considerations, not from the molecular symmetry.

Let us consider the important example of an octahedral ($\mathbf{O}_h$) molecule $AB_6$. The hybrid bonds are oriented along the $C_4$ axes, i.e. along the $AB_i$ directions. Applying the $\mathbf{O}_h$ group operations to the six vectors $\boldsymbol{\sigma}_i$ ($AB_i$ bonds), we obtain the following:

| $O_h$ | $\hat{E}$ | $8\hat{C}_3$ | $6\hat{C}_2$ | $6\hat{C}_4$ | $3\hat{C}_2$ | $\hat{I}$ | $6\hat{S}_4$ | $8\hat{S}_6$ | $3\hat{\sigma}_h$ | $6\hat{\sigma}_d$ |
|---|---|---|---|---|---|---|---|---|---|---|
| $\chi_\sigma$ | 6 | 0 | 0 | 2 | 2 | 0 | 0 | 0 | 4 | 2 |

We reduce $\Gamma_\sigma$: $\Gamma_\sigma \doteq A_{1g} + E_g + T_{1u}$. The following functions correspond to these representations:

$A_1$ orbital : s;
$E_g$ orbitals : $d_{z^2}, d_{x^2-y^2}$;
$T_{1u}$ orbitals : $p_x, p_y, p_z$.

The only possible hybridization is $sd^2p^3$.

## 7.3 SITE-SYMMETRY METHOD

This section describes an alternative approach to the classification of hybrid orbitals. It is based on the method of the so-called site symmetry [21]. The *site symmetry* or *local symmetry* of an atom in a molecule is determined by the *site group* $\mathbf{G}_S$, whose elements are those elements of the full molecular symmetry group $\mathbf{G}$ that, when acting on the molecule, leave that atom in its initial position. It is easy to see, for instance, that in the tetrahedral molecule $AB_4$ ($\mathbf{T}_d$) the site-group of each B atom is $\mathbf{C}_{3v}$ while that of the A atom is $\mathbf{T}_d$. The atomic site symmetry groups in nonlinear molecules can be divided into four types:

(1) the molecule's point group, for atoms situated at the intersections of all the symmetry elements;

(2) $\mathbf{C}_{nv}$, for the atoms lying on the axes;

(3) $\mathbf{C}_s$, for the atoms lying in the symmetry planes;

(4) $\mathbf{C}_1$, for the atoms that are not located on symmetry elements.

It should be noted that the site group $\mathbf{G}_S$ is always a subgroup of the full molecular symmetry group $\mathbf{G}$, and indeed the latter can always be expressed as the *direct product of the site group* and *a certain permutation group* $\mathbf{G}'$, whose operations interchange equivalent atoms (a more rigorous description is given in [21]). For $\mathbf{T}_d$, for instance, such a $\mathbf{G}'$ is the $\mathbf{D}_2(\hat{E}, 3\hat{C}_2) : \mathbf{T}_d = \mathbf{C}_{3v} \times \mathbf{D}_2$. Data on site symmetries for some molecules are given in Table 7.1.

It is worth noting that the hybrid $\sigma$ orbital oriented towards an atom B is invariant under the operations of the site group of B. In other words, the hybrid orbital is transformed according to the totally symmetric representation $A_1$ of the site group. This simple statement provides a way of determining the

**Table 7.1** Site symmetry groups of some molecules [21].

| Molecular point group **G** | Molecule | Atom type | Site group $\mathbf{G_S}$ | Permutation group $\mathbf{G'}$ |
|---|---|---|---|---|
| $\mathbf{C_{3v}}$ | $AB_3$ | B | $\mathbf{C_s}$ | $\mathbf{C_3}$ |
| | $AB_3C$ | C | $\mathbf{C_{3v}}$ | $\mathbf{C_1}$ |
| $\mathbf{C_{4v}}$ | $AB_4$ | B | $\mathbf{C_s}$ | $\mathbf{C_4}$ |
| $\mathbf{D_{2h}}$ | $AB_2C_2D_2$ | B, C, D | $\mathbf{C_{2v}}$ | $\mathbf{C_2}$ |
| $\mathbf{D_{3h}}$ | $AB_3$ | B | $\mathbf{C_{2v}}$ | $\mathbf{C_3}$ |
| $\mathbf{D_{3d}}$ | $AB_6$ | B | $\mathbf{C_s}$ | $\mathbf{S_6}$ |
| $\mathbf{O_h}$ | $AB_6$ | B | $\mathbf{C_{4v}}$ | $\mathbf{S_6}$ |

irreducible representations whose basis functions are contained in the hybrid orbitals:

(1) find the point group **G** of the molecule;

(2) determine the site groups $\mathbf{G_S}$ of the equivalent atoms;

(3) reduce all irreducible representations $\Gamma$ of **G** in the subgroup $\mathbf{G_S}$ ($\Gamma \doteq \Sigma\Gamma_S$);

(4) find the totally symmetric representations $\Gamma_{1S}$ among representations $\Gamma_S$; the irreducible representations $\Gamma$ producing the representations $\Gamma_{1S}$ are the required representations of **G**;

(5) find the atomic orbital types, from which one can construct the $\Gamma$ basis.

**Example 7.1** Consider the tetrahedral molecule $AB_4$.

(1) the point group is $\mathbf{T_d}$;

(2) the site groups of the atoms $B_i$ are $\mathbf{C_{3v}}$;

(3) the reduction of the irreducible representations of $\mathbf{T_d}$ in the subgroup is given by the following diagram:

| $\mathbf{T_d}$ | $A_1$ | $A_2$ | E | $T_1$ | $T_2$ |
|---|---|---|---|---|---|
| $\mathbf{C_{3v}}$ | $A_1$ | $A_2$ | E | $A_2 + E$ | $A_1 + E$ |

(4) the totally symmetric representations of $\mathbf{C_{3v}}$ are derived from the representations $A_1$ and $T_2$ of $\mathbf{T_d}$; therefore four $\sigma$ orbitals form the basis $A_1 + T_2$;

(5) as in Section 7.2.1, we find the atomic basis is $sp^3d^3$.

It should be noted that the method of site symmetry takes account of the electron density distribution within the hybrid states in a more satisfactory way than the method of $\boldsymbol{\sigma}$ vector transformation. For example the electron density of tetrahedral hybrid bonding is invariant under rotation $\hat{C}_3$, while the $\boldsymbol{\sigma}$ vector possesses complete axial symmetry.

## 7.4 CLASSIFICATION OF HYBRID $\pi$ BONDS

The group-theoretical methods of $\sigma$-orbital classification involve the general concept of matrix character. This can also be applied to more complicated problems. Since the axis of the $\pi$-orbital electron density is perpendicular to the bond axis, we shall represent it by a vector, also perpendicular to this axis. For definiteness we shall take the $\pi$-bond vector $\boldsymbol{\pi}_i$ as being oriented towards the positive lobe of the $\pi$ orbital (Fig. 7.4). Then, for instance, a double $\pi$ bond is represented by two pairs of orthogonal vectors (Fig. 7.7). Symmetry operations transform the atoms into another, simultaneously transforming the $\pi$ vectors. As in the case of the $\sigma$ bonds, we can find the matrices and characters of the group operations. Expansion of the reducible representation $\Gamma_\pi$ into irreducible representations gives the type of the atomic orbitals forming the hybrid $\pi$ bonds.

Let us see how this method works, taking as an example the octahedral molecule $AB_6$. Each ligand $B_i$ has two orthogonal $\pi$ orbitals, to which there correspond orthogonal $\boldsymbol{\pi}$ vectors (Fig. 7.8). $\mathbf{O}_h$ group operations transform the atoms $B_i$ into one other; to the new location of each atom there corresponds a new orientation of the $\boldsymbol{\pi}$-vector pair belonging to this atom. Thus the 12 vectors $\boldsymbol{\pi}$ of the molecule $AB_6$ form the basis of the 12-dimensional reducible representation $\Gamma_\pi$. This representation in matrix form is easily found by applying the group operations to the $[\pi_{i\alpha}]$ $(i = 1, 2, \ldots, 6;\ \alpha = x, y, z)$. The character of the transformation matrix is given by the number of atoms left unchanged by this transformation. The sign of the character is positive if the symmetry operation does not alter the direction of the $\pi$ vector, and negative if it does. For example, $\chi(\hat{E}) = 12$, since the identity transformation leaves all $\boldsymbol{\pi}$ vectors unchanged. The operation $\hat{C}_2(x)$ does not affect atoms 1 and 4 (Fig. 7.8); however, it rotates the vectors $\pi_x$ and $\pi_y$ of these atoms through 180°. Hence $\chi(\hat{C}_2(x)) = -4$, which holds for the class of three

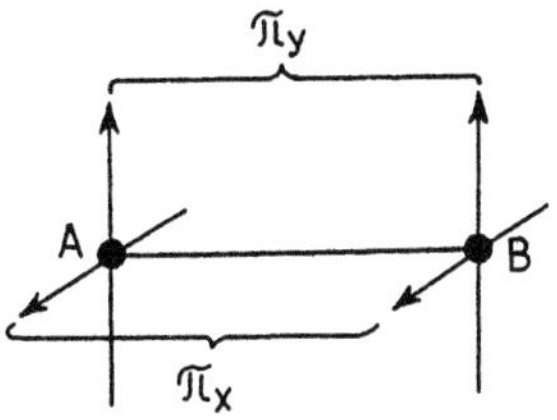


**Fig. 7.7** Vectorial representation of a double $\pi$ bond.

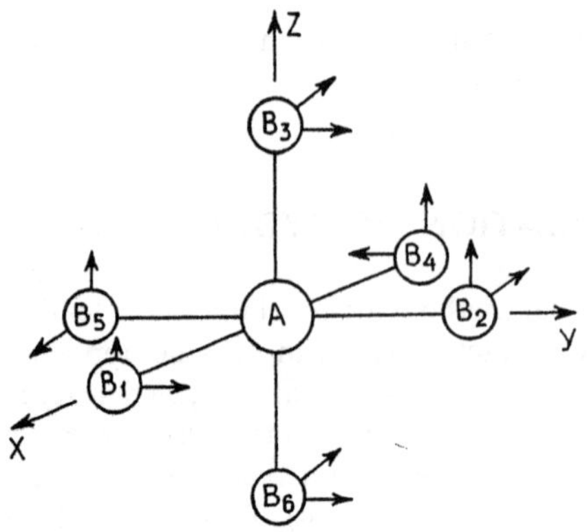


**Fig. 7.8** Vector representation of $\pi$ orbitals of the ligands of an octahedral $AB_6$ molecule.

rotations $\hat{C}_2$. The other operations of $\mathbf{O}_h$ change the positions of all the atoms; therefore their characters are zero.

Taking the above into account, we find the characters of the representation $\Gamma_\pi$:

| $\mathbf{O}_h$ | $\hat{E}$ | $8\hat{C}_3$ | $6\hat{C}_2$ | $6\hat{C}_4$ | $3\hat{C}_4^2$ | $\hat{I}$ | $6\hat{S}_4$ | $8\hat{S}_6$ | $3\hat{\sigma}_h$ | $6\hat{\sigma}_d$ |
|---|---|---|---|---|---|---|---|---|---|---|
| $\chi_\pi$ | 12 | 0 | 0 | 0 | −4 | 0 | 0 | 0 | 0 | 0 |

The reduction of the representation $\Gamma_\pi$ leads to the result $\Gamma_\pi \doteq T_{1g} + T_{2g} + T_{1u} + T_{2u}$. Confining attention to the s, p and d orbitals of the central atom, it is easy to see how they relate to the orbitals involved in the $\pi$ bonds:

$T_{1g}$ orbitals : nonexistent;
$T_{2g}$ orbitals : $d_{yz}, d_{xz}, d_{xy}$;
$T_{1u}$ orbitals : $p_x, p_y, p_z$;
$T_{2u}$ orbitals : nonexistent.

Hence all 12 $\pi$ bonds cannot be realized in the molecule $AB_6$, since the central atom (taking into account only the s, p and d states) does not have the appropriate orbitals. It should also be noted that not all the d orbitals can be involved in $\pi$ bonding: two orbitals of the type $E_g(d_{z^2}, d_{x^2-y^2})$ are involved only in $\sigma$ bonds (Section 7.2.2).

Let us now consider an example illustrating other features of $\pi$ bonding; namely, the square-planar molecule $AB_4$ ($\mathbf{D}_{4h}$). From Fig. 7.9 it can be seen that the 12 $\pi$ vectors are divided into 2 sets: 4 of them are perpendicular to the molecular plane while 8 lie within it. These sets (we shall call them *perpendicular* and *parallel orbitals* respectively) are independent, since symmetry operations do not transform the vectors belonging to different sets

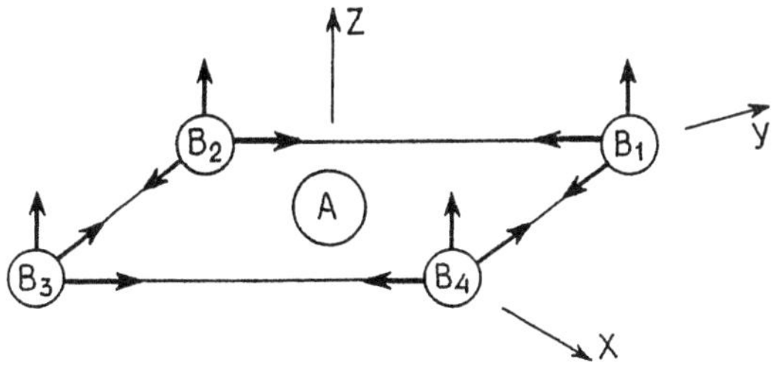


**Fig. 7.9** Vector representation of $\pi$ orbitals in a square-planar $AB_4$ molecule.

into one another. It is therefore sufficient to consider two representations whose bases are vectors from the two different sets.

We denote by $\chi_\pi^\perp$ and $\chi_\pi^\parallel$ the characters of these representations; the character of the representation realized by all the $\pi$ orbitals is $\chi_\pi = \chi_\pi^\perp + \chi_\pi^\parallel$. Applying the $\mathbf{D}_{4h}$ group operations it is easy to obtain the following table:

| $\mathbf{D}_{4h}$ | $\hat{E}$ | $2\hat{C}_4$ | $\hat{C}_2$ | $2\hat{C}_2'$ | $\hat{C}_2''$ | $\hat{I}$ | $2\hat{S}_4$ | $\hat{\sigma}_h$ | $2\hat{\sigma}_v$ | $2\hat{\sigma}_d$ |
|---|---|---|---|---|---|---|---|---|---|---|
| $\chi_\pi^\perp$ | 4 | 0 | 0 | −2 | 0 | 0 | 0 | −4 | 2 | 0 |
| $\chi_\pi^\parallel$ | 4 | 0 | 0 | −2 | 0 | 0 | 0 | 4 | −2 | 0 |

with

$$\Gamma_\pi^\perp \doteq A_{2u} + B_{2u} + E_g, \qquad \Gamma_\pi^\parallel \doteq A_{2g} + B_{2g} + E_u.$$

The orbitals of the central atom that can form $\pi$ bonds are obtained as usual. Reducing the representations $D^{(0)}$, $D^{(1)}$ and $D^{(2)}$ in $\mathbf{D}_{4h}$ (Table 4.3), we find

$$D^{(0)} = A_{1g}, \qquad D^{(1)} = A_{2u} + E_u, \qquad D^{(2)} = A_{1g} + B_{1g} + B_{2g} + E_g.$$

From the tables of functions belonging to irreducible representations of $\mathbf{D}_{4h}$ we obtain the following types of orbital:

$$\underbrace{A_{2u}(p_z), \quad B_{2u}(\text{nonexistent}), \quad E_g(d_{xz}, d_{yz})}_{\text{"perpendicular" } \pi \text{ bonds}}$$

$$\underbrace{A_{2g}(\text{nonexistent}), \quad B_{2g}(d_{xy}), \quad E_u(p_x, p_y)}_{\text{"parallel" } \pi \text{ bonds}}$$

Thus in the molecule $AB_4$ ($\mathbf{D}_{4h}$) not all $\pi$ orbitals of the atoms $B_i$ are involved in $\pi$ bonds with the central atom A. We see that the $A_{2g}$ and $B_{2u}$ orbitals of A are nonexistent in the s, p and d atomic state sets. Figure 7.10 shows that $B_{2g}(d_{xy})$ orbital of the central atom and the p orbitals of the atoms $B_i$ forming "perpendicular" $\pi$ bonds, and the $A_{2u}(p_z)$ orbitals of A and the $p_z$ orbitals of $B_i$ forming "parallel" $\pi$ bonds.

The above group-theoretical classification of directed $\pi$ and $\sigma$ valence bonds

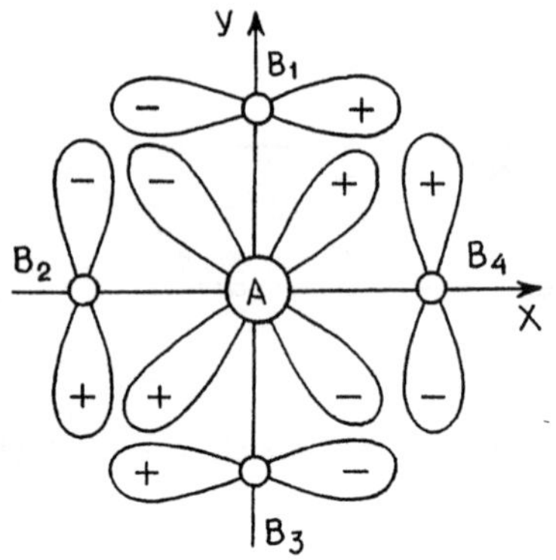


(a)

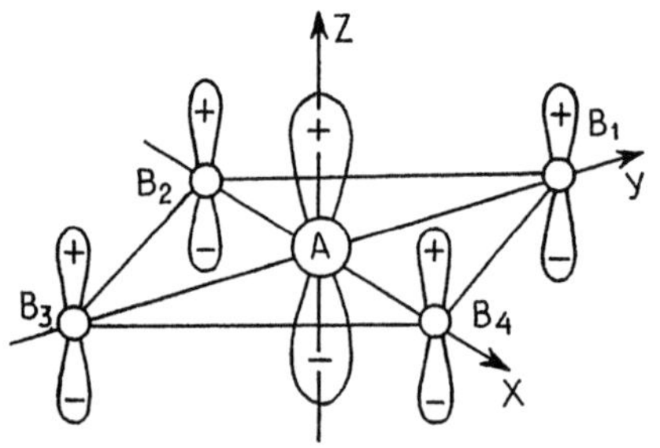


(b)

**Fig. 7.10** Electron density of $\pi$ orbitals of a square $AB_4$ molecule: perpendicular $\pi$ orbitals, $d_{xy}$ orbital of A atom ($B_{2g}$) and $p_x$, $p_y$ orbitals of B atoms; (b) $p_z$ orbital of A atoms ($A_{2u}$) and $p_z$ orbitals of B atoms.

was applied by Kimbal [58] to the study of many important spatial bonding situations and different coordination numbers.

**Example 7.2** Let us consider the bonds in the triatomic molecules $AB_2$ and BAC of $\mathbf{C}_{2v}$ symmetry. The classification of s, p and d orbitals as well as $\sigma$ and $\pi$ hybrid orbitals of the atom is given in Table 7.2 [58]. It can be seen that in $\mathbf{C}_{2v}$ the *s* function constitutes the basis of the totally symmetric representation $A_1$; the three p functions are transformed according to the representations $A_1$, $B_1$ and $B_2$; and the *d* functions constitute the basis of two representations $A_1$ as well as of $A_2$, $B_1$ and $B_2$. The $\sigma$ bonds form the bases of the representations $A_1$ and $B_2$ and the $\pi$ bonds form the bases of $A_1$, $A_2$, $B_1$ and $B_2$.

**Example 7.3** Table 7.3 gives the results for molecules with coordination number 3 of $AB_3$ type and $\mathbf{C}_{3v}$ symmetry, having a trigonal-pyramidal structure. Thus $\sigma$ bonds of $A_1$ and E types and $\pi$ bonds of $A_1$, $A_2$ and E types are possible.

**Table 7.2** Symmetry of directed valence of a three-atom molecule.

| $\mathbf{C}_{2v}$ | $A_1$ | $A_2$ | $B_1$ | $B_2$ |
|---|---|---|---|---|
| s | 1 | 0 | 0 | 0 |
| p | 1 | 0 | 1 | 1 |
| d | 2 | 1 | 1 | 1 |
| $\sigma$ | 1 | 0 | 0 | 1 |
| $\pi$ | 1 | 1 | 1 | 1 |

**Table 7.3** Directed valence symmetry in $AB_3$ ($\mathbf{C}_{3v}$) molecules.

| $\mathbf{C}_{3v}$ | $A_1$ | $A_2$ | E |
|---|---|---|---|
| s | 1 | 0 | 0 |
| p | 1 | 0 | 1 |
| d | 1 | 0 | 2 |
| $\sigma$ | 1 | 0 | 1 |
| $\pi$ | 1 | 1 | 2 |

**Table 7.4** Directed valence symmetry in a trigonal prismatic molecule.

| $\mathbf{D}_{3h}$ | $A_1'$ | $A_1''$ | $A_2'$ | $A_2''$ | $E'$ | $E''$ |
|---|---|---|---|---|---|---|
| s | 1 | 0 | 0 | 0 | 0 | 0 |
| p | 0 | 0 | 0 | 1 | 1 | 0 |
| d | 1 | 0 | 0 | 0 | 1 | 1 |
| $\sigma$ | 1 | 0 | 0 | 1 | 1 | 1 |
| $\pi$ | 1 | 1 | 1 | 1 | 2 | 2 |

**Table 7.5** Directed valence symmetry in a molecule with trigonal antiprismatic bonds.

| $\mathbf{D}_{3d}$ | $A_{1g}$ | $A_{1u}$ | $A_{2g}$ | $A_{2u}$ | $E_g$ | $E_u$ |
|---|---|---|---|---|---|---|
| s | 1 | 0 | 0 | 0 | 0 | 0 |
| p | 0 | 0 | 0 | 1 | 0 | 1 |
| d | 1 | 0 | 0 | 0 | 2 | 0 |
| $\sigma$ | 1 | 0 | 0 | 1 | 1 | 1 |
| $\pi$ | 1 | 1 | 1 | 1 | 2 | 2 |

**Table 7.6** Hybrid bonding symmetry in a seven-coordinate ($\mathbf{C}_{3v}$) ion.

| $\mathbf{C}_{3v}$ | $A_1$ | $A_2$ | E |
|---|---|---|---|
| s | 1 | 0 | 0 |
| p | 1 | 0 | 1 |
| d | 1 | 0 | 2 |
| f | 2 | 1 | 2 |
| $\sigma$ | 3 | 0 | 2 |
| $\pi$ | 2 | 2 | 5 |

**Example 7.4** High coordination numbers admit a wide diversity of structural types. The most frequently occurring octahedral complexes have already been analysed. Several other examples of coordination number 6 are given in Tables 7.4 and 7.5.

Large heavy atoms lead to complicated coordinating polyhedra because of the diversity of hybrid bondings, involving not only s, p and d, but also f orbitals. Consider for example the seven-coordinate ion $ZrF_7^-$, whose octahedral coordination is complemented by the ion $F^-$ in the face centre. The hybrid bonding characteristics are given in Table 7.6 [58].

It should again be emphasized that the possibility of realizing certain types of hybrid bonds and the corresponding stereochemical configurations in a molecule is determined not only by symmetry and coordination number but also by energetic considerations. The latter fall outside the scope of group theory and require quantum-chemical calculations.

## 7.5 CONSTRUCTION OF HYBRID ORBITALS

Assuming known atomic orbital types and the corresponding irreducible representations, which play an active part in hybridization, we can proceed to construct linear combinations of atomic orbitals. The general approach to the construction of hybrid state can be illustrated by the tetrahedral molecule $AB_4$. Hybrid $\sigma$ bonds can be constructed from $s(A_1)$ and $p(T_2)$ functions (Section 7.2.1). Let us write the four hybrid $sp^3$ orbitals as linear combinations:

$$\Psi_i = a_i s + b_i p_x + c_i p_y + d_i p_z, \tag{7.3}$$

where $i = 1, \ldots, 4$, and $s$, $p_x$, $p_y$ and $p_z$ are atomic wavefunctions. The electron density distribution in the four states $\Psi_i$ must be elongated towards the four B atoms, i.e. directed along the $\boldsymbol{\sigma}$ vectors. The wavefunctions $\Psi_i$ are therefore transformed by $\mathbf{T}_d$ group operations in the same way as the $\boldsymbol{\sigma}$ vectors. Since the $s$ function has spherical symmetry, all the coefficients $a_i$ are identical: $a_i = a$. From the three $p$ functions one can easily construct four functions, oriented along the $\sigma_i$ bonds (Fig. 7.6):

$$\left.\begin{array}{ll} p_1 = p_x - p_y + p_z, & p_2 = -p_x + p_y + p_z, \\ p_3 = -p_x - p_y - p_z, & p_4 = p_x + p_y - p_z, \end{array}\right\} \qquad (7.4)$$

where the signs correspond to the directions of the coordinate axes. Hence $b_i$, $c_i$ and $d_i$ are either $+1$ or $-1$, and (7.3) takes the form

$$\left.\begin{array}{ll} \Psi_1 = as + bp_1, & \Psi_2 = as + bp_2, \\ \Psi_3 = as + bp_3, & \Psi_4 = as + bp_4. \end{array}\right\} \qquad (7.5)$$

The $\Psi_i$ orbitals should be normalized; therefore we require $a^2 + 3b^2 = 1$. If, as usual, we require different hybrids to be orthogonal, then we must take

$$\langle \Psi_1 | \Psi_2 \rangle = a^2 \langle s|s \rangle + b^2 \langle p_1 | p_2 \rangle = a^2 - b^2 = 0, \qquad (7.6)$$

so that $a = b = \frac{1}{2}$, or $a = -b = \frac{1}{2}$. In the former case the electron density is elongated along the positive directions of the $\boldsymbol{\sigma}$ vectors; that is, towards the B atoms. Hence the hybrid orbitals are†

$$\left.\begin{array}{l} \Psi_1 = \frac{1}{2}(s + p_x - p_y + p_z), \\ \Psi_2 = \frac{1}{2}(s - p_x + p_y + p_z), \\ \Psi_3 = \frac{1}{2}(s - p_x - p_y - p_z), \\ \Psi_4 = \frac{1}{2}(s + p_x + p_y - p_z). \end{array}\right\} \qquad (7.7)$$

The above example leads us to formulate the following general rules:

(1) the hybrid orbitals can be constructed as linear combinations of atomic orbitals, whose types are prescribed by the symmetry rules enunciated in Section 7.2:

$$\Psi_i = a_i \varphi_1 + b_i \varphi_2 + c_i \varphi_3 + \ldots; \qquad (7.8)$$

(2) the orbitals $\Psi_i$ will be transformed by the symmetry operations in the same way as equivalent atoms;

(3) the hybrid orbitals $\Psi_i$ can be normalized; by taking,

$$a_i^2 + b_i^2 + c_i^2 + \ldots + 1; \qquad (7.9)$$

(4) the hybrid orbitals can be orthogonalized, so that $\langle \psi_i | \psi_j \rangle = 0$ by requiring the condition

$$a_i a_j + b_i b_j + c_i c_j + \ldots = 0 \qquad (7.10)$$

These conditions completely determine the coefficients of the linear combinations (7.8).

† It is usual but not essential to make the hybrid functions orthogonal.

**Example 7.5** Let us find the hybrid $sp^2$ orbitals of an $AB_3$ ($\mathbf{D}_3$) molecule. One of these, say $sp^2$, is written as

$$\Psi_1 = as + b_1 p_x + b_2 p_y. \tag{7.11}$$

Let the vector $\boldsymbol{\sigma}_1$ be oriented along the $x$ axis; then $b_2 = 0$ and therefore

$$\Psi_1 = a_1 s + b_1 p_x. \tag{7.12}$$

The orbitals $\Psi_2$ and $\Psi_3$ are obtained from $\Psi_3$ by the application of the operations $\hat{C}_3$ and $\hat{C}_3^2 = \hat{C}_3^{-1}$. Since $p_x$, $p_y$ transform like basis vectors $i$, $j$ under rotations (Section 3.5.2) we can use (3.18) and (3.19):

$$\left.\begin{aligned} \hat{C}_3 p_x &= p_x \cos\vartheta + p_y \sin\vartheta = -\tfrac{1}{2}p_x - \tfrac{1}{2}\sqrt{3}p_y, \\ \hat{C}_3^{-1} p_x &= -\tfrac{1}{2}p_x - \tfrac{1}{2}\sqrt{3}p_y. \end{aligned}\right\} \tag{7.13}$$

Therefore, since $\hat{C}_3 s = s$, we obtain

$$\left.\begin{aligned} \Psi_1 &= as + bp_x, \\ \Psi_2 &= as + b(-\tfrac{1}{2}p_x + \tfrac{1}{2}\sqrt{3}p_y), \\ \Psi_3 &= as + b(-\tfrac{1}{2}p_x - \tfrac{1}{2}\sqrt{3}p_y). \end{aligned}\right\} \tag{7.14}$$

We now apply the normalization and orthogonality conditions

$$\langle \Psi_1 | \Psi_1 \rangle = a^2 + b^2 = 1,$$

$$\langle \Psi_1 | \Psi_2 \rangle = a^2 - \tfrac{1}{2}b^2 = 0,$$

from which $a = \sqrt{1/3}$, $b = 2\sqrt{1/6}$. The final expressions for the hybrid $sp^2$ orbitals are

$$\left.\begin{aligned} \Psi_1 &= \sqrt{1/3}s + 2\sqrt{1/6}p_x, \\ \Psi_2 &= \sqrt{1/3}s - \sqrt{1/6}p_x + \sqrt{1/2}p_y, \\ \Psi_3 &= \sqrt{1/3}s - \sqrt{1/6}p_x - \sqrt{1/2}p_y. \end{aligned}\right\} \tag{7.15}$$

**Example 7.6** Consider hybrid orbitals in the square planar molecule $AB_4$ ($\mathbf{D}_{4h}$). The $dsp^2$ hybrid bonds are formed by the s, $p_x$, $p_y$ and $d_{x^2-y^2}$ orbitals. If the $x$ axis is oriented towards atom 1 then the $p_y$ function does not contribute to the $\Psi_1$ orbital, so that

$$\Psi_1 = as + bp_x + cd_{x^2-y^2}. \tag{7.16}$$

The hybrid orbitals $\Psi_2$, $\Psi_3$ and $\Psi_4$ are obtained from $\Psi_1$ by applying the operations $\hat{C}_4$, $\hat{C}_4^2$ and $\hat{C}_4^3$ respectively:

$$\left.\begin{aligned} \Psi_2 &= as + bp_y - cd_{x^2-y^2}, \\ \Psi_3 &= as - bp_x + cd_{x^2-y^2}, \\ \Psi_4 &= as - bp_y - cd_{x^2-y^2}. \end{aligned}\right\} \tag{7.17}$$

The coefficients a, b, c are determined from the conditions of orthogonality and normalization: $a = \tfrac{1}{2}$, $b = 1/\sqrt{2}$, $c = \tfrac{1}{2}$.

**Example 7.7** Consider the octahedral complex $AB_6$ ($\mathbf{O}_h$). The $\sigma$-vector classification leads us to the conclusion that there must exist $sp^3d^2$ bonds (Section 7.2.2). The action of $\mathbf{O}_h$ group operations on the $E_g$ basis $d_{x^2-y^2}$, $d_{z^2}$ was considered in Section 3.5.3. Using the method described there, we obtain the $sp^3d^2$ hybrid orbitals oriented along the $z$, $-z$, $x$, $-x$, $y$ and $-y$ axes:

$$\left.\begin{aligned}
\Psi_1 &= \sqrt{1/6}s + \sqrt{1/2}p_z + \sqrt{1/3}d_{z^2},\\
\Psi_2 &= \sqrt{1/6}s - \sqrt{1/2}p_z + \sqrt{1/3}d_{z^2},\\
\Psi_3 &= \sqrt{1/6}s + \sqrt{1/2}p_x - \tfrac{1}{2}\sqrt{1/3}d_{z^2} + \tfrac{1}{2}d_{x^2-y^2},\\
\Psi_4 &= \sqrt{1/6}s - \sqrt{1/2}p_x - \tfrac{1}{2}\sqrt{1/3}d_{z^2} + \tfrac{1}{2}d_{x^2-y^2},\\
\Psi_5 &= \sqrt{1/6}s + \sqrt{1/2}p_y - \tfrac{1}{2}\sqrt{1/3}d_{z^2} - \tfrac{1}{2}d_{x^2-y^2},\\
\Psi_6 &= \sqrt{1/6}s - \sqrt{1/2}p_y - \tfrac{1}{2}\sqrt{1/3}d_{z^2} - \tfrac{1}{2}d_{x^2-y^2}.
\end{aligned}\right\} \qquad (7.18)$$

Inequivalent bonds can be analysed in a similar way.

## PROBLEMS

**7.1** Obtain the character table of irreducible representation $\Gamma_\sigma$ for five vectors $\mathbf{a}_i$, corresponding to the $\sigma$ bonds in the molecule $PF_5$ (Section 7.2.2).

**7.2** Prove that for the plane trigonal molecule $AB_3$ possessing $\mathbf{D}_{3h}$ symmetry the $\sigma$ orbitals are transformed according to the representations $\Gamma_6 \doteq A_1' + E'$, and that the following wavefunction types are possible: (1) $s$, $p_x$, $p_y$ ($sp^2$); (2) $s$, $d_{xy}$, $d_{x^2-y^2}$ ($sd^2$); (3) $d_{z^2}$, $p_x$, $p_y$ ($dp^2$); (4) $d_{z^2}$, $d_{xy}$, $d_{x^2-y^2}$ ($d^3$).

**7.3** Prove that for the square-planar molecule $AB_4$ ($\mathbf{D}_{4h}$) $\Gamma_\sigma \doteq A_{1g} + B_{1g} + E_u$ and the hybrid orbitals are formed according to one of the following schemes: (1) $s$, $d_{x^2-y^2}$, $p_x$, $p_y$ ($dsp^2$); (2) $d_{z^2}$, $d_{x^2-y^2}$, $p_x$, $p_y$ ($d^2p^2$).

**7.4** Find the representation (transformation matrices) for the 12 $\pi$ vectors of the octahedral molecule $AB_6$ (Fig. 7.7).

**7.5** Determine $\sigma$ and $\pi$ hybrid orbital types in a molecule of $AB_4$ type having a tetragonal pyramidal structure ($\mathbf{C}_{4v}$).

**7.6** Determine the hybrid orbital types of a three-atom molecule $AB_2$ (Table 7.2).

**7.7** Construct the hybrid orbitals for an $AB_3$ type molecule (Table 7.3).

# 8 Molecular Orbital Method

## 8.1 GENERAL BACKGROUND

In the previous chapters we have examined the group-theoretical aspects of two methods used for the investigation of electronic structure—the crystal field approximation and hybrid bonding theory—both of which are based on several essentially restricting assumptions. A more effective tool in quantum chemistry is the *molecular orbital method.* This takes account of the electronic structure of all atoms in the molecule or complex, removing the principal restriction of crystal field theory—the approximation of pointlike and structureless ligands. The molecular orbital method also provides a natural way of describing coordination bonds, removing the need to assume pair bondings that is adopted in hybrid orbital theory. A detailed discussion of the many aspects of chemical bonding can be found in various textbooks (e.g. [59]). A most modern and comprehensive presentation of molecular quantum mechanics is given in the fundamental book by McWeeny [60].

In the first approximation *molecular orbitals* are usually constructed as *linear combinations of atomic orbitals (LCAO)*:

$$\Psi = c_1\psi_1 + c_2\psi_2 + \ldots + c_n\psi_n, \tag{8.1}$$

where $\psi_i$ are the atomic wavefunctions and $c_i$ are unknown coefficients. In (8.1) the sum is over all the atoms of the molecule, and over different atomic orbitals of each atom. The collection of $\psi_i$ orbitals is called the *atomic basis.* In the framework of the molecular orbital method the exact many-electron molecular Hamiltonian is replaced by an approximate one-electron Hamiltonian in which the interelectron interaction is supposed to be taken into account as an 'effective field' in the sense of the Hartree–Fock method [60]. In a given atomic basis the best approximation to the molecular orbitals are found by solving the secular equation

$$\begin{vmatrix} H_{11}-\varepsilon S_{11} & H_{12}-\varepsilon S_{12} & \dots & H_{1n}-\varepsilon S_{1n} \\ H_{21}-\varepsilon S_{21} & H_{22}-\varepsilon S_{22} & \dots & H_{2n}-\varepsilon S_{2n} \\ \vdots & \vdots & & \vdots \\ H_{n1}-\varepsilon S_{n1} & H_{n2}-\varepsilon S_{n2} & \dots & H_{nn}-\varepsilon S_{nn} \end{vmatrix} = 0, \qquad (8.2)$$

where $H_{ij} = \int \psi_i^* \hat{H} \psi_j \, d\tau$ are the matrix elements of a certain effective Hamiltonian $\hat{H}$ between the atomic functions $\psi_i$, $S_{ij} = \int \psi_i^* \psi_j \, d\tau$ are the overlap integrals (which are nonzero for orbitals of different atoms) and $\varepsilon$ are the required energy levels for the individual electrons. There are $n$ solutions $(\epsilon_1, \epsilon_2, \dots \epsilon_n)$ and with each $\epsilon$ there is a corresponding set of coefficients $c_i$ $(i = 1, 2, \dots, n)$, determining the molecular orbital (MO). The $c_i$ are the solutions of the system of linear homogeneous equations (for given $\epsilon$)

$$\sum_k c_k (H_{ik} - \varepsilon S_{ik}) = 0. \qquad (8.3)$$

Equation (8.2) generalizes the system (4.21) for the case of *nonorthogonal atomic orbitals (AO)*.

The group-theoretical approach used in setting up approximate wavefunctions of LCAO MO form is based on similar concepts to those developed in crystal field theory (Section 4.3.5). According to the Wigner theorem (Section 4.2), each energy level $\varepsilon_\alpha$ corresponds to a certain irreducible representation $\Gamma_\alpha$ of the molecule symmetry point group **G**. Therefore the level $\varepsilon_\alpha$ eigenfunctions form the $\Gamma_\alpha$ basis. Denoting them by $\Psi_{\Gamma_\alpha \gamma_\alpha} \equiv |\Gamma_\alpha \gamma_\alpha\rangle$, it can be shown that $\langle \Gamma_\alpha \gamma_\alpha | \hat{H} | \Gamma_\beta \gamma_\beta \rangle = 0$ for different representations $\Gamma_\alpha$ and $\Gamma_\beta$, and for $\gamma_2 \neq \gamma_\beta$, while $\langle \Gamma_\alpha \gamma_\alpha | \hat{H} | \Gamma_\alpha \gamma_\alpha \rangle = \varepsilon_\alpha$. It follows from the properties of identical (repeating) representations $\Gamma_\alpha$ and $\Gamma_\alpha$ (Section 5.6.2) that the configuration mixing matrix elements $\langle \Gamma_\alpha \gamma_\alpha | \hat{H} | \Gamma'_\alpha \gamma'_\alpha \rangle = H_{\alpha\alpha'}$ are nonzero. Thus in a symmetry adapted basis $\Psi_{\Gamma_\gamma}$ the secular equation (8.2) of $n$th order is split into blocks. For nonrepeating representations the blocks are one-dimensional, i.e. $\varepsilon_\alpha = H_{\alpha\alpha}$. To repeating representations there correspond blocks whose dimension is equal to the number of representations (Section 5.6.2).

The above considerations indicate the range of problems that can be solved by group-theoretical methods in LCAO MO theory:

(1) choice of atomic basis and determination of the irreducible representations according to which the MOs constructed from the chosen basis AOs are transformed;

(2) construction of LCAO MOs that are transformed according to the irreducible representations of the molecular symmetry group.

Section 8.2 is concerned with the solution of the first problem.

## 8.2 GROUP-THEORETICAL CLASSIFICATION OF MOLECULAR ORBITALS

### 8.2.1 Illustrative example

Let us begin with an illustration, considering the $H_3$ molecule and taking for the atomic basis three $s$ functions $s_1$, $s_2$ and $s_3$. The $\mathbf{D}_{3h}$ group operations transform the functions into one another according to the general law (3.64) of transformation under the group operations:

$$\left.\begin{aligned} \hat{E}[s_1\, s_2\, s_3] &= [s_1\, s_2\, s_3]\begin{bmatrix} 1 & 0 & 0 \\ 0 & 1 & 0 \\ 0 & 0 & 1 \end{bmatrix}, \\ \hat{C}_3[s_1\, s_2\, s_3] &= [s_1\, s_2\, s_3]\begin{bmatrix} 0 & 0 & 1 \\ 1 & 0 & 0 \\ 0 & 1 & 0 \end{bmatrix} \quad \text{etc.} \end{aligned}\right\} \tag{8.4}$$

The set of matrices in (8.4) forms a three-dimensional reducible representation of the group. The basis of this representation comprises the three functions $s_1$, $s_2$ and $s_3$. From these, it is possible to construct three MOs of the form

$$\Psi_i = c_{1i}s_1 + c_{2i}s_2 + c_{3i}s_3 \quad (i = 1, 2, 3). \tag{8.5}$$

The matrices $\mathbf{D}(\hat{R})$ in the $\Psi_i$ basis will, of course, be different; however, as it follows from Section 3.1.4, the transformation (8.5) does not change their characters. Therefore, in calculating the characters of the matrices $\mathbf{D}(\hat{R})$ in the atomic basis, (8.4), it is possible to obtain at the same time the characters of the irreducible representation $\Gamma$ according to which the MOs transform. In the present case we obtain the following:

| $\mathbf{D}_{3h}$ | $\hat{E}$ | $2\hat{C}_3$ | $3\hat{C}_2$ | $2\hat{S}_3$ | $3\hat{\sigma}_v$ | $\hat{\sigma}_h$ |
|---|---|---|---|---|---|---|
| $\chi_{MO}$ | 3 | 0 | 1 | 0 | 1 | 3 |

Reducing this representation, we obtain

$$\Gamma \doteq A_1' + E'.$$

Thus the basis of three $s$ functions leads to two MOs: one nondegenerate, of $A_1'$ type, and another twofold-degenerate, of $E'$ type. The energy spectrum of the $H_3$ molecule contains the singlet $\varepsilon_A$ and the doublet $\varepsilon_E$. These are *one-electron* levels, and for the construction of many-electron states, taking account of spin as well as the Pauli principle (i.e. of *molecular terms*) the Wigner and Clebsch–Gordan coefficients must be used. We shall return to this question below.

The example considered above illustrates the main concept of group-theoretical MO classification: after the choice of the AO basis $\psi_1, \psi_2, \ldots, \psi_n$, it is necessary to determine the characters $\chi$ of the irreducible representation $\mathbf{D}(\hat{R})$:

$$\hat{R}[\psi_1, \psi_2 \ldots \psi_n] = [\psi_1 \psi_2 \ldots \psi_n]\, \mathbf{D}\,(\hat{R}) \tag{8.6}$$

Expanding this into irreducible representations provides the MO symmetry types constructed using the chosen MO basis.

The practical implementation of this concept for the case of the $H_3$ molecule has proved so easy because of the spherical symmetry of the $s$ orbitals. There are contributions to the character $\chi_{MO}(\hat{R})$ only from the s orbitals of the atoms left unchanged by the operations $\hat{R}$. It is obvious that atomic functions with $l > 0$ do not contribute to the character $\chi_{MO}(\hat{R})$ if they belong to an atom that is shifted under the operation $\hat{R}$. That is why the total character $\chi_{MO}(\hat{R})$ can differ from zero only for those operations $\hat{R}$ that constitute the site group of the given atom.

### 8.2.2 Ammonia molecule

Before stating the site-symmetry technique in general form, we shall utilize the above considerations to solve a more complicated problem: the determination of the MO LCAO symmetry of the ammonia molecule $NH_3$ ($\mathbf{C}_{3v}$). In the AO basis we include the 2s and 2p orbitals of nitrogen and the 1s orbitals of hydrogen. Since the AO basis comprises seven functions, the matrices $\mathbf{D}(\hat{R})$ are seven-dimensional; they can be represented schematically as

$$\mathbf{D}(\hat{R}) = \begin{bmatrix} \times & \times & \times & \times & \times & \times & \times \\ \times & \times & \times & \times & \times & \times & \times \\ \times & \times & \times & \times & \times & \times & \times \\ \times & \times & \times & \times & \times & \times & \times \\ \times & \times & \times & \times & \times & \times & \times \\ \times & \times & \times & \times & \times & \times & \times \\ \times & \times & \times & \times & \times & \times & \times \end{bmatrix} \begin{matrix} s^{N} \\ p_x^{N} \\ p_y^{N} \\ p_z^{N} \\ s_1^{H} \\ s_2^{H} \\ s_3^{H} \end{matrix} \tag{8.7}$$

The nitrogen atom possesses $\mathbf{C}_{3v}$ site symmetry, while the hydrogen atoms within the $\sigma_v$ planes possess $\mathbf{C}_s$ symmetry. Let us consider the $4 \times 4$ blocks of the matrices $\mathbf{D}(\hat{R})$, related to the nitrogen atom. The identity transformation provides $\chi_s^N(\hat{E}) = 1$ and $\chi_p^N(\hat{E}) = 3$ (the superscript denotes the atom and the subscript the AO type). For the $\hat{C}_3$ rotation $\chi_s^N(\hat{C}_3) = 1$. The character of the $p^N$-orbital transformation block for $C_3$ rotations can be found by inserting $\varphi = \frac{2}{3}\pi$ into the expression for the character $\chi^{(l)}(\varphi)$, (4.16), for $l = 1$: $\chi_p^N(\hat{C}_3) = 0$. Finally, for $\hat{\sigma}_v$ reflections $\chi_p^N(\hat{\sigma}_v) = 1$. The characters $\chi^N$ are shown in Table 8.1.

Reducing the representations $\Gamma_s^N$ and $\Gamma_p^N$, we obtain

$$\Gamma_s^N \doteq A_1, \quad \Gamma_p^N \doteq A_1 + E.$$

**Table 8.1** Characters of nitrogen AO transformation.

| $\mathbf{C}_{3v}$ | $\hat{E}$ | $2\hat{C}_3$ | $3\hat{\sigma}_v$ |
|---|---|---|---|
| $\chi_s^N$ | 1 | 1 | 1 |
| $\chi_p^N$ | 3 | 0 | 1 |

**Table 8.2** Characters of AO transformation for three hydrogen atoms.

| $\mathbf{C}_{3v}$ | $\hat{E}$ | $2\hat{C}_3$ | $3\hat{\sigma}_v$ |
|---|---|---|---|
| $\chi_s^H$ | 3 | 0 | 1 |

This shows that the nitrogen 2s orbital is involved in the $A_1$ symmetry MO, while its p orbitals permit the construction of one $A_1$-type MO and two MOs for the E represention involved in the MO transformed according to the representations $A_1$ and E.

In the $\mathbf{C}_s$ site groups the $s^H$ orbitals are totally symmetric ($A_1$); therefore only the characters $\chi_s^H(\hat{\sigma}_v) = 1$ and $\chi_s^H(\hat{E}) = 3$ are nonzero, and we arrive at the results shown in Table 8.2, from which it follows that $\Gamma_s^H \doteq A_1 + E$. Summarizing the results, we obtain the irreducible representations according to which the MOs are transformed in the assumed AO basis:

$$\Gamma_{MO} \doteq 3A_1 + 2E.$$

This result indicates that for $NH_3$ there are five one-electron levels: three nondegenerate and two comprising a doublet.

### 8.2.3 Tetrahedral molecules: formulation of method

Let us consider a more complicated example—a tetrahedral metal complex $ML_4$, where the use of site symmetry is more efficient. The group $\mathbf{G}_S$ for the metal is $\mathbf{T}_d$, while that for the ligands, $\mathbf{G}_S$, is $\mathbf{C}_{3v}$ (Table 7.1). In the AO basis we shall include the s, p and d orbitals. Since the site group of the central atom coincides with the full group, irreducible representations $\Gamma^M$ are obtained on reducing the rotation group representation $D^{(l)}$ ($l = 0, 1, 2$) in $\mathbf{T}_d$ (Table 4.3):

$$\Gamma_s^M \doteq A_1, \quad \Gamma_p^M \doteq T_2, \quad \Gamma_d^M \doteq E + T_2.$$

Thus the metal AOs are involved in MOs of $A_1$ type, in two MOs of E type and in one MO of $T_2$ type.

In the site symmetry subgroup ($\mathbf{C}_{3v}$) of each ligand, by means of (4.16) for $l = 0, 1, 2$, we obtain the characters $\chi^L(\hat{R})$ (Table 8.3).

**Table 8.3** Characters of ligand s, p and d AO transformation for the system $ML_4$ in the site group $\mathbf{C}_{3v}$.

| $\mathbf{C}_{3v}$ | $\hat{E}$ | $2\hat{C}_3$ | $3\hat{\sigma}_v$ |
|---|---|---|---|
| $\chi_s^L$ | 1 | 1 | 1 |
| $\chi_p^L$ | 3 | 0 | 1 |
| $\chi_d^L$ | 5 | −1 | 1 |

Now we must return to the initial full group $\mathbf{T}_d$. Since the site group $\mathbf{C}_{3v}$ is one of its subgroups, the site group representation characters are identical with those of the same operations of $\mathbf{T}_d$. The characters of $\mathbf{T}_d$ group operations not included in site groups are zero. To obtain the characters $\chi^L$ of the full group, the characters $\chi^L(\hat{R})$ from Table 8.3 must be multiplied by the numbers of ligands left unchanged by the $\mathbf{T}_d$ group operation $\hat{R}$: namely 4, 1 and 2 for $\hat{E}$, $\hat{C}_3$ and $\hat{\sigma}_v$ respectively. This gives the results shown in Table 8.4.

Expanding the ligand representations into their irreducible components, we obtain

$$\Gamma_s^L \doteq A_1 + T_2, \qquad \Gamma_p^L \doteq A_1 + E + T_1 + 2T_2, \qquad \Gamma_d^L \doteq A_1 + 2E + 2T_1 + 3T_2.$$

Thus the ligand s orbitals are involved in $A_1$ and $T_2$ type MOs, and the p orbitals in $A_1$, E and $T_1$ and two $T_2$ type MOs. Proceeding from the classification of the metal orbitals, we can obtain all LCAO MO types of the $ML_4$ tetrahedral system:

$$\Gamma_{MO} \doteq 4A_1 + 4E + 3T_1 + 8T_2.$$

The assumed basis thus provides 19 one-electron levels. The $45 \times 45$ matrix of the Hamiltonian is split into blocks whose dimensions are equal to the multiplicity of the repeating identical representations—blocks of dimension 4 ($A_1$ and E), and blocks of dimension 3 and 8 ($T_1$ and $T_2$ respectively).

Let us recapitulate the main stages of MO classification according to the irreducible representations of the point group:

(1) choose the AO basis for each atom $A_i$ $(i = 1, \ldots, n)$ of the molecule,

**Table 8.4** Characters of ligand s, p and d orbitals for $ML_4$ in $\mathbf{T}_d$.

| $\mathbf{T}_d$ | $\hat{E}$ | $8\hat{C}_3$ | $3\hat{C}_2$ | $6\hat{S}_4$ | $6\hat{\sigma}_d$ |
|---|---|---|---|---|---|
| $\chi_s^L$ | 4 | 1 | 0 | 0 | 2 |
| $\chi_p^L$ | 12 | 0 | 0 | 0 | 2 |
| $\chi_d^L$ | 20 | −1 | 0 | 0 | 2 |

$\psi_0(i), \psi_1(i), \ldots, \psi_k(i)$, where $l = 0, 1, \ldots, k$ are the values of the orbital angular momentum;

(2) determine the point group **G** of the molecule;

(3) find the site group $\mathbf{G}_S(i)$ of each atom $A_i$;

(4) compute the characters of the $\mathbf{G}_S(i)$ group operations $R_S(i)$ for each atom $A_i$ in the basis of its AOs $\psi_0(i), \psi_1(i), \ldots, \psi_k(i)$;

(5) multiply the characters of each site-group operation $\hat{R}_S(i)$ by the number of atoms of the molecule left unchanged by the operation;

(6) the numbers thus obtained are the characters $\chi^{A_i}_{\psi_l(i)}(\hat{R})$ of the full group **G** with reference to the operation $\hat{R}$ for the $\psi_l(i)$ orbital; the characters of operations not contained in site groups are zero;

(7) expand the reducible representations $\Gamma^{A_i}_{\psi_l(i)}$ corresponding to the $i$th atom or to a group of equivalent atoms into irreducible representations $\Gamma$: the irreducible representations thus obtained provide a classification of the MOs as well as the AOs involved in MOs of different symmetries.

## 8.3 CYCLIC $\pi$ SYSTEMS

MO classification by symmetry type gives irreducible representations whose basis functions can be constructed from an AO set. The LCAO MO construction may be carried out by using a projection operator (Section 6.3.2). We begin with the simple example of the system $H_3$, which we shall use for detailed exposition of the construction procedure for cyclic $\pi$ MO systems. The features of these systems will be discussed below. We introduce an initial simplification, considering instead of the point group $\mathbf{D}_{3h}$ its subgroup $\mathbf{C}_3$. Since $\mathbf{C}_3$ is Abelian, its irreducible representations are one-dimensional and the quantities $D^{(\Gamma)}_{\gamma\gamma}(\hat{R})^*$ in the projection operator are equal to the characters of the operations $\hat{R} = \hat{E}$, $\hat{C}_3$ and $\hat{C}_3^2$ in the representations $\Gamma = $ A, $\varepsilon$ and $\varepsilon^*$. These characters are given in Section 3.7.3. Let us apply the projection operator (6.13) to one of the basis functions, say $s_1$:

$$\left.\begin{aligned} \hat{P}^{A} s_1 &= \tfrac{1}{3}(1 \cdot \hat{E}s_1 + 1 \cdot \hat{C}_3 s_1 + 1 \cdot \hat{C}_3^2 s_1), \\ \hat{P}^{\varepsilon} s_1 &= \tfrac{1}{3}(1 \cdot \hat{E}s_1 + \omega \hat{C}_3 s_1 + \omega^* \hat{C}_3^2 s_1), \\ \hat{P}^{\varepsilon^*} s_1 &= \tfrac{1}{3}(1 \cdot \hat{E}s_1 + \omega^* \hat{C}_3 s_1 + \omega \hat{C}_3^2 s_1). \end{aligned}\right\} \tag{8.8}$$

Since $\hat{C}_3 s_1 = s_2$ and $\hat{C}_3^2 s_1 = s_3$ we obtain, after normalizing,

$$\left.\begin{aligned} \Psi(\mathrm{A}) &= \sqrt{1/3}(s_1 + s_2 + s_3), \\ \Psi(\varepsilon) &= \sqrt{1/3}(s_1 + \omega s_2 + \omega^* s_3), \\ \Psi(\varepsilon^*) &= \sqrt{1/3}(s_1 + \omega^* s_2 + \omega s_3). \end{aligned}\right\} \tag{8.9}$$

where we have assumed all AOs to be orthonormal. From the character table (Section 3.7.3) it can be seen that $\Psi(\varepsilon)$ and $\Psi(\varepsilon^*)$ are the bases of one-dimensional complex-conjugate representations; these functions are transformed as $x + \mathrm{i}y$ and $x - \mathrm{i}y$ respectively. *Two* MOs, $\Psi(\varepsilon)$ and $\Psi(\varepsilon^*)$, correspond to *one* twofold-degenerate level; therefore it is convenient to introduce the real $x$ and $y$ basis of the *two-dimensional* representation E (Section 3.7.3):

$$\left.\begin{aligned} \Psi(\mathrm{E}x) &= \sqrt{1/2}[\Psi(\varepsilon) + \Psi(\varepsilon^*)] = \sqrt{1/6}(2s_1 - s_2 - s_3), \\ \Psi(\mathrm{E}y) &= (-\mathrm{i}\sqrt{1/2})[\Psi(\varepsilon) - \Psi(\varepsilon^*)] = \sqrt{1/2}(s_2 - s_3). \end{aligned}\right\} \tag{8.10}$$

We now return to the initial group $\mathbf{D}_{3h}$, whose characters are given in Table 8.5. It can be seen that $\Psi_{\mathrm{A}}(\mathbf{C}_3)$ is transformed in the $\mathbf{D}_{3h}$ group according to the totally symmetric representation $\mathrm{A}_1'$, while the functions $\Psi(\mathrm{E}x)$ and $\Psi(\mathrm{E}y)$ correspond to the $\mathbf{D}_{3h}$ basis $\mathrm{E}'$.

Let us now consider homocyclic systems of the $(\mathrm{CH})_n$ type, assuming that there is an $\sigma$-bonded "core" involving hydrogen s orbitals overlapping with carbon $sp^2$ hybrids. The problem that remains is the construction of $n$ MOs from the carbon $p_z$ orbitals, perpendicular to the molecular plane.

A typical example is that of benzene $C_6H_6$ (Fig. 8.1), with group $\mathbf{D}_{6h}$. Let us determine the types of $\mathbf{C}_6$ subgroup MOs. All the rotations interchange the carbon-atom positions; therefore their characters are zero ($\chi_{\mathrm{AO}}(\hat{C}_6^k) = 0$). For the operation $\hat{E}$ the character $\chi_{\mathrm{AO}}(\hat{E}) = 6$. Since the transformation from AO to MO leaves the characters unchanged the table for $\chi_{\mathrm{MO}}$ is obtained as

| $\mathbf{C}_6$ | $\hat{E}$ | $\hat{C}_6$ | $\hat{C}_6^2$ | $\hat{C}_6^3$ | $\hat{C}_6^4$ | $\hat{C}_6^5$ |
|---|---|---|---|---|---|---|
| $\chi_{\mathrm{MO}}$ | 6 | 0 | 0 | 0 | 0 | 0 |

**Table 8.5** $\mathbf{D}_{3h}$ group characters.

| $\mathbf{D}_{3h}$ | $\hat{E}$ | $2\hat{C}_3$ | $2\hat{C}_2$ | $\hat{\sigma}_h$ | $2\hat{S}_3$ | $3\hat{\sigma}_v$ | $f^\Gamma$ |
|---|---|---|---|---|---|---|---|
| $\mathrm{A}_1'$ | 1 | 1 | 1 | 1 | 1 | 1 | $x^2+y^2$, $z^2$ |
| $\mathrm{A}_2'$ | 1 | 1 | −1 | 1 | 1 | −1 | $R_z$ |
| $\mathrm{A}_1''$ | 1 | 1 | 1 | −1 | −1 | −1 | |
| $\mathrm{A}_2''$ | 1 | 1 | −1 | −1 | −1 | 1 | $z$ |
| $\mathrm{E}'$ | 2 | −1 | 0 | 2 | −1 | 0 | $x, y$; $x^2-y^2$, $xy$ |
| $\mathrm{E}''$ | 2 | −1 | 0 | −2 | 1 | 0 | $R_x, R_y$; $yz$, $xz$ |

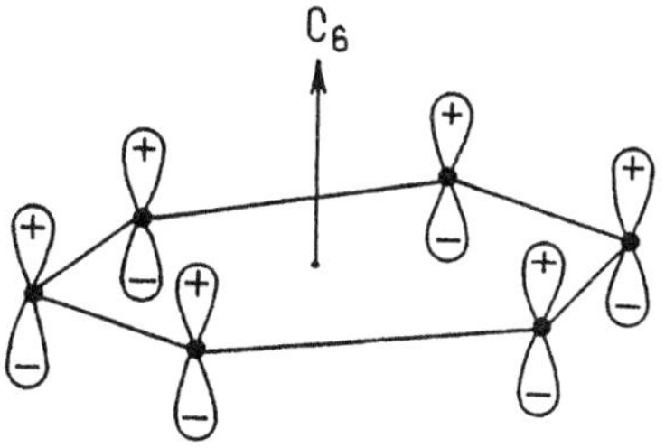


**Fig. 8.1** The $p_\pi$ orbitals of the benzene molecule.

On reducing the reducible representation $\Gamma_{MO}$, we find that $\Gamma_{MO} \doteq A + B + \varepsilon_1 + \varepsilon_1^* + \varepsilon_2 + \varepsilon_2^* \equiv A + B + E_1 + E_2$. As shown in Section 3.7.3, in $\mathbf{C}_6$ there are two pairs of one-dimensional complex-conjugate representations $\varepsilon_1$, $\varepsilon_1^*$ and $\varepsilon_2$, $\varepsilon_2^*$. Let us denote by $p_i$ the $p_z$ orbital of the $i$th atom. The representations are one-dimensional; therefore, in order to apply the projection operator, only the $\mathbf{C}_6$ group characters are required (see the table in Section 3.7.3 for $\mathbf{C}_6$). As in the previous example, we obtain

$$\Psi(A) = \sqrt{1/6}(p_1 + p_2 + p_3 + p_4 + p_5 + p_6), \tag{8.11}$$

$$\Psi(B) = \sqrt{1/6}(p_1 - p_2 + p_3 - p_4 + p_5 - p_6), \tag{8.12}$$

$$\left.\begin{aligned} \Psi(\varepsilon_1) &= \sqrt{1/6}(p_1 - \omega p_2 - \omega^* p_3 - p_4 - \omega p_5 + \omega^* p_6), \\ \Psi(\varepsilon_1^*) &= \sqrt{1/6}(p_1 - \omega^* p_2 - \omega p_3 - p_4 - \omega^* p_5 + \omega p_6), \end{aligned}\right\} \tag{8.13}$$

$$\left.\begin{aligned} \Psi(\varepsilon_2) &= \sqrt{1/6}(p_1 - \omega^* p_2 - \omega p_3 + p_4 - \omega^* p_5 - \omega p_6, \\ \Psi(\varepsilon_2^*) &= \sqrt{1/6}(p_1 - \omega p_2 - \omega^* p_3 + p_4 - \omega p_5 - \omega^* p_6, \end{aligned}\right\} \tag{8.14}$$

where $\omega = e^{2\pi i/3} = \cos\frac{2}{3}\pi + i\sin\frac{2}{3}\pi = \frac{1}{2} + \frac{1}{2}\sqrt{3}i$.

On passing to the real basis, instead of (8.13) and (8.14), we obtain MOs forming the basis of the two-dimensional representations $E_1$ and $E_2$ of $\mathbf{C}_6$:

$$\left.\begin{aligned} \Psi(E_1x) &= \tfrac{1}{2}\sqrt{1/3}(2p_1 + p_2 - p_3 - 2p_4 - p_5 + p_6), \\ \Psi(E_1y) &= \tfrac{1}{2}(p_1 + p_2 - p_4 - p_5), \end{aligned}\right\} \tag{8.15}$$

$$\left.\begin{aligned} \Psi(E_2a) &= \tfrac{1}{2}\sqrt{1/3}(2p_1 - p_2 - p_3 + 2p_4 - p_5 - p_6), \\ \Psi(E_2b) &= \tfrac{1}{2}(p_1 - p_2 + p_3 - p_4). \end{aligned}\right\} \tag{8.16}$$

From the character tables in [1] of $\mathbf{C}_6$ (Section 3.7.3) we find that the simplest complex basis of $E_1$ comprises the functions $\mp(x \pm iy)$, while that of $E_2$ comprises the functions $u_+ = (x + iy)^2$ and $u_- = (x - iy)^2$. The real basis is $x$, $y$ for $E_1$ and $a = x^2 - y^2$, $b = 2xy$ for $E_2$, which corresponds to the notation in (8.15) and (8.16).

It is now easy to pass to $\mathbf{D}_{6h}$. The $p_z$-type orbitals change sign under a reflection $\hat{\sigma}_h$. In $\mathbf{D}_{6h}$ the characters $\chi(\hat{\sigma}_h)$ in the representations $E_{1g}$ and $E_{2u}$ are negative. Therefore the MO pairs (8.15) and (8.16) are the respective basis functions of these $\mathbf{D}_{6h}$ group representations. Also,

$$A(\mathbf{C}_6) \to A_{2u}(\mathbf{D}_{6h}), \qquad B(\mathbf{C}_6) \to B_{2u}(\mathbf{D}_{6h}).$$

It should be noted that in the examples considered above all MOs belong to *different* irreducible representations. Nonrecurrence of representations is a general feature of homonuclear cyclic $\pi$ systems. This feature makes it unnecessary to solve the secular equation (8.2), which decomposes into *one-dimensional* blocks. In other words, the energy levels are expressed through diagonal matrix elements in the MO $\Psi(\Gamma\gamma)$ basis. For example, for benzene

$$\left.\begin{aligned} &\varepsilon(A_{2u}) = \langle A_{2u}|\hat{H}|A_{2u}\rangle, \qquad \varepsilon(E_{1g}) = \langle E_{1g}\gamma|\hat{H}|E_{1g}\gamma\rangle, \\ &\varepsilon(E_{2u}) = \langle E_{2u}\gamma|\hat{H}|E_{2u}\gamma\rangle, \end{aligned}\right\} \tag{8.17}$$

where $\gamma = x, y$ or $a, b$. In the Hückel approximation (see for example [59]) integrals of the type $\int p_i \hat{H} p_i \, d\tau$ are denoted by $\alpha$, while those of the type $\int p_i \hat{H} p_j \, d\tau$ for neighbouring carbon atoms are denoted by $\beta$. With neglect of the overlap integrals $\int p_i p_j \, d\tau$, the energy is written as

$$\left.\begin{aligned} &\varepsilon(A_{2u}) = \alpha + 2\beta, \qquad \varepsilon(E_{1g}) = \alpha + \beta, \\ &\varepsilon(E_{2u}) = \alpha - \beta, \qquad \varepsilon(B_{2g}) = \alpha - 2\beta. \end{aligned}\right\} \tag{8.18}$$

Other cyclic $\pi$ systems are treated similarly; a detailed discussion is given in [13]. In more complex $\pi$ systems reducible representations $\Gamma_\pi$ contain repeated irreducible representations. An example is the tetramethylenecyclobutane molecule [13] (Fig. 8.2). The representation $\Gamma_\pi$ of $\mathbf{D}_{4h}$ is reduced by the scheme $\Gamma_\pi \doteq 2A_{2u} + 2B_{1u} + 2E_g$ so that, instead of an eighth-order secular equation, we have to solve three second-order equations. The results are given in detail in [13].

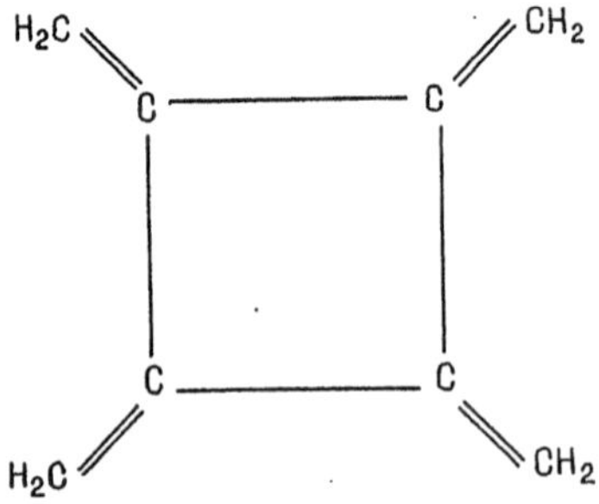


**Fig. 8.2** Tetramethylenecyclobutane.

## 8.4 TRANSITION METAL COMPLEXES

Many transition metal complexes possess symmetrical octahedral or tetrahedral structures. Ligand substitution in the first coordination sphere results in less symmetrical structures. A common feature of the electronic structure of such systems is the presence of ramified coordinate bonds involving s, p and d orbitals of the transition metal [59] (for iron-group ions these are the 3d, 4s and 4p AOs), and s and p orbitals of the ligands.

Let us apply a projection operator in order to construct MOs from ligand $\sigma$ functions of the octahedral ($\mathbf{O}_h$) complex $ML_6$. The ligand $\sigma$ orbitals transform like six $\boldsymbol{\sigma}$ vectors, and the corresponding reducible representation $\Gamma_\sigma$ decomposes into three irreducible representations $A_{1g}$, $E_g$ and $T_{1u}$ (Section 7.2.2). To construct the $\Psi(A_{1g})$ MO, we apply the operator $\hat{P}^{A_{1g}}$, (6.13), to one of the initial basis functions, for instance to $\sigma$ (the ion numbering is shown in Fig. 7.8). Group operations transform the $\boldsymbol{\sigma}$ vectors into one another:

$$\hat{E}\sigma_1 = \sigma_1, \qquad \hat{C}_3(xyz)\sigma_1 = \sigma_2, \qquad \hat{C}_4\sigma_1 = \sigma_2, \qquad \hat{C}_4^2(z)\sigma_1 = \sigma_4.$$

For $A_{1g}$ representations $D^{(A_{1g})*} = 1$, and, after normalization (assuming orthonormal AOs), we obtain

$$\Psi(A_{1g}) = \sqrt{1/6}(\sigma_1 + \sigma_2 + \sigma_3 + \sigma_4 + \sigma_5 + \sigma_6). \tag{8.19}$$

To construct MOs from ligand functions transforming according to *degenerate* (two- and three-dimensional) representations $E_g$ and $T_{1u}$, the *matrices* of these representations are required [7, 15]. The final result is

$$\left.\begin{aligned} \Psi(E_g u) &= \tfrac{1}{2}\sqrt{1/3}(2\sigma_3 + 2\sigma_6 - \sigma_1 - \sigma_2 - \sigma_4 - \sigma_5), \\ \Psi(E_g v) &= \tfrac{1}{2}(\sigma_1 + \sigma_4 - \sigma_2 - \sigma_5), \end{aligned}\right\} \tag{8.20}$$

$$\left.\begin{aligned} \Psi(T_{1u}x) &= \sqrt{1/2}(\sigma_1 - \sigma_4), \\ \Psi(T_{1u}y) &= \sqrt{1/2}(\sigma_2 - \sigma_5), \\ \Psi(T_{1u}z) &= \sqrt{1/2}(\sigma_3 - \sigma_6). \end{aligned}\right\} \tag{8.21}$$

The total basis of the AO collection also contains metal-ion functions. For iron-group ions one may usually confine attention to the 3d, 4s and 4p AOs, which carry the following irreducible representations of $\mathbf{O}_h$ (see the basis notation in Section 3.5):

$$\underbrace{d_u, d_v;}_{E_g} \qquad \underbrace{d_\xi, d_\eta, d_\zeta;}_{T_{2g}} \qquad \underbrace{s\;\;;}_{A_{1g}} \qquad \underbrace{p_x, p_y, p_z}_{T_{1u}}.$$

Within this total basis the representations $A_{1g}$, $E_g$, $T_{1u}$ and $T_{2g}$ appear twice—the total basis is formed only by $d$ functions of the metal ion. Let $\Psi^L$ and $\Psi^M$ be the MOs of the ligands and the metal-ion functions respectively. According to the configuration mixing rules (Section 5.6.2), the matrix elements $\langle \Psi^{L}(\Gamma\gamma)|\hat{H}|\Psi^{M}(\Gamma\gamma)\rangle$ are nonzero and identical for all basis functions $\gamma$ of the irreducible representation $\Gamma$. In the present case the secular equation (8.2) of 15th order (with $A_{1g}$, $E_g$ and $T_{1u}$ ligand functions, and $E_g$, $T_{2g}$ and $A_{1g}$ metal functions) decomposes into second-order blocks and one one-dimensional block, the latter corresponding to the nonrepeating representation $T_{2g}$. The block for the two $E_g$ representations is

$$\begin{vmatrix} \langle d_u|\hat{H}|d_u\rangle - \varepsilon & \langle d_u|\hat{H}|\sigma_u\rangle - \varepsilon\langle d_u|\sigma_u\rangle \\ \langle \sigma_u|\hat{H}|d_u\rangle - \varepsilon\langle \sigma_u|d_u\rangle & \langle \sigma_u|\hat{H}|\sigma_u\rangle - \varepsilon \end{vmatrix} = 0, \qquad (8.22)$$

where the $\sigma$ function, of the ligand MO $\Psi^{L}(E_g u)$ is denoted for brevity as $|\sigma_u\rangle$, $\hat{H}$ is the one-electron (effective) Hamiltonian of the complex, $\langle d_u|\sigma_u\rangle$ is the *group overlap integral* describing the overlap of the central-ion function $d_u$ with the ligand MO function $\sigma_u$. The same configuration mixing equation is obtained for the functions $d_v$ and $\sigma_v$.

The two solutions of (8.22) reveal that atomic and ligand levels having the same symmetry "repel each other", as shown schematically in Fig. 8.3. The ground state is called the *bonding state*, while the excited state is the *antibonding state* [37, 59, 60]. When an electron goes into the bonding state, this leads to a decrease in energy relative to the separate metal and ligand orbitals, i.e. to chemical bonding. The corresponding wavefunctions $\psi_u^-$, $\psi_v^-$ (bonding) and $\psi_u^+$, $\psi_v^+$ (antibonding) also form bases of $E_g$ representations and in the general case are linear combinations of $d_u$, $d_v$ and $\sigma_u$, $\sigma_v$:

$$\left.\begin{aligned} \psi_\gamma^+ &= (1-c^2)^{1/2} d_\gamma - c\sigma_\gamma, \\ \psi_\gamma^- &= c d_\gamma - (1-c^2)^{1/2}\sigma_\gamma \end{aligned}\right\} \quad (\gamma = u \text{ or } v), \qquad (8.23)$$

where the constant $c$ is determined from the solution of the secular equation (8.2). It should be stressed that this constant cannot be obtained using group-theoretical procedures. In lower-symmetry complexes the number of repeating representations increases, which results in higher-order secular equations,

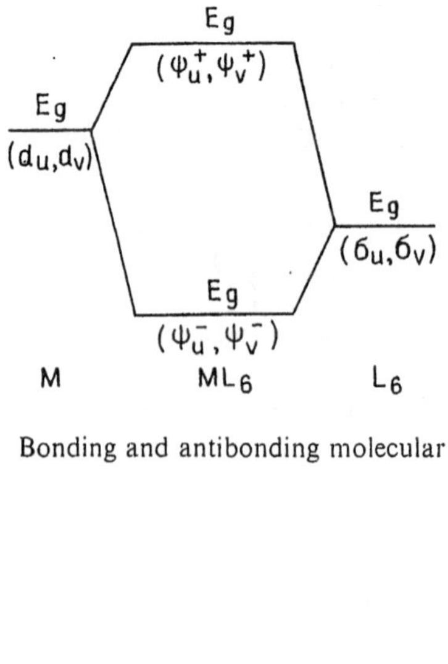


**Fig. 8.3** Bonding and antibonding molecular orbitals.

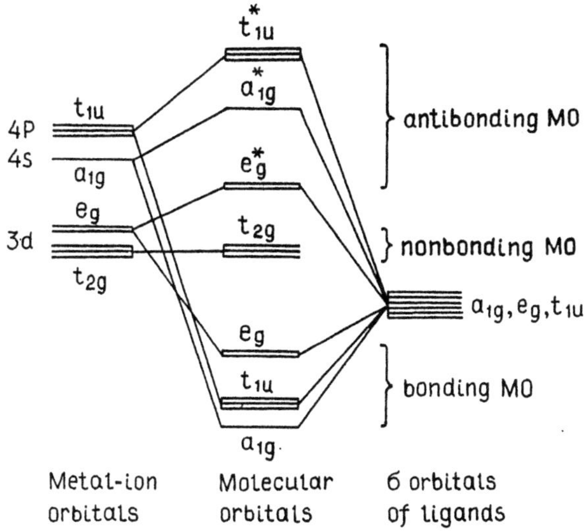


**Fig. 8.4** The energy diagram of an octahedral complex of a transition metal, taking into account $\sigma$ bonds; the antibonding states are indicated by asterisks.

whose solutions must be found by numerical methods. Similar complications result from expansion of the atomic basis.

The energy level $\varepsilon(T_{2g})$ is equal to the metal-ion energy and corresponds to a *nonbonding orbital.* Thus chemical bonding can occur only when there are identical representations in the basis function sets of the metal ion and the ligands.

The energy diagram of the $\sigma$ orbitals of an octahedral complex is sketched in Fig. 8.4. For each pair of repeating representations the complex has two corresponding orbitals—one bonding and one antibonding; the nonbonding orbital ($t_{2g}$) corresponds to the central ion level. It should be noted that the irreducible representations in Fig. 8.4 are denoted by lower-case letters, this is the usual convention to indicate that they are one-electron MOs.

The construction of $\pi$-type MOs can also be performed by the projection-operator method. Results for octahedral, tetrahedral and tetragonal six-coordinate complexes are given in [56, 59].

## 8.5 SANDWICH-TYPE COMPOUNDS

Compounds $(C_nH_n)_2M$ containing a metal ion situated at the centre of symmetry between two parallel $C_nH_n$ rings are called *sandwich* compounds. Examples are ferrocene $(C_5H_5)_2Fe$ and dibenzenechromium $(C_6H_6)_2Cr$. Sandwich-complex symmetry depends on the relative positions of the $C_nH_n$

rings in planes perpendicular to the $C_n$ axis. The dibenzenechromium symmetry group is $\mathbf{D}_{6h}$, since the benzene rings form an "eclipsed" hexagonal prism configuration (Fig. 8.5a). Ferrocene, whose cyclopentadienyl rings are staggered, has $\mathbf{D}_{5d}$ symmetry (Fig. 8.5b).

Sandwich compound MOs can be constructed in a standard way using the projection operator method. However, the specific structure of these compounds allows a simpler approach. The main idea consists in using as a basis not the AOs of individual atoms but the MOs of entire atomic clusters.

Let us illustrate this, taking dibenzenechromium as an example [13]. The atomic basis comprises the chromium 3d, 4s and 4p orbitals, as well as the 12 $p_z$ orbitals of the two benzene rings. Instead of the 12 $p_z$ orbitals, we take the MOs of the rings, found in Section 8.3. Reflection of the whole molecule in the $\sigma_h$ plane transforms the MOs of the benzene rings I and II (Fig. 8.5a) into one another. We now introduce the site groups $\mathbf{C}_6$ for each benzene ring, and construct the symmetric and antisymmetric combinations of the benzene I and II ring MOs, taking into account their classification in the ring site groups $\mathbf{C}_6$. The symmetric combinations are transformed according to the even representations of the full group $\mathbf{D}_{6h}$, while the antisymmetric combinations are transformed according to its odd representations. Using (8.11), (8.12), (8.15) and (8.16), we obtain the dibenzenechromium MO, composed of benzene ring $\pi$ orbitals:

$$\left.\begin{aligned}\Psi(A_{1g}) &= \sqrt{1/2}\,[\Psi_I(A) + \Psi_{II}(A)],\\ \Psi(A_{2u}) &= \sqrt{1/2}\,[\Psi_I(A) - \Psi_{II}(A)],\end{aligned}\right\} \tag{8.24}$$

$$\left.\begin{aligned}\Psi(B_{2g}) &= \sqrt{1/2}\,[\Psi_I(B) + \Psi_{II}(B)],\\ \Psi(B_{1u}) &= \sqrt{1/2}\,[\Psi_I(B) - \Psi_{II}(B)],\end{aligned}\right\} \tag{8.25}$$

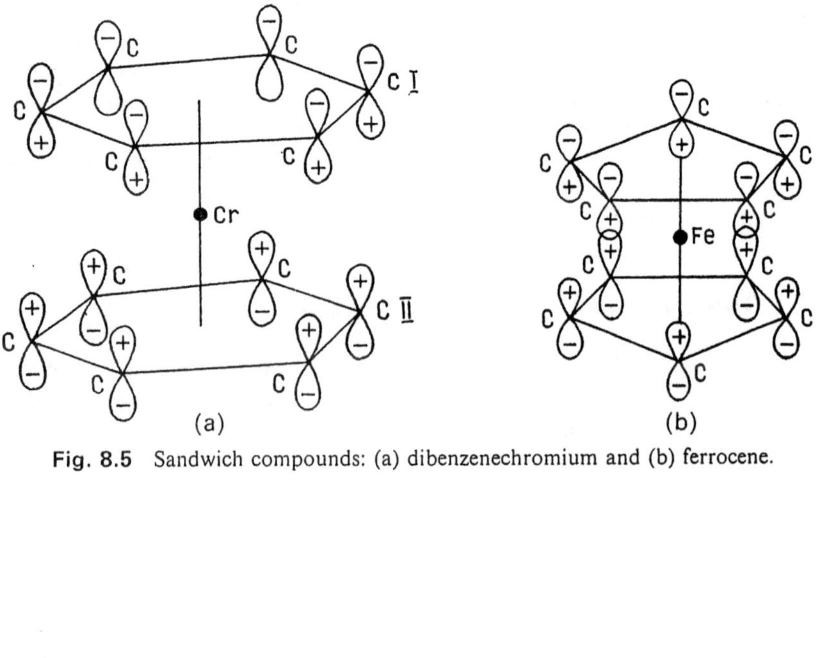


**Fig. 8.5** Sandwich compounds: (a) dibenzenechromium and (b) ferrocene.

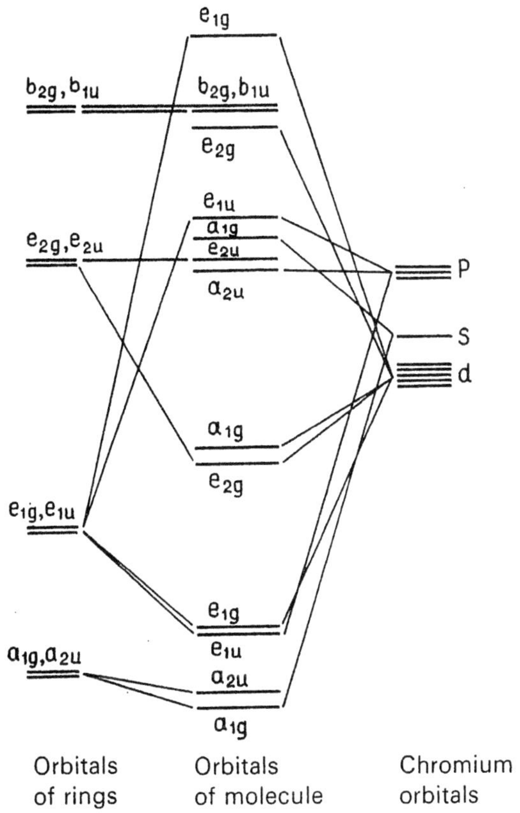


**Fig. 8.6** The energy diagram of dibenzenechromium.

$$\left.\begin{aligned} \Psi(\mathrm{E_{1g}}xz) &= \sqrt{1/2}\,[\Psi_{\mathrm{I}}(\mathrm{E_1}x) + \Psi_{\mathrm{II}}(\mathrm{E_1}x)], \\ \Psi(\mathrm{E_{1u}}x) &= \sqrt{1/2}\,[\Psi_{\mathrm{I}}(\mathrm{E_1}x) - \Psi_{\mathrm{II}}(\mathrm{E_1}x)], \end{aligned}\right\} \tag{8.26}$$

$$\left.\begin{aligned} \Psi(\mathrm{E_{2g}}x^2 - y^2) &= \sqrt{1/2}\,[\Psi_{\mathrm{I}}(\mathrm{E_2}a) + \Psi_{\mathrm{II}}(\mathrm{E_2}a)], \\ \Psi(\mathrm{E_{2u}}a) &= \sqrt{1/2}\,[\Psi_{\mathrm{I}}(\mathrm{E_2}a) - \Psi_{\mathrm{II}}(\mathrm{E_2}a)], \end{aligned}\right\} \tag{8.27}$$

where the subscripts I and II indicate the benzene ring MOs. It follows from the tables in [1] that the $\mathrm{E_{1g}}$ basis is $yz$, $xz$ (or $R_x$, $R_y$), and the $\mathrm{E_{2g}}$ basis is $x^2 - y^2$, $xy$. Only the first basis functions of the doublets are given. It should be noted that two A type MOs (in site groups) provide different irreducible representations of the full group $\mathbf{D}_{6h}$. The correlation diagram is

$$2\mathrm{A} \rightarrow \mathrm{A_{1g}} + \mathrm{A_{2u}}, \quad 2\mathrm{B} \rightarrow \mathrm{B_{2g}} + \mathrm{B_{1u}}, \quad 2\mathrm{E_1} \rightarrow \mathrm{E_{1g}} + \mathrm{E_{1u}}, \quad 2\mathrm{E_2} \rightarrow \mathrm{E_{2g}} + \mathrm{E_{2u}}.$$

We must now classify the metal-ion AOs with $l = 0, 1, 2$ according to the irreducible representations of $\mathbf{D}_{6h}$. Reducing the representations $D^{(l)}$ in $\mathbf{D}_{6h}$, we obtain

$$D^{(0)} \doteq A_{1g}, \quad D^{(1)} \doteq A_{2u} + E_{u}, \quad D^{(2)} \doteq A_{1g} + E_{1g} + E_{2g},$$

where the parity is determined by the factor $(-1)^l$. The metal orbitals are distributed over these irreducible representations as follows:

$$\underbrace{s, dz^2}_{A_{1g}}; \quad \underbrace{d_{xz}, d_{yz}}_{E_{1g}}; \quad \underbrace{d_{xy}, d_{x^2-y^2}}_{E_g}; \quad \underbrace{p_z}_{A_{2u}}; \quad \underbrace{p_x, p_y}_{E_{1u}}.$$

We see that the metal s and $d_{z^2}$ AOs are mixed with the benzene core $A_{1g}$ MO, resulting in a third-order secular equation. The $d_{xz}$ and $d_{yz}$ ($E_{1g}$) AOs mix with the $E_{1g}$ MO, giving a second-order secular equation. Another three second-order equations are obtained for the representations $E_{2g}$, $A_{2u}$ and $E_{1u}$. Benzene ring MOs of types $B_{2g}$, $B_{1u}$ and $E_{2u}$ lead to nonbonding states. The energy diagram calculated in [61] is sketched in Fig. 8.6.

## 8.6 SUPEREXCHANGE IN CLUSTERS

Ion groupings forming magnetic fragments of polynuclear coordination compounds, and of some inorganic salts, are usually called *exchange clusters.* Magnetic properties of a cluster are determined by the individual properties of the electron shells of the ion involved, and by those of the ion interactions. The interactions occurring in the case of direct ion contact at short distances are called *direct exchange* [62, 63]. This is related to the overlap of ion electron shells. Ions with unpaired electron shells are usually far enough apart to ensure neglible orbital overlap. In this case the interaction between unpaired (magnetic) electrons occurs through intermediate diamagnetic atoms of the bridging ligands. Such an exchange is called *superexchange* or *indirect exchange* [64–66].

Group-theoretical aspects of exchange interaction problems in polynuclear complexes are discussed in Chapter 13. For now, we shall consider the application of the MO method in explaining the character of the exchange interaction between two bridging-ligand-bound metal ions. The principal results in this area are due to W. E. Hatfield (see the reviews [64, 65]). The simplest structure is a square-planar fragment of a dimeric (binuclear) copper(II) cluster (Fig. 8.7), whose copper ions are bound by oxygen bridge atoms [64]. It is assumed that the copper–copper distance exceeds the sum of the covalent radii, so that d-electron delocalization is due to their partial transfer to oxygen atoms. To describe superexchange, it is necessary to construct the cluster MOs. We include in the atomic basis eight functions:

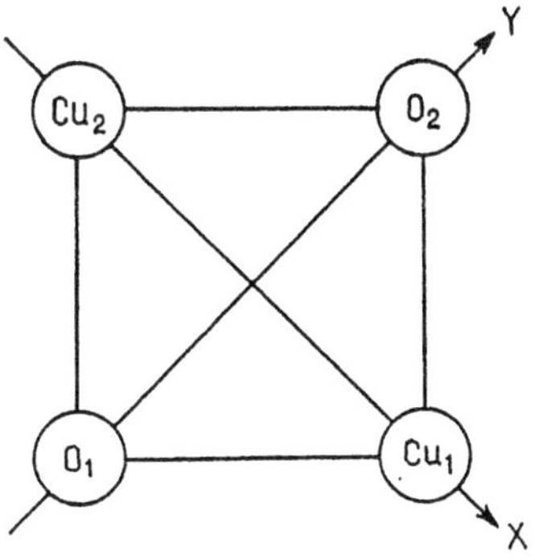


**Fig. 8.7** The structure of a binuclear copper fragment.

copper $d_{xy}$ and $d_{x^2-y^2}$ orbitals and oxygen $p_x$ and $p_y$ orbitals. Other AOs overlap much less, and they can be omitted in a qualitative analysis.

The cluster point group is $\mathbf{D}_{2h}$. The LCAO MO classification is easily performed using the method described in Section 8.2. The result is

$$\Gamma_{\mathrm{MO}} \doteq 2\mathrm{A_g} + 2\mathrm{B_{1g}} + 2\mathrm{B_{2u}} + 2\mathrm{B_{3u}}.$$

Since the one-dimensional representations of $\mathbf{D}_{2h}$ occur only twice, the eighth-order secular equation for the energy is reduced to four second-order equations. Let us use the procedure of separating fragments employed in Section 8.5. Since the fragments must include equivalent atoms, we divide the cluster into two fragments, one containing two copper ions, the other one consisting of two oxygen ions. We construct separately the MOs of the two copper ions, each having site group $\mathbf{C}_{2v}$. Application of the projection operator technique gives

$$\left.\begin{aligned}
\Psi(\mathrm{A_g}) &= \sqrt{1/2}\,[d_{x^2-y^2}(1) + d_{x^2-y^2}(2)],\\
\Psi(\mathrm{B_{1g}}) &= \sqrt{1/2}\,[d_{xy}(1) + d_{xy}(2)],\\
\Psi(\mathrm{B_{2u}}) &= \sqrt{1/2}\,[d_{xy}(1) - d_{xy}(2)],\\
\Psi(\mathrm{B_{3u}}) &= \sqrt{1/2}\,[d_{x^2-y^2}(1) - d_{x^2-y^2}(2)],
\end{aligned}\right\} \qquad (8.28)$$

where the numbers in parentheses are the atom numbers (Fig. 8.7). For the oxygen fragment MO we obtain in the same way

$$\left.\begin{aligned}
\psi(\mathrm{A_g}) &= \sqrt{1/2}\,[p_y(1) - p_y(2)],\\
\psi(\mathrm{B_{1g}}) &= \sqrt{1/2}\,[p_x(1) - p_x(2)],\\
\psi(\mathrm{B_{2u}}) &= \sqrt{1/2}\,[p_y(1) + p_y(2)],\\
\psi(\mathrm{B_{3u}}) &= \sqrt{1/2}\,[p_x(1) + p_x(2)].
\end{aligned}\right\} \qquad (8.29)$$

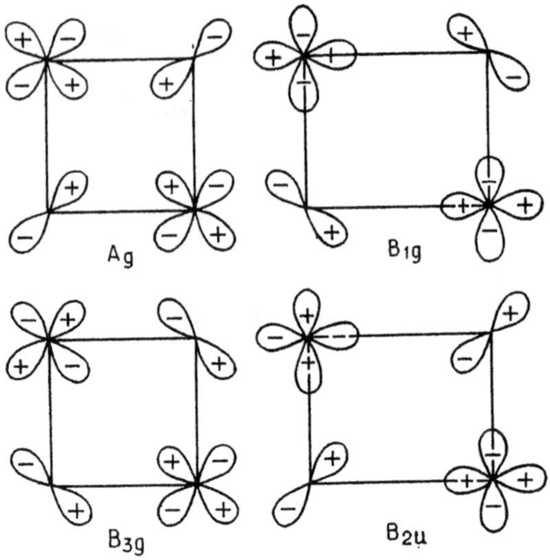


**Fig. 8.8** The distribution of electron density in one-electron states of a binuclear copper cluster with oxygen bridges.

The spatial forms of these MOs is shown in Fig. 8.8 [64]. It is clear that MOs of identical symmetry, belonging to different fragments, overlap and this leads to delocalization of unpaired electrons. This is expressed mathematically by the fact that MOs transforming according to identical representations mix. For example for the representation $A_g$ the two MOs of the entire cluster are

$$\left.\begin{aligned}\Phi(A_g) &= c(A_g)\Psi(A_g) + [1 - c^2(A_g)]^{1/2}\psi(A_g),\\ \Phi^*(A_g) &= [1 - c^2(A_g)]^{1/2}\Psi(A_g) - c(A_g)\psi(A_g),\end{aligned}\right\} \tag{8.30}$$

where $\Phi$ and $\Phi^*$ are the MOs of the bonding and antibonding states respectively. The constant $c(A_g)$ is determined from the solution of the secular equation in the basis of the two $A_g$ MOs. According to quantum mechanical rules, $|c(A_g)|^2$ is the probability of electron localization on copper ions in the bonding state and on oxygen ions in the antibonding state. Thus the MOs of type (8.30) describe the superexchange interaction in transition metal clusters by means of oxygen bridges.

## 8.7 MANY-ELECTRON STATES IN THE MOLECULAR ORBITAL METHOD

### 8.7.1 Molecular terms

The construction of many-electron states when using the LCAO MO method does not differ in principle from that described in Section 5.5 for strong-crystal-field terms. Let us assume that two electrons occupy two MOs: $\Psi(\Gamma_1\gamma_1)$ and $\Psi(\Gamma_2\gamma_2)$. Multiplying these by the spin functions $\chi(\frac{1}{2}m)$ (Section 5.5), we obtain two spin orbitals $\varphi(\Gamma_i\gamma_i m_i) = \Psi(\Gamma_i\gamma_i)\chi(\frac{1}{2}m_i)$. According to the Pauli principle, a two-electron wavefunction must be antisymmetric under electronic permutation; it may therefore be constructed from the determinants $|\varphi(\Gamma_1\gamma_1 m_1) \quad \varphi(\Gamma_2\gamma_2 m_2)|$. The states of the system are classified by the total spin $S = 0$ or 1 and the irreducible representation $\Gamma$ contained in the direct product $\Gamma_1 \times \Gamma_2$. The method of wavefunctions construction does not depend on the specific character of the spin orbitals $\varphi(\Gamma\gamma m)$; therefore we can use the properties of Wigner and Clebsch–Gordan coefficients and apply (5.37):

$$\Psi(\alpha S\Gamma M\gamma) = \sum_{m_1 m_2}\sum_{\gamma_1\gamma_2} |\varphi(\Gamma_1\gamma_1 m_1)\cdot\varphi(\Gamma_2\gamma_2 m_2)|\langle\Gamma_1\gamma_1\Gamma_2\gamma_2|\Gamma\gamma\rangle\langle\tfrac{1}{2}m_1\tfrac{1}{2}m_2|SM\rangle, \tag{8.31}$$

where $\alpha$ is an additional quantum number, enumerating repeating molecular terms $S\Gamma$.

The "annexation" of successive electrons in a many-electron system is performed in the same way as in the strong-crystal-field theory (Section 5.7.3). All the electrons whose AOs are employed in constructing the MOs must then be distributed over one-electron *molecular* spin orbitals $\varphi(\Gamma\gamma m)$ according to the Pauli principle. Different ways of filling these lead to different sets of possible molecular terms.

### 8.7.2 Cyclic $\pi$-system terms

Let us consider the many-electron states of the benzene molecule. The diagram of one-electron levels, as found in Section 8.3, (8.18), is shown in Fig. 8.9. Since

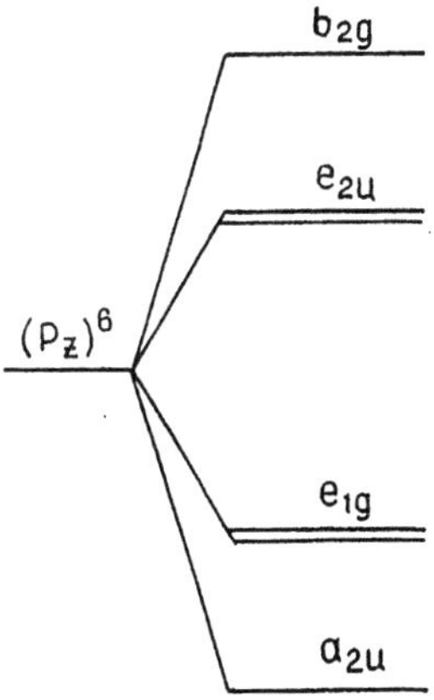


**Fig. 8.9** The energy diagram of one-electron levels of a benzene molecule.

the resonance integral $\beta$ is negative, the ground level is $\varepsilon(a_{2u})$. Six $\pi$ electrons fill the one-electron MOs $a_{2u}$ and $e_{1g}$, forming the configuration $(a_{2u})^2(e_{1g})^4$ (Fig. 8.10a). This is a *closed-shell configuration*, since it accommodates the maximum number of electrons allowed by the Pauli principle. It is clear that for a closed-shell the total spin $S = 0$, the only possible wavefunction being the *Slater determinant*

$$|(a_{1g})\ (\bar{a}_{1g})\ (e_{1g}x)\ (\bar{e}_{1g}x)\ (e_{1g}y)\ (\bar{e}_{1g}y)|.$$

The corresponding representation is $A_{1g}$; therefore the closed shell provides the term $^1A_{1g}$. This is a general conclusion. The many-electron function of a closed shell is transformed according to the totally symmetric representation, and the total spin is $S = 0$.

The first group of excited states is obtained when the electron passes from the $e_{1g}$ MO to the energetically closest MO, $e_{2u}$. The terms of the configuration $(a_{2u})^2(e_{1g})^3e_{2u}$ (Fig. 8.10b) can be found if we construct the many-electron wavefunctions. For one-electron excitations from a closed shell the problem can be simplified, bearing in mind that $(a_{2u})^2$ gives $^1A_{1g}$ and considering the configuration $(e_{1g})^3$ as a "hole" in the closed configuration $(e_{1g})^4$. Multiplying the "hole" representation $e_{1g}$ by $e_{2u}$, we obtain

$$e_{1g} \times e_{2u} = b_{1u} + b_{2u} + e_{1u}.$$

Taking into account that $S = 0$ or 1, we obtain the terms of the configuration $(a_{2u})^2(e_{1g})^3e_{2u}$:

$$^1B_{1u},\quad ^3B_{1u},\quad ^1B_{2u},\quad ^3B_{2u},\quad ^1E_{1u},\quad ^3E_{1u}.$$

The corresponding energy levels (Fig. 8.11) are separated by gaps determined by the interelectronic interaction.

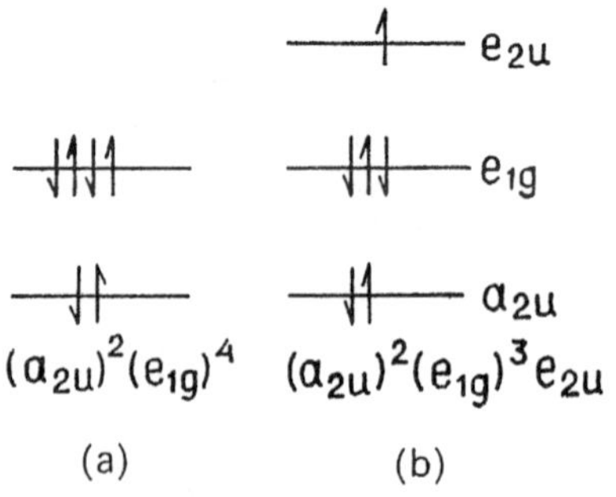


**Fig. 8.10** Electronic configurations of the ground and first excited groups of benzene levels.

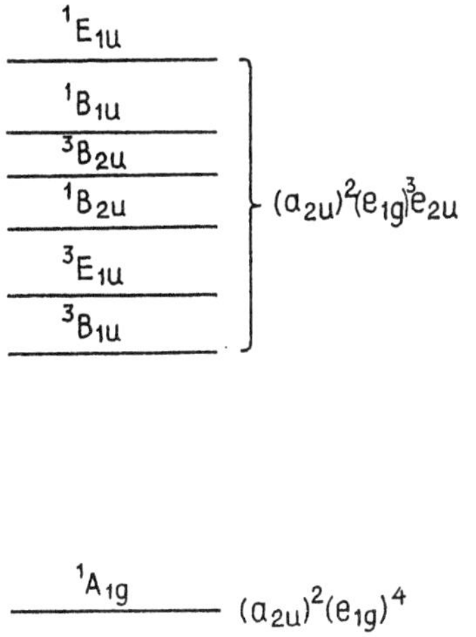


**Fig. 8.11** Terms of a benzene molecule.

## 8.7.3 Terms of transition metal complexes

Transition metal complexes have unfilled electronic shells. The bonding MOs (Fig. 8.4) are completely filled, while the nonbonding level $t_{2g}$ and the antibonding level $e_g^*$ are partially filled. The electronic configuration in the $\sigma$ approximation (Fig. 8.4) can be represented as $(a_{1g})^2(t_{1u})^6(e_g)^4(t_{2g})^k e_g^{n-k}$, where $n$ is the number of transition metal d electrons. Denoting the closed-shell configuration $(a_{1g})^2(t_{1u})^6(e_g)^4$ by [A], we can represent the electronic configuration of the $d^1$ ion as $[A]t_{2g}$, that of the $d^2$ ion as $[A]t_{2g}^2$, etc. Thus for the LCAO MO method, all of the conclusions of Section 5.5–5.7 concerning term classification are valid, as are the expressions for the many-electron wavefunctions. Therefore the $t_{2g}^2$ configuration wavefunctions for the $d^2$ ion are given by (5.41). Of course, it should be kept in mind that in the LCAO MO approximation $\xi$, $\eta$, $\zeta$ and $\bar{\xi}$, $\bar{\eta}$, $\bar{\zeta}$ are one-electron MOs, not $d$ functions of a metal ion, as in the case of (5.41). That is why the term energies differ from those obtained in the ionic model of the crystal field. A discussion of the calculation methods is given in [59] the detailed presentation of the present day state of many-electron theory is given in [60].

## 8.7.4 Magnetic states of dimeric clusters

Copper(II) complexes usually exhibit the paramagnetism of the unpaired hole in the d shell ($d^9$). In dimeric clusters interaction of the holes results in a principal spin-paired state ($S = 0$) and an excited spin triplet ($S = 1$). This leads to cluster diamagnetism at zero temperature and an increase in the magnetic moment with temperature (for a review see [63]).

The one-electron MOs constructed above (Section 8.6) for the magnetic fragment

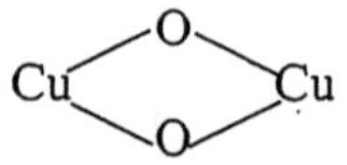


account for the origin of the superexchange itself, being related to electron delocalization in the absence of direct metal–metal interaction. Cluster magnetic properties depend on the position of levels with certain total spins, i.e. terms. Figure 8.12(a) shows a diagram of the one-electron levels [64, 65] of a symmetric ($\mathbf{D}_{2h}$) $Cu_2O_2$ cluster having Cu–O–Cu bond angles equal to 90°. The one-electron energies of the $b_{1g}$ and $a_g$ levels are equal to the energies of the $b_{2u}$ and $b_{3u}$ levels, respectively. This conclusion, drawn from calculations, is illustrated by the distribution of electron density for these MOs (Fig. 8.8).

Eight electrons occupy the bonding MO, and the antibonding orbitals $a_g^*$ and

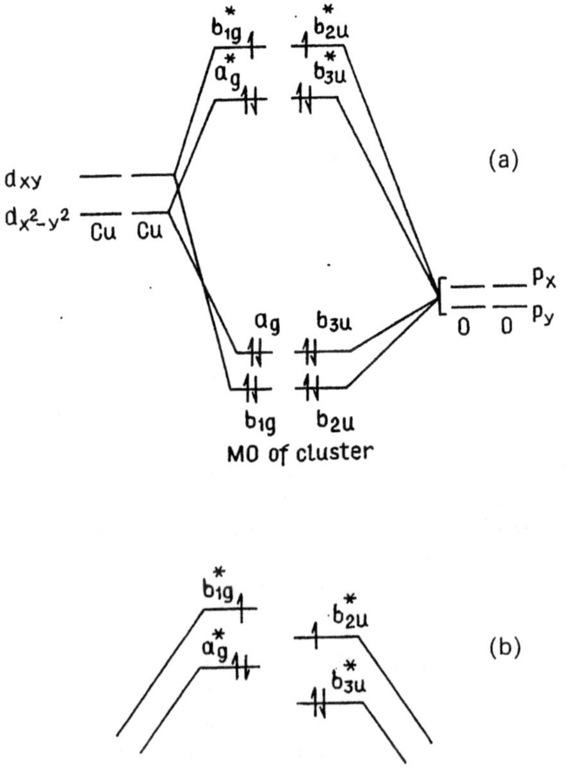


**Fig. 8.12** One-electron levels of a dimeric $Cu_2O_2$ cluster: (a) $\mathbf{D}_{2h}$ symmetry; (b) antibonding states of a distorted cluster (after [64]).

$b_{3u}^*$ also constitute closed configurations. Finally, each of the antibonding MOs $b_{1g}^*$ and $b_{2u}^*$ is occupied by one electron. The configuration $[A](b_{1g})(b_{2u})$ can result either in a triplet term or a singlet term (relative to the spin) $^3B_{3u}$ or $^1B_{3u}$ $(b_{1g} \times b_{2u} = b_{3u})$. Terms of the configurations $[A](b_{1g})^2$ and $[A](b_{2u})^2$, for which $S = 0$, are also possible; these are the $^1A_g$ terms. Calculations show that the ground term is the triplet term; in practice this manifests itself in the ferromagnetism of dimers with 90° Cu–O–Cu bond angles.

The cluster behaviour as the Cu–O–Cu angle increases is important. When this angle changes, there appears a difference between the energies $\varepsilon(b_{1g}^*)$ and $\varepsilon(b_{2u}^*)$. If this difference $\varepsilon(b_{1g}^*) - \varepsilon(b_{2u}^*)$ exceeds the electron interaction energy and the ground term is $^1B_{3u}$ of configuration $(b_{2u}^*)^2$ then cluster diamagnetism occurs at low temperatures; that is, there is an antiferromagnetic super-exchange interaction. The triplet terms are excited and consequently an increase in temperature should lead to an increase in magnetic moment; this is indeed observed experimentally.

## PROBLEMS

**8.1** Determine the MO types and construct the MOs for the cyclic $\pi$ systems $C_4H_4$ and $C_8H_8$.

**8.2** Classify the tetramethylenecyclobutane MOs (Fig. 8.2) and construct them using the projection operator method.

**8.3** Construct the $\sigma$ MOs for an octahedral complex.

**8.4** Construct the $\sigma$ MOs for a tetrahedral complex.

**8.5** Construct $\pi$-type LCAO MOs for the octahedral complex $ML_6$.

**8.6** Construct $\pi$-type LCAO MOs for the tetrahedral complex $ML_4$.

**8.7** Using the site-symmetry method, classify LCAO MOs of the dimeric copper cluster shown in Fig. 8.7, including in the AO basis the $d_{xy}$ and $d_{x^2-y^2}$ copper orbitals and the $p_x$ and $p_y$ oxygen orbitals.

**8.8** Using the projection operator method, construct the LCAO MOs of the dimeric copper cluster of Problem 8.7.

# 9 Intensities of Optical Lines

## 9.1 SELECTION RULES FOR OPTICAL TRANSITIONS

### 9.1.1 Interaction with an electromagnetic field

Interaction of molecules or atoms with an electromagnetic field (photons) leads to light absorption or emission. On absorbing light, a molecule is excited, and its energy is increased by an amount $\varepsilon(\Gamma_2) - \varepsilon(\Gamma_1) = h\nu$, where $\nu$ is the light frequency, $h$ is Planck's constant, and the irreducible representations $\Gamma_1$ and $\Gamma_2$ refer to the initial and final states respectively. To each possible transition $\Gamma_1 \to \Gamma_2$ there corresponds a spectral line, characterized by the frequency $\nu$ of the absorbed light (i.e. by the line position) and by its intensity. The line intensity $I(\Gamma_1, \Gamma_2)$ is proportional to the sum of the squared matrix elements of the type

$$I(\Gamma_1, \Gamma_2) = \frac{1}{g(\Gamma_1)} \sum_{\gamma_1, \gamma_2} |\langle \Gamma_2\gamma_2 | \hat{V}_{\mathrm{ep}} | \Gamma_1\gamma_1 \rangle|^2, \tag{9.1}$$

where $\gamma_i$ labels the basis functions of the irreducible representation $\Gamma_i$, $\hat{V}_{\mathrm{ep}}$ is the operator associated with the interaction of the molecule's electrons with the electromagnetic field of the light wave (i.e. the electron–photon interaction). The observed line intensity is due to transitions to *all* final states $|\Gamma_2\gamma_2\rangle$; therefore (9.1) includes a summation over final states ($\Sigma_{\gamma_2} \ldots$). Since the transition starts from a certain state, the intensity $I$ should be averaged over the initial states; that is, it should be summed over $\gamma_1$ and divided by the degree of degeneracy $g(\Gamma_1)$ of the level $\varepsilon(\Gamma_1)$. There are three likely electron–radiation interaction mechanisms, and accordingly the following optical transition types [67].

(1) *electric dipole transitions.* These are due to interaction of the electron dipole moment with the electric field of the light wave. These transitions are known as *E1 transitions*. The associated operator $\hat{V}_{\mathrm{ep}}$ (electron photon) is (up to a factor)

$$\hat{V}_{\mathrm{ep}} = \mathbf{u} \cdot \hat{\mathbf{d}}, \tag{9.2}$$

where $\hat{\mathbf{d}} = -e\Sigma_i \mathbf{r}_i$ is the electric (dipole) moment and $\mathbf{u}$ is the (electric) *polarization vector* of the light, i.e. the unit vector along the direction of oscillation of the (light-wave) electric field (its components $u_x$, $u_y$ and $u_z$ are conventionally denoted by $l$, $m$ and $n$ respectively).

(2) *magnetic dipole (M1) transitions.* The operator of electron interaction with the magnetic field of the light wave (electron-photon interaction, $\hat{V}_{\text{ep}}$) is given (up to a factor) by

$$\hat{V}_{\text{ep}} = \mathbf{u} \cdot \hat{\mathbf{M}}, \tag{9.3}$$

where the electron magnetic moment operator

$$\hat{\mathbf{M}} = -\frac{e\hbar}{2mc} \sum_i (\hat{\mathbf{l}}_i + 2\hat{\mathbf{s}}_i); \tag{9.4}$$

$\hat{\mathbf{l}}_i$ is the orbital angular momentum operator of the $i$th electron, the sum is over all electrons, and $\mathbf{u}$ is the unit vector along the direction of oscillation of the (light-wave) magnetic field (the magnetic polarization vector);

(3) *Electric quadrupole (E2) transitions.* These are induced by the operator

$$\hat{V}_{\text{ep}} = \sum_i (l\hat{Q}_{yz} + m\hat{Q}_{xz} + n\hat{Q}_{xy}), \tag{9.5}$$

where $\hat{Q}_{\alpha\beta}$ are the components of the electron quadrupole moment operator,

$$\hat{Q}_{yz} = -eyz, \qquad \hat{Q}_{xz} = -exz, \qquad \hat{Q}_{xy} = -exy, \tag{9.6}$$

and $\mathbf{u}$ (with components $l$, $m$, $n$) is the (electric) polarization vector.

To calculate the optical line intensity for a given polarization along the $x$, $y$ or $z$ axis, only the term containing the respective component ($l$, $m$ or $n$) of $\mathbf{u}$ is retained. In the general case, the equation (9.1) for the line intensity depends on all the components of $\mathbf{u}$; this dependence is called the *angular* or *polarization dependence* of the intensity.

## 9.1.2 Selection rules

Let us consider the matrix elements $\langle \Gamma_2\gamma_2 | \hat{V}_{\text{ep}} | \Gamma_1\gamma_1 \rangle$ giving the intensity of the optical lines. For each pair of levels $\Gamma_1$ and $\Gamma_2$ these form a matrix $\mathbf{V}_{\text{ep}}^{\Gamma_1\Gamma_2}$ of dimension $g(\Gamma_1) \times g(\Gamma_2)$ (where the $g$s are the degrees of degeneracy), with $\gamma_1$ and $\gamma_2$ labelling matrix rows and columns. If $\mathbf{V}_{\text{ep}}^{\Gamma_1\Gamma_2}$ contains nonzero elements, the optical transition is called *allowed*, and its intensity $I(\Gamma_1\Gamma_2)$ is finite and nonzero. For *forbidden* transitions all the matrix elements $\langle \Gamma_2\gamma_2 | \hat{V}_{\text{ep}} | \Gamma_1\gamma_1 \rangle$ are zero, and $I(\Gamma_1\Gamma_2) = 0$. The concept of allowed and forbidden transitions applies to each of the three above mechanisms. For example, there may be an allowed electric dipole transition $\Gamma_1 \to \Gamma_2$; however, this transition may be forbidden when the operator describes a magnetic dipole interaction, or an electric quadrupole interaction.

The fact that a transition is allowed or forbidden is revealed by the selection rules for the matrix elements of the operator $\hat{V}_{ep}$. The selection rules can be established without computing the matrix $\mathbf{V}_{ep}^{\Gamma_1\Gamma_2}$ itself.

We consider the ensemble of matrix elements of the operator $\hat{\mathbf{V}}_{\Gamma\gamma}$ that transform according to the representation $\Gamma$:

$$\langle\Gamma_2\gamma_2|\hat{\mathbf{V}}_{\Gamma\gamma}|\Gamma_1\gamma_1\rangle \equiv \int \varphi^*_{\Gamma_2\gamma_2}(\mathbf{r})\hat{\mathbf{V}}_{\Gamma\gamma}\varphi_{\Gamma_1\gamma_1}(\mathbf{r})\,d\tau. \tag{9.7}$$

We state the following important rule: some of the matrix elements (9.7) are nonzero only if the direct product $\Gamma_1 \times \Gamma$ contains the irreducible representation $\Gamma_2$.

The fundamental theorem establishing the selection rule for an operator $\hat{V}_{\Gamma\gamma}$ [15] can be formulated in an equivalent way: the ensemble of matrix elements $\langle\Gamma_2\gamma_2\,|\,\hat{V}_{\Gamma\gamma}\,|\,\Gamma_1\gamma_1\rangle$ contains non-vanishing elements if the direct product of three irreducible representations $\Gamma_2 \times \Gamma \times \Gamma_1$ contains the full-symmetric irreducible representations of the point symmetry group. The result of reduction of a direct product of three irreducible representations does not depend on the order of multiplication. Since the totally symmetric representation is contained only in a direct product of two identical representations (Section 5.4.3) the condition stated above proves to be fulfilled if (1) $\Gamma \times \Gamma_1$ contains $\Gamma_2$, (2) $\Gamma_2 \times \Gamma_1$ contains $\Gamma$ or (3) $\Gamma_2 \times \Gamma$ contains $\Gamma_1$.† This selection rule has a general character and can be applied to any operators, particularly to $\hat{\mathbf{d}}$, $\hat{\mathbf{M}}$ and $\hat{Q}_{\alpha\beta}$.

Let us consider as an example the E1 transitions between the levels $E_g$ and $T_{2u}$ of the group $\mathbf{O}_h$. The matrix elements of the operator $\hat{V}_{ep}$ are represented as

$$\langle T_{2u}\gamma_2|\hat{V}_{ep}|E_g\gamma_1\rangle = l\langle T_{2u}\gamma_2|\hat{d}_x|E_g\gamma_1\rangle + m\langle T_{2u}\gamma_2|\hat{d}_y|E_g\gamma_1\rangle + n\langle T_{2u}\gamma_2|\hat{d}_z|E_g\gamma_1\rangle. \tag{9.8}$$

Referring to the table of basis functions for $\mathbf{O}_h$, we find that the three components of the polar vector $\hat{\mathbf{d}}$ transform according to the irreducible representation $T_{1u}$. Taking the direct product $E_g \times T_{2u} \doteq T_{1u} + T_{2u}$, we note that it contains $T_{1u}(\hat{\mathbf{d}})$. Therefore the electric dipole transitions $E_g \leftrightarrow T_{2u}$ (direct and inverse) in systems with $\mathbf{O}_h$ symmetry are allowed.

The components of the magnetic moment $\hat{\mathbf{M}}$ in $\mathbf{O}_h$ form a basis of $T_{1g}$. Since in the reduction scheme of $E_g \times T_{1g}$ the irreducible representation $T_{2u}$ is missing, the M1 transition $E_g \leftrightarrow T_{2u}$ is forbidden. In the same way it can easily

† When the representations are real the order of coupling is unimportant. Consideration of complex representations is given in [20].

**Table 9.1** Selection rules for electric dipole transitions in $\mathbf{O}_h$ (only half of the symmetrical table is given).

| $\mathbf{O}_h$ | $A_{1g}$ | $A_{2g}$ | $E_g$ | $T_{1g}$ | $T_{2g}$ |
|---|---|---|---|---|---|
| $A_{1u}$ | — | — | — | E1 | — |
| $A_{2u}$ | | — | — | — | E1 |
| $E_u$ | | | — | E1 | E1 |
| $T_{1u}$ | | | | E1 | E1 |
| $T_{2u}$ | | | | | E1 |

be seen that the E2 transitions are forbidden, since the three components $\hat{Q}_{\alpha\beta}$ form the basis of the representation $T_{2g}$.

The components of the electric dipole moment are odd under inversion. Therefore in centrosymmetric groups they are transformed according to odd representations ($\Gamma_u$). This gives a selection rule common to all centrosymmetric groups: *electric dipole transitions between states of identical parity* ($g \leftrightarrow g$, $u \leftrightarrow u$) *are forbidden.* E1 transitions can be allowed only when the parity changes (u $\leftrightarrow$ g). The magnetic dipole and electric quadrupole transitions are induced by the even operators $\hat{\mathbf{M}}$ and $\hat{Q}_{\alpha\beta}$. Therefore M1 and E2 transitions between states of different parity are forbidden. Thus in centrosymmetric groups E1 and M1 transitions cannot be allowed at the same time, while M1 and E2 transitions can. The results for $\mathbf{O}_h$ are given in Tables 9.1 and 9.2.

In noncentrosymmetric groups the selection rules relating to parity are absent since the basis functions of irreducible representations do not have a particular parity. In $\mathbf{T}_d$ the components of the polar vector $\hat{\mathbf{d}}$ form the basis of $T_2$, as well as three components $\hat{Q}_{\alpha\beta}$. The quantities $\hat{M}_\alpha$ form the basis of $T_1$. Using the multiplication table for $\mathbf{T}_d$ group representations [1], we obtain the permitted transitions (Table 9.3). Table 9.3 shows that in $\mathbf{T}_d$ all of the above-mentioned optical transitions can be permitted at the same time.

The selection rules based on group theory do not provide the possibility of comparing the intensities of lines induced by different mechanisms of interaction with radiation. It should be pointed out that in transition metal

**Table 9.2** Selection rules for magnetic dipole and electric quadrupole transitions in $\mathbf{O}_h$.

| $\mathbf{O}_h$ | $A_{1g}$ | $A_{2g}$ | $E_g$ | $T_{1g}$ | $T_{2g}$ |
|---|---|---|---|---|---|
| $A_{1g}$ | — | — | — | M1 | E2 |
| $A_{2g}$ | | — | — | E2 | M1 |
| $E_g$ | | | — | M1, E2 | M1, E2 |
| $T_{1g}$ | | | | M1, E2 | M1, E2 |
| $T_{2g}$ | | | | | M1, E2 |

**Table 9.3** Selection rules for optical transitions in $\mathbf{T}_d$.

| $\mathbf{T}_d$ | $A_1$ | $A_2$ | E | $T_1$ | $T_2$ |
|---|---|---|---|---|---|
| $A_1$ | — | — | — | M1 | E1, E2 |
| $A_2$ | | — | — | E1, E2 | M1 |
| E | | | — | E1, M1, E2 | E1, M1, E2 |
| $T_1$ | | | | E1, M1, E2 | E1, M1, E2 |
| $T_2$ | | | | | E1, M1, E2 |

complexes allowed E1 transitions are far more intense than M1 and E2 transitions. Of course, if E1 transitions are forbidden, the observed lines are due to other mechanisms. M1 transitions are more typical of lanthanide and actinide spectra. It should be also stressed that the selection rules deduced above do not take account of spin; transitions are allowed for $\Delta S = 0$.

### 9.1.3 Optical line polarization for allowed transitions

Let us consider the selection rules for noncubic groups with $C_n$ $(n \geq 3)$ symmetry axes. From tables of basis functions [1] (see Table 5.5) it is easy to see that in $\mathbf{C}_{4v}$ the pairs $(\hat{d}_x, \hat{d}_y)$, $(\hat{M}_x, \hat{M}_y)$ and $(\hat{Q}_{yz}, \hat{Q}_{xz})$ are the basis functions for the representation E, while $\hat{d}_z$, $\hat{M}_z$ and $\hat{Q}_{xy}$ are the bases for $A_1$, $A_2$ and $B_2$ respectively. The transition $A_1 \rightarrow A_1$ between terms of the same symmetry is allowed in the E1 approximation. Since $A_1 \times A_1 \doteq A_1$, in this transition only the $z$ component of the dipole moment is active, i.e. the operator is $u_z d_z$. This means that the absorption and emission optical line $A_1 \leftrightarrow A_1$ is polarized along the $C_4(z)$ axis. In point groups with $C_n$ $(n \geq 3)$ axes such a polarization is called *parallel* ($\|$) or $\pi$ *polarization.*

The selection rules for the singlet–doublet $B_1 \rightarrow E$ transition are determined by the direct product $B_1 \times E \doteq E$. The transverse components of the vectors $\hat{\mathbf{d}}$, $\hat{\mathbf{M}}$ and $\hat{Q}_{yz}$, $\hat{Q}_{xz}$ are transformed according to this representation. Hence the optical $B_1 \rightarrow E$ line is polarized in the $\sigma(xy)$ plane perpendicular to the $C_4(z)$ axis. Such a polarization is called *perpendicular* ($\perp$) or $\sigma$ *polarization.*

Finally, if both the initial and final terms are degenerate, *mixed polarization* is possible. In $\mathbf{C}_{3v}$ the E1 transition of the $E \rightarrow E$ type is allowed for both $\sigma$ and $\pi$ polarizations. Indeed, this group has $E \times E \doteq A_1 + A_2 + E$, while the components of $\hat{\mathbf{d}}$ are transformed according to the representations $A_1(d_z)$ and $E(d_x, d_y)$. Therefore the $E \rightarrow E$ transition optical line is *partially* $\pi$- and $\sigma$-*polarized.* The light absorption is anisotropic and depends on the direction of the polarization vector, the absorption coefficients for $\sigma$ and $\pi$ polarizations being different. It should be kept in mind that mixed polarization does not always result from multiplet–multiplet transitions. For example, in tetragonal groups the $E \rightarrow E$ transitions are $\pi$-polarized. In $\mathbf{C}_{4v}$ we have

**Table 9.4** Selection rules for optical transitions in $\mathbf{C}_{4v}$.

| $\mathbf{C}_{4v}$ | $A_1$ | $A_2$ | $B_1$ | $B_2$ | E |
|---|---|---|---|---|---|
| $A_1$ | E1($\pi$) | M1($\pi$) | — | E2($\pi$) | E1($\sigma$), E2($\sigma$), M1($\sigma$) |
| $A_2$ | | E1($\pi$) | E2($\pi$) | | E1($\sigma$), E2($\sigma$), M1($\sigma$) |
| $B_1$ | | | | M1($\pi$) | E1($\sigma$), E2($\sigma$), M1($\sigma$) |
| $B_2$ | | | | | E1($\sigma$), E2($\sigma$), M1($\sigma$) |
| E | | | | | E1($\pi$), E2($\pi$), M1($\pi$) |

$E \times E \doteq A_1(\hat{d}_z) + A_2(\hat{M}_z) + B_1 + B_2(\hat{Q}_{xy})$; therefore the selection rules permit only $\pi$-polarized transitions of E1, M1 and E2 types (Table 9.4).

Let us now consider a rhombic crystal symmetry group, such as $\mathbf{D}_2$, having three $C_2$ axes. The common feature of such groups is that different components of $\hat{\mathbf{d}}$ and $\hat{\mathbf{M}}$, as well as $\hat{Q}_{\alpha\beta}$, belong to different one-dimensional representations. In $\mathbf{D}_2$ $\hat{d}_x$, $\hat{d}_y$ and $\hat{d}_z$ are bases for $B_1$, $B_2$ and $B_3$ respectively, as are $\hat{M}_x$, $\hat{M}_y$ and $\hat{M}_z$ and $\hat{Q}_{xy}$, $\hat{Q}_{xz}$ and $\hat{Q}_{yz}$. Therefore the $A \leftrightarrow B_1$, $A \leftrightarrow B_2$ and $A \leftrightarrow B_3$ transition lines are polarized along the $C_2(z)$, $C_2(y)$ and $C_2(x)$ axes, while the transition $B_1 \rightarrow B_2$ is polarized along the $C_2(x)$ axis ($B_1 \times B_2 \doteq B_3(\hat{d}_x)$). The three-axial rhombic symmetry induces polarization of each optical line along one of the $C_2$ axes.

The selection rules thus allow the determination of allowed and forbidden transitions and of optical line polarization. The results for the point groups $\mathbf{O}_h$, $\mathbf{T}_d$ and $\mathbf{C}_v$ have been given in Tables 9.1–9.4 above; those for $\mathbf{C}_{3v}$ and $\mathbf{D}_2$ are shown in Tables 9.5 and 9.6. The general polarization behaviour resulting from the allowed transitions considered above is as follows:

(1) if the components of $\hat{\mathbf{d}}$ form the basis of one irreducible representation of the point group (*cubic group*), the optical E1 lines are unpolarized; similarly, if the components of $\hat{\mathbf{M}}$ and $\hat{\mathbf{Q}}_{\alpha\beta}$ form the basis of one irreducible representation each, the M1 and E2 transitions are unpolarized;

(2) if the components of $\hat{\mathbf{d}}$, $\hat{\mathbf{M}}$ and $\hat{Q}_{\alpha\beta}$ are transformed according to two representations (groups with $C_3$ axes, $n \geq 3$), $\pi$ or $\sigma$ polarization is possible;

**Table 9.5** Selection rules for optical transitions in $\mathbf{C}_{3v}$.

| $\mathbf{C}_{3v}$ | $A_1$ | $A_2$ | E |
|---|---|---|---|
| $A_1$ | E1($\pi$) | M1($\pi$) | E1($\sigma$), M1($\sigma$), E2($\pi$), E2($\sigma$) |
| $A_2$ | | E1($\pi$) | E1($\sigma$), M1($\sigma$), E2($\pi$), E2($\sigma$) |
| E | | | E1($\pi$), M1($\pi$), E1($\sigma$), M1($\sigma$), E2($\pi$) |

**Table 9.6** Selection rules for optical transitions in $\mathbf{D}_2$.

| $\mathbf{D}_2$ | A | $B_1$ | $B_2$ | $B_3$ |
|---|---|---|---|---|
| A | — | E1($z$), E2($z$), M1($z$) | E1($y$), E2($y$), M1($y$) | E1($x$), E2($x$), M1($x$) |
| $B_1$ | | — | E1($x$), E2($x$), M1($x$) | E1($y$), E2($y$), M1($y$) |
| $B_2$ | | | — | E1($z$), E2($z$), M1($z$) |
| $B_3$ | | | | — |

(3) if each component of $\hat{\mathbf{d}}$, $\hat{\mathbf{M}}$ or $\hat{Q}_{\alpha\beta}$ forms the basis of a one-dimensional representation (groups with three $C_2$ axes), the lines are polarized along one of the $C_2(x)$, $C_2(y)$ or $C_2(z)$ axes.

### 9.1.4 Polarization dichroism in low-symmetry fields

Let us consider the allowed electric dipole transition $A_{1g} \leftrightarrow T_{1u}$ in a cubic ($\mathbf{O}_h$) complex $AX_6$. The optical line for this transition is unpolarized. Let us further assume that the complex is subject to the action of a $\mathbf{D}_{4h}$ low-symmetry crystal field, brought about by axial ligand substitution or induced by remote coordination spheres. The symmetry reduction $\mathbf{O}_h \rightarrow \mathbf{D}_{4h}$ leads to $T_{1u}$ splitting according to $T_{1u} \doteq A_{2u} + E_u$ (Fig. 4.2). The corresponding optical E1 line is also split (Fig. 9.1). Proceeding as in Section 9.1.3, it is easy to establish that the transitions $A_{1g} \leftrightarrow A_{2u}$ and $A_{1g} \leftrightarrow E_u$ give $\pi$- and $\sigma$-polarized lines respectively. Thus a low-symmetry crystal field both splits the energy lines and leads to a polarized spectrum. Since lines polarized parallel and perpendicularly have different corresponding wavelengths, the crystal acquires a colour depending on the light polarization. In the above example (see Fig. 9.1) the $\pi$-polarized

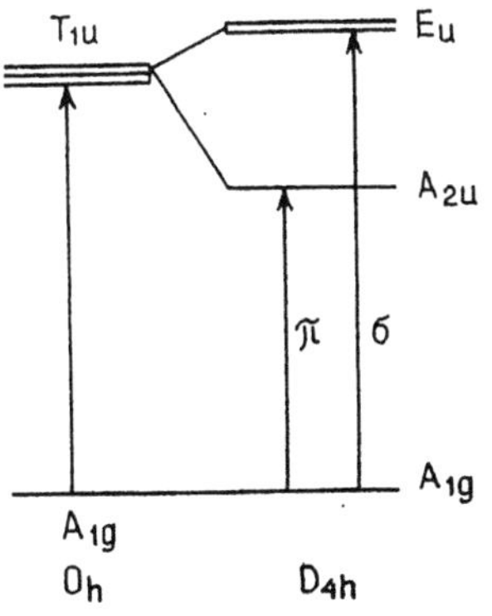


**Fig. 9.1** The origin of polarization dichroism under the action of a tetragonal deformation of the cubic centre.

spectrum is shifted towards the red, while the $\sigma$-polarized spectrum is shifted towards the violet. This effect is called *polarization dichroism (dichromatism).*

A similar polarization dichroism effect occurs in a $\mathbf{D}_{3d}$ trigonal field, which splits the $T_{1u}$ level according to $T_{1u} \doteq A_{2u} + E_u$. In this case $\pi$ and $\sigma$ polarizations are related to the $C_3$ axis and the plane perpendicular to it. Finally, for rhombic symmetry, the crystal colour changes when the light polarization changes relative to the three $C_2$ axes. This effect, characteristic of three-axis systems, is called *polychroism.* Analysis of monocrystal polarized spectra provides information on the point symmetry of unfilled-shell ions [68].

The calculational procedure for all level splittings of cubic crystals and for polarization characteristics can be found in [69], forming the foundation of *crystal piezospectroscopy.*

## 9.2 WIGNER–ECKART THEOREM FOR POINT GROUPS

Let $\varphi_x$ and $\varphi_y$ be two basis functions of the representation $T_{1u}$ in $\mathbf{O}_h$ (e.g. $f(r)x$ and $f(r)y$), with $\hat{V}_{xy} \equiv \hat{V}_{T_{2g}\zeta}$ being the operator transformed as the $\zeta$ component of $T_{2g}$. The matrix element

$$\langle T_{1u}x|\hat{V}_{T_2xy}|T_{1u}y\rangle \equiv \int \varphi_x \hat{V}_{xy}\varphi_y \, d\tau \tag{9.9}$$

remains unchanged if the same symmetry operation is applied to the functions and to the operator. Therefore it is equal to the matrix elements $\langle T_{1u}y|\hat{V}_{T_{2g}yz}|T_{1u}z\rangle$ and $\langle T_{1u}z|\hat{V}_{T_{2g}zx}|T_{1u}x\rangle$, obtained from (9.9) as the result of the $\mathbf{O}_h$ group operation $\hat{C}_3$. This simple example shows that there are relations between the matrix elements of the tensor operators $\hat{V}_{\Gamma\gamma}$. These relations follow from the invariance of the matrix elements under point-group operations. The relations are expressed by the *Wigner–Eckart theorem for point groups*:

$$\langle \alpha_1\Gamma_1\gamma_1|\hat{V}_{\Gamma\gamma}|\alpha_2\Gamma_2\gamma_2\rangle = [g(\Gamma_1)]^{-1/2}\langle \alpha_1\Gamma_1\|\hat{V}_\Gamma\|\alpha_2\Gamma_2\rangle\langle \Gamma_1\gamma_1|\Gamma_2\gamma_2\Gamma\gamma\rangle. \tag{9.10}$$

As in the rotation group (Section 6.2), the matrix element of the point-group tensor operator $\hat{V}_{\Gamma\gamma}$ involves two factors: the *reduced matrix element* $\langle \alpha_1\Gamma_1\|\hat{V}_\Gamma\|\alpha_2\Gamma_2\rangle$ and the Clebsch–Gordan coefficient $\langle \Gamma_1\gamma_1|\Gamma_2\gamma_2\Gamma\gamma\rangle$. The reduced matrix element is determined only by the *irreducible* representations $\Gamma_1$, $\Gamma_2$ and $\Gamma$, together with the supplementary quantum numbers $\alpha_1$ and $\alpha_2$. Therefore reduced matrix elements are identical for all matrices $\mathbf{V}_{\Gamma\gamma}$ of the operator $\hat{V}_{\Gamma\gamma}$. The dependence on the row and columns labels $\gamma_1$ and $\gamma_2$ of the matrix $\mathbf{V}_{\Gamma\gamma}$ is contained in the Clebsch–Gordan coefficients. These are nonzero if $\Gamma_1$ is contained in the expansion of the direct product $\Gamma \times \Gamma_2$; that is, if the selection rules are satisfied. This makes it possible to represent the Wigner–Eckart theorem in matrix form as

$$\mathbf{V}_{\Gamma\gamma}^{\Gamma_1\Gamma_2} = \frac{1}{[g(\Gamma_1)]^{1/2}}\langle \alpha_1\Gamma_1\|\hat{V}_\Gamma\|\alpha_2\Gamma_2\rangle \mathbf{O}_{\Gamma\gamma}^{\Gamma_1\Gamma_2} \tag{9.11}$$

where, as in Section 6.2, $\mathbf{O}_{\Gamma\gamma}^{\Gamma_1\Gamma_2}$ is the matrix of Clebsch–Gordan coefficients.

Thus, in order to calculate the set of $\hat{\mathbf{V}}_{\Gamma\gamma}$ operator matrices only one parameter $[g(\Gamma_1)]^{-1/2}\langle\alpha_1\Gamma_1\|\hat{\mathbf{V}}_\Gamma\|\alpha_2\Gamma_2\rangle$ need be known; it is denoted for brevity by $V_\Gamma^{\Gamma_1\Gamma_2}$. This single parameter can be determined by calculating the simplest matrix element $\langle\alpha_1\Gamma_1\gamma_1|\hat{\mathbf{V}}_{\Gamma\gamma}|\alpha_2\Gamma_2\gamma_2\rangle$ and dividing it by $\langle\Gamma_1\gamma_1|\Gamma_2\gamma_2\Gamma\gamma\rangle$. Within the scope of the *semiempirical* or *phenomenological* theory the $V_\Gamma^{\Gamma_1\Gamma_2}$ are parameters obtained by comparison of theoretical and experimental data. In this case the Wigner–Eckart theorem provides a criterion of parameter independence.†

It should be emphasized that the *Wigner–Eckart theorem is applicable only if the operator $\hat{V}$ belongs to one irreducible representation.* It cannot be applied, for instance, to the operator $\hat{V}_{xx} = x^2$, which in **O** enters in the basis functions $A_1(\hat{V}_{xx}+\hat{V}_{yy}+\hat{V}_{zz})$ and $Ev(x^2-y^2)$ (Table 3.4). In this sense the operator $\hat{V}_{xx}$ is not an irreducible tensor.

**Example 9.1** We shall construct the dipole moment ($T_{1u}$) component matrix for the transition $E_g \leftrightarrow T_{1u}$ in $\mathbf{O}_h$. We apply the Wigner–Eckart theorem in matrix form for the $x$ component of the dipole moment

$$\mathbf{d}_{T_{1u}x} = d_{T_{1u}}^{E_gT_{1u}}\mathbf{O}_{T_{1u}x}^{E_gT_{1u}}, \tag{9.12}$$

where

$$\mathbf{O}_{T_{1u}x}^{E_gT_{1u}} = \begin{matrix} & x & y & z \\ u & \langle E_gu|T_{1u}xT_{1u}x\rangle & \langle E_gu|T_{1u}yT_{1u}x\rangle & \langle E_gu|T_{1u}xT_{1u}z\rangle \\ v & \langle E_gv|T_{1u}xT_{1u}x\rangle & \langle E_gv|T_{1u}yT_{1u}x\rangle & \langle E_gu|T_{1u}xT_{1u}z\rangle \end{matrix}. \tag{9.13}$$

In order to obtain the matrix (9.13) in the explicit form we must use the Clebsch–Gordan coefficients for the $T_{1u}\times T_{1u}$ product in the $\mathbf{O}_h$ group which coincide with those for the $T_1\times T_1$ product in **O** [1, 14] (Table 9.7). We obtain from Table 9.7:

$$\mathbf{O}_{T_{1u}x}^{E_gT_{1u}} = \begin{bmatrix} \sqrt{1/6} & 0 & 0 \\ -\sqrt{1/2} & 0 & 0 \end{bmatrix}. \tag{9.14}$$

Similarly, we obtain

$$\mathbf{O}_{T_{1u}y}^{E_gT_{1u}} = \begin{bmatrix} 0 & \sqrt{1/6} & 0 \\ 0 & \sqrt{1/2} & 0 \end{bmatrix}, \quad \mathbf{O}_{T_{1u}z}^{E_gT_{1u}} = \begin{bmatrix} 0 & 0 & -\sqrt{2/3} \\ 0 & 0 & 0 \end{bmatrix}. \tag{9.15}$$

The $\hat{V}_{ep}$ operator matrix for the transition $T_{1u} \leftrightarrow E_g$ is

$$V_{ep} = V_{T_{1u}}^{E_gT_{1u}}(l\mathbf{O}_{T_{1u}x}^{E_gT_{1u}} + m\mathbf{O}_{T_{1u}y}^{E_gT_{1u}} + n\mathbf{O}_{T_{1u}z}^{E_gT_{1g}}). \tag{9.16}$$

**Example 9.2** Let us obtain the matrix representation of the E2 operator for transition between the $T_{1g}$ and $T_{2g}$ levels. Using the Wigner–Eckart theorem, we find

$$\mathbf{V}_{ep} = V_{T_{2g}}^{T_{1g}T_{2g}}(l\mathbf{O}_{T_{2g}yz}^{T_{1g}T_{2g}} + m\mathbf{O}_{T_{2g}xz}^{T_{1g}T_{2g}} + n\mathbf{O}_{T_{2g}xy}^{T_{1g}T_{2g}}), \tag{9.17}$$

† The simple form of the theorem applies only for simply reducible groups, where no irreducible representation appears more than once in the reduction of a direct product $\Gamma_2\times\Gamma$–generally another index is then required [15]. A more complicated case of repeating representation $\Gamma_1$ in $\Gamma_2\times\Gamma$ will be considered in Section 11.2.2.

Now we must use the Clebsch–Gordan coefficient for the product $T_{2g} \times T_{2g}$ in $\mathbf{O}_h$, i.e. for $T_2 \times T_2$ in $\mathbf{O}$ (Table 9.8). From Table 9.8 we obtain:

$$\mathbf{O}^{T_{1g}T_{2g}}_{T_{2g}yz} = \frac{1}{\sqrt{2}}\begin{bmatrix} 0 & 0 & 0 \\ 0 & 0 & -1 \\ 0 & 1 & 0 \end{bmatrix}, \quad \mathbf{O}^{T_{1g}T_{2g}}_{T_{2g}xz} = \frac{1}{\sqrt{2}}\begin{bmatrix} 0 & 0 & 1 \\ 0 & 0 & 0 \\ -1 & 0 & 0 \end{bmatrix}, \quad \mathbf{O}^{T_{1g}T_{2g}}_{T_{2g}yz} = \frac{1}{\sqrt{2}}\begin{bmatrix} 0 & -1 & 0 \\ 1 & 0 & 0 \\ 0 & 0 & 0 \end{bmatrix}. \tag{9.18}$$

**Table 9.7** Clebsch–Gordan coefficients for the $\mathbf{O}_h$ group product $T_{1u} \times T_{1u}$.

| $T_{1u}\times T_{1u}$ | $\Gamma$ | $A_{1g}$ | $E_g$ | | $T_{1g}$ | | | $T_{2g}$ | | |
|---|---|---|---|---|---|---|---|---|---|---|
| $\gamma_1$ | $\gamma_2$ \ $\gamma$ | $e_1$ | u | v | $\alpha$ | $\beta$ | $\gamma$ | $\xi$ | $\eta$ | $\zeta$ |
| x | x | $-1/\sqrt{3}$ | $1/\sqrt{6}$ | $-1/\sqrt{2}$ | 0 | 0 | 0 | 0 | 0 | 0 |
| x | y | 0 | 0 | 0 | 0 | 0 | $-1/\sqrt{2}$ | 0 | 0 | $-1/\sqrt{2}$ |
| x | z | 0 | 0 | 0 | 0 | $1/\sqrt{2}$ | 0 | 0 | $-1/\sqrt{2}$ | 0 |
| y | x | 0 | 0 | 0 | 0 | 0 | $1/\sqrt{2}$ | 0 | 0 | $-1/\sqrt{2}$ |
| y | y | $-1/\sqrt{3}$ | $1/\sqrt{6}$ | $1/\sqrt{2}$ | 0 | 0 | 0 | 0 | 0 | 0 |
| y | z | 0 | 0 | 0 | $-1/\sqrt{2}$ | 0 | 0 | $-1/\sqrt{2}$ | 0 | 0 |
| z | x | 0 | 0 | 0 | 0 | $-1/\sqrt{2}$ | 0 | 0 | $-1/\sqrt{2}$ | 0 |
| z | y | 0 | 0 | 0 | $1/\sqrt{2}$ | 0 | 0 | $-1/\sqrt{2}$ | 0 | 0 |
| z | z | $-1/\sqrt{3}$ | $-2/\sqrt{6}$ | 0 | 0 | 0 | 0 | 0 | 0 | 0 |

**Table 9.8** Glebsch–Gordan coefficients for $T_{2g} \times T_{2g}$ in $\mathbf{O}_h$.

| $T_{2g}\times T_{2g}$ | $\Gamma$ | $A_{1g}$ | $E_g$ | | $T_{1g}$ | | | $T_{2g}$ | | |
|---|---|---|---|---|---|---|---|---|---|---|
| $\gamma_1$ | $\gamma_2$ \ $\gamma$ | $e_1$ | u | v | $\alpha$ | $\beta$ | $\gamma$ | $\xi$ | $\eta$ | $\zeta$ |
| $\xi$ | $\xi$ | $1/\sqrt{3}$ | $-1/\sqrt{6}$ | $1/\sqrt{2}$ | 0 | 0 | 0 | 0 | 0 | 0 |
| $\xi$ | $\eta$ | 0 | 0 | 0 | 0 | 0 | $1/\sqrt{2}$ | 0 | 0 | $1/\sqrt{2}$ |
| $\xi$ | $\zeta$ | 0 | 0 | 0 | 0 | $-1/\sqrt{2}$ | 0 | 0 | $1/\sqrt{2}$ | 0 |
| $\eta$ | $\xi$ | 0 | 0 | 0 | 0 | 0 | $-1/\sqrt{2}$ | 0 | 0 | $1/\sqrt{2}$ |
| $\eta$ | $\eta$ | $1/\sqrt{3}$ | $-1/\sqrt{6}$ | $-1/\sqrt{2}$ | 0 | 0 | 0 | 0 | 0 | 0 |
| $\eta$ | $\zeta$ | 0 | 0 | 0 | $1/\sqrt{2}$ | 0 | 0 | $1/\sqrt{2}$ | 0 | 0 |
| $\zeta$ | $\xi$ | 0 | 0 | 0 | 0 | $1/\sqrt{2}$ | 0 | 0 | $1/\sqrt{2}$ | 0 |
| $\zeta$ | $\eta$ | 0 | 0 | 0 | $-1/\sqrt{2}$ | 0 | 0 | $1/\sqrt{2}$ | 0 | 0 |
| $\zeta$ | $\xi$ | $1/\sqrt{3}$ | $2/\sqrt{6}$ | 0 | 0 | 0 | 0 | 0 | 0 | 0 |

In agreement with the general conclusions, the operators $\hat{V}_{ep}$ in matrix form for the cubic group contain only one parameter each.

The Wigner–Eckart theorem explains the configuration mixing rules, stated without proof in Section 5.6.2. The application of this theorem to the scalar operator $\hat{\mathbf{V}}_{A_1}$ shows that the matrix $\mathbf{O}_{A_1}^{\Gamma'\Gamma''}$ (where $\Gamma'$ and $\Gamma''$ are the same representations) is a unit matrix; in other words, $\mathbf{V}_{A_1} = V_{A_1}^{\Gamma'\Gamma''}\mathbf{1}$.

## 9.3 POLARIZATION DEPENDENCE OF SPECTRA FOR ALLOWED TRANSITIONS

After determining the matrices and independent parameters of the operator $\hat{V}_{ep}$ by the help of the Wigner–Eckart theorem, we can substitute them into (9.1) to get the optical line intensity. The intensity thus obtained depends on the polarization-vector components and certain phenomenological parameters (reduced matrix elements). The dependence of the line intensity on the orientation of the polarization vector with respect to the symmetry elements of the system is called the *angular* or *polarization dependence of the spectrum.* To use matrix concepts, we represent (9.1) as a matrix product. According to the matrix multiplication rules (3.6), we obtain

$$\begin{aligned} I(\Gamma_1\Gamma_2) &= \frac{1}{g(\Gamma_1)}\sum_{\gamma_1\gamma_2}\langle\Gamma_2\gamma_2|\hat{V}_{ep}|\Gamma_1\gamma_1\rangle\langle\Gamma_1\gamma_1|\hat{V}_{ep}|\Gamma_2\gamma_2\rangle \\ &= \frac{1}{g(\Gamma_1)}\sum_{\gamma_2}\langle\Gamma_2\gamma_2|\hat{V}_{ep}^{\dagger}\hat{V}_{ep}|\Gamma_2\gamma_2\rangle \\ &\equiv \sum_{\gamma_2}(\mathbf{V}_{ep}^{\dagger}\mathbf{V}_{ep})_{\gamma_2\gamma_2}. \end{aligned} \tag{9.19}$$

where $\mathbf{V}_{ep}$ is the matrix with elements $\langle\Gamma_1\gamma_1\,|\,\hat{V}_{ep}\,|\,\Gamma_2\gamma_2\rangle$ and $\mathbf{V}_{ep}^{\dagger}$ is its Hermitian transpose with elements $\langle\Gamma_2\gamma_2\,|\,\hat{V}_{ep}\,|\,\Gamma_1\gamma_1\rangle$. The sum over $\gamma_2$ on the right-handside of (9.19) is the character of the square matrix $\mathbf{V}_{ep}^{\dagger}\mathbf{V}_{ep}$, so that

$$I(\Gamma_1\Gamma_2) = \frac{1}{g(\Gamma_1)}\chi(\mathbf{V}_{ep}^{\dagger}\mathbf{V}_{ep}). \tag{9.20}$$

Using (9.20) let us now establish the angular dependence of the spectrum for the simple case of the E1 transition $A_{1g} \rightarrow T_{1u}(\mathbf{O}_h)$, whose matrix is

$$\mathbf{V}_{ep} = d_{T_{1u}}^{A_{1g}T_{1u}}\left(l\begin{bmatrix}1\\0\\0\end{bmatrix} + m\begin{bmatrix}0\\1\\0\end{bmatrix} + n\begin{bmatrix}0\\0\\1\end{bmatrix}\right). \tag{9.21}$$

**O** matrices (transposed and complex-conjugate) are represented by rows, e.g.

[100]. Therefore, for instance,

$$[1 \quad 0 \quad 0]\begin{bmatrix}1\\0\\0\end{bmatrix} = 1, \qquad [1 \quad 0 \quad 0]\begin{bmatrix}0\\1\\0\end{bmatrix} = 0, \qquad \text{etc.,}$$

and we find

$$I(\Gamma_1\Gamma_2) = |d_{\mathrm{T_{1u}}}^{\mathrm{A_{1g}T_{1u}}}|^2(l^2 + m^2 + n^2). \tag{9.22}$$

Since **u** is a unit vector, the expression in parentheses is equal to unity, so that

$$I(\Gamma_1\Gamma_2) = |d_{\mathrm{T_{1u}}}^{\mathrm{A_{1g}T_{1u}}}|^2. \tag{9.23}$$

This result shows that the spectral line intensity is given by the square of the reduced matrix element and does not depend on the orientation of the polarization vector; in other words, the spectrum is *isotropic*. In the example of the $\mathrm{T_{1u}} \rightarrow \mathrm{E_g}$ transition (Section 9.2) the **O** matrices are also orthogonal (superscripts are omitted below):

$$\mathbf{O}^{\dagger}_{\mathrm{T_{1u}}x}\mathbf{O}_{\mathrm{T_{1u}}y} = 0,$$

$$\mathbf{O}^{\dagger}_{\mathrm{T_{1u}}x}\mathbf{O}_{\mathrm{T_{1u}}x} = \begin{bmatrix}\sqrt{1/6} & -\sqrt{1/2}\\ 0 & 0\\ 0 & 0\end{bmatrix}\begin{bmatrix}\sqrt{1/6} & 0 & 0\\ -\sqrt{1/2} & 0 & 0\end{bmatrix} = \begin{bmatrix}2/3 & 0 & 0\\ 0 & 0 & 0\\ 0 & 0 & 0\end{bmatrix}, \quad \text{etc.}$$

We have again arrived at (9.23), including $|d_{\mathrm{T_{1u}}}^{\mathrm{E_gT_{1u}}}|^2$; that is, at the isotropic spectrum.

The result obtained has a general character: *the allowed optical lines of cubic crystal systems are isotropic*. This statement can be proved using the orthogonality relations (5.18) and (5.19) for Clebsch–Gordan coefficients (we shall not give the proof here). It should be stressed that optical line isotropy, as an obvious attribute of free atoms belonging to the rotation group, is also characteristic of cubic systems, whose symmetry, although lower, is still very high.

Using the above procedure, it is easy to show that in groups with $C_n$ $(n \geq 3)$ axes the angular dependence of the spectrum is expressed by

$$I(\Gamma_1\Gamma_2) \sim \alpha n^2 + \beta(l^2 + m^2), \tag{9.24}$$

where $\alpha$ and $\beta$ are two semiempirical parameters. For transitions between singlet states $\alpha \neq 0$, $\beta = 0$, for singlet–doublet transitions $\alpha = 0$, $\beta \neq 0$, and, finally, for doublet–doublet transitions $\alpha, \beta \neq 0$. This corresponds to the selection rules established in Section 9.1.3.

Finally, triaxial symmetry admits an angular dependence of $\alpha n^2$, $\beta m^2$ or $\gamma l^2$ type, according to the electronic terms.

## 9.4 APPROXIMATE SELECTION RULES

Group-theoretical selection rules are based on exact characteristics of many-electron systems; that is, they are based on irreducible representations, related to wavefunctions and to the operator $\hat{V}_{ep}$. As well as these rules, one should bear in mind some other considerations, which can be termed *approximate selection rules*. These imply that their intensities are weak.

Let us consider a tetrahedral complex of a transition metal. In the crystal field the atomic $d^1$ level splits into the triplet $T_2$ and the doublet E (see Table 4.3). Components of $\hat{\mathbf{d}}$ vector form the basis of the irreducible representation $T_1$ in $\mathbf{T}_d$. The E1 transition $E \leftrightarrow T_2$ proves to be allowed since the direct product $T_1(\hat{\mathbf{d}}) \times T_2 \doteq A_2 + E + T_1 + T_2$, i.e. contains E. In particular, parity rules do not impose any restrictions in the noncentrosymmetrical group $\mathbf{T}_d$. On the other hand, in the crystal field approximation the wavefunctions $\psi(E\gamma)$ and $\psi(T_2\gamma)$ are composed of *d* functions ($l = 2$) of even type. Therefore, although formally not having definite parity in $\mathbf{T}_d$, these functions do not give nonzero probability of light absorption. Such supplementary restrictions (*exclusions*) are eliminated when the basis functions are calculated more precisely, for example taking covalence into account. Indeed, for functions of the form

$$\Psi(\Gamma\gamma) = a\psi_{\Gamma\gamma}(3d) + b\psi_{\Gamma\gamma}(MO)$$

where $\psi_{\Gamma\gamma}(MO)$ is the molecular orbital of ligands not having specific parity, the d–d transition intensity becomes greater as the covalence factor $b$ increases.

Another example of this kind concerns interconfiguration transitions. It can be shown that optical transitions are possible only between terms resulting from one configuration $t_2^n e^m$ or those characterized by a one-electron transition ($t_2^n e^m \rightarrow t_2^{n\pm1} e^{m\mp1}$). Transitions of the type $\Gamma_1(t_2^n e^m) \rightarrow \Gamma_2(t_2^{n\pm2} e^{m\mp2})$ are forbidden, if we are dealing with the pure $t_2^n e^m$ configurations. This selection rule is only approximate, since it no longer applies when configuration mixing is taken into account. Now we can introduce the "weakly allowed" and "strongly allowed" transitions. Transitions of the type $t_2^n e^m \leftrightarrow t_2^{n\pm2} e^{m\pm2}$ (accompanied by two-electron jumps) can be referred to as "weakly allowed". The $A_{1g}(s^2) \leftrightarrow T_{1u}(sp)$ transition in a cubic centre ($\mathbf{O}_h$) represents the example of "strongly allowed" transition.

It should be noted that the true selection rules are determined purely by symmetry characteristics. The approximate selection rules, illustrated above by examples, are not related to point symmetry and determine only the magnitude of the operator's reduced matrix element.

## 9.5 TWO-PHOTON SPECTRA

### 9.5.1 Selection rules for two-photon transitions

The application of lasers as optical excitation sources has made it possible to study the optical lines of two-photon light absorption. The selection rules and angular dependences in two-photon absorption spectra are substantially different from those for one-photon processes. Two-photon spectroscopy has therefore considerably expanded the scope of studies of energy levels in crystals and coordination compounds.

Let us start by deducing the equation for the $\Gamma_1 \to \Gamma_2$ absorption intensity of two photons with frequencies $\Omega_1$ and $\Omega_2$ and polarization vectors $\mathbf{u}_1$ and $\mathbf{u}_2$. Two light beams having these characteristics can be produced, for example, by two lasers. Since the electric dipole mechanism of electron–photon interaction is the most effective, we shall assume that the absorption is primarily due to the action of two excitation operators $\hat{V}_{1ep} = \mathbf{u}_1 \cdot \hat{\mathbf{d}}$ and $\hat{V}_{2ep} = \mathbf{u}_2 \cdot \hat{\mathbf{d}}$. To calculate the matrix element $A(\Gamma_1\gamma_1\Gamma_2\gamma_2)$ of the transition $\Gamma_1\gamma_1 \to \Gamma_2\gamma_2$ we can use the quantum mechanical equation of second-order perturbation theory (see e.g. [39, 67]):

$$A(\Gamma_1\gamma_1\Gamma_2\gamma_2) = \sum_{\Gamma_i\gamma_i}\left[\frac{\langle\Gamma_2\gamma_2|\mathbf{u}_1\cdot\hat{\mathbf{d}}|\Gamma_i\gamma_i\rangle\langle\Gamma_i\gamma_i|\mathbf{u}_2\cdot\hat{\mathbf{d}}|\Gamma_1\gamma_1\rangle}{\varepsilon(\Gamma_i)-\varepsilon(\Gamma_1)-\hbar\Omega_2} + \frac{\langle\Gamma_2\gamma_2|\mathbf{u}_2\cdot\hat{\mathbf{d}}|\Gamma_i\gamma_i\rangle\langle\Gamma_i\gamma_i|\mathbf{u}_1\cdot\hat{\mathbf{d}}|\Gamma_1\gamma_1\rangle}{\varepsilon(\Gamma_i)-\varepsilon(\Gamma_1)-\hbar\Omega_1}\right]. \quad (9.25)$$

The $\Gamma_i\gamma_i$ states are called *intermediate* or *virtual*, being involved only in the second-order matrix element products. The diagrams of processes corresponding to two possible transitions via virtual states

$$\Gamma_1 \xrightarrow{\Omega_2} \Gamma_i \xrightarrow{\Omega_1} \Gamma_2$$

and

$$\Gamma_1 \xrightarrow{\Omega_1} \Gamma_i \xrightarrow{\Omega_2} \Gamma_2$$

are represented in Fig. 9.2. Energy balance is satisfied only relative to the initial and final states:

$$\varepsilon(\Gamma_2) - \varepsilon(\Gamma_1) = \hbar\Omega_2 - \hbar\Omega_1 \quad (\Omega_2 > \Omega_1).$$

We define the *Green's function*

$$\Lambda(\Omega_\alpha) = \sum_{\Gamma_i\gamma_i}\frac{|\Gamma_i\gamma_i\rangle\langle\Gamma_i\gamma_i|}{\varepsilon(\Gamma_i)-\varepsilon(\Gamma_1)-\hbar\Omega_\alpha} \quad (\alpha = 1, 2), \quad (9.26)$$

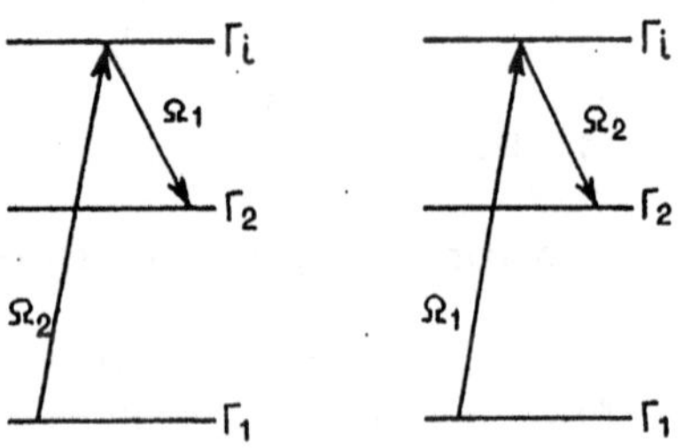


**Fig. 9.2** Diagrams of two-photon transitions through intermediate states.

together with the quantities

$$\Lambda_+ = \Lambda(\Omega_2) + \Lambda(\Omega_1), \qquad \Lambda_- = \Lambda(\Omega_2) - \Lambda(\Omega_1), \tag{9.27}$$

which are *symmetric* and *antisymmetric* respectively under interchange of the photon frequencies $\Omega_1$ and $\Omega_2$ [70]. It should be stressed that the Green function $\Lambda(\Omega)$ (as well as $\Lambda_+$ and $\Lambda_-$) is invariant under any transformation of the point group. Summing over $\gamma_i$ within one representation $\Gamma_i$, we obtain the scalar quantity

$$\sum_{\gamma_i} |\Gamma_i\gamma_i\rangle\langle\Gamma_i\gamma_i| \equiv \sum_{\gamma_1} \psi^*(\Gamma_i\gamma_i)\psi(\Gamma_i\gamma_i).$$

Since $[\varepsilon(\Gamma_i) - \varepsilon(\Gamma_1) - \hbar\Omega_\alpha]^{-1}$ are numerical coefficients, summing over $\Gamma_i$ does not affect the invariance of (9.26) as a whole.

The matrix element (9.26) can now be written as

$$A = \langle\Gamma_2\gamma_2|[\mathbf{u}_1\cdot\hat{\mathbf{d}}\Lambda(\Omega_2)\hat{\mathbf{d}}\cdot\mathbf{u}_2 + \mathbf{u}_2\cdot\hat{\mathbf{d}}\Lambda(\Omega_1)\hat{\mathbf{d}}\cdot\mathbf{u}_1]|\Gamma_2\gamma_2\rangle. \tag{9.28}$$

Inserting $\Lambda_+$ and $\Lambda_-$, we transform the two-photon transition operator as [71]

$$\tfrac{1}{2}(\mathbf{u}_1\cdot\hat{\mathbf{d}}\Lambda_+\hat{\mathbf{d}}\cdot\mathbf{u}_2 + \mathbf{u}_2\cdot\hat{\mathbf{d}}\Lambda_+\hat{\mathbf{d}}\cdot\mathbf{u}_1) + \tfrac{1}{2}(\mathbf{u}_1\cdot\hat{\mathbf{d}}\Lambda_-\hat{\mathbf{d}}\cdot\mathbf{u}_2 - \mathbf{u}_2\cdot\hat{\mathbf{d}}\Lambda_-\hat{\mathbf{d}}\cdot\mathbf{u}_1)$$

$$= \mathbf{u}_1\cdot[(\hat{\mathbf{d}}\Lambda_+\hat{\mathbf{d}})_\mathrm{s} + (\hat{\mathbf{d}}\Lambda_-\hat{\mathbf{d}})_\mathrm{a}]\cdot\mathbf{u}_2 \equiv \mathbf{u}_1\cdot\hat{\mathbf{T}}\cdot\mathbf{u}_2. \tag{9.29}$$

where subscripts s and a denote the symmetric and antisymmetric parts of the tensor operator $\hat{\mathbf{T}}$ within the square brackets. The operator $\hat{\mathbf{T}}$ can be written in matrix form as

$$\hat{\mathbf{T}} = \begin{bmatrix} \hat{A}_{xx} & \hat{A}_{xy} & \hat{A}_{xz} \\ \hat{A}_{yx} & \hat{A}_{yy} & \hat{A}_{yz} \\ \hat{A}_{zx} & \hat{A}_{zy} & \hat{A}_{zz} \end{bmatrix} + \begin{bmatrix} 0 & \hat{B}_z & -\hat{B}_y \\ -\hat{B}_z & 0 & \hat{B}_x \\ \hat{B}_y & -\hat{B}_x & 0 \end{bmatrix}, \tag{9.30}$$

where

$$\left.\begin{aligned} \hat{B}_x &= \tfrac{1}{2}(\hat{d}_y\Lambda_-\hat{d}_z - \hat{d}_z\Lambda_-\hat{d}_y), \\ \hat{B}_y &= \tfrac{1}{2}(\hat{d}_z\Lambda_-\hat{d}_x - \hat{d}_x\Lambda_-\hat{d}_z), \\ \hat{B}_z &= \tfrac{1}{2}(\hat{d}_x\Lambda_-\hat{d}_y - \hat{d}_y\Lambda_-\hat{d}_x). \end{aligned}\right\} \tag{9.31}$$

Since $\Lambda_-$ is a scalar, the quantities $B_i$ transform as the components of an *axial vector* (Section 3.7.2). Finally, the quantities

$$\hat{A}_{ik} = \tfrac{1}{2}(\hat{d}_i\Lambda_+\hat{d}_k + \hat{d}_k\Lambda_+\hat{d}_i) \tag{9.32}$$

transform in the same way as coordinate products ($A_{xy} \sim xy$, etc.).

The expression $\mathbf{u}_1\cdot\hat{\mathbf{T}}\cdot\mathbf{u}_2$ becomes

$$\begin{aligned} \mathbf{u}_1\cdot\hat{\mathbf{T}}\cdot\mathbf{u}_2 = {} & l_1l_2\hat{A}_{xx} + m_1m_2\hat{A}_{yy} + n_1n_2\hat{A}_{zz} + (l_1m_2 + l_2m_1)\hat{A}_{xy} \\ & + (n_1l_2 + n_2l_1)\hat{A}_{xz} + (m_1n_2 + m_2n_1)\hat{A}_{yz} + (l_1m_2 - l_2m_1)\hat{B}_z \\ & + (n_1l_2 - n_2l_1)\hat{B}_y + (m_1n_2 - m_2n_1)\hat{B}_x, \end{aligned} \tag{9.33}$$

where $l_1, m_1, n_1$ and $l_2, m_2, n_2$ are the components of the polarization vectors $\mathbf{u}_1$ and $\mathbf{u}_2$. From this representation of the two-photon transition operator it is obvious that the set of nine quantities $\hat{A}_{ik}$ and $\hat{B}_i$, dependent on the coordinates of the electrons of the absorbing centre, are transformed according to the direct product of vector representations $\Gamma(\hat{\mathbf{d}})$ (Section 5.4). Calculation of the matrix element $A(\Gamma_1\gamma_1\Gamma_2\gamma_2)$ involves only calculation of the quantities $\langle\Gamma_2\gamma_2|\hat{A}_{ik}|\Gamma_1\gamma_1\rangle$ and $\langle\Gamma_2\gamma_2|\hat{B}_i|\Gamma_1\gamma_1\rangle$. We thus arrive at the following selection rule: *the two-photon transition $\Gamma_1 \leftrightarrow \Gamma_2$ is allowed if the direct product of $\Gamma_1$ and the representation $\Gamma(\mathbf{d}) \times \Gamma(\mathbf{d})$ contains $\Gamma_2$ at least once. $\Gamma_1 \times \Gamma_2$ involves at least one irreducible representation from the direct product $\Gamma(\hat{\mathbf{d}}) \times \Gamma(\hat{\mathbf{d}})$*. In groups having an inversion centre $\Gamma(\hat{\mathbf{d}})$ is an odd (u) representation and the direct product $\Gamma(\hat{\mathbf{d}}) \times \Gamma(\hat{\mathbf{d}})$ consists of even (g) representations. Therefore *two-photon electric dipole transitions are allowed only if parity is conserved* (g $\leftrightarrow$ g, u $\leftrightarrow$ u). Parity conservation makes it possible to observe two-photon spectra for transitions forbidden in one-photon optics, for instance for d–d transitions in transition metal complexes.

It should be noted that if the photon frequencies coincide ($\Omega_1 = \Omega_2$), the antisymmetric part of the Green function $\Lambda_- = 0$, all the $B_i$ being zero, and, in order to deduce the selection rules, the representations whose bases comprise the components $B_x$, $B_y$ and $B_z$ should be excluded from the direct product $\Gamma(\hat{\mathbf{d}}) \times \Gamma(\hat{\mathbf{d}})$. This part of the direct product is called the *antisymmetric part* and is discussed in detail in Section 11.3. The above supplementary selection rule must be taken into consideration if, for instance, the photon excitation is provided by a single laser.

**Example 9.3** In $\mathrm{O_h}$, $\Gamma(\hat{\mathbf{d}}) = \mathrm{T_{1u}}$. Then $\Gamma(\hat{\mathbf{d}}) \times \Gamma(\hat{\mathbf{d}}) \doteq \mathrm{T_{1u}} \times \mathrm{T_{1u}} \doteq \mathrm{A_{1g}} + \mathrm{E_g} + \mathrm{T_{1g}} + \mathrm{T_{2g}}$. Two-photon transitions are allowed if the direct product $\Gamma_1 \times \Gamma_2$ involves at least one of the representations $\mathrm{A_{1g}}$, $\mathrm{E_g}$, $\mathrm{T_{1g}}$ and $\mathrm{T_{2g}}$. From the $\Gamma_1 \times \Gamma_2$ multiplication table (Table 5.7) we find that $\mathrm{A_{2g}} \times \mathrm{E_g} \doteq \mathrm{E_g}$. Hence the transition $\mathrm{A_{2g}} \leftrightarrow \mathrm{E_g}$ is allowed. Likewise the transitions $\mathrm{E_g} \leftrightarrow \mathrm{T_{1g}}$ and $\mathrm{E_g} \leftrightarrow \mathrm{T_{2g}}$ are also allowed ($\mathrm{E_g} \times \mathrm{T_{1g}} = \mathrm{E_g} \times \mathrm{T_{2g}} \doteq \mathrm{T_{1g}} + \mathrm{T_{2g}}$).

In $\mathbf{O}_\mathrm{h}$ the components of the axial vector $\mathbf{B}$ form the basis of $\mathrm{T_{1g}}$. Therefore for one-frequency excitation ($\Omega_1 = \Omega_2$) from the matrix elements of the transitions $\mathrm{E_g} \leftrightarrow \mathrm{T_{1g,2g}}$ the terms containing $B_i$ (i.e. the last line in (9.9)) vanish. The transitions $\mathrm{A_{1g}} \leftrightarrow \mathrm{T_{1g}}$ and $\mathrm{A_{2g}} \leftrightarrow \mathrm{T_{2g}}$ are allowed for $\Omega_1 \neq \Omega_2$ and forbidden for $\Omega_1 = \Omega_2$. For these transitions $\mathrm{A_{1g}} \times \mathrm{T_{1g}} = \mathrm{A_{2g}} \times \mathrm{T_{2g}} \doteq \mathrm{T_{1g}}$, and the intensity results from the contributions of the terms containing $B_i$. Taking this rule and Table 5.7 into consideration, we find that in $\mathbf{O}_\mathrm{h}$ all two-photon $\Gamma_{1\mathrm{g(u)}} \leftrightarrow \Gamma_{2\mathrm{g(u)}}$ transitions are allowed, except for $\mathrm{A_{1g(u)}} \leftrightarrow \mathrm{A_{2g(u)}}$.

### 9.5.2 Polarization dependence of two-photon spectra

To find the polarization-dependence of two-photon spectra, the intensity equation

$$I(\Gamma_1\Gamma_2) = \frac{1}{g(\Gamma_1)}\sum_{\gamma_1\gamma_2} |\langle\Gamma_2\gamma_2|\mathbf{u}_1 \cdot \hat{\mathbf{T}} \cdot \mathbf{u}_2|\Gamma_1\gamma_1\rangle|^2 \tag{9.34}$$

must be expressed as a function of certain parameters (reduced matrix elements of the operator $\hat{\mathbf{T}}$) and of the components of the vectors $\mathbf{u}_1$ and $\mathbf{u}_2$. As in the case of one-photon transitions (Section 9.3), this problem is solved using the Wigner–Eckart theorem. However, since not all the operators $\hat{A}_{ik}$ are irreducible tensors, it is impossible to use this theorem immediately. Elementary rearrangement of the terms in (9.33) reveals that it is the sum of the scalar combinations of the irreducible tensors $\hat{\mathbf{V}}_{\Gamma\gamma}(\hat{\mathbf{d}}_1\hat{\mathbf{d}}_2) \equiv \hat{\mathbf{V}}_{\Gamma\gamma}(\hat{\mathbf{d}})$ in the space of electronic coordinates and tensors $V_{\Gamma\gamma}(\mathbf{u}_1\mathbf{u}_2) \equiv V_{\Gamma\gamma}(\mathbf{u})$, formed from the components of polarization vectors $\mathbf{u}_1$ and $\mathbf{u}_2$:

$$\mathbf{u}_1 \cdot \hat{\mathbf{T}} \cdot \mathbf{u}_2 = \sum_{\Gamma\gamma} \hat{\mathbf{V}}_{\Gamma\gamma}(\hat{\mathbf{d}}_1\hat{\mathbf{d}}_2) V_{\Gamma\gamma}(\mathbf{u}_1\mathbf{u}_2). \tag{9.35}$$

To construct $\hat{\mathbf{V}}_{\Gamma\gamma}(\hat{\mathbf{d}})$ and $V_{\Gamma\gamma}(\mathbf{u})$, we shall use the definition (5.15) of the Clebsch–Gordan coefficients. For $\mathbf{O}_h$ we obtain the expressions already found in Section 5.4.2:

$$\left.\begin{aligned} \hat{\mathbf{V}}_{A_{1g}}(\hat{\mathbf{d}}) &= \sqrt{1/3}\,(\hat{A}_{xx} + \hat{A}_{yy} + \hat{A}_{zz}),\\ V_{A_{1g}}(\mathbf{u}) &= \sqrt{1/3}\,(l_1 l_2 + m_1 m_2 + n_1 n_2) \end{aligned}\right\} (A_{1g}), \tag{9.36}$$

$$\left.\begin{aligned} \hat{\mathbf{V}}_{Eu}(\hat{\mathbf{d}}) &= \sqrt{1/6}\,(2\hat{A}_{zz} - \hat{A}_{xx} - \hat{A}_{yy}),\\ V_{Eu}(\mathbf{u}) &= \sqrt{1/6}\,(2n_1 n_2 - l_1 l_2 - m_1 m_2),\\ \hat{\mathbf{V}}_{Ev}(\hat{\mathbf{d}}) &= \sqrt{1/2}\,(\hat{A}_{xx} - \hat{A}_{yy}),\\ V_{Ev}(\mathbf{u}) &= \sqrt{1/2}\,(l_1 l_2 - m_1 m_2) \end{aligned}\right\} (E_g), \tag{9.37}$$

$$\left.\begin{aligned} \hat{\mathbf{V}}_{T_{1g}\alpha}(\hat{\mathbf{d}}) &= \hat{B}_x, & V_{T_{1g}\alpha}(\mathbf{u}) &= \sqrt{1/2}\,(m_1 n_2 - n_1 m_2),\\ \hat{\mathbf{V}}_{T_{1g}\beta}(\hat{\mathbf{d}}) &= \hat{B}_y, & V_{T_{1g}\beta}(\mathbf{u}) &= \sqrt{1/2}\,(n_1 l_2 - l_1 n_2),\\ \hat{\mathbf{V}}_{T_{1g}\gamma}(\hat{\mathbf{d}}) &= \hat{B}_z, & V_{T_{1g}\gamma}(\mathbf{u}) &= \sqrt{1/2}\,(l_1 m_2 - m_1 l_2) \end{aligned}\right\} (T_{1g}), \tag{9.38}$$

$$\left.\begin{aligned} \hat{\mathbf{V}}_{T_{2g}\xi}(\hat{\mathbf{d}}) &= \hat{A}_{yz}, & V_{T_{2g}\xi}(\mathbf{u}) &= \sqrt{1/2}\,(m_1 n_2 + n_1 m_2),\\ \hat{\mathbf{V}}_{T_{2g}\eta}(\hat{\mathbf{d}}) &= \hat{A}_{xz}, & V_{T_{2g}\eta}(\mathbf{u}) &= \sqrt{1/2}\,(n_1 l_2 + l_1 n_2),\\ \hat{\mathbf{V}}_{T_{2g}\zeta}(\hat{\mathbf{d}}) &= \hat{A}_{xy}, & V_{T_{2g}\zeta}(\mathbf{u}) &= \sqrt{1/2}\,(l_1 m_2 + m_1 l_2) \end{aligned}\right\} (T_{2g}). \tag{9.39}$$

We can now apply the Wigner–Eckart theorem to the operator $\mathbf{u}_1 \cdot \hat{\mathbf{T}} \cdot \mathbf{u}_2$, expressed in terms of irreducible tensors,

$$\langle \Gamma_2\gamma_2 | \mathbf{u}_1 \cdot \hat{\mathbf{T}} \cdot \mathbf{u}_2 | \Gamma_1\gamma_1 \rangle = \sum_{\Gamma\gamma} V_{\Gamma}^{\Gamma_2\Gamma_1}(\hat{\mathbf{d}}_1\hat{\mathbf{d}}_2) V_{\Gamma\gamma}(\mathbf{u}_1\mathbf{u}_2) \langle \Gamma_2\gamma_2 | \Gamma_1\gamma_1\Gamma\gamma \rangle, \tag{9.40}$$

or, in matrix form,

$$\mathbf{u}_1 \cdot \hat{\mathbf{T}} \cdot \mathbf{u}_2 = \sum_{\Gamma} V_{\Gamma}^{\Gamma_2\Gamma_1}(\hat{\mathbf{d}}_1\hat{\mathbf{d}}_2) \sum_{\gamma} \mathbf{O}_{\Gamma\gamma}^{\Gamma_1\Gamma_2} V_{\Gamma\gamma}(\mathbf{u}_1\mathbf{u}_2), \tag{9.41}$$

where $V_\Gamma^{\Gamma_2\Gamma_1} = [g(\Gamma_2)]^{-1/2}\langle\Gamma_2\|\mathbf{V}_\Gamma(\hat{\mathbf{d}}_1\hat{\mathbf{d}}_2)\|\Gamma_1\rangle$ is a reduced matrix element, playing the role of an independent parameter. The number of independent parameters is given by the number of irreducible representations $\Gamma$ for which $V_\Gamma^{\Gamma_1\Gamma_2} \neq 0$.

Using $\mathbf{O}_{\Gamma\gamma}$ matrix orthogonality and the properties of the Clebsch–Gordan coefficients, we can perform the summation in (9.10) and obtain [72]

$$I(\Gamma_1\Gamma_2) = \sum_\Gamma |V_\Gamma^{\Gamma_1\Gamma_2}(\hat{\mathbf{d}})|^2 \sum_\gamma |V_{\Gamma\gamma}(\mathbf{u})|^2. \tag{9.42}$$

This equation can be used to determine the polarization dependence of any two-photon transition $\Gamma_1 \leftrightarrow \Gamma_2$ in any point group. The results are given for some groups in Table 9.9, where a shorthand notation ($\lambda_i$ and $c_i$) is adopted for the parameters $|V_\Gamma^{\Gamma_1\Gamma_2}|^2$. By measuring line intensities for different orientations of $\mathbf{u}_1$ and $\mathbf{u}_2$ it is possible to determine whether this line is related to a certain transition. It should be mentioned that, unlike one-photon transitions, two-photon transitions are strongly anisotropic. Of course, group theory does not provide any relationship between the parameters $\lambda$ and $c$. Because they are taken to be independent, their values can be obtained from the solution of a quantum mechanical problem within the framework of a given model.

## 9.6 EFFECTIVE DIPOLE MOMENT METHOD

### 9.6.1 Effective dipole moment

As has been mentioned already, electric dipole transitions between equal-parity levels are forbidden. However, if there is a weak odd crystal field in the system, this prohibition is removed. The selection rules for allowed transitions can be established using the method described in Section 9.1.

**Example 9.4** Consider an octahedral complex ($\mathbf{O}_h$) with a weak field $\mathbf{C}_{4v}$ (Section 6.4.3). Such a field may be either the crystal field itself or an external electric field oriented along the $C_4(z)$ axis (see Table 4.4). From the $\mathbf{O}_h \to \mathbf{C}_{4v}$ reduction tables [1] we find $T_{2g} \doteq B_2 + E$ and $E_g \doteq A_1 + B_1$. For the components of $\hat{\mathbf{d}}$ we obtain $T_{1u}(\hat{\mathbf{d}}) \doteq A_1(\hat{d}_z) + E(\hat{d}_x, \hat{d}_y)$. This leads to splitting and to the selection rules for the transition $T_{2g} \leftrightarrow E_g$ in an octahedral complex of $\mathbf{C}_{4v}$ conformation, shown in Fig. 9.3:

$$(B_2 \times A_1 \doteq B_2, \quad B_2 \times B_1 \doteq A_2, \quad E \times A_1 \doteq E, \quad E \times B_1 \doteq E).$$

The selection rules thus obtained do not provide information on the relation between the intensities of different transitions. This problem can be solved using the method of the effective dipole moment [14, 46]. Let $\Gamma_1$ and $\Gamma_2$ be two levels of a centrosymmetric group (e.g. $\mathbf{O}_h$) with the same parity and let $\hat{V}_u$ be the odd component of the crystal field (Section 6.3.1) or the interaction with the external electric field. Making use of the second-order equation of

**Table 9.9** Polarization dependences $I(\mathbf{u}_1\mathbf{u}_2)$ of two-photon transition intensities [71].

| Groups and transitions | | Polarization dependence |
|---|---|---|
| **O, T$_d$** | **O$_h$** | $I(\mathbf{u}_1\mathbf{u}_2)$ |
| $A_1 \leftrightarrow A_1$<br>$A_2 \leftrightarrow A_2$ | $A_{1g(u)} \leftrightarrow A_{1g(u)}$<br>$A_{2g(u)} \leftrightarrow A_{2g(u)}$ | $(l_1 l_2 + m_1 m_2 + n_1 n_2)^2 \equiv A_1(\mathbf{u}_1\mathbf{u}_2)$ |
| $A_1 \leftrightarrow A_2$ | $A_{1g(u)} \leftrightarrow A_{2g(u)}$ | Forbidden |
| $A_1 \leftrightarrow E$<br>$A_2 \rightarrow E$ | $A_{1g(u)} \leftrightarrow E_{g(u)}$<br>$A_{2g(u)} \leftrightarrow E_{g(u)}$ | $(l_1^2 l_2^2 + m_1^2 m_2^2 + n_1^2 n_2^2) - (l_1 l_2 m_1 m_2 + m_1 m_2 n_1 n_2$<br>$+ n_1 n_2 l_1 l_2) \equiv E(\mathbf{u}_1\mathbf{u}_2)$ |
| $A_1 \leftrightarrow T_1$<br>$A_2 \leftrightarrow T_2$ | $A_{1g(u)} \leftrightarrow T_{1g(u)}$<br>$A_{2g(u)} \leftrightarrow T_{2g(u)}$ | $(m_1 n_2 - m_2 n_1)^2 + (l_2 n_1 - l_1 n_2)^2$<br>$+ (l_1 m_2 - l_2 m_1)^2 \equiv T_1(\mathbf{u}_1\mathbf{u}_2)$ |
| $A_1 \leftrightarrow T_2$<br>$A_2 \leftrightarrow T_1$ | $A_{1g(u)} \leftrightarrow T_{2g(u)}$<br>$A_{2g(u)} \leftrightarrow T_{1g(u)}$ | $(l_1 m_2 + l_2 m_1)^2 + (l_1 n_2 + l_2 n_1)^2 + (m_1 n_2 + m_2 n_1)^2$<br>$\equiv T_2(\mathbf{u}_1\mathbf{u}_2)$ |
| $E \leftrightarrow E$ | $E_{g(u)} \leftrightarrow E_{g(u)}$ | $A_1(\mathbf{u}_1\mathbf{u}_2) + cE(\mathbf{u}_1\mathbf{u}_2)$ |
| $E \rightarrow T_1$<br>$E \rightarrow T_2$ | $E_{g(u)} \leftrightarrow T_{1g(u)}$<br>$E_{g(u)} \leftrightarrow T_{2g(u)}$ | $T_1(\mathbf{u}_1\mathbf{u}_2) + cT_2(\mathbf{u}_1\mathbf{u}_2)$ |
| $T_1 \leftrightarrow T_1$<br>$T_2 \leftrightarrow T_2$ | $T_{1g(u)} \leftrightarrow T_{1g(u)}$<br>$T_{2g(u)} \leftrightarrow T_{2g(u)}$ | $A_1(\mathbf{u}_1\mathbf{u}_2) + c_1 E(\mathbf{u}_1\mathbf{u}_2) + c_2 T_1(\mathbf{u}_1\mathbf{u}_2) + c_3 T_2(\mathbf{u}_1\mathbf{u}_2)$ |
| $T_1 \leftrightarrow T_2$ | $T_{1g(u)} \leftrightarrow T_{2g(u)}$ | $E(\mathbf{u}_1\mathbf{u}_2) + c_1 T_1(\mathbf{u}_1\mathbf{u}_2) + c_2 T_2(\mathbf{u}_1\mathbf{u}_2)$ |
| **D$_4$, C$_{4v}$, D$_{2d}$** | **D$_{4h}$** | $I(\mathbf{u}_1\mathbf{u}_2)$ |
| $A_1 \leftrightarrow A_1$<br>$A_1 \leftrightarrow A_2$<br>$B_1 \leftrightarrow B_2$<br>$B_2 \leftrightarrow B_2$ | $A_{1g(u)} \leftrightarrow A_{1g(u)}$<br>$A_{1g(u)} \leftrightarrow A_{2g(u)}$<br>$B_{1g(u)} \leftrightarrow B_{1g(u)}$<br>$B_{2g(u)} \leftrightarrow B_{2g(u)}$ | $[(l_1 l_2 + m_1 m_2) + \lambda n_1 n_2]^2 \equiv A_1(\mathbf{u}_1\mathbf{u}_2)$ |
| $A_1 \leftrightarrow A_2$<br>$B_1 \leftrightarrow B_2$ | $A_{1g(u)} \leftrightarrow A_{2g(u)}$<br>$B_{1g(u)} \leftrightarrow B_{2g(u)}$ | $(l_1 m_2 - l_2 m_1)^2 \equiv A_2(\mathbf{u}_1\mathbf{u}_2)$ |
| $A_1 \leftrightarrow B_1$<br>$A_2 \rightarrow B_2$ | $A_{1g(u)} \leftrightarrow B_{1g(u)}$<br>$A_{2g(u)} \leftrightarrow B_{2g(u)}$ | $(l_1 l_2 - m_1 m_2)^2 \equiv B_1(\mathbf{u}_1\mathbf{u}_2)$ |
| $A_1 \leftrightarrow B_2$<br>$A_2 \leftrightarrow B_1$ | $A_{1g(u)} \leftrightarrow B_{2g(u)}$<br>$A_{2g(u)} \leftrightarrow B_{1g(u)}$ | $(l_1 m_2 + l_2 m_1)^2 \equiv B_2(\mathbf{u}_1\mathbf{u}_2)$ |
| $A_1 \leftrightarrow E$<br>$A_2 \leftrightarrow E$<br>$B_1 \leftrightarrow E$<br>$B_2 \leftrightarrow E$ | $A_{1g(u)} \leftrightarrow E_{g(u)}$<br>$A_{2g(u)} \leftrightarrow E_{g(u)}$<br>$B_{1g(u)} \leftrightarrow E_{g(u)}$<br>$B_{2g(u)} \leftrightarrow E_{g(u)}$ | $[(m_1 n_2 + m_2 n_1) + \lambda_2(m_1 n_2 - m_2 n_1)]^2$<br>$+ [(n_1 l_2 + n_2 l_1) - \lambda_2(n_1 l_2 - n_2 l_1)]^2$<br>$\equiv E(\mathbf{u}_1\mathbf{u}_2)$ |
| $E \leftrightarrow E$ | $E_{g(u)} \leftrightarrow E_{g(u)}$ | $A_1(\mathbf{u}_1\mathbf{u}_2) + c_1 A_2(\mathbf{u}_1\mathbf{u}_2) + c_2 B_1(\mathbf{u}_1\mathbf{u}_2) + c_2 B_2(\mathbf{u}_1\mathbf{u}_2)$ |

perturbation theory, we can write the $\Gamma_1 \to \Gamma_2$ transition matrix element as

$$\langle \Gamma_2\gamma_2|\hat{V}_{\mathrm{ep}}|\Gamma_1\gamma_1\rangle = \sum_i \left[\frac{\langle \Gamma_2\gamma_2|\hat{V}_{\mathrm{u}}|\Gamma_i\gamma_i\rangle\langle \Gamma_i\gamma_i|\mathbf{u}\cdot\hat{\mathbf{d}}|\Gamma_1\gamma_1\rangle}{\varepsilon(\Gamma_2)-\varepsilon(\Gamma_i)} + \frac{\langle \Gamma_2\gamma_2|\mathbf{u}\cdot\hat{\mathbf{d}}|\Gamma_i\gamma_i\rangle\langle \Gamma_i\gamma_i|\hat{V}_{\mathrm{u}}|\Gamma_1\gamma_1\rangle}{\varepsilon(\Gamma_1)-\varepsilon(\Gamma_i)}\right], \tag{9.43}$$

where $\Gamma_i$ are the odd representations of $\mathbf{O}_{\mathrm{h}}$, related, for example, to $3\mathrm{d}^{n-1}4\mathrm{p}$ states. Since both $\hat{V}_{\mathrm{u}}$ and $\mathbf{u}\cdot\hat{\mathbf{d}}$ are odd, the matrix elements of (9.43) are nonzero in general. If the energy gap between the groups of levels relating to the $3\mathrm{d}^n$ and $3\mathrm{d}^{n-1}4\mathrm{p}$ configurations is large as compared with the crystal field splittings within each group, the denominators in (9.43) can be approximately replaced by a certain mean energy $\Delta E$, which can be taken outside the summation. Then performing the matrix multiplication, we obtain

$$\langle \Gamma_2\gamma_2|\hat{V}_{\mathrm{ep}}|\Gamma_1\gamma_1\rangle = \frac{2}{\Delta E}\langle \Gamma_2\gamma_2|\hat{V}_{\mathrm{u}}\mathbf{u}\cdot\hat{\mathbf{d}}|\Gamma_1\gamma_1\rangle \tag{9.44}$$

The operator

$$\hat{P}_{\mathrm{eff}} = \frac{2}{\Delta E}\hat{V}_{\mathrm{u}}\mathbf{u}\cdot\hat{\mathbf{d}}$$

is called the *effective dipole moment*. It takes into account both the odd field and the E1 electron–photon interaction. Being even, the effective dipole moment connects identical-parity states of $\mathbf{O}_{\mathrm{h}}$.

### 9.6.2 Intensities of spectral lines

Let us transform $\hat{P}_{\mathrm{eff}}$, expressing it in terms of the irreducible tensors of $\mathbf{O}_{\mathrm{h}}$. The odd field $\hat{V}_{\mathrm{u}}(\mathbf{r})$ is invariant under the operations of the low-symmetry group ($\mathbf{O}_{\mathrm{h}}$ subgroup); that is, it transforms according to the totally symmetric representation $\mathrm{A}_1$ of this group. Therefore $\hat{V}_{\mathrm{u}}$ is a tensor operator $\hat{\mathbf{V}}_{\Gamma'\gamma'}$ of $\mathbf{O}_{\mathrm{h}}$, forming the $\mathrm{A}_1$ basis of the subgroup. From this there follows a simple procedure for determining the operators $\hat{\mathbf{V}}_{\Gamma'\gamma'}$:

(1) from the reduction tables we determine the representations $\Gamma'(\mathbf{O}_{\mathrm{h}})$, containing $\mathrm{A}_1$ of low-symmetry group;

(2) from the tables of basis functions $f(\Gamma')$ we find the function $\hat{\mathbf{V}}_{\Gamma'\gamma'}(\mathbf{O}_{\mathrm{h}})$ forming the $\mathrm{A}_1$ basis of the subgroup.

Inserting $\hat{V}_{\mathrm{u}} = \hat{\mathbf{V}}_{\Gamma'\gamma'}$ into $\hat{P}_{\mathrm{eff}}$ and using the transformation (5.20), we obtain

$$\hat{P}_{\mathrm{eff}} = \frac{2}{\Delta E}\hat{\mathbf{V}}_{\Gamma'\gamma'}\sum_{\Gamma\gamma} u_{\Gamma\gamma}\hat{\mathbf{d}}_{\Gamma\gamma} = \frac{2}{\Delta E}\sum_{\Gamma\gamma} u_{\Gamma\gamma}\sum_{\bar{\Gamma}\bar{\gamma}} \hat{\mathbf{w}}_{\bar{\Gamma}\bar{\gamma}}(\Gamma'\Gamma)\langle\bar{\Gamma}\bar{\gamma}|\Gamma'\gamma'\Gamma\gamma\rangle. \tag{9.45}$$

Here $\mathbf{u} \cdot \hat{\mathbf{d}}$ is represented as the scalar product of the basis functions $\Gamma(\hat{\mathbf{d}})$, $\hat{\mathbf{w}}(\bar{\Gamma}\bar{\gamma})$ is a function of the electron coordinates of the $\bar{\Gamma}\bar{\gamma}$ basis, and $\bar{\Gamma}$ is contained in $\Gamma' \times \Gamma$. We can now apply the Wigner–Eckart theorem to the operator $\hat{P}_{\text{eff}}$:

$$\langle \Gamma_2\gamma_2|\hat{P}_{\text{eff}}|\Gamma_1\gamma_1\rangle = \frac{2}{\Delta E[g(\Gamma_2)]^{1/2}} \sum_{\Gamma\gamma} \mathbf{u}_{\Gamma\gamma} \sum_{\bar{\Gamma}} w_{\bar{\Gamma}}^{\Gamma_2\Gamma_1}(\Gamma'\Gamma) \times \sum_{\bar{\gamma}} \langle \Gamma_2\gamma_2|\Gamma_1\gamma_1\bar{\Gamma}\bar{\gamma}\rangle\langle\bar{\Gamma}\bar{\gamma}|\Gamma'\gamma'\Gamma\gamma\rangle. \tag{9.46}$$

The reduced matrix elements $w_{\bar{\Gamma}}^{\Gamma_2\Gamma_1}(\Gamma'\Gamma) \equiv \langle\Gamma_2\|\hat{\mathbf{w}}_{\bar{\Gamma}}(\Gamma'\Gamma)\|\Gamma_1\rangle$ are semi-empirical parameters. To calculate the line intensities for a given polarization, we must select the corresponding term from the sum over $\Gamma\gamma$.

**Example 9.5** We consider the $\mathbf{C}_{4v}$ conformation of an octahedral complex, for which $\Gamma(\hat{\mathbf{d}}) = \mathrm{T}_{1u}$ and $\gamma = x, y, z$ (polarization direction). From the $\Gamma'(\mathbf{O}_h)$ reduction schemes (see Table 4.4) we find that $\mathrm{A}_1(\mathbf{C}_{4v})$ results only from $\mathrm{T}_{1u}(\mathbf{O}_h)$ reduction, whereby $f(A_1) = f(\mathrm{T}_{1u}z)$. Hence

$$\hat{P}_{\text{eff}} = \frac{2}{\Delta E}\hat{\mathbf{V}}_{\mathrm{T}_{1u}z} \sum_{\gamma=x,y,z} u_{\mathrm{T}_{1u}\gamma}\hat{\mathbf{d}}_{\mathrm{T}_{1u}\gamma} = \frac{2}{\Delta E}\sum_{\gamma=x,y,z} u_{\mathrm{T}_{1u}\gamma} \sum_{\bar{\Gamma}\bar{\gamma}} \hat{\mathbf{w}}_{\bar{\Gamma}\bar{\gamma}}(\mathrm{T}_{1u}\mathrm{T}_{1u})\langle\bar{\Gamma}\bar{\gamma}|\mathrm{T}_{1u}z\mathrm{T}_{1u}\gamma\rangle, \tag{9.47}$$

where $\bar{\Gamma} = \mathrm{T}_{1u} \times \mathrm{T}_{1u} \doteq \mathrm{A}_{1g} + \mathrm{E}_g + \mathrm{T}_{1g} + \mathrm{T}_{2g}$. Inserting the Clebsch–Gordan coefficients, we obtain the components of the effective dipole moment:

$$\left.\begin{aligned} \hat{P}_{\text{eff}} &= l\hat{P}_x + m\hat{P}_y + n\hat{P}_z, \\ \hat{P}_x &= -\frac{1}{\sqrt{2}\Delta E}(\hat{\mathbf{w}}_{\mathrm{T}_{1g}\beta} + \hat{\mathbf{w}}_{\mathrm{T}_{2g}\beta}), \\ \hat{P}_y &= -\frac{1}{\sqrt{2}\Delta E}(\hat{\mathbf{w}}_{\mathrm{T}_{1g}\alpha} - \hat{\mathbf{w}}_{\mathrm{T}_{1g}\zeta}), \\ \hat{P}_z &= -\frac{2}{\sqrt{3}\Delta E}(\hat{\mathbf{w}}_{\mathrm{A}_{1g}} + \sqrt{2}\hat{\mathbf{w}}_{\mathrm{E}_g u}). \end{aligned}\right\} \tag{9.48}$$

For the transition $\mathrm{T}_{2g} \leftrightarrow \mathrm{E}_g$ only the $x$ and $y$ components of $\hat{P}_{\text{eff}}$ are involved, including the operators of $\mathrm{T}_{1g}$ and $\mathrm{T}_{2g}$ type ($\mathrm{E}_g \times \mathrm{T}_{2g} \doteq \mathrm{T}_{1g} + \mathrm{T}_{2g}$). This corroborates our conclusion about the possibility of having only $\sigma$-polarized lines (Fig. 9.3), and shows that their intensities depend on two parameters: $w_{\mathrm{T}_{1g}}$ and $w_{\mathrm{T}_{2g}}$. Applying the Wigner–Eckart theorem, we obtain the matrix $\hat{\mathbf{P}}_x$ as

$$\mathbf{P}_x = w_1\begin{bmatrix} 0 & \sqrt{1/2} & 0 \\ 0 & -\sqrt{1/6} & 0 \end{bmatrix} + w_2\begin{bmatrix} 0 & -\sqrt{1/6} & 0 \\ 0 & -\sqrt{1/2} & 0 \end{bmatrix}, \tag{9.49}$$

where

$$w_1 = \frac{-w_{\mathrm{T}_{1g}}}{\sqrt{2}\Delta E}, \qquad w_2 = \frac{-w_{\mathrm{T}_{2g}}}{\sqrt{2}\Delta E}.$$

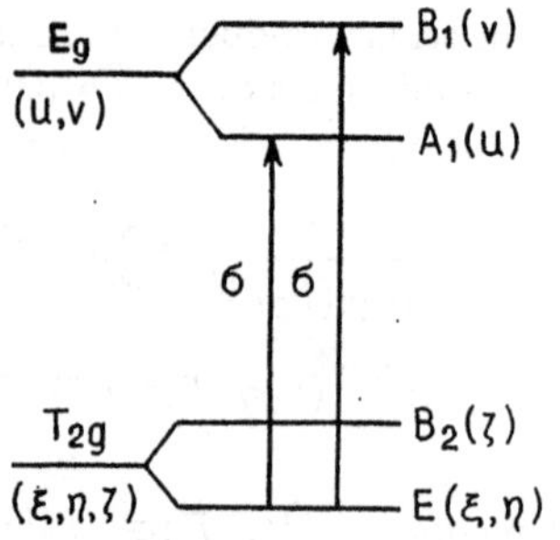


**Fig. 9.3** Levels and allowed transitions in an octahedral complex of $\mathbf{C}_{4v}$ conformation (the basis functions are given in parentheses).

The parameters $w_1$ and $w_2$ are independent. In approximate models (e.g. using crystal field theory) these parameters can be related by certain equations [6].

## PROBLEMS

**9.1** Find the selection rules for optical transitions in $\mathbf{O}_h$ (Tables 9.1 and 9.2).

**9.2** Find the selection rules for optical transitions in $\mathbf{T}_d$, $\mathbf{C}_{4v}$, $\mathbf{C}_{3v}$ and $\mathbf{D}_2$ (Tables 9.4–9.6).

**9.3** Determine the dipole moment matrices for the transition $T_{1u} \leftrightarrow T_{2g}$ in $\mathbf{O}_h$.

**9.4** Determine the matrices of the magnetic dipole moment operator for the $E_g \leftrightarrow T_{2g}$ transition in $\mathbf{O}_h$.

**9.5** Show that in cubic systems the optical lines resulting from M1 and E2 transitions are isotropic.

**9.6** Find the angular dependences of the intensities of all allowed transitions in $\mathbf{C}_{2h}$, $\mathbf{D}_{4h}$, $\mathbf{D}_3$ and $\mathbf{D}_{3h}$.

**9.7** Find the selection rules for two-photon transitions in $\mathbf{D}_{3h}$, $\mathbf{D}_{4h}$ and $\mathbf{D}_{2h}$.

**9.8** Construct the angular dependences of two-photon transitions in $\mathbf{C}_{2v}$, $\mathbf{C}_{3v}$, $\mathbf{D}_{3d}$ and $\mathbf{D}_{3h}$.

**9.9** Construct the effective dipole moment for the $\mathbf{C}_{2v}$ conformation of an octahedral complex.

# 10 Double Groups

## 10.1 SPIN–ORBIT INTERACTION

In an atom the spin of an electron interacts with its orbital angular momentum. The spin–orbit interaction operator may be written as [14, 39, 73]

$$\hat{H}_{\mathrm{so}} = \lambda(\mathbf{r})\hat{\mathbf{l}}\cdot\hat{\mathbf{s}} \equiv \lambda(r)\,(\hat{l}_x\hat{s}_x + \hat{l}_y\hat{s}_y + \hat{l}_z\hat{s}_z), \tag{10.1}$$

where $\hat{\mathbf{l}}$ and $\hat{\mathbf{s}}$ are the orbital angular momentum and spin operators and $\lambda(r)$ is a function of the electron coordinates:

$$\lambda(r) = -\frac{e\hbar^2}{2m^2c^2}\,\frac{1}{r}\,\frac{\mathrm{d}U(r)}{\mathrm{d}r}; \tag{10.2}$$

$U(r)$ is the spherically symmetric potential in which the electron moves in the atom.

Below we determine the electronic energy levels, taking the spin–orbit interaction into account. It should be noted that the atomic levels in this case are characterized by the total angular momentum quantum number. The quantum numbers $j$ (for one electron) and $J$ (for a many-electron atom) are determined according to the coupling rules given in Section 5.1.

For one electron ($s = \frac{1}{2}$)

$$j = l + \tfrac{1}{2} \quad \text{and} \quad l - \tfrac{1}{2};$$

that is, the total angular momentum quantum number takes half-integer values. In many-electron atoms the total spin $S$, as well as $J = L + S, \ldots, |L - S|$ may also be half-integers. We apply (4.16) for the operation $\hat{R}_\alpha$ of rotation through an angle $\alpha$ for wavefunctions $Y_{JM}$ having half-integer angular momentum:

$$\chi^{(J)}(\alpha) = \frac{\sin (J + \frac{1}{2})\alpha}{\sin \frac{1}{2}\alpha}. \tag{10.3}$$

Let us determine $\chi^{(J)}(\alpha)$ for a rotation angle of $2\pi$:

$$\chi^{(J)}(\alpha+2\pi) = \frac{\sin[(J+\frac{1}{2})(\alpha+2\pi)]}{\sin[\frac{1}{2}(\alpha+2\pi)]} = \frac{\sin[(J+\frac{1}{2})\alpha+2\pi]}{\sin(\frac{1}{2}\alpha+\pi)}$$

$$= -\chi^{(J)}(\alpha). \qquad (10.4)$$

Thus when $J$ is a half-integer, a rotation through $2\pi$ changes the sign of the character. Since a $2\pi$ rotation is an identity transformation, wavefunctions $\psi_{JM}$ of half-integer angular momentum are said to form the basis of a *double-valued representation*. To each group operation there corresponds not one but two matrices, whose characters have different signs.

## 10.2 DOUBLE-VALUED REPRESENTATIONS

### 10.2.1 The concept of a double group

It is possible to obtain double-group representations using the methods developed by Bethe [20, 39]. Let us introduce a new operation $\hat{Q}$ of the group, namely rotation about an axis through an angle $2\pi$, so that $\hat{Q}^2 = \hat{E}$. Now $\hat{C}_n$ rotations provide an identity transformation when applied $2n$ times:

$$\hat{C}_n^n = \hat{Q}, \qquad \hat{C}_n^{2n} = \hat{E}.$$

Also

$$\hat{I}^2 = \hat{I},$$

but

$$\hat{\sigma}^2 = \hat{Q}, \qquad \hat{\sigma}^4 = \hat{E}.$$

The operation $\hat{Q}$ commutes with all point-group operations $\hat{R}$. Multiplying the operations $\hat{R}$ by $\hat{Q}$, we obtain a set of operations: $\hat{R}$ and $\hat{R}\hat{Q}$. The new set of operations thus obtained is called a *double group*. The order of the double group is twice that of the original group. Double-valued representations of the original point group are single-valued (i.e. conventional) representations of the double group.

### 10.2.2 Classes of double groups

In order to determine double-group irreducible representations, it is necessary to distribute the operations over the double-group classes. Let us apply the rules for determining classes (Section 2.3) to double-group operations. The operation $\hat{Q}$ commutes with all the operations and is a one-dimensional class itself. The operations $\hat{C}_n^k$ and $\hat{C}_n^{n-k}\hat{Q} \equiv \hat{C}_n^{2n-k}$ are conjugate when the $C_n$ axis is two-sided. Therefore the rotation groups include two one-dimensional classes $\hat{E}$ and $\hat{Q}$, and the two-dimensional classes $\hat{C}_2$ and $\hat{C}_2\hat{Q}$, $\hat{C}_n$ and $\hat{C}_n^{n-1}\hat{Q}$, $\hat{C}_n^m$ and $\hat{C}_n^{n-m}\hat{Q}$. As in Section 3.6, it is possible to determine the number of irreducible representations and their dimensions. Double groups are indicated by a prime. Thus a double **O** group is denoted by **O**′, and a double $\mathbf{D}_4$ by $\mathbf{D}_4'$.

**Example 10.1** The group $\mathbf{D}_4$ consists of eight operations: $\hat{E}$, $\hat{C}_4$, $\hat{C}_4^s$, $\hat{C}_2$, $2\hat{C}_2'$, $2\hat{C}_2''$. By multiplying these operations by $\hat{Q}$ we obtain 16 operations of the double group $\mathbf{D}_4'$ distributed over 7 classes:

$$\begin{array}{ccccccc} \hat{E} & \hat{Q} & \hat{C}_4 & \hat{C}_4^3 & \hat{C}_2 & 2\hat{C}_2' & 2\hat{C}_2'' \\ & & \hat{C}_4^3\hat{Q} & \hat{C}_4\hat{Q} & \hat{C}_2\hat{Q} & 2\hat{C}_2'\hat{Q} & 2\hat{C}_2''\hat{Q} \end{array}$$

The dimensions of the irreducible representations satisfy the equation

$$g_1^2 + g_2^2 + g_3^2 + g_4^2 + g_5^2 + g_6^2 + g_7^2 = 16$$

That is, $g_i$ are the numbers

$$1, \quad 1, \quad 1, \quad 1, \quad 2, \quad 2, \quad 2.$$

The group $\mathbf{D}_4'$ contains four one-dimensional and three two-dimensional irreducible representations. The one-dimensional representations ($A_1$, $A_2$, $B_1$, $B_2$) and one (of the three) two-dimensional representation (E) belong to the original group $\mathbf{D}_4$ (Section 4.3.2). The new (i.e. the double-valued) representations are two two-dimensional representations.

**Example 10.2** The operations of the double group $\mathbf{D}_3'$ are distributed in classes as follows:

$$\hat{E} \quad \hat{Q} \quad \hat{C}_3 \quad \hat{C}_3^2 \quad 3\hat{C}_2 \quad 3\hat{C}_2\hat{Q} \quad \hat{C}_3^2\hat{Q}$$

The equation for the dimensions,

$$g_1^2 + g_2^2 + g_3^2 + g_4^2 + g_5^2 + g_6^2 = 12,$$

provides the numbers 1, 1, 1, 1, 2, 2.

Since $\mathbf{D}_3$ contains three representations $A_1$, $A_2$ and E (Section 3.7.4), there are one two-dimensional and two one-dimensional double-valued representations (we shall see below that the latter are complex-conjugate).

### 10.2.3 Character tables of the double groups

The characters of all double-valued representations of point groups are given in the tables in [1], the general structure of which is described in Section 3.7. Explanations are needed only for certain pieces of notations whose meaning is obvious from the given examples.

**Example 10.3** The characters of the original group $\mathbf{D}_4$ are given in the example 4.1 and for the double group $\mathbf{D}_4'$ they are listed in Table 10.1. Similarly to the original

**Table 10.1** Characters of $\mathbf{D}_4'$.

| $\mathbf{D}_4'$ | $\hat{E}$ | $\hat{Q}$ | $\hat{C}_4$ $\hat{C}_4^3\hat{Q}$ | $\hat{C}_4^3$ $\hat{C}_4\hat{Q}$ | $\hat{C}_2$ $\hat{C}_2\hat{Q}$ | $2\hat{C}_2'$ $2\hat{C}_2'\hat{Q}$ | $2\hat{C}_2''$ $2\hat{C}_2''\hat{Q}$ | $f^{(\Gamma)}$ |
|---|---|---|---|---|---|---|---|---|
| $A_1\ \Gamma_1$ | 1 | 1 | 1 | 1 | 1 | 1 | 1 | $R$ |
| $A_2\ \Gamma_2$ | 1 | 1 | 1 | 1 | 1 | $-1$ | $-1$ | $z;\ R_z$ |
| $B_1\ \Gamma_3$ | 1 | 1 | $-1$ | $-1$ | 1 | 1 | $-1$ | $x^2-y^2$ |
| $B_2\ \Gamma_4$ | 1 | 1 | $-1$ | $-1$ | 1 | $-1$ | 1 | $xy$ |
| $E\ \Gamma_5$ | 2 | 2 | 0 | 0 | $-2$ | 0 | 0 | $R_x,\ R_y$ |
| $E_{1/2}\ \Gamma_6$ | 2 | $-2$ | $\sqrt{2}$ | $-\sqrt{2}$ | 0 | 0 | 0 | $\Phi(\frac{1}{2},\frac{1}{2}),\ \Phi(\frac{1}{2},-\frac{1}{2})$ |
| $E_{3/2}\ \Gamma_7$ | 2 | $-2$ | $-\sqrt{2}$ | $\sqrt{2}$ | 0 | 0 | 0 | $\Gamma_3\times\Gamma_6$ |

group $\mathbf{D}_4$, the column headings show operations divided into seven classes of the double group $\mathbf{D}_4'$. The characters of the $\mathbf{D}_4$ group operations $\hat{R}$ are the same as in Table of characters for original group $\mathbf{D}_4$ (Section 4.3.2), and the characters of the operations $\hat{R}\hat{Q}$ for conventional (single-valued) representations coincide with the characters $\chi(\hat{R})$. The characters of the two double-valued representations occupy the two bottom rows separated by a horizontal line. These representations are denoted by $E_{1/2}$ and $E_{3/2}$ in the Mulliken system and by $\Gamma_6$ and $\Gamma_7$ in the Bethe system. The basis functions of $E_{1/2}$ are $\Phi(\frac{1}{2},\frac{1}{2})$ and $\Phi(\frac{1}{2},-\frac{1}{2})$, where $\Phi(sm)$ are the spin functions of a particle of spin $s=\frac{1}{2}$, determined in Section 5.4.5, (5.24), and denoted by $\alpha(m=\frac{1}{2})$ and $\beta(m=-\frac{1}{2})$ (Section 5.3). The basis $E_{3/2}$ is the direct product $B_1\times E_{1/2}$ ($\Gamma_3\times\Gamma_6$); this basis can be formed by two functions $\Phi(\frac{3}{2}m)$ with $m=\frac{3}{2}$ and $m=-\frac{3}{2}$ (a more detailed description is given later).

**Example 10.4** Table 10.2 gives only the double-valued representations of $\mathbf{D}_3'$. The $E_{1/2}$ basis (an alternative notation is $\bar{E}$) is formed by the spin functions $\alpha$ and $\beta$; $E_{3/2}$ ($\bar{A}_1$ and $\bar{A}_2$) are two one-dimensional complex-conjugate representations with bases of $\Phi(\frac{3}{2}m)$ functions. Physically they correspond to the same energy level.

The cubic groups $\mathbf{O}$ and $\mathbf{T}_d$ contain three double-valued representations $\Gamma_6(E_{1/2})$, $\Gamma_7(E_{5/2})$ and $\Gamma_8(G_{3/2})$ each. The representation $\Gamma_8$ is four-

**Table 10.2** Characters of double-valued representations of $\mathbf{D}_3$.

| $\mathbf{D}_3'$ | $\hat{E}$ | $\hat{Q}$ | $\hat{C}_3$ $\hat{C}_3^2\hat{Q}$ | $\hat{C}_3^2$ $\hat{C}_3\hat{Q}$ | $3\hat{C}_2'$ | $3\hat{C}_2'\hat{Q}$ | $f^{(\Gamma)}$ |
|---|---|---|---|---|---|---|---|
| . . | . | . | . | . | . | . | . |
| $E_{1/2}\ \bar{E}\ \Gamma_4$ | 2 | $-2$ | 1 | $-1$ | 0 | 0 | $\Phi(\frac{1}{2},\frac{1}{2}),\ \Phi(\frac{1}{2},-\frac{1}{2})$ |
| $E_{3/2}$ $\bar{A}_1\ \Gamma_5$ | 1 | $-1$ | $-1$ | 1 | i | $-$i | $\Phi(\frac{3}{2},-\frac{3}{2})-\mathrm{i}\Phi(\frac{3}{2},\frac{3}{2})$ |
| $E_{3/2}$ $\bar{A}_2\ \Gamma_6$ | 1 | $-1$ | $-1$ | 1 | $-$i | i | $-\Phi(\frac{3}{2},\frac{3}{2})-\mathrm{i}\Phi(\frac{3}{2},-\frac{3}{2})$ |

dimensional, with basis functions $\Phi(\frac{3}{2}m)$ $(m = \pm\frac{3}{2}, \pm\frac{1}{2})$, the basis functions of $\Gamma_6$ are $\Phi(\frac{1}{2}, -\frac{1}{2})$, while $\Gamma_7 = \Gamma_6 \times A_2$.

Thus, taking account of the spin in cubic groups, a fourfold degeneracy of the levels is possible. In double groups having an inversion centre the same representations are divided into even and odd ones ($E_{1/2u}$, $E_{1/2g}$, $E_{5/2u}$, $E_{5/2g}$ and $G_{3/2u}$, $G_{3/2g}$).

## 10.3 REDUCTION OF DOUBLE-VALUED REPRESENTATIONS. THE KRAMERS THEOREM

As in the case of conventional representations, double-valued representations are reduced when the symmetry decreases. Reduction is performed by means of (4.12), where $\hat{R}$ are the operations of a double low-symmetry group **G**, $\Gamma_i$ are their double-valued irreducible representations, and the characters $\chi(\Gamma)$ of the high-symmetry group $\mathbf{G}_0$ and $\chi(\Gamma_i)$ are determined from tables [1]. We shall illustrate the calculation methods by way of examples.

**Example 10.5** Let us consider the splitting of a one-electron d level in a cubic field **O**. For one d electron $l = 2$ and $s = \frac{1}{2}$. Hence the total angular momentum $j = l + \frac{1}{2}$ and $l - \frac{1}{2}$; that is, $j = \frac{5}{2}$ and $\frac{3}{2}$. The spin–orbit interaction splits the $^2D$ term of a free atom into two fine-structure sublevels $^2D_{5/2}$ and $^2D_{3/2}$ (Fig. 10.1). These sublevels are sixfold and fourfold-degenerate respectively. The characters of the double-valued irreducible representations **O**$'$ and the characters of the reducible representations $\Gamma_{5/2}$ and $\Gamma_{3/2}$ are given in Table 10.3, from which it can be seen that $\Gamma_{3/2}$ is the irreducible representation of $G_{3/2}$ ($\Gamma_8$), while $\Gamma_{5/2}$ is reduced by means of (4.12) according to the scheme

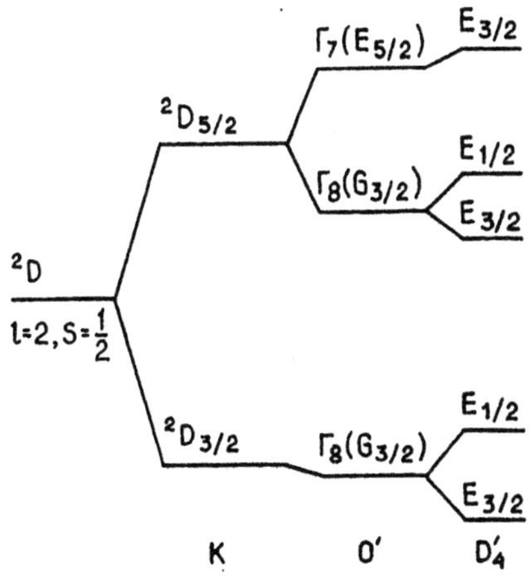


**Fig. 10.1** Splitting of an atomic $^2D$ term by a strong spin–orbit interaction and a crystal field.

**Table 10.3** Characters of the double-valued representations of **O** and the characters of $\Gamma_{5/2}$ and $\Gamma_{3/2}$.

| $\mathbf{O}'$ | | $\hat{E}$ | $\hat{Q}$ | $4C_3$<br>$4C_3^2 Q$ | $4\hat{C}_3^2$<br>$4\hat{C}_3 \hat{Q}$ | $3\hat{C}_4^2$<br>$3\hat{C}_4^2 \hat{Q}$ | $3\hat{C}_4$<br>$3\hat{C}_4^3 \hat{Q}$ | $3\hat{C}_4^3$<br>$3\hat{C}_4 \hat{Q}$ | $6\hat{C}_2'$<br>$6\hat{C}_2' \hat{Q}$ |
|---|---|---|---|---|---|---|---|---|---|
| . | . | . | . | . | . | . | . | . | . |
| $E_{1/2}$ | $\Gamma_6$ | 2 | −2 | 1 | −1 | 0 | $\sqrt{2}$ | $-\sqrt{2}$ | 0 |
| $E_{5/2}$ | $\Gamma_7$ | 2 | −2 | 1 | −1 | 0 | $-\sqrt{2}$ | $\sqrt{2}$ | 0 |
| $G_{3/2}$ | $\Gamma_8$ | 4 | −4 | −1 | 1 | 0 | 0 | 0 | 0 |
| | $\Gamma_{3/2}$ | 4 | −4 | −1 | 1 | 0 | 0 | 0 | 0 |
| | $\Gamma_{5/2}$ | 6 | −6 | 0 | 0 | 0 | $-\sqrt{2}$ | $\sqrt{2}$ | 0 |

$\Gamma_{5/2} \doteq E_{5/2} + G_{3/2}(\Gamma_7 + \Gamma_8)$. Therefore in a cubic field the atomic quadruplet $^2D_{3/2}$ remains degenerate, while $^2D_{5/2}$ is split into a doublet $E_{5/2}(\Gamma_7)$ and a quadruplet $G_{3/2}(\Gamma_8)$ (Fig. 10.1).

**Example 10.6** Let us consider the splitting of a one-electron d level in a tetragonal $\mathbf{D}_4$ field. The reducible $\Gamma_{3/2}$ and $\Gamma_{5/2}$ representations of the double group $\mathbf{D}_4'$ are given in Table 10.4. Decomposing these into irreducible representations, we obtain

$$\Gamma_{3/2} \doteq E_{1/2} + E_{3/2} \equiv \Gamma_6 + \Gamma_7, \qquad \Gamma_{5/2} \doteq E_{1/2} + 2E_{3/2} \equiv \Gamma_6 + 2\Gamma_7.$$

The $^2D_{3/2}$ level is split into two doublets, while the $^2D_{5/2}$ level is split into three doublets.

Since $\mathbf{D}_4'$ is a subgroup of $\mathbf{O}'$, reduction of the rotation group representations $D^{(3/2)}$ and $D^{(5/2)}$ can be performed sequentially in two steps ($\mathbf{K} \to \mathbf{O}' \to \mathbf{D}_4'$). Indeed, applying the reduction formula to the representations $E_{5/2}$ and $G_{3/2}$ of $\mathbf{O}'$, we obtain

$$E_{5/2} \doteq E_{3/2}, \qquad G_{3/2} \doteq E_{1/2} + E_{3/2}.$$

Hence the step-by-step reduction scheme has the following form (corresponding energy levels are sketched in Fig. 10.1).

| | | | | | |
|---|---|---|---|---|---|
| $\mathbf{K}$ | $D^{(3/2)}$ | | $D^{(5/2)}$ | | |
| | ↓ | | ↙ | ↘ | |
| $\mathbf{O}'$ | $G_{3/2}$ | | $E_{5/2}$ | $G_{3/2}$ | |
| | ↙ | ↘ | ↓ | ↙ | ↘ |
| $\mathbf{D}_4'$ | $E_{1/2}$ | $E_{3/2}$ | $E_{3/2}$ | $E_{1/2}$ | $E_{3/2}$ |

**Table 10.4**

| $\mathbf{D}_4'$ | $\hat{E}$ | $\hat{Q}$ | $2\hat{C}_4$ | $2\hat{C}_4\hat{Q}$ | $2\hat{C}_2$ | $4C_2'$ | $4C_2''$ |
|---|---|---|---|---|---|---|---|
| $\Gamma_{3/2}$ | 4 | −4 | 0 | 0 | 0 | 0 | 0 |
| $\Gamma_{5/2}$ | 6 | −6 | $-\sqrt{2}$ | $\sqrt{2}$ | 0 | 0 | 0 |

In practice, the reduction of double-valued representations does not require any calculation (the examples of this section simply illustrating the method), since all the results are collected in the tables in [1]. Table 10.5 shows the results for the reduction of the representations $D_J$ (in the notation of [1]) with the half-integer $J$ for the group $\mathbf{T}_d$. In contrast with Table 5.4 (for integer $J$) this reduction of $D_J$ leads to double-valued representations. Since $\mathbf{T}_d$ is a non-centrosymmetric group, these representations do not contain a parity symbol. On the other hand, the original representations $D_J$ of the full group $\mathbf{K}_h$ may be either even ($D_J^+$) or odd ($D_J^-$), the result of reduction being dependent on the parity of the original representation ($D_{5/2}^+ \doteq \Gamma_7 + \Gamma_8$, but $D_{5/2}^- \doteq \Gamma_6 + \Gamma_8$). In double groups with an inversion centre the $D_J^+$ reduction gives even double-valued representations, while $D_J^-$ gives odd representations of the same type. Thus, for example, in $\mathbf{O}_h$ we have $D_{5/2}^{\pm} \doteq \Gamma_7^{\pm} + \Gamma_8^{\pm}$.

Similar tables (Section 4.37) are used for the reduction of double-valued representations of the 32 point groups on decrease in symmetry. As an example we shall take (from [1]) the branching table of the double-valued representations $\mathbf{O}_h$, which is a continuation of Table 4.4. As for conventional representations, we can determine from Table 10.5 the energy-level splitting corresponding to double-valued representations on decrease in symmetry, in particular, on application of an external field. However, application of the Wigner theorem (Section 4.2) in the case of double-valued representations calls for special treatment. In this connection we shall present without proof the following statements:

(1) The electric field can remove degeneracy entirely only in a system with an even number of electrons. Here the spin $S$ and the total angular momentum $J$ are integers, while the irreducible representations of the point group are conventional (i.e. single-valued).

(2) A system with an odd number of electrons has a half-integer spin as well as a total angular momentum. The wavefunctions transform according to double-valued representations, and any electric field (either the crystal

**Table 10.5** Representation reduction of the full rotation group $\mathbf{K}_h$ in $\mathbf{T}_d$ for half-integer $J$.

| | | | |
|---|---|---|---|
| $D_{1/2}^+$ | $\Gamma_6$ | $D_{1/2}^-$ | $\Gamma_7$ |
| $D_{3/2}^+$ | $\Gamma_8$ | $D_{3/2}^-$ | $\Gamma_8$ |
| $D_{5/2}^+$ | $\Gamma_7 + \Gamma_8$ | $D_{5/2}^-$ | $\Gamma_6 + \Gamma_8$ |
| $D_{7/2}^+$ | $\Gamma_6 + \Gamma_7 + \Gamma_8$ | $D_{7/2}^-$ | $2\Gamma_6 + \Gamma_7 + \Gamma_8$ |
| $D_{9/2}^+$ | $\Gamma_6 + 2\Gamma_8$ | $D_{9/2}^-$ | $2\Gamma_7 + 2\Gamma_8$ |
| $D_{11/2}^+$ | $\Gamma_6 + \Gamma_7 + 2\Gamma_8$ | $D_{11/2}^-$ | $2\Gamma_6 + \Gamma_7 + 2\Gamma_8$ |
| $D_{13/2}^+$ | $\Gamma_6 + 2\Gamma_7 + 2\Gamma_8$ | $D_{13/2}^-$ | $2\Gamma_6 + \Gamma_7 + 2\Gamma_8$ |

field or an arbitrary oriented external electric field) will maintain the twofold degeneracy of energy levels. This statement (proved by Kramers and Wigner [74]) is called the *Kramers theorem* and the corresponding twofold degeneracy the *Kramers degeneracy*.

Let us consider Table 10.6 from this point of view. The electric field $\mathscr{E}(z)$ along the tetragonal axis reduces the symmetry $\mathbf{O} \to \mathbf{C}_{4v}$, while the twofold representations $\Gamma_6^{\pm}$, $\Gamma_7^{\pm}(\mathbf{O})$ remain degenerate ($\Gamma_6$, $\Gamma_7(\mathbf{C}_{4v})$) and the quadruplet $\Gamma_8^{\pm}$ is split into two doublets $\Gamma_6$ and $\Gamma_7$. In a trigonal field $\mathscr{E}(w)$ (group $\mathbf{C}_{3v}$) $\Gamma_8^{\pm}$ is split according to the scheme $\Gamma_4 + \Gamma_5 + \Gamma_6$, with $\Gamma_4$ being twofold and $\Gamma_5$, $\Gamma_6$ being complex-conjugate representations (see the characters of $\mathbf{C}_{3v}$). To the latter, according to the Kramers theorem, there corresponds a doublet energy level (Sections 3.7.3 and 4.2).

(3) One-dimensional double-valued representations should be doubled even if their characters are real. This means that, in the absence of a magnetic field, two independent wavefunctions constituting the basis of a single one-dimensional representation belong to a twofold-degenerate energy level. These one-dimensional representations occur in the rotation groups $\mathbf{C}_n'$. Thus $\mathbf{C}_3'$ includes one-dimensional complex-conjugate representations $\Gamma_4(\Phi(\frac{1}{2}, \frac{1}{2}))$ and $\Gamma_5(\Phi(\frac{1}{2}, -\frac{1}{2}))$ and a one-dimensional real representation $\Gamma_6$. The basis of the real representation is a spin function $\Phi(\frac{3}{2}, \frac{3}{2})$ or $\Phi(\frac{3}{2}, -\frac{3}{2})$ (see the characters in [1]). It is clear that, in the absence of a magnetic field, the spin states with $m = \frac{3}{2}$ and $m = -\frac{3}{2}$ have equal energy.

**Table 10.6** Branching scheme of the double-valued representations of $\mathbf{O}_h'$.

| $\mathbf{O}_h'$ | $\Gamma_6^{\pm}$ | $\Gamma_7^{\pm}$ | $\Gamma_8^{\pm}$ |
|---|---|---|---|
| $\mathbf{O}$ | $\Gamma_6$ | $\Gamma_7$ | $\Gamma_8$ |
| $\mathbf{T}_d$ | $\Gamma_6$ | $\Gamma_7$ | $\Gamma_8$ |
| $\mathbf{T}_h$ | $\Gamma_5^{\pm}$ | $\Gamma_5^{\pm}$ | $\Gamma_6 + \Gamma_7$ |
| $\mathbf{D}_{4h}$ | $\Gamma_6^{\pm}$ | $\Gamma_7^{\pm}$ | $\Gamma_6^{\pm} + \Gamma_7^{\pm}$ |
| $\mathbf{D}_{3d}$ | $\Gamma_4^{\pm}$ | $\Gamma_4^{\pm}$ | $\Gamma_4^{\pm} + \Gamma_5^{\pm} + \Gamma_6^{\pm}$ |
| $\mathbf{C}_{4h} : \mathscr{H}(z)$ | $\Gamma_5^{\pm} + \Gamma_6^{\pm}$ | $\Gamma_7^{\pm} + \Gamma_8^{\pm}$ | $\Gamma_5^{\pm} + \Gamma_6^{\pm} + \Gamma_7^{\pm} + \Gamma_8^{\pm}$ |
| $\mathbf{C}_{2h} : \mathscr{H}(v)$ | $\Gamma_3^{\pm} + \Gamma_4^{\pm}$ | $\Gamma_3^{\pm} + \Gamma_4^{\pm}$ | $2\Gamma_3^{\pm} + 2\Gamma_4^{\pm}$ |
| $\mathbf{C}_{3i} : \mathscr{H}(w)$ | $\Gamma_4^{\pm} + \Gamma_5^{\pm}$ | $\Gamma_4^{\pm} + \Gamma_5^{\pm}$ | $\Gamma_4^{\pm} + \Gamma_5^{\pm} + 2\Gamma_6^{\pm}$ |
| $\mathbf{C}_{4v} : \mathscr{E}(z)$ | $\Gamma_6$ | $\Gamma_7$ | $\Gamma_6 + \Gamma_7$ |
| $\mathbf{C}_{2v} : \mathscr{E}(v)$ | $\Gamma_5$ | $\Gamma_5$ | $2\Gamma_5$ |
| $\mathbf{C}_{3v} : \mathscr{E}(w)$ | $\Gamma_4$ | $\Gamma_4$ | $\Gamma_4 + \Gamma_5 + \Gamma_6$ |

(4) A magnetic field removes the degeneracy completely, and to each one-dimensional representation (double-valued or conventional) there corresponds a single nondegenerate energy level. This also applies to one-dimensional real double-valued representations. In other words, the Kramers degeneracy is removed only by a magnetic field.

It follows from Table 10.5, for example, that in a magnetic field $\mathscr{H}(z)$ the $\Gamma_8^+(\mathbf{O}')$ level is split into four sublevels: $\Gamma_5^+$, $\Gamma_6^+$, $\Gamma_7^+$ and $\Gamma_8^+(\mathbf{C}'_{4h})$. From tables of the group characters and $D^+_{3/2}$ reduction for $\mathbf{C}'_{4h}$ it is easy to determine that the basis functions of these representations are the functions $\Phi(\frac{3}{2}, \frac{1}{2})$, $\Phi(\frac{3}{2}, -\frac{1}{2})$, $\Phi(\frac{3}{2}, -\frac{3}{2})$ and $\Phi(\frac{3}{2}, \frac{3}{2})$ respectively. This result corresponds to ordering the spin energy levels in a magnetic field according to the spin projection $m$. In contrast to the case of a magnetic field, in an electric field the energies of the states $\Phi(\frac{3}{2}, m)$ and $\Phi(\frac{3}{2}, -m)$ are equal; therefore in a field $\mathscr{E}(z)$ $(\mathbf{C}'_{4v})$ two Kramers doublets, $\Gamma_6(\Phi(\frac{3}{2}, \pm\frac{1}{2}))$ and $\Gamma_7(\Phi(\frac{3}{2}, \pm\frac{3}{2}))$ are formed.

## PROBLEMS

**10.1** Distribute the $\mathbf{D}'_{3d}$ double-group operations over classes.

**10.2** Determine the total angular momentum values $j$ for one f electron and classify states in crystal fields with symmetry $\mathbf{O}_h$, $\mathbf{D}_{4h}$, $\mathbf{D}_{3d}$ and $\mathbf{D}_{2h}$.

**10.3** Using the equations for rotation-group characters and reduction of reducible representations, determine the crystal states in $\mathbf{D}_{4h}$ and $\mathbf{C}_{3v}$ for $J = \frac{1}{2}, \frac{3}{2}, \frac{5}{2}, \frac{7}{2}$ and $\frac{9}{2}$.

**10.4** Determine the branching scheme of the double-valued representations of $\mathbf{D}_{4h}$.

# 11 Spin–Orbit Interaction in Crystal Fields

## 11.1 CLASSIFICATION OF FINE-STRUCTURE LEVELS

### 11.1.1 One-electron terms in a cubic field

So far, in classifying ion levels in crystal fields (Sections 4.2 and 4.3) the spin and spin–orbit interaction have been neglected. Let us consider the spin–orbit splitting in the simplest case of one d electron in an **O** cubic field (Section 4.3.5). If the cubic field is stronger than the spin–orbit interaction, it should be taken into account at the first stage. We then obtain two levels: $^2T_2$ (the ground level) and $^2E$ (the excited level). With allowance for spin, the orbital triplet $^2T_2$ is sixfold-degenerate—it has six corresponding wavefunctions of the form (see the notation in Section 5.5)

$$\xi\alpha, \quad \eta\alpha, \quad \zeta\alpha, \quad \xi\beta, \quad \eta\beta, \quad \zeta\beta.$$

This notation shows clearly that the six functions of the $^2T_2$ term form a direct product basis comprising two sets: a cubic $T_2$ basis and two functions $\alpha$ and $\beta$ from the $D^{(1/2)}$ basis of the rotation group. This direct product is denoted by $T_2 \times D^{(1/2)}$. Since the double group **O**$'$ has no six-dimensional irreducible representations, the direct product $T_2 \times D^{(1/2)}$ is reducible. Let us decompose the reducible representation $T_2 \times D^{(1/2)}$ into its irreducible components. It should be noted here that in **O**$'$ the $D^{(1/2)}$ basis gives the irreducible representation $E_{1/2}(\Gamma_6)$ (Section 10.2). It is now easy to construct the direct product $T_2 \times E_{1/2}$. The characters of this direct product are obtained according to (5.12):

$$\chi^{(T_2 \times E_{1/2})}(\hat{R}) = \chi^{(T_2)}(\hat{R})\chi^{(E_{1/2})}(\hat{R}),$$

where $\hat{R}$ are *double-group* operations. From the character tables (Section 4.3.4 and Table 10.3) we obtain Table 11.1.

**Table 11.1** Characters of the direct product $T_2 \times E_{1/2}$.

| $\mathbf{O}'$ | $\hat{E}$ | $\hat{Q}$ | $4\hat{C}_3$ $4\hat{C}_3^2\hat{Q}$ | $4\hat{C}_3^2$ $4\hat{C}_3\hat{Q}$ | $3\hat{C}_4^2$ $3\hat{C}_4^2\hat{Q}$ | $3\hat{C}_4$ $3\hat{C}_4^3\hat{Q}$ | $3\hat{C}_4^3$ $3\hat{C}_4\hat{Q}$ | $6\hat{C}_2'$ $6\hat{C}_2'\hat{Q}$ |
|---|---|---|---|---|---|---|---|---|
| $T_2$ | 3 | 3 | 0 | 0 | −1 | −1 | −1 | 1 |
| $E_{1/2}$ | 2 | −2 | 1 | −1 | 0 | $\sqrt{2}$ | $-\sqrt{2}$ | 0 |
| $T_2 \times E_{1/2}$ | 6 | −6 | 0 | 0 | 0 | $-\sqrt{2}$ | $\sqrt{2}$ | 0 |

Using the reduction equation (4.12), the number of times $\Gamma_i$ appears is

$$a(\Gamma_i) = \frac{1}{g}\sum_{\hat{R}} \chi^{(T_2\times E_{1/2})}(\hat{R})\chi^{(\Gamma_i)}(\hat{R}), \tag{11.2}$$

where $g = 48$ is the order of $\mathbf{O}'$. In this way we find

$$T_2 \times E_{1/2} \doteq E_{1/2} + G_{3/2}. \tag{11.3}$$

According to the Wigner theorem each irreducible representation is related to a certain energy level; in other words, *the spin–orbit interaction splits the* $^2T_2$ *term into a doublet* $E_{1/2}(\Gamma_6)$ *and a quadruplet* $G_{3/2}(\Gamma_8)$ (see Table 10.3), called *fine-structure levels* (Fig. 11.1). The fine-structure levels in a free atom are enumerated by the total angular momentum $J$ (or $j$), and in a crystal field by point-group irreducible representations (which are double-valued for odd numbers of electrons).

Similarly, we can consider the excited term $^2E$, for which we obtain

$$E \times E_{1/2} \doteq G_{3/2}(\Gamma_8). \tag{11.4}$$

The existence of one irreducible representation of the double group indicates that the cubic field term $^2E$ is not split by the spin–orbit interaction. Generalization of these conclusions for a real cubic system with inversion

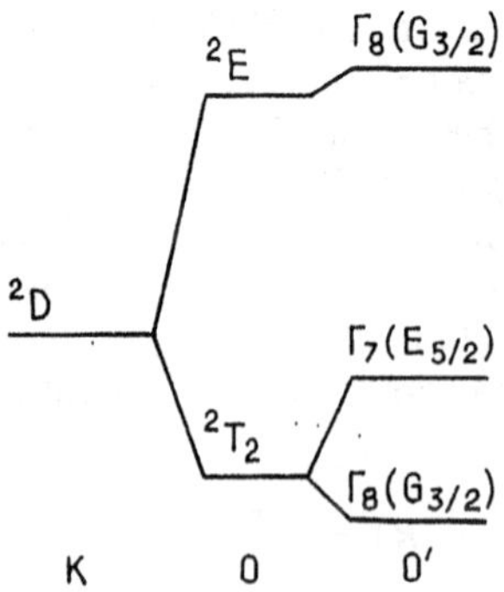


**Fig. 11.1** Fine structure of the one-electron cubic term.

centre ($\mathbf{O}'_h$) is simple:

$$T_{2g} \times E_{1/2g} = E_{1/2g} + G_{3/2g}(\Gamma_6 + \Gamma_8), \qquad E_g \times E_{1/2g} = G_{3/2g}(\Gamma_8).$$

### 11.1.2 One-electron terms in low-symmetry fields

Before formulating general rules, let us consider more complicated versions of the above examples. Let us imagine that the octahedral complex ($\mathbf{O}'$) is distorted tetragonally ($\mathbf{D}'_4$). If the tetragonal field is weaker than the spin–orbit interaction, it is possible to consider splitting of the fine-structure levels (produced by spin–orbit interaction in strong cubic crystal field) by the weak tetragonal field. For multiplets $^2T_2$ and $^2E$ (Fig. 11.1), under the reduction $\mathbf{O}' \rightarrow \mathbf{D}'_4$, we obtain

$$^2T_2 \begin{cases} \Gamma_7 \doteq \Gamma_7 \\ \Gamma_8 \doteq \Gamma_6 + \Gamma_7 \end{cases} \qquad {}^2E \;\{\, \Gamma_8 \doteq \Gamma_6 + \Gamma_7$$

If tetragonal components of the crystal field (reducing symmetry from $\mathbf{O}$ to $\mathbf{D}_4$) are strong in comparison with the spin–orbit interaction, it seems to be reasonable to take into account the crystal field and not focusing attention on the spin and spin–orbit interaction as the first consideration. In other words, in a strong tetragonal field it is convenient to proceed from the tetragonal components of $^2T_2$ following to the reduction scheme ($\mathbf{O} \rightarrow \mathbf{D}_4$) only of orbital ($T_2$) components of the full wavefunction in terms of the conventional representations

$$^2T_2 \doteq {}^2B_2 + {}^2E.$$

Spin wavefunctions $\phi(\frac{1}{2}, \frac{1}{2}), \phi(\frac{1}{2}, -\frac{1}{2})$ are basis functions of the irreducible representation $\Gamma_6$ in $\mathbf{D}'_4$ ($D^{(1/2)} \doteq \Gamma_6$ in $\mathbf{D}'_4$). Decomposing the direct products $B_2 \times \Gamma_6$ and $E \times \Gamma_6$ in $\mathbf{D}'_4$ we obtain the irreducible representations labelling the fine structure level of strong tetragonal crystal field components

$$^2B_2 \rightarrow \Gamma_7, {}^2E \rightarrow \Gamma_6 + \Gamma_7.$$

This result of course, coincides with the result already obtained. The transition from a weak to a strong spin–orbit interaction is expressed by the *correlation diagram* given in Fig. 11.2.

As was shown earlier (e.g. (11.4)) spin–orbit interaction does not split the cubic $^2E$ term since only one irreducible representation $\Gamma_8$ of $\mathbf{O}'$ exists in the direct product $E \times D^{(1/2)}$. Let us consider the behaviour of the cubic $^2E(\Gamma_8)$ term in a trigonal (say, $\mathbf{D}_3$ symmetry) crystal field. To find the number and specification of the fine structure level in a trigonal field we must reduce the cubic ($\mathbf{O}'$) irreducible representation $\Gamma_8$ in the trigonal ($\mathbf{D}'_3$) group. Using the general equation (4.12) for decomposition of representation and characters of

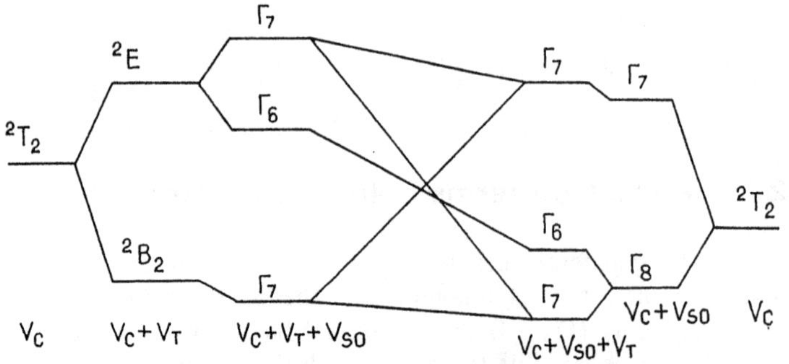


**Fig. 11.2** Correlation diagram of the fine structure of the $^2T_2$ term in a tetragonal distorted complex ($V_c$ is a cubic field, $V_t$ is a tetragonal field and $V_{So}$ is the spin–orbit interaction).

$\mathbf{O}'$ (Table 10.3) and $\mathbf{D}_3'$ (Table 10.2) we can easily find

$$\Gamma_8 \doteq \Gamma_4 + \Gamma_5 + \Gamma_6 (\equiv \bar{E} + \bar{A}_1 + \bar{A}_2)$$

Hence the $^2$E term in the trigonal field is split into two fine-structure doublets $\Gamma_4$ and $\Gamma_5$, $\Gamma_6$ (complex-conjugate $\mathbf{D}_3'$ group representations). This result is interesting in that, on the one hand, the trigonal field itself does not split the cubic orbital doublet E (see Section 4.3.6), and, on the other, the spin–orbit interaction does not split the cubic term $^2$E. Therefore, it is clear that fine structure is provided by the combined effect of the trigonal field and the spin–orbit interaction. We shall return to this problem below.

### 11.1.3 Many-electron terms

Classification of many-electron terms is carried out according to the same scheme. In a two-electron system the total spin $S = 0$ and 1; that is, $^1\Gamma$ and $^3\Gamma$ terms are possible, which are singlet and triplet according to their spin. For integer spin $S$ the representation $D^{(S)}$ consists of simple (not double-valued) point-group representations. Thus the $^3T_2$ term ($\mathbf{O}$) of a $d^2$ ion is split according to the scheme

$$D^{(1)} \times T_2 = T_1 \times T_2 \doteq A_1 + E + T_1 + T_2,$$

with the fine-structure spectrum containing the $A_2$ singlet prohibited by the Kramers theorem for half-integer spins. Similarly, in a $d^3$ ion the $^4T_2$ term gives

$$D^{(3/2)} \times T_2 = \Gamma_8 \times T_2 \doteq \Gamma_6 + \Gamma_7 + 2\Gamma_8.$$

It should be noted that many-electron wavefunctions are formed from determinants and cannot be represented directly, as at the beginning of Section 11.1.1, as products of coordinate and spin functions. However, under the action of the group operations on the electron and spin coordinates, they transform as the direct product $D^{(S)} \times \Gamma$. In fact, the symmetric (i.e. permutation) and rotation group operations *commute*. It is therefore possible to build a wavefunction from products of orbitals and spin functions and to antisymmetrize *afterwards*. This truns the products into determinants, without changing spatial symmetry properties. This makes it possible to apply the present classification method to many-electron ions. For higher spins or lower symmetry $D^{(S)}$ is reduced into several representations. Thus in **O**

$$D^{(5/2)} \doteq \Gamma_7 + \Gamma_8.$$

This means that even an orbital singlet term $^6A_1$ ($d^5$ ion, $^6S$ atomic state or $t_2^3e^2$ strong-field configuration) may have fine structure. In noncubic complexes fine structure appears for all terms with $S > \frac{1}{2}$. The orbital singlets $^2A$ (or $^2B$) retain Kramers degeneracy.

Therefore classification of fine-structure levels of $S\Gamma$-terms is performed as follows:

(1) using the tables from [1], reduce $D^{(S)}$ for the point group;

(2) then form the direct product $D^{(S)} \times \Gamma$;

(3) decomposition of this direct product into irreducible representation gives the desired answer.

## 11.2 SPIN–ORBIT SPLITTING IN ONE-ELECTRON IONS

### 11.2.1 Wavefunctions of fine-structure levels

In order to calculate the fine-structure levels of the $S\Gamma$ term, it is necessary to use the secular equation for the operator $\hat{H}_{so}$ (10.1). The basis functions of $\hat{H}_{so}$ are the wavefunctions $\Psi(\alpha S\Gamma M\gamma)$, and hence calculation of the spin–orbit interaction leads to a secular equation of dimension $(2S+1)g(\Gamma)$:

$$|\langle\alpha S\Gamma M\gamma|\hat{H}_{so}|\alpha S\Gamma M'\gamma'\rangle - \epsilon\delta_{MM'}\delta_{\gamma\gamma'}| = 0. \tag{11.5}$$

This can be substantially simplified if we use the considerations given in Chapter 5 regarding basis selection. Indeed, the basis $\Psi(\alpha S\Gamma M\gamma)$ is the direct product $D^{(S)} \times \Gamma$. By reducing it for the point group, we obtain the irreducible representations $\Gamma_i$ describing the fine-structure levels. The corresponding wavefunctions $\Phi(\Gamma_i\gamma_i)$ can be constructed by means of the Clebsch–Gordan decomposition

$$\Phi(\alpha S\Gamma\Gamma_i\gamma_i) = \sum_{\gamma_S,\gamma} \Psi(\alpha S\Gamma M\gamma)\langle\Gamma_S\gamma_S\Gamma\gamma|\Gamma_i\gamma_i\rangle. \tag{11.6}$$

where $\Gamma_S$ labels the irreducible representations (double-valued for an odd number of electrons) into which $D^{(S)}$ is reduced in the point group, and $\alpha S\Gamma$

shows the origin of the function $\Phi$. After constructing $\Phi(\alpha S\Gamma\Gamma_i\gamma_i)$ we obtain the correct wavefunctions in the zeroth-order approximation of the operator $\hat{H}_{\rm so}$ belonging to the $S\Gamma$ term. In the $\Phi(\Gamma_i\gamma_i)$ basis the secular equation (11.5) is diagonal, the fine-structure levels being expressed by the simple equation

$$\epsilon(\Gamma_i) = \langle \alpha S\Gamma\Gamma_i\gamma_i|\hat{H}_{\rm so}|\alpha S\Gamma\Gamma_i\gamma_i\rangle, \tag{11.7}$$

which holds for nonrepeating representations $\Gamma_i$.

**Example 11.1** Consider a one-electron ${}^2{\rm T}_{1{\rm u}}$ state in $\mathbf{O}_{\rm h}$:

$$D^{(1/2)} \doteq \Gamma_6{}^+({\rm E}_{1/2{\rm g}}) \equiv \Gamma_S, \qquad \Gamma_i = {\rm T}_{1{\rm u}} \times \Gamma_6{}^+ \doteq \Gamma_6{}^- + \Gamma_8{}^-.$$

In the case under consideration (11.6) has the form

$$\Phi({}^2{\rm T}_{1{\rm u}}, \Gamma_i\gamma_i) = \sum_{\gamma_S=\pm 1/2}\ \sum_{\gamma=x,y,z} \Psi({}^2{\rm T}_{1{\rm u}}, \pm\tfrac{1}{2}, \gamma)\langle \Gamma_s^+\gamma_s {\rm T}_{1{\rm u}}\gamma|\Gamma_i\gamma_i\rangle. \tag{11.8}$$

The basis vectors of $\Gamma_6{}^+$ ($\alpha \equiv \phi(\frac{1}{2},\frac{1}{2})$ and $\beta \equiv \phi(\frac{1}{2},-\frac{1}{2})$) are enumerated by the spin-projection values $\gamma_S = \pm\frac{1}{2}$ and the ${\rm T}_{1{\rm u}}$ basis is formed by three functions $\varphi_x$, $\varphi_y$ and $\varphi_z$. To construct the wavefunctions $\phi({}^2{\rm T}_{1{\rm u}}, \Gamma_i\gamma_i)$ ($\Gamma_i = \Gamma_6^-, \Gamma_8^-$) we can use the Clebsch–Gordan coefficient for the direct product ${\rm T}_1 \times \Gamma_6$ in $\mathbf{O}'$ [1] listed in Table 11.2. In the Table 11.2 the basis functions of $\Gamma_6$ and $\Gamma_8$ contained in the direct product ${\rm T}_1 \times \Gamma_6$ are denoted as $\phi(\Gamma_6, \pm\frac{1}{2})$ and $\phi(\Gamma_8, \pm\frac{3}{2})$, $\phi(\Gamma_8, \pm\frac{1}{2})$. Two functions $\phi(\Gamma_6, \pm\frac{1}{2})$ transform like spin-functions $\phi(\frac{1}{2}, \pm\frac{1}{2})$, while four functions $\phi(\Gamma_8, m)$ ($m = \pm\frac{3}{2}, \pm\frac{1}{2}$) transform like $\phi(\frac{3}{2}, m)$. Clebsch–Gordan coefficients for ${\rm T}_{1{\rm u}} \times \Gamma_6^+$ in $\mathbf{O}_{\rm h}$ are the same and we can use Table 11.2 when taking into account that the resulting wavefunctions are odd (superscript "—"). We now obtain the basis functions that reduce the direct product:

$$\left.\begin{aligned} \Phi({}^2{\rm T}_{1{\rm u}}, \Gamma_6{}^-, -\tfrac{1}{2}) &= -\sqrt{\tfrac{1}{3}}\,{\rm i}\,\varphi_x\alpha - \sqrt{\tfrac{1}{3}}\,\varphi_y\alpha + \sqrt{\tfrac{1}{3}}\,{\rm i}\,\varphi_z\beta, \\ \Phi({}^2{\rm T}_{1{\rm u}}, \Gamma_6{}^-, +\tfrac{1}{2}) &= -\sqrt{\tfrac{1}{3}}\,{\rm i}\,\varphi_x\beta + \sqrt{\tfrac{1}{3}}\,\varphi_y\beta - \sqrt{\tfrac{1}{3}}\,{\rm i}\,\varphi_z\alpha, \end{aligned}\right\}\ (\Gamma_6{}^-) \tag{11.9}$$

**Table 11.2** Clebsch–Gordan coefficients for the direct product $\Gamma_6 \times {\rm T}_1$ in $\mathbf{O}'$.

| $\Gamma_6 \times {\rm T}_1$ $\gamma_1$ | $\gamma_2$ | $\phi(\Gamma_6, -\frac{1}{2})$ | $\phi(\Gamma_6, \frac{1}{2})$ | $\phi(\Gamma_8, -\frac{3}{2})$ | $\phi(\Gamma_8, -\frac{1}{2})$ | $\phi(\Gamma_8, \frac{1}{2})$ | $\phi(\Gamma_8, -\frac{3}{2})$ |
|---|---|---|---|---|---|---|---|
| $\phi(\frac{1}{2}, -\frac{1}{2})$ | $x$ | 0 | $-{\rm i}/\sqrt{3}$ | ${\rm i}/\sqrt{2}$ | 0 | $-{\rm i}/\sqrt{6}$ | 0 |
| | $y$ | 0 | ${\rm i}/\sqrt{3}$ | ${\rm i}/\sqrt{2}$ | 0 | $1/\sqrt{6}$ | 0 |
| | $z$ | ${\rm i}/\sqrt{3}$ | 0 | 0 | ${\rm i}\sqrt{2/3}$ | 0 | 0 |
| $\phi(\frac{1}{2}, \frac{1}{2})$ | $x$ | $-{\rm i}/\sqrt{3}$ | 0 | 0 | ${\rm i}/\sqrt{6}$ | 0 | $-{\rm i}/\sqrt{2}$ |
| | $y$ | $-1/\sqrt{3}$ | 0 | 0 | $1/\sqrt{6}$ | 0 | $1/\sqrt{2}$ |
| | $z$ | 0 | $-{\rm i}/\sqrt{3}$ | 0 | 0 | ${\rm i}\sqrt{2/3}$ | 0 |

$$\left.\begin{aligned}
\Phi({}^2\mathrm{T}_{1u}, \Gamma_8^-, -\tfrac{3}{2}) &= \sqrt{\tfrac{1}{2}}\,\mathrm{i}\,\varphi_x\beta + \sqrt{\tfrac{1}{2}}\,\varphi_y\beta,\\
\Phi({}^2\mathrm{T}_{1u}, \Gamma_8^-, -\tfrac{1}{2}) &= \sqrt{\tfrac{1}{6}}\,\mathrm{i}\,\varphi_x\alpha + \sqrt{\tfrac{1}{6}}\,\varphi_y\alpha + \sqrt{\tfrac{2}{3}}\,\mathrm{i}\,\varphi_z\beta,\\
\Phi({}^2\mathrm{T}_{1u}, \Gamma_8^-, +\tfrac{1}{2}) &= -\sqrt{\tfrac{1}{6}}\,\mathrm{i}\,\varphi_x\beta + \sqrt{\tfrac{1}{6}}\,\varphi_y\beta + \sqrt{\tfrac{2}{3}}\,\mathrm{i}\,\varphi_z\alpha,\\
\Phi({}^2\mathrm{T}_{1u}, \Gamma_8^-, +\tfrac{3}{2}) &= -\sqrt{\tfrac{1}{2}}\,\mathrm{i}\,\varphi_x\alpha + \sqrt{\tfrac{1}{2}}\,\varphi_y\alpha.
\end{aligned}\right\}\ (\Gamma_8^-) \qquad (11.10)$$

For calculation of the energies $\epsilon(\Gamma_6)$ and $\epsilon(\Gamma_8)$ we may take any of the functions of (11.9) and (11.10) respectively.

**Example 11.2** For the one-electron ${}^2\mathrm{T}_{2g}(\mathrm{d}^1)$ term in $\mathbf{O}_h$ we have $\Gamma_6^+ \times \mathrm{T}_{2g} = \Gamma_7^+ + \Gamma_8^+$. Similarly, using the table of Clebsch–Gordan coefficients in [1] we obtain (we give one basis function from each of $\Gamma_7^+$ and $\Gamma_8^+$):

$$\Phi({}^2\mathrm{T}_{2g}, \Gamma_7^-, -\tfrac{1}{2}) = -\sqrt{\tfrac{1}{3}}\,\mathrm{i}\,\varphi_\xi\alpha - \sqrt{\tfrac{1}{3}}\,\varphi_\eta\alpha + \sqrt{\tfrac{1}{3}}\,\mathrm{i}\,\varphi_\zeta\beta, \qquad (11.11)$$

$$\Phi({}^2\mathrm{T}_{2g}, \Gamma_8^+, +\tfrac{3}{2}) = -\sqrt{\tfrac{1}{6}}\,\mathrm{i}\,\varphi_\xi\beta + \sqrt{\tfrac{1}{6}}\,\varphi_\eta\beta + \sqrt{\tfrac{2}{3}}\,\mathrm{i}\,\varphi_\zeta\alpha. \qquad (11.12)$$

In certain cases $\mathrm{D}^{(S)} \times \Gamma$ contains repeating representations. Thus for the cubic ${}^4\mathrm{T}_{1g}$ term of a $\mathrm{d}^3$ ion in an octahedral $\mathbf{O}$ field $\mathrm{D}^{(3/2)} \doteq \Gamma_8^+$ and $\Gamma_8^+ \times \mathrm{T}_{2g} = \Gamma_6^+ + \Gamma_7^+ + 2\Gamma_8^+$. The table of Clebsch–Gordan coefficients [1] for the product $\Gamma_8 \times \mathrm{T}_{2g}$ contains two sets, giving the functions $\Psi(\Gamma_8\gamma)$ (see Table 11.3 for $\mathbf{O}$). The basis functions of the repeating $\Gamma_8$ representations, calculated from the general equation (11.6), are orthogonal.

The existence of *repeating irreducible representations* in some direct products for double groups requires some comments about the Wigner–Eckart theorem (Section 9.2) as it relates to such cases. Suppose, for example, that we need the matrix of an operator $\hat{\mathrm{V}}_{\mathrm{T}_2\gamma}(\gamma = \xi, \eta, \zeta)$ of $\mathrm{T}_2$ type in $\mathbf{O}$ (say, components of an electric quadrupole moment, Section 9.11) connecting two basis sets of $\Gamma_8$: $\phi(\Gamma_8, m)$ and $\phi'(\Gamma_8, m')$. Matrix elements

$$\langle \Gamma_8 m | \hat{V}_{\mathrm{T}_2\gamma} | \Gamma_8' m' \rangle \equiv \int \phi^*(\Gamma_8, m)\, \hat{V}_{\mathrm{T}_2\gamma}\, \phi'(\Gamma_8, m')\, \mathrm{d}\tau$$

are proportional to the Clebsch–Gordan coefficients $\langle \Gamma_8 m | \Gamma_8 m' \mathrm{T}_2\gamma \rangle$ appearing in the expression for the Wigner–Eckart theorem (Equation (9.10)). In the case under consideration we have *two sets of coupling coefficients and two reduced matrix elements* $\langle \Gamma_8 \| \hat{V}_{\mathrm{T}_2} \| \Gamma_8 \rangle$. It should be stressed that these two reduced matrix elements are independent from the point of view of symmetry, i.e. there are no relations between them, resulting from the group theory. Remembering that Table 11.3 contains Clebsch–Gordan coefficients

$$\langle \Gamma_8 m' \mathrm{T}_2\gamma | \Gamma_8 m \rangle = \langle \Gamma_8 m | \Gamma_8 m' \mathrm{T}_2\gamma \rangle^*$$

**Table 11.3** Clebsch–Gordan coefficients for the direct product $T_2 \times \Gamma_8 \doteq \Gamma_6 \times \Gamma_7 + 2\Gamma_8$ in $\mathbf{O}'$ (basis functions belonging to the two basis sets of the resulting $\Gamma_8$ are denoted as $\phi(\Gamma_8, m)$ and $\phi'(\Gamma_8, m)$.

| $T_2 \times \Gamma_8$ | | $\Gamma_8$ | | | | $\Gamma_8$ | | | | $\Gamma_7$ | | $\Gamma_6$ | |
|---|---|---|---|---|---|---|---|---|---|---|---|---|---|
| $\gamma_1$ | $\gamma_2$ | $\phi(\Gamma_8,-\frac{3}{2})$ | $\phi(\Gamma_8;-\frac{1}{2})$ | $\phi(\Gamma_8,\frac{1}{2})$ | $\phi(\Gamma_8,\frac{3}{2})$ | $\phi'(\Gamma_8,-\frac{3}{2})$ | $\phi'(\Gamma_8,-\frac{1}{2})$ | $\phi'(\Gamma_8,\frac{1}{2})$ | $\phi'(\Gamma_8,\frac{3}{2})$ | $\phi(\Gamma_7,-\frac{1}{2})$ | $\phi(\Gamma_7,\frac{1}{2})$ | $\phi(\Gamma_6,-\frac{1}{2})$ | $\phi(\Gamma_6;\frac{1}{2})$ |
| $\xi$ | $\phi(\frac{3}{2},-\frac{3}{2})$ | 0 | 0 | 0 | $i/\sqrt{5}$ | 0 | $i\sqrt{5/12}$ | 0 | $i/\sqrt{20}$ | $-i/2$ | 0 | $i/\sqrt{12}$ | 0 |
| | $\phi(\frac{3}{2},-\frac{1}{2})$ | $i\sqrt{4/15}$ | 0 | $-i/\sqrt{5}$ | 0 | $-i\sqrt{3/20}$ | 0 | $-i/\sqrt{20}$ | 0 | 0 | $-i/\sqrt{12}$ | 0 | $-i/2$ |
| | $\phi(\frac{3}{2},\frac{1}{2})$ | 0 | $-i/\sqrt{5}$ | 0 | $i\sqrt{4/15}$ | 0 | $-i/\sqrt{20}$ | 0 | $-i\sqrt{3/20}$ | $i\sqrt{12}$ | 0 | $i/2$ | 0 |
| | $\phi(\frac{3}{2},\frac{3}{2})$ | $i/\sqrt{5}$ | 0 | 0 | 0 | $i/\sqrt{20}$ | 0 | $i\sqrt{5/12}$ | 0 | 0 | $i/2$ | 0 | $-i/\sqrt{12}$ |
| $\eta$ | $\phi(\frac{3}{2},-\frac{3}{2})$ | 0 | 0 | 0 | $-1/\sqrt{5}$ | 0 | $\sqrt{5/12}$ | 0 | $-1/\sqrt{20}$ | $1/2$ | 0 | $1/\sqrt{12}$ | 0 |
| | $\phi(\frac{3}{2},-\frac{1}{2})$ | $-\sqrt{4/15}$ | 0 | $-1/\sqrt{5}$ | 0 | $\sqrt{3/20}$ | 0 | $-1/\sqrt{20}$ | 0 | 0 | $1/\sqrt{12}$ | 0 | $-1/2$ |
| | $\phi(\frac{3}{2},\frac{1}{2})$ | 0 | $-1/\sqrt{5}$ | 0 | $\sqrt{4/15}$ | 0 | $1/\sqrt{20}$ | 0 | $-\sqrt{3/20}$ | $1\sqrt{12}$ | 0 | $-1/2$ | 0 |
| | $\phi(\frac{3}{2},\frac{3}{2})$ | $1/\sqrt{5}$ | 0 | 0 | 0 | $1/\sqrt{20}$ | 0 | $-\sqrt{5/12}$ | 0 | 0 | $1/2$ | 0 | $1/\sqrt{12}$ |
| $\zeta$ | $\phi(\frac{3}{2},-\frac{3}{2})$ | 0 | 0 | $i\sqrt{3/5}$ | 0 | 0 | 0 | $-i/\sqrt{15}$ | 0 | 0 | 0 | 0 | $-i/\sqrt{3}$ |
| | $\phi(\frac{3}{2},-\frac{1}{2})$ | 0 | 0 | 0 | $-i/\sqrt{15}$ | 0 | 0 | 0 | $-i/\sqrt{3/5}$ | $-i\sqrt{3}$ | 0 | 0 | 0 |
| | $\phi(\frac{3}{2},\frac{1}{2})$ | $i/\sqrt{15}$ | 0 | 0 | 0 | $i\sqrt{3/5}$ | 0 | 0 | 0 | 0 | $-i/\sqrt{3}$ | 0 | 0 |
| | $\phi(\frac{3}{2},\frac{3}{2})$ | 0 | $-i\sqrt{3/5}$ | 0 | 0 | 0 | $i/\sqrt{15}$ | 0 | 0 | 0 | 0 | $-i/\sqrt{3}$ | 0 |

(unitarity condition) we can write the matrix of $\hat{V}_{T_2\xi}$ in the following form:

$$\begin{array}{c} \phi'(\Gamma_8,-\tfrac{3}{2})\ \phi'(\Gamma_8,-\tfrac{1}{2})\ \phi'(\Gamma_8,\tfrac{1}{2})\ \phi'(\Gamma_8,\tfrac{3}{2}) \end{array}$$

$$\mathbf{V}_{T_2\xi}^{\Gamma_8\Gamma_8'} = c_1 \begin{bmatrix} 0 & -i\sqrt{4/15} & 0 & -i/\sqrt{5} \\ 0 & 0 & i/\sqrt{5} & 0 \\ 0 & i/\sqrt{5} & 0 & 0 \\ -i/\sqrt{5} & 0 & -i\sqrt{4/15} & 0 \end{bmatrix} \begin{array}{c} \phi(\Gamma_8,-\tfrac{3}{2}) \\ \phi(\Gamma_8,-\tfrac{1}{2}) \\ \phi(\Gamma_8,\tfrac{1}{2}) \\ \phi(\Gamma_8,\tfrac{3}{2}) \end{array} +$$

$$+c_2 \begin{bmatrix} 0 & i\sqrt{3/20} & 0 & -i/\sqrt{20} \\ -i\sqrt{5/12} & 0 & i\sqrt{20} & 0 \\ 0 & i/\sqrt{20} & 0 & -i\sqrt{5/12} \\ -i/\sqrt{20} & 0 & i\sqrt{3/20} & 0 \end{bmatrix}$$

where $c_1$ and $c_2$ are the two independent reduced matrix elements (including factor $[g(\Gamma_8)]^{-1/2} = \frac{1}{2}$). Matrices of $\hat{V}_{T_2\eta}$ and $\hat{V}_{T_2\zeta}$ are obtained in the same way; these matrices contain two parameters $c_1$ and $c_2$ only. It should be emphasized that no relation can be established between $c_1$ and $c_2$ in consideration of group theory. To find the parameters $c_1$ and $c_2$ we must calculate two (but not one as was the case with simply reducible groups; see Section 9.2) matrix elements; for example:

$$\langle \Gamma_8, -\tfrac{3}{2} | \hat{V}_{T_2\xi} | \Gamma_8', -\tfrac{1}{2} \rangle = -ic_1\sqrt{4/15} + ic_2\sqrt{3/20},$$
$$\langle \Gamma_8, -\tfrac{1}{2} | \hat{V}_{T_2\xi} | \Gamma_8', -\tfrac{3}{2} \rangle = -ic_2\sqrt{5/12}.$$

Solving this system of equation we find $c_1$ and $c_2$.

### 11.2.2 Spin–orbit splitting of p and d levels in a cubic field

Knowing the wavefunctions of spin–orbit states, it is possible to calculate the fine-structure energy levels according to (11.7). For this purpose it is necessary to use the well-known quantum mechanical relations

$$\left.\begin{aligned} \hat{l}_z \varphi_{nlm}(\mathbf{r}) &= m\varphi_{nlm}(\mathbf{r}) \quad (m = -l, \ldots, l), \\ \hat{l}_\pm \varphi_{nlm}(\mathbf{r}) &= [l(l+1) - m(m \pm 1)]^{1/2} \varphi_{nl,m\pm1}(\mathbf{r}), \end{aligned}\right\} \quad (11.13)$$

where $\hat{l}_\pm = \hat{l}_x \pm i\hat{l}_y$ are raising and lowering operators, under whose action the hydrogen-like wavefunction is transformed into $\varphi_{nl,m+1}$ and $\varphi_{nl,m-1}$. Equations (11.13) allow the construction of the matrices of the operators $\hat{l}_x$, $\hat{l}_y$

and $\hat{l}_z$ for the p level ($l = 1$; see Table 3.1):

$$\begin{array}{c} \quad\; p_x \;\; p_y \;\; p_z \\ \mathbf{l}_x = \begin{bmatrix} 0 & 0 & 0 \\ 0 & 0 & -i \\ 0 & i & 0 \end{bmatrix} \end{array}, \quad \mathbf{l}_y = \begin{bmatrix} 0 & 0 & i \\ 0 & 0 & 0 \\ -i & 0 & 0 \end{bmatrix}, \quad \mathbf{l}_z = \begin{bmatrix} 0 & -i & 0 \\ i & 0 & 0 \\ 0 & 0 & 0 \end{bmatrix}. \qquad (11.14)$$

We have also used the well-known matrices of spin operators for $s = \frac{1}{2}$:

$$\begin{array}{c} \qquad \alpha \;\; \beta \\ \hat{\mathbf{s}}_x = \dfrac{1}{2}\begin{bmatrix} 0 & 1 \\ 1 & 0 \end{bmatrix} \end{array}, \quad \begin{array}{c} \qquad \alpha \;\; \beta \\ \hat{\mathbf{s}}_y = \dfrac{1}{2}\begin{bmatrix} 0 & -i \\ i & 0 \end{bmatrix} \end{array}, \quad \begin{array}{c} \qquad \alpha \;\; \beta \\ \hat{\mathbf{s}}_z = \dfrac{1}{2}\begin{bmatrix} 1 & 0 \\ 0 & -1 \end{bmatrix} \end{array}. \qquad (11.15)$$

We now determine the energy $\epsilon(\Gamma_8^-)$:

$$\epsilon(\Gamma_8^-) = \langle {}^2\mathrm{T}_{1u}, \Gamma_8^-, +\tfrac{3}{2} | \hat{H}_{so} | {}^2\mathrm{T}_{1u}, \Gamma_8^-, +\tfrac{3}{2} \rangle$$

$$= \langle \sqrt{\tfrac{1}{2}}\, i\varphi_x\alpha + \sqrt{\tfrac{1}{2}}\varphi_y\alpha | \lambda(r)(\hat{l}_x\hat{s}_x + \hat{l}_y\hat{s}_y + \hat{l}_z\hat{s}_z) | -\sqrt{\tfrac{1}{2}}\, i\varphi_x\alpha + \sqrt{\tfrac{1}{2}}\varphi_y\alpha \rangle.$$

On substituting the matrix elements from (11.14), we obtain

$$\epsilon(\Gamma_8^-) = \tfrac{1}{2}\lambda, \qquad (11.16)$$

where

$$\lambda = \int_0^\infty \lambda(r) |R_{nl}(r)|^2 r^2 \, dr \quad (l = 1)$$

is the *spin–orbit interaction constant*; it depends upon the electron term: $\lambda \equiv \lambda({}^2\mathrm{T}_{1u})$. Similarly, we find

$$\epsilon(\Gamma_6^-) = -\lambda. \qquad (11.17)$$

Hence the spin–orbit splitting $\epsilon(\Gamma_8^-) - \epsilon(\Gamma_6^-) = \frac{3}{2}\lambda$. The sign of $\lambda$ determines the order of the levels.

Using (11.13), it is possible to determine the matrices of $\mathbf{l}_x$, $\mathbf{l}_y$ and $\mathbf{l}_z$ in the $t_{2g}$ basis $d_\xi$, $d_\eta$, $d_\zeta$ (Table 3.1). It turns out that these matrices differ from those in (11.14) only in sign ($\mathbf{l}_i(t_{2g}) = -\mathbf{l}_i(\mathrm{p})$), which makes it possible to speak about the so-called T–P equivalence [14]. The energies of the ${}^2\mathrm{T}_{2g}$ sublevels are expressed as

$$\epsilon(\Gamma_7^+) = \lambda, \quad \epsilon(\Gamma_8^+) = \tfrac{1}{2}\lambda, \quad \lambda \equiv \lambda({}^2\mathrm{T}_{2g}) \qquad (11.18)$$

and differ in sign from those in (11.16) and (11.17).

### 11.2.3 Selection rules for mixing of *SΓ* terms

The level splittings calculated above appear in first order of perturbation theory; that is, they are proportional to $\lambda$. In many cases it is necessary to find the matrix elements $\hat{H}_{so}$ connecting different $S\Gamma$ terms. Let us confine ourselves to a summary of selection rules for matrix elements for the mixing of different many-electron $S\Gamma$ terms. The matrix elements $\langle \alpha_1 S_1 \Gamma_1 M_1 \gamma_1 | \hat{H}_{so} | \alpha_2 S_2 \Gamma_2 M_2 \gamma_2 \rangle$ are nonzero provided the following two conditions are satisfied.

(1) The direct product $\Gamma_1 \times \Gamma_2$ must contain representations $\Gamma(\mathbf{l})$ according to which the three components of the orbital moment $\hat{\mathbf{l}}$ transform. The meaning of this rule is clear without a rigorous proof: since the operators $\hat{l}_x$, $\hat{l}_y$ and $\hat{l}_z$ affect the coordinate parts of the $S\Gamma$ states, they should satisfy conventional selection rules (Section 9.12, (9.7)).

(2) The direct product $D^{(S_1)} \times D^{(S_2)}$ in the rotation group must contain $D^{(1)}(\hat{S}_x, \hat{S}_y, \hat{S}_z)$; in other words, the "triangle condition" $S_1 + S_2 \geq 1 \geq |S_1 - S_2|$ must be satisfied. In practice this means that the states being mixed must differ in spin by not more than 1. The only exception is the case $S_1 = S_2 = 0$, for which there is no spin–orbit mixing.

**Example 11.3** Consider one-electron d states $^2T_{2g}$ and $^2E_g$. In $\mathbf{O}_h$ we have $\Gamma(\mathbf{l}) = T_{1g}$, $\Delta S = 0$ and $T_{2g} \times E_g \doteq T_{1g} + T_{2g}$. Therefore $\langle ^2T_{2g} SM | \hat{H}_{so} | ^2E_g SM \rangle \neq 0$.

**Example 11.4** Consider the term $^4T_{1g}$ and $^4T_{2g}$ of the $d^3$ ion: $T_{1g} \times T_{2g} \doteq A_{2g} + E_g + T_{1g}(\mathbf{l}) + T_{2g}$ and $\Delta S = 1$. Hence these terms are mixing.

**Example 11.5** Consider the terms $^4B_{1g}$ and $^2E_g$ of $\mathbf{D}_{4h}$. The vector $\hat{\mathbf{l}}$ is transformed according to two irreducible representations: $A_{2g}(\hat{l}_z)$ and $E_g(\hat{l}_x, \hat{l}_y)$. Since $B_{1g} \times E_g \doteq E_g(\hat{l}_x, \hat{l}_y)$ and $\Delta S = 1$, the terms $^4B_{1g}$ and $^2E_g$ are mixed.

It should be noted that the above selection rules refer to different $S\Gamma$ levels, while in the basis of a single $S\Gamma$ term other rules are valid (Section 11.3.3).

### 11.2.4 Shifts in the fine-structure levels

Let us apply the selection rules for $\hat{H}_{so}$ to the calculation of the shifts of fine-structure levels. Crystal $S\Gamma$ terms for which $D^{(S)} \times \Gamma$ is irreducible (Section 11.1.1) are not split by the spin–orbit interaction. These levels are subjected to shifts that can be calculated in second order of perturbation theory. The shift is calculated from the quantum mechanical expression for the correction $\Delta\epsilon$ to the $\epsilon_0$ energy level due to a perturbation $V$:

$$\Delta\epsilon = \sum_{i \neq 0} \frac{|\langle i | \hat{V} | 0 \rangle|^2}{\epsilon_0 - \epsilon_i}, \tag{11.19}$$

where $i$ runs over the other unperturbed states.

Let us consider a cubic $^2$E term as an example to illustrate the group-theoretical principles for calculating energy corrections. As already mentioned, $D^{(1/2)} \times E = \Gamma_6 \times E \doteq \Gamma_8$. Since $D^{(S)} \times \Gamma$ is irreducible, the absence of spin–orbit splitting is an exact result. In calculating the shift we shall take into account only the "mixing" with the nearest $^2T_2$ term:

$$\Delta\epsilon(^2E) = \sum_{\gamma=\xi,\eta,\zeta} \sum_{m=\pm 1/2} \frac{|\langle ^2T_2 m\gamma|\hat{H}_{so}|^2Em'\gamma'\rangle|^2}{10Dq}, \tag{11.20}$$

where $10Dq = \epsilon(^2E) - \epsilon(^2T_2) = \epsilon_0 - \epsilon_i$ for one electron in the **O** crystal field ($\epsilon_0 - \epsilon_i = -10Dq$ for one hole; that is, for the $d^8$ shell); the shift $\Delta\epsilon$ does not depend on $\gamma' = u, v$ or $m' = +\frac{1}{2}, -\frac{1}{2}$.

Let us digress for a while from this example and formulate three important general statements.

(1) The spin–orbit interaction operator is transformed in any double group according to the totally symmetric representation $A_1$. Indeed, $\hat{l}_x$, $\hat{l}_y$ and $\hat{l}_z$, for example, in $\mathbf{O}_h$ transform according to the representation $T_{1g}$, as are $\hat{S}_x$, $\hat{S}_y$, $\hat{S}_z$. The operator $\hat{H}_{so}$ constitutes the totally symmetric part; that is, the scalar part of the direct product $T_{1g} \times T_{1g}$. Since it is a scalar in the rotation group (and in $\mathbf{O}_h$), $\hat{H}_{so}$ is also a scalar in all point groups.

(2) Assume $\Gamma'$ and $\Gamma''$ label the fine structure levels of $S_1\Gamma_1$ and $S_2\Gamma_2$ terms (i.e. the irreducible representations belonging to $D^{(S_1)} \times \Gamma_1$ and $D^{(S_2)} \times \Gamma_2$). The spin–orbit interaction can mix only the wavefunctions belonging to identical (repeating) $\Gamma' = \Gamma''$ representations and basis function $\gamma$ of these representations. In other words, only matrix elements of the form $\langle\Gamma'\gamma|\hat{H}_{so}|\Gamma''\gamma\rangle$ are nonzero. These matrix elements do not depend on $\gamma$; that is, they are equal to each other.

This follows from the Wigner–Eckart theorem (Section 9.2 and 11.2.1) and therefore duplicates the configuration mixing rules (Section 5.6.2), valid for any scalar operator.

(3) Independently of these rules, $\langle\Gamma'\gamma|\hat{H}_{so}|\Gamma''\gamma\rangle \neq 0$ only when $\Gamma_1 \times \Gamma_2$ contains $\Gamma(\hat{1})$, and $\Delta S = 0$ or 1 (Section 11.2.3).

Now it is clear that the $^2E(\Gamma_8)$ level is "repelled" only by the $\Gamma_8$ component of $^2T_2$, and (11.20) may be rewritten as

$$\Delta\epsilon(^2E) = \frac{|\langle ^2T_2, \Gamma_8\gamma|\hat{H}_{so}|^2E, \Gamma_8\gamma\rangle|^2}{10Dq}, \tag{11.21}$$

where owing to the above property, only the diagonal elements are taken into account. On substituting (11.10) and (11.12) ($\gamma = \frac{3}{2}$), we obtain

$$\Delta\epsilon(^2E) = \frac{3\lambda^2}{20Dq}. \tag{11.22}$$

The same shift, but in the other direction, applies also to the sublevel $\Gamma_8(^2T_2)$.

Let us consider a many-electron system. In the absence of spin–orbit interaction each energy level belongs to a certain total spin $S$ and irreducible representation $\Gamma$ and we have a set of states $\alpha_1 S_1 \Gamma_1, \alpha_2 S_2 \Gamma_2, \ldots \alpha_n S_n \Gamma_n$ ($\alpha$ numbers repeating $S\Gamma$-terms). Within the scope of the first order approximation of the perturbation theory we can consider the spin–orbit splitting of each $\alpha S\Gamma$ term, neglecting by spin–orbit mixing of different $\alpha S\Gamma$ terms. If spin–orbit interaction can not be considered as a weak perturbation we must solve the problem more precisely, also taking into account spin–orbit mixing of different $\alpha S\Gamma$ terms. In other words we must solve the secular equation in which all n $S\Gamma$ terms are involved:

$$|\langle \alpha_i S_i \Gamma_i | \hat{H}_{so} | \alpha_j S_j \Gamma_j \rangle - \epsilon \sigma_{ij}| = 0.$$

The general statements (1) – (3) given above allow us to considerably simplify this equation. In fact only the states belonging to the same irreducible representations $\Gamma'$ contained in $D^{(S_i)} \times \Gamma_i$ are mixed by $\hat{H}_{so}$. Hence the order of the secular equation we have to solve is equal to the total number of the repeating $\Gamma'$ irreducible representations arising from all $\alpha S\Gamma$ terms (but not to the total dimension of n $\alpha S\Gamma$-terms or to the total dimension of $\Gamma'$ terms).

## 11.3 FINE STRUCTURE OF MANY-ELECTRON TERMS

### 11.3.1 Effective spin–orbit interaction

For many-electron ions the operator $\hat{H}_{so}$, (10.1), is generalized by summation over all the electrons:

$$\hat{H}_{so} = \sum_i \lambda(r_i) \hat{\mathbf{l}} \cdot \hat{\mathbf{s}}_i \tag{11.23}$$

The matrix elements of $\hat{H}_{so}$ are calculated using many-electron determinantal functions. For cubic terms ${}^{2S+1}T_1$ and ${}^{2S+1}T_2$, the solution of the spin–orbit splitting problem is simplified in practice by using the results of many-electron theory [14]. It turns out to be possible to introduce an *effective* spin–orbit interaction operator of the form

$$\hat{H}_{so} = \lambda(\alpha S\Gamma) \hat{\mathbf{S}} \cdot \hat{\mathbf{T}}. \tag{11.24}$$

The quantities involved here are as follows: $\hat{\mathbf{S}}$ is the total spin operator of the electron shell; and $\hat{\mathbf{T}} \equiv \hat{\mathbf{T}}(T_1)$ is an operator whose components $\hat{\mathbf{T}}_x$, $\hat{\mathbf{T}}_y$, and $\hat{\mathbf{T}}_z$ are transformed according to the representation $T_1$ (or $T_{1g}$) of the cubic groups (i.e. like the orbital angular momentum components). The matrices $\mathbf{T}(T_1)$ are obtained through use of the Wigner–Eckart theorem: they coincide with those in (11.14) in the $T_1$ basis and differ from them in sign in the $T_2$ basis. Hence the effective operator matrix $\hat{\mathbf{S}} \cdot \hat{\mathbf{T}}$ may be written for any ${}^{2S+1}T_1$ and ${}^{2S+1}T_2$ term using only the definitions of the operators $\hat{\mathbf{S}}$ and $\hat{\mathbf{T}}$, without using the actual wavefunction of $|S\Gamma M\gamma\rangle$. On the other hand, the constant $\lambda$ depends on $\alpha S\Gamma$; that is, it takes into account the nature of the term (the "internal structure"). Table 11.4 gives splittings of ${}^{2S+1}$T levels of cubic systems. The many-electron parameters $\lambda(\alpha S\Gamma)$ are given in [14]; for some terms they have the forms

$$\left.\begin{array}{c}\lambda(t_2^2\,{}^3T_1) = -\tfrac{1}{2}\lambda, \quad \lambda(t_2e\,{}^3T_1) = \tfrac{1}{4}\lambda, \quad \lambda(t_2e\,{}^3T_2) = -\tfrac{1}{4}\lambda, \\ \lambda(t_2^2({}^3T_1)e\,{}^4T_1) = \tfrac{1}{6}\lambda, \quad \lambda(t_2({}^3T_1)e^4T_1) = -\tfrac{1}{6}\lambda, \quad \lambda(t_2e^2({}^3A_2)\,{}^4T_1) = -\tfrac{1}{3}\lambda,\end{array}\right\} \tag{11.25}$$

where $\lambda$ is the one-electron parameter (Section 11.2.2). From Table 11.2 it can be seen that the first-order spin–orbit interaction leaves an "accidental" degeneracy of states with $S \geq \frac{3}{2}$ ($\Gamma_7 + \Gamma_8$ in ${}^4T_1$; $A_2 + T_1 + T_2$ in ${}^5T_1$). This degeneracy is removed in second order of perturbation theory (the splitting $\sim \lambda^2$) and is consistent with the Wigner theorem.

It should be emphasized that the *effective operator (11.24) works only within the limits of the $\alpha S\Gamma$ basis*. It cannot be used for calculating the "repulsion" between different $\alpha S\Gamma$ terms, in particular for calculating second-order splittings. The group-theoretical apparatus necessary for these calculations is described in detail in [14, 46, 47], which give all the matrix elements $\langle \alpha_1 S_1 \Gamma_1 | \hat{H}_{so} | \alpha_2 S_2 \Gamma_2 \rangle$ for $d^n$ ions necessary for complete diagonalization of the configuration mixing matrix of the spin–orbit interaction. The wavefunctions of spin–orbit components are calculated by means of the Clebsch–Gordan decomposition.

**Table 11.4** Spin–orbit splitting of $2^{S+1}$T terms (the upper sign refers to $T_1$ and the lower to $T_2$) [14].

| Terms | Levels | Classification |
|---|---|---|
| ${}^3T_1(T_2)$ | $\mp 2\lambda(\alpha S\Gamma)$ | $A_1$ ($A_2$) |
| | $\mp\lambda(\alpha S\Gamma)$ | $T_1(T_2)$ |
| | $\pm\lambda(\alpha S\Gamma)$ | E, $T_2$ (E, $T_1$) |
| ${}^4T_1(T_2)$ | $\mp\frac{5}{2}\lambda(\alpha S\Gamma)$ | $\Gamma_6$ ($\Gamma_7$) |
| | $\mp\lambda(\alpha S\Gamma)$ | $\Gamma_8$ ($\Gamma_8$) |
| | $\pm\frac{3}{2}\lambda(\alpha S\Gamma)$ | $\Gamma_7\,\Gamma_8$ ($\Gamma_6$, $\Gamma_8$) |
| ${}^5T_1(T_2)$ | $\mp 3\lambda(\alpha S\Gamma)$ | $T_1$ ($T_2$) |
| | $\mp\lambda(\alpha S\Gamma)$ | E, $T_2$ (E, $T_1$) |
| | $\pm 2\lambda(\alpha S\Gamma)$ | $A_2$, $T_1$, $T_2$ ($A_1$, $T_2$, $T_1$) |

### 11.3.2 Symmetric and antisymmetric parts of the direct product

The selection rules given in Section 9.1.2 refer to transitions between different levels $\Gamma_1$ and $\Gamma_2$. Matrix elements of the form $\langle \Gamma\gamma_1 | \hat{\mathbf{V}}_{\Gamma'\gamma'} | \Gamma\gamma_2 \rangle$ connecting states *within* one electron configuration obey special selection rules. In order to determine these rules, it is necessary to deal with some new concepts.

We shall start with an example. Assume that $\varphi(\mathrm{E}u)$, $\varphi(\mathrm{E}v)$ and $\psi(\mathrm{E}u)$, $\psi(\mathrm{E}v)$ (briefly $\varphi_u$, $\varphi_v$ and $\psi_u$, $\psi_v$) are two basis sets of the irreducible representation E (group **O**). The direct product $\mathrm{E} \times \mathrm{E}$ consists of four functions $\varphi_\gamma \psi_{\gamma'}$. Let us construct the basis functions $\Phi(\Gamma\gamma)$ of the irreducible representations contained in $\mathrm{E} \times \mathrm{E}$, namely $A_1 + A_2 + E$ using the Clebsch–Gordan decomposition:

$$\left.\begin{aligned} \Phi(\mathrm{A}_1) &= \sqrt{\tfrac{1}{2}}(\varphi_u \psi_u + \varphi_v \psi_v); \quad & \Phi(\mathrm{E}u) &= \sqrt{\tfrac{1}{2}}(-\varphi_u \psi_u + \varphi_v \psi_v), \\ \Phi(\mathrm{A}_2) &= \sqrt{\tfrac{1}{2}}(-\varphi_u \psi_v + \varphi_v \psi_u), \quad & \Phi(\mathrm{E}v) &= \sqrt{\tfrac{1}{2}}(\varphi_u \psi_v + \varphi_v \psi_u). \end{aligned}\right\} \tag{11.26}$$

From the appearance of the functions $\Phi$ it is obvious that $\Phi(\mathrm{A}_1)$ and the two functions $\Phi(\mathrm{E}\gamma)$ have one property in common: they are both symmetric; that is, they do not change their form under the interchange $\varphi \leftrightarrow \psi$. If $\varphi$ and $\psi$ are functions of one basis ($\varphi_\gamma = \psi_\gamma$), then $\Phi(\mathrm{A}_1)$ and $\Phi(\mathrm{E}\gamma)$ are simplified: taking normalization into account, they have the form

$$\left.\begin{aligned} \Phi(\mathrm{A}_1) &= \sqrt{\tfrac{1}{2}}(\varphi_u^2 + \varphi_v^2) \quad & \Phi(\mathrm{E}u) &= \sqrt{\tfrac{1}{2}}(\varphi_v^2 - \varphi_u^2), \\ & & \Phi(\mathrm{E}v) &= \varphi_u \varphi_v. \end{aligned}\right\} \tag{11.27}$$

The function $\Phi(\mathrm{A}_2)$ is antisymmetric, since it reverses sign under $\varphi \leftrightarrow \psi$. This function becomes zero if the direct product $\mathrm{E} \times \mathrm{E}$ is constructed from one basis set, that is, for eigenfunctions belonging to the same energy level.

Therefore the representations $\mathrm{A}_1$ and E belong to the so-called *symmetric part of the direct product* $\mathrm{E} \times \mathrm{E}$, while $\mathrm{A}_2$ is the *antisymmetric part*. These parts are denoted by $[\mathrm{E} \times \mathrm{E}]$ and $\{\mathrm{E} \times \mathrm{E}\}$ respectively, or $[\mathrm{E}^2]$ and $\{\mathrm{E}^2\}$; of course, $[\mathrm{E}^2] + \{\mathrm{E}^2\} = \mathrm{E} \times \mathrm{E}$.

For the direct product $\mathrm{T}_{1u} \times \mathrm{T}_{1u}$ (Section 5.4.2) it is easy to find

$$[\mathrm{T}_{1u}^2] \doteq \mathrm{A}_{1g} + \mathrm{E}_g + \mathrm{T}_{2g}, \qquad \{\mathrm{T}_{1u}^2\} \doteq \mathrm{T}_{1g}.$$

The systematic decomposition procedure $\Gamma \times \Gamma \doteq [\Gamma \times \Gamma] + \{\Gamma \times \Gamma\}$ is based on reduction.

We give here without proof the equations for the characters of symmetric, $\chi^{[\Gamma^2]}(\hat{R})$, and antisymmetric, $\chi^{\{\Gamma^2\}}(\hat{R})$, parts of the direct product $\Gamma \times \Gamma$:

$$\chi^{[\Gamma^2]}(\hat{R}) = \tfrac{1}{2}\{[\chi^{(\Gamma)}(\hat{R})]^2 + \chi^{(\Gamma)}(\hat{R}^2)\}, \tag{11.28}$$

$$\chi^{\{\Gamma^2\}}(\hat{R}) = \tfrac{1}{2}\{[\chi^{(\Gamma)}(R)]^2 - \chi^{(\Gamma)}(\hat{R}^2)\}. \tag{11.29}$$

By substituting from the character tables the squares $[\chi^{(\Gamma)}(\hat{R})]^2$and $\chi^{(\Gamma)}(\hat{Q})$ ($\hat{Q} = \hat{R}^2$), it is easy to calculate $\chi^{[\Gamma\times\Gamma]}$ and $\chi^{\{\Gamma\times\Gamma\}}$. Decomposing them into irreducible parts according to the reduction formulae

$$\left.\begin{aligned} a(\Gamma_i) &= \frac{1}{g}\sum_{\hat{R}} \chi^{[\Gamma^2]}(\hat{R})\chi^{(\Gamma_i)}(\hat{R}), \\ a(\Gamma_i) &= \frac{1}{g}\sum_{\hat{R}} \chi^{\{\Gamma^2\}}(\hat{R})\chi^{(\Gamma_i)}(\hat{R}), \end{aligned}\right\} \tag{11.30}$$

we obtain symmetric and antisymmetric-type representations.

For practical purposes it is enough to use the tables compiled by Jahn and Teller (see [74]), containing both the symmetric products of ordinary representations and antisymmetric products of double-valued representations for all the point groups (see Appendix III). For example, for $\mathbf{T}_d$ and $\mathbf{O}$

$$[\mathrm{E}^2] \doteq \mathrm{A}_1 + \mathrm{E}, \quad [\mathrm{T}_1^2] = [\mathrm{T}_2^2] \doteq \mathrm{A}_1 + \mathrm{E} + \mathrm{T}_2,$$

$$\{\mathrm{E}^2\} \doteq \mathrm{A}_2, \quad \{\mathrm{T}_1^2\} = \{\mathrm{T}_2^2\}, \quad \{\Gamma_8^2\} \doteq \mathrm{A}_1 + \mathrm{E} + \mathrm{T}_2.$$

In centrosymmetric groups all $[\Gamma^2]$ and $\{\Gamma^2\}$ are even ($g$) under spatial inversion.

### 11.3.3 Selection rules for real and imaginary operators

Let us first consider the time-dependent Schrödinger equation including no spin–orbit interaction and interaction with a magnetic field:

$$i\hbar\frac{\partial\psi(t)}{\partial t} = \hat{H}\psi(t). \tag{11.31}$$

Let us replace $t \to -t$ and write a complex conjugate equation bearing in mind that $\hat{H}^* = \hat{H}$

$$i\hbar\frac{\partial\psi^*(-t)}{\partial t} = \hat{H}\psi^*(-t). \tag{11.32}$$

Comparing (11.31) and (11.32) we see that the time-dependent wavefunctions $\psi(t)$ and $\psi^*(-t)$ satisfy the same Schrödinger equation. The state $\hat{K}_0\psi(t) = \psi^*(-t)$ is called the time reversed state of $\psi(t)$ and $\hat{K}$ is the time reversal operator. In the stationary (time-independent) case, the time-reversed state of $\psi(\mathbf{r})$ is simply a complex conjugate state

$$\hat{K}_0\psi(\mathbf{r}) = \psi_0^*(\mathbf{r}) \tag{11.33}$$

and hence for the time-dependent problems the time-reversal operator for the orbital functions is the complex conjugation operator. This reversal is called Wigner's time reversal [12].

Now we give without proof (the proof is given for example in [14] and [74]) the following important statement:

> Wigner's time reversal operator $\hat{K}_0$ changes the sign of the orbital momentum operator $\hat{\mathbf{l}}$
>
> $$\hat{K}_0 \hat{\mathbf{l}} \hat{K}_0^{-1} = -\hat{\mathbf{l}} \tag{11.34}$$

The equation (11.34) means that each component of the vector operator $\hat{\mathbf{l}}$ changes its sign under time inversion.

The spin operator $\hat{S}$ possesses the same property

$$\hat{K} \hat{\mathbf{S}} \hat{K}^{-1} = -\mathbf{S} \tag{11.35}$$

where $K$ is the time reversal operator acting in the spin space.

Let us introduce the important concepts of *real* and *imaginary operators*. The orbital angular momentum operator $\hat{\mathbf{l}}$ is an imaginary operator; it may be represented as a vector product

$$-i\hbar \mathbf{r} \times \nabla \quad \left( \nabla_x = \frac{\partial}{\partial x} \text{ etc.} \right).$$

In contrast, interaction with crystal fields is described by real operators. *Imaginary operators reverse sign under time inversion*, while *real operators are invariant* (imaginary operators are said to be *T-odd*). The operators $\hat{\mathbf{l}}$ and $\hat{\mathbf{s}}$ are *T*-odd. The irreducible representations of a group can be related to both real and imaginary representations. In **O** the real representations are $A_1$, E and $T_2$, while the imaginary ones are $A_2$ and $T_1$; in $\mathbf{O}_h$ the real representations are $A_{1g}$, $A_{2u}$, $E_g$, $E_u$, $T_{1u}$, $T_{2g}$ and $T_{2u}$, while the imaginary ones are $A_{2g}$ and $T_{1g}$. The *T*-odd operators $\hat{\mathbf{s}}$ and $\hat{\mathbf{l}}$ transform according to the imaginary representation $T_1(\mathbf{O})$ or $T_{1g}(\mathbf{O}_h)$. We confine ourselves to this short resumé, since a complete description of the time-inversion problem requires some new concepts (a comprehensive description of which is given in [12], [14] and [75]).

We can now return to a discussion of the selection rules for the matrix elements $\langle \Gamma\gamma_1 | \hat{\mathbf{V}}_{\Gamma'\gamma'} | \Gamma\gamma_2 \rangle$ relating to one energy level $\epsilon(\Gamma)$. The rules depend on the number of electrons and properties of the representation $\Gamma'$ under time inversion (Table 11.5). Proofs of these results are given in [39].

Let us apply the selection rules to spin–orbit interaction. The $\alpha S\Gamma$ term is split in first order of perturbation theory if $\{\Gamma^2\}$ contains at least one of the representations $\Gamma(\hat{\mathbf{l}})$ and $S \neq 0$. As an example we can take triplets for which $\{T_1^2\} = \{T_2^2\} \doteq T_{1g}(\hat{\mathbf{l}})$. This condition means that the mean value of the orbital angular momentum is nonzero. In contrast, $\{E^2\} \doteq A_2$; hence the doublet $^2E$ does not split (which does not contradict the conclusion about the irreducibility of $D^{(1/2)} \times E = \Gamma_8$). For $S > \frac{1}{2}$ splitting can occur even when $\Gamma(\hat{\mathbf{l}})$

**Table 11.5** Selection rules for the matrix elements $\langle \Gamma\gamma_1 | \hat{V}_{\Gamma'\gamma'} | \Gamma\gamma_2 \rangle$.

| $\Gamma$ \ $\Gamma'$ | Number of electrons | |
|---|---|---|
| | Even<br>$\Gamma$ single-valued | Odd<br>$\Gamma$ double-valued |
| Real | $[\Gamma^2]$ contains $\Gamma'$ | $\{\Gamma^2\}$ contains $\Gamma'$ |
| Imaginary | $\{\Gamma^2\}$ contains $\Gamma'$ | $[\Gamma^2]$ contains $\Gamma'$ |

is not contained in $\{\Gamma^2\}$. As an example we can take the $^3$E term ($D^{(1)} \times E \doteq T_1 + T_2$). In this case the degeneracy is removed by the second-order spin–orbit interaction. Finally, for $S > \frac{1}{2}$ in noncubic groups the spin degeneracy is removed (in second order) and this results in the existence of fine structure in the spin states.

## 11.4 FINE STRUCTURE OF OPTICAL LINES

### 11.4.1 Intensities and selection rules

By splitting the $S\Gamma$ terms, the spin–orbit interaction leads to fine structure of optical lines. Let us consider a one-electron impurity centre in a cubic crystal with ground state $ns(^2A_{1g})$ and excited $n'p(^2T_{1u})$ (e.g. hydrogen-like 1s and 2p levels of an F-centre vacancy having entrapped an electron). The transition $^2A_{1g} \rightarrow {^2T_{1u}}$ is electric-dipole allowed (as shown in Section 9.3). Figure 11.3 shows the spin–orbit splitting. Denoting the matrix element $\langle \varphi_\alpha | \hat{d}_\alpha | s \rangle$ ($\alpha = x, y, z$) by $d$, it is easy to calculate all the matrix elements

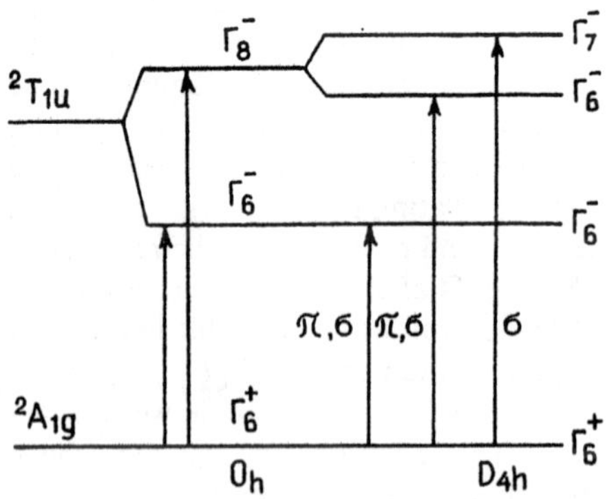


**Fig. 11.3** Fine structure of the $^2A_{1g} \rightarrow {^2T_{1u}}$ line in a cubic field and its splitting by tetragonal deformation.

$\langle {}^2T_{1u}, \Gamma_6^-\gamma|\hat{d}_\alpha|{}^2A_{1g}, \Gamma_6^+\gamma'\rangle$ and $\langle {}^2T_{1u}, \Gamma_8^-\gamma|\hat{d}_\alpha|{}^2A_{1g}, \Gamma_6^+\gamma'\rangle$. Simple calculations using (9.1) show that for any polarization ($\alpha = x, y, z$) the ratio of line intensities $I(\Gamma_6^+ \to \Gamma_6^-) : I(\Gamma_6^+ \to \Gamma_8^-)$ is 1 : 2. This example shows that the absorption $\Gamma_6^+ \to \Gamma_6^-$ and $\Gamma_6^+ \to \Gamma_8^-$ is isotropic and the given intensity ratio is maintained for any light polarization. *The spin–orbit interaction, in splitting the levels, does not lead to polarization dichroism* (as distinct from low-symmetry fields, which also split the $S\Gamma$ multiplets). Polarization dichroism occurs only when the point symmetry decreases, while the spin–orbit interaction leaves the system cubic, connecting only the spin with the orbital angular momentum.

Taking the direct products $\Gamma_6^+ \times \Gamma_6^-$ and $\Gamma_6^+ \times \Gamma_8^-$,

$$\Gamma_6^+ \times \Gamma_6^- \doteq A_{1u} + T_{1u}(\hat{\mathbf{d}}),$$

$$\Gamma_6^+ \times \Gamma_8^- \doteq E_u + T_{1u}(\hat{\mathbf{d}}) + T_{2u},$$

we see that $T_{1u}(\hat{\mathbf{d}})$ is included in both decompositions. On the other hand, transition to the fine-structure levels $\Gamma_6^-$, $\Gamma_8^-({}^2T_{1u})$ is allowed if the matrix elements $\langle {}^2A_{1g}|\hat{\mathbf{d}}|{}^2T_{1u}\gamma\rangle \equiv \langle s|\hat{\mathbf{d}}|p_\alpha\rangle$ are nonzero.

Generalizing this reasoning, we arrive at the following selection rules. The transition $\Gamma_1' \rightleftarrows \Gamma_2'$ between the fine-structure levels of $\alpha S_1\Gamma_1$ and $\beta S_2\Gamma_2$ terms is allowed if the following three conditions are fulfilled simultaneously:

(1) the direct product $\Gamma_1' \times \Gamma_2'$ contains representations $\Gamma(\hat{\mathbf{d}})$;

(2) the direct product $\Gamma_1 \times \Gamma_2$ also contains representations $\Gamma(\hat{\mathbf{d}})$;

(3) the spins of the $S\Gamma$ terms are identical ($S_1 = S_2$).

The latter two selection rules are "approximate", they are applicable to "pure" $S\Gamma$ terms without regard to their mixing by spin–orbit interaction.

Finally, for centrosymmetric groups the parity rule should be taken into account: since the spin functions are even, the parity of the states $|\alpha S\Gamma, \Gamma'\gamma'\rangle$ coincides with that of the irreducible representation $\Gamma$.

### 11.4.2 Deformation splitting, two-photon transitions

External fields split the fine-structure levels, the selection rules being determined in the usual way.

**Example 11.6** Consider a tetragonal deformation ($\mathbf{O_h} \to \mathbf{D_{4h}}$) and the transitions $\Gamma_6^+ \to \Gamma_6^-, \Gamma_8^-$. The doublets $\Gamma_6^+$ and $\Gamma_6^-$ are Kramers doublets, and hence they are not split by deformation. According to the reduction tables, $\Gamma_8^- \doteq \Gamma_6^- + \Gamma_7^-$ and the $\Gamma_6^-$ basis is $\Phi({}^2T_{1u}, \Gamma_8^-, \pm\frac{1}{2})$, while the $\Gamma_7^-$ basis is $\Phi({}^2T_{1u}, \Gamma_8^-, \pm\frac{3}{2})$. In $\mathbf{D_{4h}}$

$$\Gamma_6^+ \times \Gamma_6^- \doteq A_{1u} + A_{2u}(\hat{d}_z) + E_u(\hat{d}_x, \hat{d}_y), \qquad \Gamma_6^+ \times \Gamma_7^- \doteq B_{1u} + B_{2u} + E_u(\hat{d}_x, \hat{d}_y).$$

Thus tetragonal deformation leads to polarization dichroism—the $\Gamma_6^+ \to \Gamma_7^-$ line is

**Table 11.6** Polarization dependences of two-photon transition intensities (double-valued representations).

| Groups and transitions | | Polarization dependences |
|---|---|---|
| $\mathbf{O}_1$, $\mathbf{T}_d$ | $\mathbf{O}_h$ | $I(\mathbf{u}_1\mathbf{u}_2)$ |
| $\Gamma_6 \leftrightarrow \Gamma_6$<br>$\Gamma_7 \leftrightarrow \Gamma_7$ | $\Gamma_6^{\pm} \leftrightarrow \Gamma_6^{\pm}$<br>$\Gamma_7^{\pm} \leftrightarrow \Gamma_7^{\pm}$ | $A(\mathbf{u}_1\mathbf{u}_2) + c_1 T_1(\mathbf{u}_1\mathbf{u}_2)$ |
| $\Gamma_6 \leftrightarrow \Gamma_8$<br>$\Gamma_7 \leftrightarrow \Gamma_8$ | $\Gamma_6^{\pm} \leftrightarrow \Gamma_8^{\pm}$<br>$\Gamma_7^{\pm} \leftrightarrow \Gamma_8^{\pm}$ | $E(\mathbf{u}_1\mathbf{u}_2) + c_1 T_1(\mathbf{u}_1\mathbf{u}_2) + c_2 T_2(\mathbf{u}_1\mathbf{u}_2)$ |
| $\Gamma_8 \leftrightarrow \Gamma_8$ | $\Gamma_8^{\pm} \leftrightarrow \Gamma_8^{\pm}$ | $A(\mathbf{u}_1\mathbf{u}_2) + c_1 T_1(\mathbf{u}_1\mathbf{u}_2) + c_2 T_1(\mathbf{u}_1\mathbf{u}_2) + c_3 T_2(\mathbf{u}_1\mathbf{u}_2)$ |
| $\mathbf{D}_4$, $\mathbf{C}_{4v}$, $\mathbf{D}_{2d}$ | $\mathbf{D}_{4h}$ | $I(\mathbf{u}_1\mathbf{u}_2)$ |
| $\Gamma_6 \leftrightarrow \Gamma_6$<br>$\Gamma_7 \leftrightarrow \Gamma_7$ | $\Gamma_6^{\pm} \leftrightarrow \Gamma_6^{\pm}$<br>$\Gamma_7^{\pm} \leftrightarrow \Gamma_7^{\pm}$ | $A_1(\mathbf{u}_1\mathbf{u}_2) + c_1 A_2(\mathbf{u}_1\mathbf{u}_2) + c_2 E(\mathbf{u}_1\mathbf{u}_2)$ |
| $\Gamma_6 \leftrightarrow \Gamma_7$ | $\Gamma_6^{\pm} \leftrightarrow \Gamma_7^{\pm}$ | $B_1(\mathbf{u}_1\mathbf{u}_2) + c_1 B_2(\mathbf{u}_1\mathbf{u}_2) + c_2 E(\mathbf{u}_1\mathbf{u}_2)$ |

$\sigma$-polarized, while the two lines $\Gamma_6^+ \to \Gamma_6^-$ have mixed $\pi$- and $\sigma$-polarization. The mixed polarization is caused by the tetragonal field and by spin–orbit interaction. Using the basis functions (11.9) and (11.10), it is easy to obtain the line-intensity ratios for transition to the levels $\Gamma_6^-$, $\Gamma_6^-$ and $\Gamma_7^-$ (in increasing order of energy; see Fig. 11.3):

$$I_\pi(\Gamma_6^-) : I_6(\Gamma_6^-) : I_\pi(\Gamma_6^-) : I_\sigma(\Gamma_6^-) : I_6(\Gamma_7^-) = 2:2:4:1:3.$$

The angular dependence of the fine-structure levels is axially symmetric, (9.24).

The polarization dependences of the two-photon transition between states relating to the double-valued representations are calculated according to the method given in Section 9.5. The results for the 32 point groups have been obtained in [71], some are given in Table 11.6 (in the notation of Section 9.5).

## PROBLEMS

**11.1** Classify the fine-structure levels of the $^2T_2$ term in a trigonal ($\mathbf{D}_3$) crystal field.

**11.2** Classify the fine-structure levels of the $^5T_1$ term in a cubic field, and of the $^4T_2$ term in a tetragonal field.

**11.3** Construct $\Gamma_8(^2E)$ wavefunctions and calculate the matrix elements of $\hat{H}_{so}$, and show that the cubic $^2E$ term is not split by spin–orbit interaction.

**11.4** Determine the selection rules for mixing the $S\Gamma$ terms of $d^2$ and $d^3$ ions (Tables 5.3, 5.13 and 5.17) in an $\mathbf{O}_h$ crystal field.

**11.5** Perform the same calculations for a $d^3$ ion in a $\mathbf{D}_{4h}$ field (Table 5.3).

**11.6** Calculate all matrix elements connecting $^2T_2$ and $^2E$ terms.

**11.7** Calculate the spin–orbit splittings of $^{2S+1}$T-terms (Table 11.4).

**11.8** Determine the symmetric and antisymmetric parts of all cubic-group products $\Gamma \times \Gamma$.

**11.9** Determine the intensity distribution of the fine-structure optical lines for the magnetic dipole transition $^2E_g \rightarrow {}^2T_{2g}(\mathbf{O}_h)$.

**11.10** Determine the symmetries of the $\Gamma_6^-$ and $\Gamma_8^-$ components of the $^2T_{1u}(\mathbf{O}_h)$ term in a rhombic field $\mathbf{D}_{2h}$ and calculate the intensity ratios of the optical lines arising from transitions $^2A_{1g} \rightarrow {}^2T_{1u}$.

# 12 Electron Paramagnetic Resonance

## 12.1 MAGNETIC RESONANCE PHENOMENA

*Electron paramagnetic resonance (EPR)* or *electron spin resonance (ESR)* is the resonant absorption of electromagnetic waves by electron spins in a magnetic field. This definition is somewhat simplified, since the spin of an electron is coupled with its orbital angular momentum in a crystal field, and interacts with the spins of other electrons, nuclear spins etc. The large amount of useful information obtained using EPR in solid state physics and in chemistry, biology and medicine is derived from analysis of these interactions and from the study of the effects of material structure and elementary processes on the observed spectra. The theory of EPR, together with its applications, is discussed in great detail in several excellent monographs (see e.g. [75, 76]). In this chapter we shall consider only the group-theoretical aspects of EPR.

The interaction of a free ion in the $LSJ$ state (where $J$ is the total angular momentum; see (Section 5.1) and a magnetic field $\mathscr{H}$ is described by the Hamiltonian

$$\hat{H}_z = g\beta\mathscr{H}\cdot\hat{\mathbf{J}}, \tag{12.1}$$

where $\hat{\mathbf{J}}$ is the total angular momentum operator, $\beta$ is the Bohr magneton and $g$ is the Landé factor (for the free electron $g = 2.0023$). In the Russell–Saunders scheme the $g$ factor depends on the quantum numbers $L$, $S$ and $J$:

$$g = 1 + \frac{J(J+1) - L(L+1) + S(S+1)}{2J(J+1)}. \tag{12.2}$$

The energy levels $\epsilon(M_J) = g\beta\mathscr{H}M_J$ are enumerated by the quantum number $M_J = -J, -(J-1), \ldots, J$ and are independent of the direction of the external field; they depend only on its magnitude $\mathscr{H}$. Figure 12.1 illustrates the splitting of the $LSJ$ term into $2J+1$ magnetic Zeeman sublevels. It is the transitions

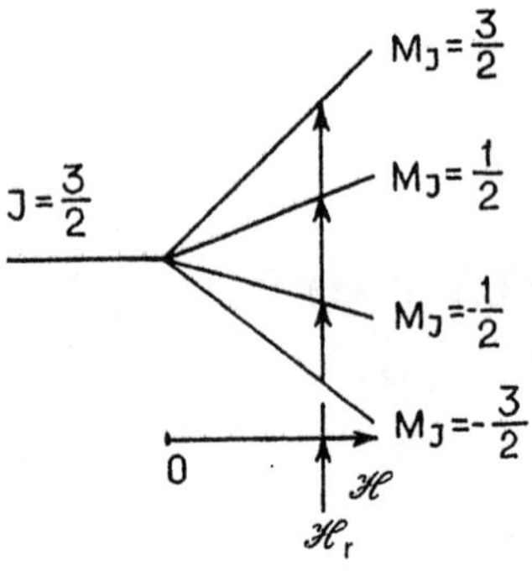


**Fig. 12.1** The Zeeman effect in a free atom.

among these levels, caused by a magnetic field, that give rise to electron paramagnetic resonance. The selection rules admit transitions between neighbouring Zeeman levels $\epsilon(M_J)$ and $\epsilon(M_J+1)$, so that the resonance condition is expressed as

$$\epsilon(M_J) - \epsilon(M_J - 1) = g\beta\mathscr{H} = h\nu \tag{12.3}$$

where $\nu$ is the external field frequency. In EPR experiments a fixed frequency $\nu$ is usually used, resonance being achieved by variation of the static field $\mathscr{H}$. The resonance fields $\mathscr{H}_r = h\nu/g\beta$ are identical for all transitions in the atomic $LSJ$ term for any field direction. Such an EPR spectrum is called *isotropic*.

The EPR spectrum of a paramagnetic ion surrounded by ligands is characterized by another type of $g$ factor and by a few lines, and is essentially anisotropic. These spectral characteristics often lead to important conclusions concerning material structure. The theoretical analysis of the spectrum is performed using the so-called effective Hamiltonian.

## 12.2 THE SPIN HAMILTONIAN

### 12.2.1 Zero-field splittings

This section deals with *orbitally nondegenerate levels*, having corresponding one-dimensional representations of a point group. Such purely spin multiplets will be denoted briefly as ${}^{2S+1}$A. In the absence of a magnetic field the spin–orbit interaction in second (or higher) order of perturbation theory splits the spin multiplets ${}^{2S+1}$A (see Section 11.3.3). This is called *zero-field splitting*. The effective Hamiltonian for the ${}^{2S+1}$A terms (spin Hamiltonian) contains only spin operators and numerical constants.

Let us first formulate the general requirements that must be satisfied by an *effective Hamiltonian* (in particular by a *spin Hamiltonian*), and then show how these requirements are to be satisfied in practice. The spin Hamiltonian must be

(1) invariant under all point-group transformations;

(2) invariant under time inversion;

(3) Hermitian (i.e. reflection of matrix elements across the main diagonal—transposition—simultaneously with complex conjugation leaves the operator matrix unchanged); this ensures that the spin Hamiltonian eigenvalues are real.

The spin Hamiltonian is constructed as a linear combination of the spin operators $\hat{S}_x$, $\hat{S}_y$, $\hat{S}_z$, their products $\hat{S}_\alpha^2$, $\hat{S}_\alpha \hat{S}_\beta$ and higher powers (the highest power being $2S$). It should be kept in mind that the spin operators reverse sign under time inversion (i.e. they are $T$-odd, while the quadratic terms $\hat{S}_\alpha \hat{S}_\beta$ are $T$-even). In most cases it is sufficient to take into account only the quadratic terms of the spin Hamiltonian, omitting higher powers. The procedure for constructing *invariant expressions* is as follows:

(1) determine the point-group symmetry;

(2) from tables in [1] find the irreducible representations $\Gamma$ whose basis is provided by the components $\hat{S}_x$, $\hat{S}_y$ and $\hat{S}_z$;

(3) for each representation $\Gamma$ form the scalar expression

$$\hat{W}_{\mathrm{A}}(\Gamma) = \sum_\gamma |\hat{S}_{\Gamma\gamma}|^2,$$

where $\hat{S}_{\Gamma\gamma}$ are the basis functions $\hat{S}_x$, $\hat{S}_y$ and $\hat{S}_z$;

(4) form a scalar sum of the form

$$\hat{H} = \sum_\Gamma c(\Gamma)\, \hat{W}_{\mathrm{A}}(\Gamma), \tag{12.4}$$

where $c(\Gamma)$ are phenomenological parameters (i.e. parameters not related by symmetry); the number of these parameters in the spin Hamiltonian $\hat{H}$ is equal to the number of irreducible representations $\Gamma$.

Before turning to this construction, we should note the following. Suppose that the Hamiltonian is expressed as a diagonal $2 \times 2$ matrix of the form

$$\mathbf{H} = \begin{bmatrix} a & 0 \\ 0 & b \end{bmatrix} \qquad (a = H_{11},\, b = H_{22})$$

having eigenfunctions $\epsilon_1 = a$ and $\epsilon_2 = b$. Only the energy gap $\epsilon_1 - \epsilon_2 = a - b$ has physical importance; it determines the frequency of the absorbed electromagnetic waves. Therefore the matrix $\mathbf{H}$ contains an excessive number of parameters (two instead of one). Elimination of the redundant parameter is possible if the energy reference level is changed. We shall take as the reference level the centre of the energy gap $\epsilon_1 - \epsilon_2$; that is, $\frac{1}{2}(a+b)$. Then $\epsilon_1 = \frac{1}{2}(a+b)$ and $\epsilon_2 = -\frac{1}{2}(a+b)$, while $\epsilon_1 - \epsilon_2$ is still equal to $a+b$. Since $a+b = \mathrm{Tr}\,\mathbf{H}$ (the trace of the matrix $\mathbf{H}$), we may eliminate the "redundant" parameter by considering instead of $\mathbf{H}$ the matrix

$$\mathbf{H}' = \mathbf{H} - \frac{1}{2}(\mathrm{Tr}\ \mathbf{H})\,\mathbf{1}, \qquad \mathbf{1} = \begin{bmatrix} 1 & 0 \\ 0 & 1 \end{bmatrix}, \tag{12.5}$$

where

$$\mathbf{H}' = \begin{bmatrix} \frac{1}{2}(a+b) & 0 \\ 0 & -\frac{1}{2}(a+b) \end{bmatrix}.$$

The new matrix $\mathbf{H}'$, containing only *one* parameter $(a+b)$, rather than two ($a$ and $b$) is traceless: $\mathrm{Tr}\ \mathbf{H}' = 0$.

Returning to the spin Hamiltonian (12.4), we impose the additional requirement that $\mathrm{Tr}\ \mathbf{H} = 0$, where $\mathbf{H}$ is the matrix of the operator (12.4) with dimension $2S+1$, calculated using the spin functions $\Phi(SM)$. The condition of zero trace eliminates one of the phenomenological parameters, expressing it in terms of the others.

**Example 12.1** Consider the term $^3$A ($^3A_1$, $^3A_2$, $^3B_1$ and $^3B_2$) in a $\mathbf{D}_4$ symmetry crystal field. The components of the vector $\hat{\mathbf{S}}$ transform according to the irreducible representations $\Gamma = A_2(\hat{S}_z)$ and $E(\hat{S}_x, \hat{S}_y)$; that is, $\hat{S}_{A_2} = \hat{S}_z$, $\hat{S}_{Ex} = \hat{S}_x$ and $\hat{S}_{Ey} = \hat{S}_y$. The scalar operators $W_A(\Gamma)$ are $W_A(A_2) = \hat{S}_z^2$ and $W_A(E) = \hat{S}_x^2 + \hat{S}_y^2$, while the full spin Hamiltonian is

$$\hat{H} = c(A_2)\hat{S}_z^2 + c(E)(\hat{S}_x^2 + \hat{S}_y^2). \tag{12.6}$$

It should be noted that $\hat{S}_x^2 + \hat{S}_y^2 = \hat{\mathbf{S}}^2 - S_z^2$, whereby $\hat{\mathbf{S}}^2\Phi(SM) = S(S+1)\Phi(SM)$. Hence the term containing $\hat{\mathbf{S}}^2$ is proportional to the unit matrix; it can be omitted since it simply shifts all the eigenvalues in the same way. After omitting $c(E)\hat{\mathbf{S}}^2$, we transform $\hat{H}$ into the form $[c(A_2) - c(E)]\hat{S}_z^2$. Since the $x$, $y$ and $z$ axes are equivalent,

$$\mathrm{Tr}\ \hat{S}_x^2 = \mathrm{Tr}\ \hat{S}_y^2 = \mathrm{Tr}\ \hat{S}_z^2 = \tfrac{1}{3}\mathrm{Tr}\hat{\mathbf{S}}^2 = \tfrac{1}{3}S(S+1)\sum_{M=-S}^{S} 1 = \tfrac{1}{3}S(S+1)(2S+1).$$

Finally, we obtain

$$\hat{H} = D[\hat{S}_z^2 - \tfrac{1}{3}S(S+1)]. \tag{12.7}$$

This spin Hamiltonian meets all of the requirements mentioned above, and $\mathrm{Tr}\ \mathbf{H} = 0$. This latter condition does not reduce the number of parameters, since $D$ is the only one.

A spin Hamiltonian of the form (12.7) is obtained for all point groups with $C_n$ axes, for $n \geq 3$, and is called an *axial spin Hamiltonian*. It is valid for any $S$, since the deduction of (12.7) did not assume that $S = 1$.

Returning to the case of the $^3$A term ($\mathbf{D}_4$), we construct the matrix $\mathbf{H}$:

$$
\left.
\begin{array}{l}
M = 1 \quad 0 \quad -1 \\
\hat{S}_z = \begin{bmatrix} 1 & 0 & 0 \\ 0 & 0 & 0 \\ 0 & 0 & -1 \end{bmatrix}, \quad \hat{S}_z^2 = \begin{bmatrix} 1 & 0 & 0 \\ 0 & 0 & 0 \\ 0 & 0 & 1 \end{bmatrix}, \quad \mathbf{H} = \begin{bmatrix} \frac{1}{3}D & 0 & 0 \\ 0 & -\frac{2}{3}D & 0 \\ 0 & 0 & \frac{1}{3}D \end{bmatrix},
\end{array}
\right\} \qquad (12.8)
$$

which leads to the eigenvalues $\epsilon_1 = \epsilon_3 = \frac{1}{3}D$ and $\epsilon_2 = -\frac{2}{3}D$, with $\epsilon_1 + \epsilon_2 + \epsilon_3 = 0$.

The $^3A_2$ splitting into the doublet $(\epsilon_1, \epsilon_3)$ and the singlet $(\epsilon_2)$ could have been predicted. Let us assume that $A = A_1$; then in $\mathbf{D}_4$ we have $A_1 \times D^{(1)} \doteq A_1 \times (A_2 + E) \doteq A_2 + E$, meaning that the fine structure of the $^3A_1$ term consists of the singlet $A_2$ and the doublet E. The basis functions $\Phi(1, 1)$, $\Phi(1, -1)$ (doublet E) and $\Phi(1, 0)$ diagonalize the spin Hamiltonian. The above reasoning shows that the preliminary group-theoretical classification of fine-structure levels (Section 11.3) gives the correct wavefunctions directly. If the basis has been chosen as shown, the spin Hamiltonian has a matrix which is either diagonal or a block diagonal (when there are repeated representations in $\Gamma \times D^{(S)}$).

**Example 12.2** Consider the $^3$A term for $\mathbf{D}_2$ triaxial symmetry. From tables in [1] we find the basis $B_1(\hat{S}_y)$, $A_2(\hat{S}_z)$, $B_2(\hat{S}_x)$. Squared one-dimensional representations provide a totally symmetric $A_1$, so that the spin Hamiltonian contains three parameters:

$$\hat{H} = c(B_1)\hat{S}_y^2 + c(A_2)\hat{S}_z^2 + c(B_2)\hat{S}_x^2. \qquad (12.9)$$

Using instead of $\hat{S}_z^2$ the traceless matrix of $\hat{S}_z^2 - \frac{1}{3}S(S+1)$, we obtain (after redefining the constants)

$$\hat{H} = D[\hat{S}_z^2 - \tfrac{1}{3}S(S+1)] + c(B_1)\hat{S}_y^2 + c(B_2)\hat{S}_x^2. \qquad (12.10)$$

We can now see that Tr $\mathbf{H} = 0$ if $c(B_2) = -c(B_1) = E$, and we obtain

$$\hat{H} = D[\hat{S}_z^2 - \tfrac{1}{3}S(S+1)] + E(\hat{S}_x^2 - \hat{S}_y^2). \qquad (12.11)$$

The spin Hamiltonian (12.11) contains two (not three!) parameters $D$ and $E$, called respectively the *axial field* and *rhombicity* parameters. For $S = 1$

$$\hat{S}_x = \frac{1}{\sqrt{2}}\begin{bmatrix} 0 & 1 & 0 \\ 1 & 0 & 1 \\ 0 & 1 & 0 \end{bmatrix}, \quad \hat{S}_y = \frac{1}{\sqrt{2}}\begin{bmatrix} 0 & -i & 0 \\ i & 0 & -i \\ 0 & i & 0 \end{bmatrix}, \quad \hat{S}_x^2 - \hat{S}_y^2 = \begin{bmatrix} 0 & 0 & 1 \\ 0 & 0 & 0 \\ 1 & 0 & 0 \end{bmatrix}. \qquad (12.12)$$

Therefore in the basis $\Phi(1M)$ we have

$$
\begin{array}{l}
M = -1 \quad 0 \quad 1 \\
\mathbf{H} = \begin{bmatrix} \frac{1}{3}D & 0 & E \\ 0 & -\frac{2}{3}D & 0 \\ E & 0 & \frac{1}{3}D \end{bmatrix},
\end{array} \qquad (12.13)
$$

which gives three nondegenerate levels:

$$\epsilon_1 = \tfrac{1}{3}D + E, \quad \epsilon_2 = \tfrac{1}{3}D - E, \quad \epsilon_3 = -\tfrac{2}{3}D. \tag{12.14}$$

Let us now obtain the eigenvalues using the diagonalizing basis. Assume, for definiteness, that $A = A_1$ in $\mathbf{D}_2$

$$D^{(1)} = B_1 + A_2 + B_2, \quad A_1 \times D^{(1)} = B_1(y) + A_z(z) + B_2(x),$$

which corresponds to the total splitting of the $^3A$ multiplet. To construct a correct basis, it is necessary to pass from the functions $\Phi(1M)$ $(M = 0, \pm 1)$ to the basis functions $\Phi(B_1)$, $\Phi(A_2)$ and $\Phi(B_2)$, which transform like $y$, $z$ and $x$ (see Table 3.1):

$$\left.\begin{aligned} &\Phi(B_1) = \sqrt{\tfrac{1}{2}}\,i\,[\Phi(1, 1) - \Phi(1, -1)], \\ &\Phi(B_2) = \sqrt{\tfrac{1}{2}}[\Phi(1, 1) + \Phi(1, -1)], \quad \Phi(A_2) = \Phi(1, 0). \end{aligned}\right\} \tag{12.15}$$

Since there are no repeating representations, the matrix $\mathbf{H}$ in the basis (12.15) is diagonal:

$$\begin{matrix} & B_2 & B_1 & A_2 \end{matrix}$$

$$\mathbf{H} = \begin{bmatrix} \tfrac{1}{3}D + E & 0 & 0 \\ 0 & \tfrac{1}{3}D - E & 0 \\ 0 & 0 & -\tfrac{2}{3}D \end{bmatrix}, \tag{12.16}$$

and its diagonal elements are the eigenvalues (12.14).

**Example 12.3** Consider the $^4A_2\,(S = \tfrac{3}{2})$ term in $\mathbf{D}_3$. The spin Hamiltonian takes the axial form (12.7) (with $z$ axis along the trigonal axis); that is, it is a diagonal Hamiltonian. The $\epsilon(|M|)$ levels are $\epsilon(\tfrac{3}{2}) = D$ and $\epsilon(\tfrac{1}{2}) = -D$. The degeneracy is not removed even if the rhombic contribution $E(\hat{S}_x^2 - S_y^2)$ is taken into account.

In conclusion, it should be emphasized that the *spin Hamiltonian operates only in the space of spin functions of an orbital singlet.* The calculation of the parameters $D$, $E$ etc. is a separate problem, which is out of the scope of symmetry consideration.

### 12.2.2 Zeeman interaction

Unlike free atoms (Section 12.1), a paramagnetic ion in a crystal field is described by an *anisotropic Zeeman Hamiltonian*, the most general form of which is

$$\begin{aligned} \hat{H}_z &= \beta \sum_{ik} g_{ik} \hat{S}_i \mathcal{H}_k \\ &= \beta(g_{xx}\hat{S}_x\mathcal{H}_x + g_{yy}\hat{S}_y\mathcal{H}_y + g_{zz}\hat{S}_z\mathcal{H}_z + g_{xy}\hat{S}_x\mathcal{H}_y \\ &\quad + g_{yx}\hat{S}_y\mathcal{H}_x + g_{xz}\hat{S}_x\mathcal{H}_z + g_{zx}\hat{S}_z\mathcal{H}_x + g_{yz}\hat{S}_y\mathcal{H}_z + g_{zy}\hat{S}_z\mathcal{H}_y), \end{aligned} \tag{12.17}$$

where $g_{xx}$, $g_{yy}$ and $g_{zz}$ are the diagonal components of the $g$ tensor, and $g_{ik}$ are its nondiagonal components. The magnetic field is produced by moving charges, whose velocity reverses sign on time inversion. Therefore the axial vector $\mathscr{H}$ is an $T$-odd vector, while the Hamiltonian $\hat{H}_Z$, containing the product of two $T$-odd quantities, is invariant under time inversion.

The requirement of $\hat{H}_Z$ invariance in a point group is satisfied as follows:

(1) the irreducible representations $\Gamma$ corresponding to the components $\hat{S}_x$, $\hat{S}_y$ and $\hat{S}_z$ are determined;

(2) $\mathscr{H}_x$, $\mathscr{H}_y$ and $\mathscr{H}_z$ are the bases of these $\Gamma$;

(3) the scalar expressions

$$\sum_{\gamma} \hat{\varphi}_{\Gamma\gamma}(\mathbf{S})\varphi_{\Gamma\gamma}(\mathscr{H}) \equiv \hat{W}_Z(\Gamma)$$

are formed;

(4) the Zeeman Hamiltonian is a linear combination of the operators $\Sigma_{\Gamma} g(\Gamma)\hat{W}_Z(\Gamma)$, where $g_{\Gamma}$ are $g$ tensor components.

**Example 12.4** Consider $\mathbf{D}_4$, and $\Gamma = \mathrm{A}_2(\hat{S}_z$ and $\mathscr{H}_Z)$, $\mathrm{E}(\hat{S}_x, \hat{S}_y$ and $\mathscr{H}_x, \mathscr{H}_y)$:

$$\hat{W}_Z(\mathrm{A}_2) = \hat{S}_z\mathscr{H}_z, \qquad \hat{W}_Z(\mathrm{E}) = \hat{S}_x\mathscr{H}_x + \hat{S}_y\mathscr{H}_y. \tag{12.18}$$

From (12.18) and (12.17) it can be seen that for $\mathbf{D}_4$ symmetry only three out of the nine components of the $g$ tensor are nonzero: $g_{zz} = g(\mathrm{A}_2)$, $g_{xx}$ and $g_{yy}$. In fact $g_{xx} = g_{yy} = g(\mathrm{E})$. The accepted notation is $g_{zz} \equiv g_{\parallel}$, $g_{xx} = g_{yy} = g_{\perp}$, using which the spin Hamiltonian becomes

$$\hat{H}_Z = \beta g_{\parallel}\hat{S}_z\mathscr{H}_z + \beta g_{\perp}(\hat{S}_x\mathscr{H}_x + \hat{S}_y\mathscr{H}_y). \tag{12.19}$$

Let us determine the energy levels for the simplest case of the Kramers doublet $^2\mathrm{A}$. Using the $S_x$, $S_y$ and $S_z$ matrices for $S = \frac{1}{2}$, (11.15), we find the matrix of the Hamiltonian (12.19):

$$\mathbf{H}_Z = \frac{\beta\mathscr{H}}{2}\begin{bmatrix} g_{\parallel}\cos\vartheta & g_{\perp}\sin\vartheta(\cos\varphi - \mathrm{i}\sin\varphi) \\ g_{\perp}\sin\vartheta(\cos\varphi + \mathrm{i}\sin\varphi) & -g_{\parallel}\cos\vartheta \end{bmatrix}. \tag{12.20}$$

Diagonalizing this matrix, we obtain the dependence of the Zeeman sublevels on the direction of $\mathscr{H}(\vartheta, \varphi)$.

$$\epsilon_{\pm} = \pm\tfrac{1}{2}\beta\mathscr{H}(g_{\parallel}^2\cos^2\vartheta + g_{\perp}^2\sin^2\vartheta)^{1/2}. \tag{12.21}$$

The angular dependence of the EPR spectrum is axial, since the level position depends on $\vartheta$ only (and not on $\varphi$).

The axial spectrum following from the Hamiltonian (12.19) is characteristic of all point groups with $C_n$ axes for $n \geq 3$.

**Example 12.5** Consider the case of $\mathbf{D}_2$ symmetry, and $\Gamma = \mathrm{B}_1\,(\hat{S}_y$ and $\mathscr{H}_y)$, $\mathrm{A}_2\,(\hat{S}_z$ and $\mathscr{H}_z)$, $\mathrm{B}_2\,(\hat{S}_x$ and $\mathscr{H}_x)$, which leads to a spin Hamiltonian of triaxial type:

$$\hat{H}_Z = \beta(g_x \hat{S}_x \mathscr{H}_x + g_y \hat{S}_y \mathscr{H}_y + g_z \hat{S}_z \mathscr{H}_z), \tag{12.22}$$

where

$$g_x = g_{xx} = g(\mathrm{B}_2), \qquad g_y = g_{yy} = g(\mathrm{B}_1), \qquad g_z = g_{zz} = g(\mathrm{A}_2).$$

The spectrum possesses triaxial symmetry; that is the level splitting for a given field magnitude $\mathscr{H}$ depends on both $\vartheta$ and $\varphi$.

**Example 12.6** Consider $\mathbf{C}_2$, in which (see the tables in [1]) there are two one-dimensional irreducible representations A and B, whereby $\hat{S}_z$ and $\mathscr{H}_z$ form the basis of A. According to the representation B, both $\hat{S}_x$ and $\hat{S}_y$ are transformed (as are $\mathscr{H}_x$ and $\mathscr{H}_y$). Therefore, along with scalars like $\hat{W}_Z(\mathrm{A}) = \hat{S}_z \mathscr{H}_z$, $\hat{W}_Z(\mathrm{B}) = \hat{S}_x \mathscr{H}_x$ and $\hat{W}_Z(\mathrm{B}) = \hat{S}_y \mathscr{H}_y$, another two scalars may be added: $\hat{W}_\Gamma(\mathrm{B}) = \hat{S}_x \mathscr{H}_y$ and $\hat{W}_Z(\mathrm{B}) = \hat{S}_y \mathscr{H}_x$. The symmetry conditions thus determine five independent components of the $g$ tensor: $g_{zz} = g(\mathrm{A})$ and four $g(\mathrm{B})$-type quantities: $g_{xx}$, $g_{yy}$, $g_{xy}$ and $g_{yx}$. The spin Hamiltonian is

$$\hat{H}_Z = g_{zz} \hat{S}_z \mathscr{H}_z + g_{xx} \hat{S}_x \mathscr{H}_x + g_{yy} \hat{S}_y \mathscr{H}_y + g_{xy} \hat{S}_x \mathscr{H}_y + g_{yx} \hat{S}_y \mathscr{H}_x. \tag{12.23}$$

Since $\mathbf{C}_2$ has a low degree of symmetry, nondiagonal components of the $g$ tensor appear in the $(x, y)$ plane.

**Example 12.7** Consider $\mathbf{C}_i$. Each $\hat{S}_i$ and $\mathscr{H}_j$ component is transformed according to the totally symmetric representation $\mathrm{A}_g$ (i.e. it is even under inversion). Therefore the Zeeman Hamiltonian includes all $\hat{S}_i \mathscr{H}_j$ and has the most general form (12.17).

The analysis of low-symmetry centres is a complex problem, owing to the great number of independent parameters in the spin Hamiltonian. In this connection one should see [77, 78]. In some cases the procedures used for the construction of invariants lead to a redundant number of phenomenological parameters. In [78] a procedure is given for eliminating inseparable combinations of parameters; such combinations are implicitly contained in the spin Hamiltonian of a low-symmetry paramagnetic centre. The general concept of a spin Hamiltonian is presented in [79].

## 12.3 HYPERFINE INTERACTION FOR SPIN MULTIPLETS

The magnetic hyperfine interaction (HFI) arises as the result of coupling between the electron and nuclear magnetic moments. In a free (spherically symmetric) atom, the HFI is proportional to the scalar product of the electron-shell total angular momentum operator $\hat{\mathbf{J}}$ and the nuclear spin operator $\hat{\mathbf{I}}$:

$$\hat{H}_{\text{hf}} = A(\hat{\mathbf{J}} \cdot \hat{\mathbf{I}}), \tag{12.24}$$

where $A$ is called the *HFI parameter*. The scalar product $\hat{\mathbf{J}} \cdot \hat{\mathbf{I}}$ is invariant under rotations, i.e. it is a scalar of the rotation group. In point groups HFI must be invariant under point-group operations, and must also satisfy the requirements of time-inversion symmetry and Hermeticity. Therefore in its most general form the HFI Hamiltonian for the orbital $^{2S+1}$A singlet of a paramagnetic ion in a crystal is

$$\hat{H}_{\text{hf}} = \sum_{ik} A_{ik} \hat{S}_i \hat{I}_k \quad (i, k = x, y, z), \tag{12.25}$$

where $A_{ik}$ is a phenomenological parameter set, called the *HFI tensor* (or A *tensor*).

The Hamiltonian (12.25) is $T$-invariant, since the operators $\hat{I}_i$ and $\hat{S}_k$ are $T$-odd. Invariance in a point group is achieved just as for the Zeeman spin Hamiltonian (Section 12.2.2). To determine the HFI operator, it is necessary to make in $\hat{H}_Z$ the substitutions $\beta g_{ik} \to A_{ik}$ and $\mathscr{H}_k \to I_k$. In cubic groups $\hat{S}_x$, $\hat{S}_y$ and $\hat{S}_z$ ($\hat{I}_x$, $\hat{I}_y$ and $\hat{I}_z$) provide one irreducible representation ($T_{1g}$ is in $\mathbf{O}_h$). The only scalar is $\hat{\mathbf{I}} \cdot \hat{\mathbf{S}}$ and the HFI operator is isotropic

$$\hat{H}_{\text{hf}} = A\,\hat{\mathbf{I}} \cdot \hat{\mathbf{S}}. \tag{12.26}$$

For point groups with $C_n$ axis ($n \geq 3$) we obtain

$$\hat{H}_{\text{hf}} = A_{\parallel} \hat{S}_z \hat{I}_z + A_{\perp}(\hat{S}_x \hat{I}_x + \hat{S}_y \hat{I}_y), \tag{12.27}$$

where $A_{\parallel} = A_{zz}$ and $A_{\perp} = A_{xx} = A_{yy}$. In the case of rhombic symmetry

$$\hat{H}_{\text{hf}} = A_x \hat{S}_x \hat{I}_x + A_y \hat{S}_y \hat{I}_y + A_z \hat{S}_z \hat{I}_z, \tag{12.28}$$

where $A_i = A_{ii}$. Finally, for lower-symmetry point groups the HFI Hamiltonian contains nondiagonal components of the A tensor. Like the Zeeman Hamiltonian, the HFI Hamiltonian is constructed proceeding only from symmetry considerations. Such an approach does not provide information concerning the quantities $A_{ik}$, it only gives the maximum number of different components of the A tensor.

The spectroscopic consequences of Zeeman interaction anisotropy and of HFI for the Kramers doublet $^2$A$(S = \frac{1}{2})$ and $I = \frac{1}{2}$ are shown in Fig. 12.2 for the case of axial symmetry. In a parallel field ($\mathscr{H} \parallel C_n$), if there is no HFI, the EPR line corresponds to the resonance field $\mathscr{H}_r = h\nu/g_{\parallel}\beta$, while in a perpendicular field ($\mathscr{H} \perp C_n$) it corresponds to the resonance field $\mathscr{H}_r = h\nu/g_{\perp}\beta$ (dashed line). When $I = \frac{1}{2}$, the HFI splits the EPR line in two, the intervals between the two lines in the field scale being $A_{\parallel}/g_{\parallel}\beta$ and $A_{\perp}/g_{\perp}\beta$ for the parallel and perpendicular fields respectively. Owing to the axial symmetry of the Zeeman Hamiltonian and HFI, the corresponding splittings

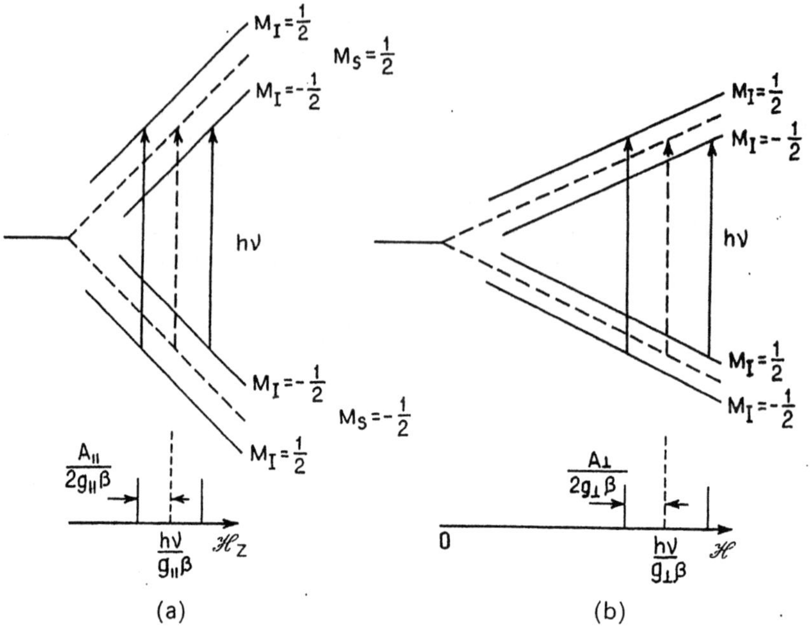


**Fig. 12.2** The EPR spectrum and its hyperfine structure for a Kramers doublet in an axial field ($I=\frac{1}{2}$): (a) $\mathscr{H} \parallel C_n$; (b) $\mathscr{H} \perp C_n$.

depend only on the angle $\vartheta$, while variation of $\varphi$ (field rotation in the plane $\sigma \perp C_n$) does not change the spectrum.

## 12.4 ELECTRIC FIELD EFFECTS

### 12.4.1 Linear electric field effect

An external electric field shifts and splits EPR lines and provides important information about the paramagnetic centre. Detailed discussions of the electric field technique and its most important results are given in [80–82]. The group-theoretical approach to the problem consists in constructing the effective Hamiltonian of a paramagnetic ion in a crystal field interacting with a magnetic field $\mathscr{H}$ and electric field $\mathscr{E}$.

The spin Hamiltonian including the field $\mathscr{E}$ must satisfy the invariance requirements mentioned above. Taking into account that $\mathscr{E}$ is a $T$-even quantity, its components $\mathscr{E}_i$ may be multiplied only by even powers of $\hat{S}_i$, while products of the type $\hat{S}_i \mathscr{H}_k \mathscr{E}_l$ are $T$-even.

The construction of an electric field spin Hamiltonian involves the following steps:

(1) allocation of the operators $\hat{S}_x$, $\hat{S}_y$ and $\hat{S}_z$ to the irreducible representations $\Gamma$ of the paramagnetic ion point group;

(2) determination of the irreducible representations $\Gamma'$ according to which the three components $\mathscr{E}_x$, $\mathscr{E}_y$, and $\mathscr{E}_z$ of the polar vector $\boldsymbol{\mathscr{E}}$ are transformed;

(3) formation of all direct products $\Gamma_1 \times \Gamma_2$, where $\Gamma_1$ and $\Gamma_2$ belong to the set $\Gamma$;

(4) use of the multiplication table of the group's irreducible representations; all $\Gamma_1 \times \Gamma_2 \doteq \bar{\Gamma}_1 + \bar{\Gamma}_2 + \ldots$ should be reduced;

(5) use of the Clebsch–Gordan expansion

$$\hat{\varphi}_{\bar{\Gamma}\bar{\gamma}}(\Gamma_1 \Gamma_2 \hat{\mathbf{S}}) = \sum_{\gamma_1 \gamma_2} \hat{\mathbf{S}}_{\Gamma_1 \gamma_1} \cdot \hat{\mathbf{S}}_{\Gamma_2 \gamma_2} \langle \Gamma_1 \gamma_1 \Gamma_2 \gamma_2 | \bar{\Gamma} \bar{\gamma} \rangle \tag{12.29}$$

to find the quadratic (with respect to spin operators) symmetry adopted combinations $\hat{\varphi}_{\bar{\Gamma}\bar{\gamma}}(\Gamma_1 \Gamma_2 \hat{\mathbf{S}})$ forming the bases of the irreducible representations $\bar{\Gamma}$;

(6) formation of the scalar

$$\hat{W}_{\mathscr{E}}(\bar{\Gamma} \Gamma_1 \Gamma_2) = \sum_{\bar{\gamma}} \hat{\varphi}_{\bar{\Gamma}\bar{\gamma}}(\Gamma_1 \Gamma_2 \hat{\mathbf{S}}) \mathscr{E}_{\bar{\Gamma}\bar{\gamma}}, \tag{12.30}$$

comprising only those $\hat{\varphi}_{\bar{\Gamma}\bar{\gamma}}$ which transform according to the same representations $\Gamma' = \bar{\Gamma}$ as the components $\mathscr{E}_i$;

(7) the linear electric field spin Hamiltonian is given as the combination

$$\hat{W}_{\mathscr{E}} = \sum_{\bar{\Gamma} \Gamma_1 \Gamma_2} c(\bar{\Gamma} \Gamma_1 \Gamma_2)\, \hat{W}_{\mathscr{E}}(\bar{\Gamma} \Gamma_1 \Gamma_2), \tag{12.31}$$

where $c$ (...) are semiempirical parameters.

The procedure is illustrated below for the cubic group.

**Example 12.8** Consider $\mathbf{T}_d$, and the linear electric effect. Three components $\mathscr{E}_x$, $\mathscr{E}_y$ and $\mathscr{E}_z$ form the basis of $T_2$, while $\hat{S}_x$, $\hat{S}_y$ and $\hat{S}_z$ ($\mathscr{H}_x$, $\mathscr{H}_y$ and $\mathscr{H}_z$) form the basis of $T_1$ [1]. The Zeeman Hamiltonian is isotropic:

$$\hat{W}_Z = g\beta \hat{\mathbf{S}} \cdot \mathscr{H}. \tag{12.32}$$

The linear (with respect to $\mathscr{E}$ ) electric field contribution $\hat{W}_{\mathscr{E}}$ is formed as a scalar from the components of $\mathscr{E}$ and the quadratic combinations $\hat{\varphi}_{T_2\gamma}(\mathrm{S})$, which transform like the components of $\mathscr{E}$ (i.e. according to the representation $T_2$). These combinations exist, since $T_1 \times T_1 \doteq A_1 + E + T_1 + T_2$ contains $T_2$. They can be written using the Clebsch–Gordan coefficients as

$$\hat{\varphi}_{T_2\gamma}(\hat{\mathbf{S}}) = \sum_{\gamma_1\gamma_2} \hat{\mathbf{S}}_{T_1\gamma_1} \cdot \hat{\mathbf{S}}_{T_1\gamma_2} \langle T_1\gamma_1 T_1\gamma_2 | T_2\gamma \rangle \quad (\gamma_1, \gamma_2 = x, y, z). \tag{12.33}$$

Using the tables in [1], we obtain

$$\left.\begin{aligned} \hat{\varphi}_{T_2x}(\hat{\mathbf{S}}) &= \sqrt{\tfrac{1}{2}}(\hat{S}_y\hat{S}_z + \hat{S}_z\hat{S}_y) \equiv \{\hat{S}_y, \hat{S}_z\}, \\ \hat{\varphi}_{T_2y}(\hat{\mathbf{S}}) &= \sqrt{\tfrac{1}{2}}(\hat{S}_x\hat{S}_z + \hat{S}_z\hat{S}_x) \equiv \{\hat{S}_x, \hat{S}_z\}, \\ \hat{\varphi}_{T_2z}(\hat{\mathbf{S}}) &= \sqrt{\tfrac{1}{2}}(\hat{S}_x\hat{S}_y + \hat{S}_y\hat{S}_x) \equiv \{\hat{S}_x, \hat{S}_y\}. \end{aligned}\right\} \tag{12.34}$$

The electric field Hamiltonian $\hat{W}_{\mathscr{E}}$ is then written as the invariant

$$\begin{aligned} \hat{W}_{\mathscr{E}} &= \alpha \sum_{\gamma} \hat{\varphi}_{T_2\gamma}(\hat{\mathbf{S}}) \mathscr{E}_{T_2\gamma} \\ &\equiv \sqrt{\tfrac{1}{2}}\alpha[\{\hat{S}_y, \hat{S}_z\}\mathscr{E}_x + \{\hat{S}_x, \hat{S}_z\}\mathscr{E}_y + \{\hat{S}_y, \hat{S}_x\}\mathscr{E}_z], \end{aligned} \tag{12.35}$$

where $\alpha$ is the single phenomenological parameter.

Let us now consider the influence of the electric field on the spin state $S = 1$ when there is no magnetic field. Using the matrices of $\hat{S}_x$, $\hat{S}_y$ and $\hat{S}_z$ (see (12.12)), we obtain

$$\left.\begin{aligned} &\{\hat{S}_x, \hat{S}_z\} = \frac{1}{\sqrt{2}}\begin{bmatrix} 0 & 1 & 0 \\ 1 & 0 & -1 \\ 0 & -1 & 0 \end{bmatrix}, \quad \{\hat{S}_y, \hat{S}_z\} = \frac{1}{\sqrt{2}}\begin{bmatrix} 0 & -\mathrm{i} & 0 \\ \mathrm{i} & 0 & \mathrm{i} \\ 0 & -\mathrm{i} & 0 \end{bmatrix}, \\ &\{\hat{S}_x, \hat{S}_y\} = \frac{1}{\sqrt{2}}\begin{bmatrix} 0 & 0 & -\mathrm{i} \\ 0 & 0 & 0 \\ \mathrm{i} & 0 & 0 \end{bmatrix}. \end{aligned}\right\} \tag{12.36}$$

Consequently $\hat{W}_{\mathscr{E}}$ can be represented in matrix form as

$$\mathbf{W}_{\mathscr{E}} = \frac{\alpha}{2}\begin{bmatrix} 0 & -\mathrm{i}(\mathscr{E}_x + \mathrm{i}\mathscr{E}_y) & -\mathrm{i}\mathscr{E}_z \\ \mathrm{i}(\mathscr{E}_x - \mathrm{i}\mathscr{E}_y) & 0 & \mathrm{i}(\mathscr{E}_x + \mathrm{i}\mathscr{E}_y) \\ \mathrm{i}\mathscr{E}_z & -\mathrm{i}(\mathscr{E}_x - \mathrm{i}\mathscr{E}_y) & 0 \end{bmatrix}. \tag{12.37}$$

Solving the scalar equation for $\mathscr{E} \parallel z$, we obtain three levels (this is the Stark effect):

$$\epsilon_1 = 0, \quad \epsilon_2 = \tfrac{1}{2}\alpha\mathscr{E}_z, \quad \epsilon_3 = -\tfrac{1}{2}\alpha\mathscr{E}_z. \tag{12.38}$$

On the other hand, the qualitative aspect of splitting can be established through reduction of the representation $D^{(1)}(S = 1)$. For $\mathbf{T}_d$ we have $D^{(1)} \doteq T_1$ ($^3A_1$ multiplet) and from the tables in [1] we find that in the field $\mathscr{E} \parallel C_4(z)$ the symmetry is lowered to $\mathbf{C}_{2v}$, which leads to the reduction $T_1 \doteq A_2(z) + B_1(y) + B_2(x)$. The basis of the three one-dimensional irreducible representations of $\mathbf{C}_{2v}$ makes it possible to write down the spin wavefunctions that diagonalize $\hat{W}_{\mathscr{E}}$. To do this, it is necessary to pass from functions with a certain spin projection $\Phi(1M)$ to linear combinations of $x$, $y$ and $z$ type. This transformation is provided by (12.15). In the general case, from the spin functions $\Phi(SM)$ it is necessary to construct linear combinations that transform according to point-group irreducible representations. The tables in [5] provide ready answers both for single-valued (Section 6.4.2) and double-valued representations, corresponding to half-integer spins.

The magnetic splitting under standard EPR conditions is usually much stronger than the Stark splitting. Therefore the Stark effect is manifested in relatively slight shifts of the Zeeman levels (Fig. 12.3) [82]. This leads to splitting of the transition frequency $\nu$ into two frequencies $\nu + \Delta\nu$ and $\nu - \Delta\nu$. In order to give a quantitative description of the effect as well as its angular dependence, it is necessary to diagonalize the $W_Z + W_{\mathscr{E}}$ matrix.

### 12.4.2 Quadratic electric field effect

In point groups with an inversion centre there is no linear electric field effect, since $\hat{\varphi}_{\bar{\Gamma}\bar{\gamma}}(\hat{\mathrm{S}})$ is even under inversion, while $\mathscr{E}$ is odd. Therefore a quadratic (with respect to the field) Stark effect is possible. Generalization of the earlier procedure is simple. The only difference is that *quadratic* combinations $\psi_{\bar{\Gamma}\bar{\gamma}}(\mathscr{E})$

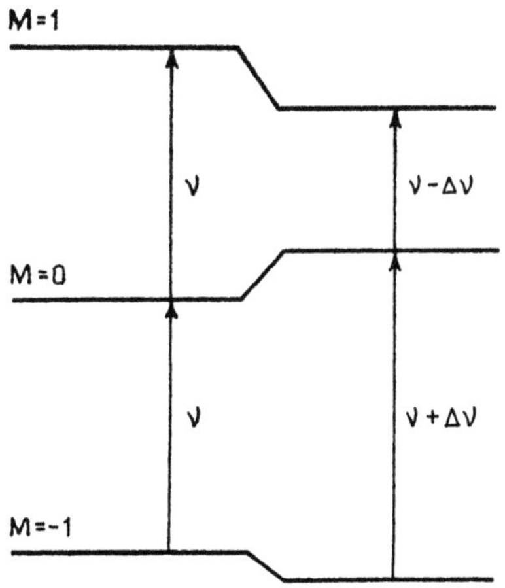


**Fig. 12.3** Electric field splitting of an EPR line ($S = 1$, $\mathbf{T}_d$).

must be constructed from the components of $\mathscr{E}$:

$$\psi_{\bar{\Gamma}\bar{\gamma}}(\mathscr{E}) = \sum_{\gamma_1 \gamma_2} \mathscr{E}_{\Gamma_1 \gamma_1} \mathscr{E}_{\Gamma_2 \gamma_2} \langle \Gamma_1 \gamma_1 \Gamma_2 \gamma_2 | \bar{\Gamma} \bar{\gamma} \rangle. \tag{12.39}$$

**Example 12.9** Consider the $^3A_{1g}$ term ($S = 1$) in $\mathbf{O}_h$. The operators $\hat{S}_\alpha$ form the basis of $T_{1g}$, while the components $\mathscr{E}_i$ form the basis of $T_{1u}$. The functions $\psi_{\bar{\Gamma}\bar{\gamma}}$ transform only according to the representations of the symmetric part of $T_{1u} \times T_{1u}$; that is as $A_{1g}$, $E_g$ and $T_{2g}$ ($T_{1g} = \{T_{1u}^2\}$; see Section 11.3.2):

$$\left.\begin{aligned}
\psi_{A_{1g}} &= \sqrt{\tfrac{1}{3}}(\mathscr{E}_x^2 + \mathscr{E}_y^2 + \mathscr{E}_z^2) = \sqrt{\tfrac{1}{3}}\mathscr{E}^2, \\
\psi_{E_g u} &= 3\mathscr{E}_z^2 - \mathscr{E}^2, \qquad \psi_{E_g v} = \sqrt{3}(\mathscr{E}_x^2 - \mathscr{E}_y^2), \\
\psi_{T_{2g}\xi} &= \mathscr{E}_y \mathscr{E}_z, \qquad \psi_{T_{2g}\eta} = \mathscr{E}_x \mathscr{E}_z, \qquad \psi_{T_{2g}\zeta} = \mathscr{E}_x \mathscr{E}_y.
\end{aligned}\right\} \tag{12.40}$$

The quadratic electric field spin Hamiltonian has the form

$$\begin{aligned}
\hat{W}_{\mathscr{E}} = \alpha\{[3\hat{S}_z^2 - S(S+1)](3\mathscr{E}_z^2 - \mathscr{E}^2) + 3(\hat{S}_x^2 - \hat{S}_y^2)(\mathscr{E}_x^2 - \mathscr{E}_y^2)\} \\
+ \beta(\{\hat{S}_y, \hat{S}_z\}\mathscr{E}_y \mathscr{E}_z + \{\hat{S}_x, \hat{S}_z\} + \mathscr{E}_x \mathscr{E}_z + \{\hat{S}_x, \hat{S}_y\}\mathscr{E}_x \mathscr{E}_y)
\end{aligned} \tag{12.41}$$

where $\alpha$ and $\beta$ are semiempirical parameters; the constant term $\mathscr{E}^2\hat{S}^2 \equiv \mathscr{E}^2 S(S+1)$ is omitted.

### 12.4.3 Combined influence of electric and magnetic fields

Let us consider in conclusion the mixed terms $\hat{W}_{\mathscr{E}\mathscr{H}} = \Sigma_{ikl}\alpha_{ikl}\hat{S}_i \mathscr{H}_k \mathscr{E}_l$, involving both electric and magnetic fields. Since the $\mathscr{H}_k$ and $S_i$ both enter linearly, their products are $T$-even. The construction procedure for the Hamiltonian $\hat{W}_{\mathscr{E}\mathscr{H}}$ is again determined from invariance considerations.

**Example 12.10** Consider $\mathbf{T}_d$. From the operators $\hat{S}_i$ and components $\mathscr{H}_k$ we construct the combinations constituting the basis of $T_2$:

$$\left.\begin{aligned}
\hat{\varphi}_{T_2 x} &= \sqrt{\tfrac{1}{2}}(\hat{S}_z \mathscr{H}_y + \hat{S}_y \mathscr{H}_z), \\
\hat{\varphi}_{T_2 y} &= \sqrt{\tfrac{1}{2}}(\hat{S}_x \mathscr{H}_z + \hat{S}_z \mathscr{H}_x), \\
\hat{\varphi}_{T_2 z} &= \sqrt{\tfrac{1}{2}}(\hat{S}_x \mathscr{H}_y + \hat{S}_y \mathscr{H}_x).
\end{aligned}\right\} \tag{12.42}$$

The Hamiltonian $\hat{W}_{\mathscr{E}\mathscr{H}}$ is then obtained as a scalar from $\hat{\varphi}_{T_2\gamma}(\hat{\mathbf{S}}, \boldsymbol{\mathscr{H}})$ and $\mathscr{E}_{T_2\gamma}$:

$$\hat{W}_{\mathscr{E}\mathscr{H}} = A(\{\hat{S}_z, \mathscr{H}_y\}\mathscr{E}_x + \{\hat{S}_x, \mathscr{H}_z\}\mathscr{E}_y + \{\hat{S}_x, \mathscr{H}_y\}\mathscr{E}_z). \tag{12.43}$$

We shall demonstrate the influence of the $\hat{W}_{\mathscr{E}\mathscr{H}}$ interaction on the EPR of the Kramers doublet $^2$A, for which $\epsilon_M = g\beta\mathscr{H}_z M$ ($M = \pm\frac{1}{2}$). An electric field $\mathscr{E} \parallel \mathscr{H}$ ($\mathscr{E}_z \neq 0$) does not affect the spectrum. On the other hand, for $\mathscr{E} \perp \mathscr{H}$ ($\mathscr{E}_x \neq 0$) we obtain $\hat{W}_{\mathscr{E}\mathscr{H}} = A\hat{S}_y \mathscr{H}_z \mathscr{E}_x$ and on solving the second-order secular equation, we obtain

$$\epsilon = g_{\mathscr{E}}\beta\mathscr{H}M, \tag{12.44}$$

where

$$g_{\mathscr{E}} = \left(g^2 - \frac{A^2}{\beta^2}\mathscr{E}_x^2\right)^{1/2}$$

is the *effective g factor*, which depends on the perpendicular electric field. The EPR line is shifted when the electric field increases, moving into the region of higher resonance magnetic fields.

A similar procedure can be used for the study of electric field effects in the hyperfine structure of EPR spectra, as well as the analysis of the influence of the electric field on the zero-field splitting parameters $D$ and $E$.

## 12.5 EFFECTIVE HAMILTONIAN FOR NON-KRAMERS DOUBLETS

We shall begin the discussion of *orbitally degenerate states* with the simplest case of *non-Kramers doublets*. The wavefunctions of such doublets form the basis of twofold-degenerate (not double-valued!) representations of symmetry groups. Among the 32 point groups there are 24 having such representations (Table 12.1). In cubic, trigonal and tetragonal groups these are the representations of E type ($E_u$, $E_g$, $E_1$, $E_2$), whose corresponding terms have zero spin ($^1$E).

The *effective Hamiltonian $\hat{H}$ gives a two-dimensional matrix*, determined by two basis functions of the representation E (we shall call them $\varphi_x$ and $\varphi_y$). Any real two-dimensional matrix can be expanded into four matrices each containing just one nonzero unit element:

$$\begin{bmatrix} a & b \\ c & d \end{bmatrix} = a\begin{bmatrix} 1 & 0 \\ 0 & 0 \end{bmatrix} + b\begin{bmatrix} 0 & 1 \\ 0 & 0 \end{bmatrix} + c\begin{bmatrix} 0 & 0 \\ 1 & 0 \end{bmatrix} + d\begin{bmatrix} 0 & 0 \\ 0 & 1 \end{bmatrix}. \tag{12.45}$$

These are called matrix units, and play the role of vectors in terms of which any vector can be expanded.

The main idea of the effective Hamiltonian procedure, developed in [83], is to make a special choice of the matrix units in terms of which the Hermitian matrix of the effective Hamiltonian is expressed. We shall demonstrate this, taking as an example $\mathbf{D}_{3h}$, for which there are two two-dimensional

**Table 12.1** Twofold-degenerate representations of the point groups (the single groups) [83].

Tetragonal series

| $D_{4h}$ | $D_4$ | $C_{4v}$ | $D_{2d}$ | $C_{4h}$ | $C_4$ | $S_4$ |
|---|---|---|---|---|---|---|
| $\Gamma_5^+$ | $\Gamma_5$ | $\Gamma_5$ | $\Gamma_5$ | $\Gamma_3^+ + \Gamma_4^+$ | $\Gamma_3 + \Gamma_4$ | $\Gamma_3 + \Gamma_4$ |
| $\Gamma_5^-$ | | | | $\Gamma_3^- + \Gamma_4^-$ | | |

Hexagonal series

| $D_{6h}$ | $D_6$ | $C_{6h}$ | $C_{6v}$ | $D_{3d}$ | $D_{3h}$ |
|---|---|---|---|---|---|
| $\Gamma_5^+$ | $\Gamma_5$ | $\Gamma_5^+ + \Gamma_6^+$ | $\Gamma_5$ | $\Gamma_3^+$ | $\Gamma_5$ |
| $\Gamma_5^-$ | | $\Gamma_5^- + \Gamma_6^-$ | | $\Gamma_3^-$ | |
| $\Gamma_6^+$ | $\Gamma_6$ | $\Gamma_2^+ + \Gamma_3^+$ | $\Gamma_6$ | $(\Gamma_3^+)$ | $\Gamma_6$ |
| $\Gamma_6^-$ | | $\Gamma_2^- + \Gamma_3^-$ | | $(\Gamma_3^-)$ | |
| $C_6$ | $D_3$ | $C_{3i}$ | $C_{3v}$ | $C_{3h}$ | $C_3$ |
| $\Gamma_5 + \Gamma_6$ | | $\Gamma_2^+ + \Gamma_3^+$ | | $\Gamma_5 + \Gamma_6$ | |
| | $\Gamma_3$ | $\Gamma_2^- + \Gamma_3^-$ | $\Gamma_3$ | $\Gamma_2 + \Gamma_3$ | $\Gamma_2 + \Gamma_3$ |
| $\Gamma_2 + \Gamma_3$ | | $(\Gamma_2^+ + \Gamma_3^+)$ | | $(\Gamma_3 + \Gamma_4)$ | |
| | | $(\Gamma_2^- + \Gamma_3^-)$ | | $(\Gamma_5 + \Gamma_6)$ | |

Cubic series

| $O_h$ | $O$ | $T_d$ | $T_h$ | $T$ |
|---|---|---|---|---|
| $\Gamma_3^+$ | $\Gamma_3$ | $\Gamma_3$ | $\Gamma_2^+ + \Gamma_3^+$ | $\Gamma_2 + \Gamma_3$ |
| $\Gamma_3^-$ | | | $\Gamma_2^- + \Gamma_3^-$ | |

representations $\Gamma_5$ and $\Gamma_6$ (E″ and E′). The characters and basis functions of $\mathbf{D}_{3h}$ are given in Table 12.2. Let us consider the procedure of constructing of the effective Hamiltonian for E″ doublet. For the doublet E″ we form the direct product

$$E'' \times E'' \doteq A_1' + A_2' + E'(\equiv \Gamma_1 + \Gamma_2 + \Gamma_6).$$

Here $\Gamma_1$ is the totally symmetric representation, $\Gamma_2$ is a one-dimensional representation, and the $z$-component of axial vector ($\hat{S}_z$ or $R_z$) is the basis of $\Gamma_2$. Finally, the two-dimensional representation $\Gamma_6$ can be represented as the product $\Gamma_6 = \Gamma_3 \times \Gamma_5$ (Table 12.2), where $\Gamma_3$ is the one-dimensional representation and $\Gamma_5$ is the two-dimensional one. The basis of $\Gamma_3$ is $z\hat{S}_z$, while the two-dimensional basis of $\Gamma_5$ is $\hat{S}_\pm = \mp(\hat{S}_x \pm i\hat{S}_y)$.

Therefore the basis of $\Gamma_6$ is $u_\pm = \mp\hat{S}_z(\hat{S}_x \pm i\hat{S}_y)$. Now we introduce

**Table 12.2** Character table and basis functions for the group $\mathbf{D}_{3h}$.

| | $\mathbf{D}_{3h}$ | $\hat{E}$ | $\hat{\sigma}_h$ | $2\hat{C}_3$ | $2\hat{S}_3$ | $3\hat{C}_2$ | $3\hat{\sigma}_v$ |
|---|---|---|---|---|---|---|---|
| $A_1'$ | $\Gamma_1$ | 1 | 1 | 1 | 1 | 1 | 1 |
| $A_2'$ | $\Gamma_2$ | 1 | 1 | 1 | 1 | −1 | −1 |
| $A_1''$ | $\Gamma_3$ | 1 | −1 | 1 | −1 | 1 | −1 |
| $A_2''$ | $\Gamma_4$ | 1 | −1 | 1 | −1 | −1 | 1 |
| $E''$ | $\Gamma_5$ | 2 | −2 | −1 | 1 | 0 | 0 |
| $E'$ | $\Gamma_6$ | 2 | 2 | −1 | −1 | 0 | 0 |

the irreducible tensor operators $\hat{\mathbf{X}}_{\gamma_i}(\Gamma_i)$ with $\Gamma_i = \Gamma_1, \Gamma_2, \Gamma_6$ belonging to the product $E'' \times E''$. The real operators $\hat{\mathbf{X}}(\Gamma_1)$ and $\hat{\mathbf{X}}_{\gamma_6}(\Gamma_6)$ are invariant under time inversion, they will be denoted by $\hat{\mathbf{V}}(\Gamma_1)$ and $\hat{\mathbf{V}}_{\gamma_6}(\Gamma_6)$, the imaginary operator $\hat{\mathbf{V}}(\Gamma_2)$ reverses its sign under time inversion and will be denoted by $\hat{\mathbf{T}}_2(\Gamma_2)$.

The main idea of the effective Hamiltonian method consists in the choice of irreducible tensors $\hat{\mathbf{X}}_{\gamma_i}(\Gamma_i)$ as matrix units instead of matrices, each containing just one non-zero unit element (Equation (12.45)). An advantage of this approach is obvious: the structure of the matrices of the irreducible tensor operators can be derived using the Wigner–Eckart theorem, the reduced matrix element playing the role of semiempirical parameters of the effective Hamiltonian. The last are independent from the point of view of symmetry, i.e. no relations between them can be derived from the group theory.

The Hermitian matrix of the effective Hamiltonian may now be expressed in terms of the operators $\hat{\mathbf{V}}$ and $\hat{\mathbf{T}}$:

$$\hat{H} = a(\mathscr{H}, \mathscr{E})\hat{\mathbf{V}}(\Gamma_1) + b(\mathscr{H}, \mathscr{E})\hat{\mathbf{T}}(\Gamma_2) + c(\mathscr{H}, \mathscr{E})\hat{\mathbf{V}}_+(\Gamma_6) + d(\mathscr{H}, \mathscr{E})\hat{\mathbf{V}}_-(\Gamma_6), \tag{12.46}$$

where $a$, $b$, $c$ and $d$ are parameters depending on the fields $\mathscr{H}$ and $\mathscr{E}$, and $\hat{\mathbf{V}}_\pm(\Gamma_6) \equiv \mathbf{V}_{u_\pm}(\Gamma_6)$. The matrices of the operators $\hat{\mathbf{X}}_\gamma(\Gamma)$ are determined by the Wigner–Eckart theorem:

$$\langle \Gamma_5\gamma_5|\hat{\mathbf{X}}_\gamma(\Gamma)|\Gamma_5\gamma_5'\rangle = \sqrt{\tfrac{1}{2}}\langle \Gamma_5\|\hat{\mathbf{X}}_\gamma(\Gamma)\|\Gamma_5\rangle\langle \Gamma_5\gamma_5|\Gamma_5\gamma_5'\Gamma\gamma\rangle, \tag{12.47}$$

where $\langle \Gamma_5\gamma_5|\Gamma_5\gamma_5'\Gamma\gamma\rangle$ are $\mathbf{D}_{3h}$ Clebsch–Gordan coefficients for the direct product $\Gamma \times \Gamma_5 (\Gamma = \Gamma_1, \Gamma_2, \Gamma_6)$. Choosing the reduced matrices as $\langle \Gamma_5\|\hat{\mathbf{V}}(\Gamma_1)\|\Gamma_5\rangle = \sqrt{2}$, $\langle \Gamma_5\|\hat{\mathbf{T}}(\Gamma_2)\|\Gamma_5\rangle = -\sqrt{2}\mathrm{i}$, and $\langle \Gamma_5\|\hat{\mathbf{V}}(\Gamma_6)\|\Gamma_5\rangle = 2$, it is easy to express the matrices $\mathbf{V}$ and $\mathbf{T}$ in terms of the Pauli matrices $\boldsymbol{\sigma}_x$, $\boldsymbol{\sigma}_y$ and $\boldsymbol{\sigma}_z$ and the unit matrix $\mathbf{1}$:

$$\mathbf{V}(\Gamma_1) = \begin{bmatrix} 1 & 0 \\ 0 & 1 \end{bmatrix} \equiv \mathbf{1}, \quad \mathbf{T}(\Gamma_2) = \begin{bmatrix} 1 & 0 \\ 0 & -1 \end{bmatrix} \equiv \boldsymbol{\sigma}_z, \tag{12.48}$$

$$\mathbf{V}_x(\Gamma_6) = \begin{bmatrix} 0 & -\mathrm{i} \\ \mathrm{i} & 0 \end{bmatrix} \equiv \boldsymbol{\sigma}_y, \qquad \mathbf{V}_y(\Gamma_6) = \begin{bmatrix} 0 & 1 \\ 1 & 0 \end{bmatrix} \equiv \boldsymbol{\sigma}_x, \tag{12.49}$$

where for $\Gamma_6$ the basis $u_x$, $u_y$ has been used.

We must now construct a Hamiltonian that satisfies the invariance requirements. Contributions to this Hamiltonian will arise as follows.

(1) The contribution $H_1$, linear in $\mathscr{H}$, can involve only the imaginary operator $\hat{\mathbf{T}}(\Gamma_2)$, which reverses sign on time inversion. The basis of $\Gamma_2$ is $\mathscr{H}_z$, hence

$$H_1 = g_{\parallel}\beta\boldsymbol{\sigma}_z\mathscr{H}_z, \tag{12.50}$$

where $g_{\parallel}\beta$ is a semiempirical constant.

(2) A contribution $H_2$ linear in the electric field will not contain $\mathscr{E}_z$, since $\Gamma_4(\mathscr{E}_z)$ does not appear in $\Gamma_5 \times \Gamma_5$. the components $\mathscr{E}_x$ and $\mathscr{E}_y$ form the basis of $\Gamma_6$, whereby the scalar from $\hat{\mathbf{V}}_+$, $\hat{\mathbf{V}}_-$ ($\hat{\mathbf{V}}_\pm = (\mp 1/\sqrt{2})(\hat{\mathbf{V}}_x \pm \mathrm{i}\hat{\mathbf{V}}_y)$† and $\mathscr{E}_+$, $\mathscr{E}_-$ ($\mathscr{E}_\pm = \mp(1/\sqrt{2})(\mathscr{E}_x \pm \mathrm{i}\mathscr{E}_y)$ is expressed as $\sqrt{\tfrac{1}{2}}\mathrm{i}[\mathscr{E}_-\hat{\mathbf{V}}_+(\Gamma_6) - \mathscr{E}_+\hat{\mathbf{V}}_-(\Gamma_6)$. Passing to the $(x, y)$ basis, we obtain

$$H_2 = G(\mathscr{E}_x\boldsymbol{\sigma}_x - \mathscr{E}_y\boldsymbol{\sigma}_y), \tag{12.51}$$

where $G$ is the parameter for the linear Stark effect

(3) For the contribution $H_3$ quadratic in $\mathscr{H}$, using a Clebsch–Gordan expansion, we obtain two combinations of the type $\varphi_{\Gamma_6\gamma_6}(\mathscr{H})$:

$$\varphi_x(\mathscr{H}) = 2\mathscr{H}_x\mathscr{H}_y, \qquad \varphi_y(\mathscr{H}) = \mathscr{H}_x^2 - \mathscr{H}_y^2$$

and two combinations of the type $\varphi_{\Gamma_1}(\mathscr{H})$,

$$\varphi_{\Gamma_1}(\mathscr{H}) = \mathscr{H}_z^2, \qquad \varphi_{\Gamma_1}(\mathscr{H}) = \mathscr{H}_x^2 + \mathscr{H}_y^2.$$

Therefore the contribution $H_3$ is

$$\begin{aligned} H_3 = {} & M_1[-2\mathscr{H}_x\mathscr{H}_y\boldsymbol{\sigma}_y + (\mathscr{H}_x^2 - \mathscr{H}_y^2)\boldsymbol{\sigma}_x] \\ & + M_2\mathscr{H}_z^2\mathbf{1} + M_3(\mathscr{H}_x^2 + \mathscr{H}_y^2)\mathbf{1}, \end{aligned} \tag{12.52}$$

and contains three parameters, $M_1$, $M_2$ and $M_3$.

(4) The contribution quadratic in $\mathscr{E}$ results from (12.52) on substituting $\mathscr{E}_\alpha$ for $\mathscr{H}_\alpha$ and contains three (other) parameters.

† Operators $\hat{\mathbf{V}}_+$, $\hat{\mathbf{V}}_-$ form the spherical basis (section 6.2.1).

The construction of an invariant expression can be applied to other doublet types in a similar way. In practice it is simplest to use the results found in [83] (see Appendix IV). These include

(1) the matrices $\mathbf{X}_\gamma(\Gamma)$ and effective Hamiltonians for all non-Kramers doublets of the 32 point groups, linear and quadratic in $\mathscr{E}$ and $\mathscr{H}$, as well as those comprising mixed terms $\mathscr{E}_\alpha\ \mathscr{H}_\beta$;

(2) the effective Hamiltonians of the hyperfine interaction between electrons in non-Kramers doublets and nuclear spins, linear ($\sim \hat{I}_\alpha$) and quadratic ($\sim \hat{I}_\alpha \hat{I}_\beta$) in the nuclear spin operators;

(3) the effective Hamiltonians for magnetic fields and nuclear spins;

(4) the selection rules for transitions between the levels of non-Kramers doublets split by electric and magnetic fields under the action of oscillating electric $\mathscr{E}_{\text{osc}}$ and magnetic $\mathscr{H}_{\text{osc}}$ fields, differently oriented with respect to the symmetry axes.

To determine the selection rules, according to the reduction tables in [1], it is necessary to establish the symmetry group, taking account of the external field, then the irreducible representations $\Gamma'$ and $\Gamma''$ corresponding to the sublevels of the doublet split in this group. The transition, polarized along the $\alpha$ axis, is allowed if $\Gamma' \times \Gamma''$ contain the representations $\Gamma_\alpha$ according to which $\mathscr{H}^\alpha_{\text{osc}}$ or $\mathscr{E}^\alpha_{\text{osc}}$ are transformed ($\alpha, \beta = x, y, z$). The selection rules for all point groups are given in Table 12.3.

**Table 12.3** Selection rules for transitions between Kramers doublets sublevels, split by electric and magnetic fields [83].

| Static field | Oscillating field | Point group |
|---|---|---|
| $\mathscr{H} \parallel z$ | $\mathscr{E}_{\text{osc}} \parallel z$ | $S_4$, $D_{2d}$ |
| | $\mathscr{E}_{\text{osc}} \perp z$ | $D_{3h}$, $D_3$, $C_{3h}$, $C_{3v}$, $C_3$ |
| $\mathscr{H} \parallel z$ | $\mathscr{H}_{\text{osc}} \parallel z$ | $D_{4h}$, $D_4$, $C_{4v}$, $D_{2d}$, $C_{4h}$, $C_4$, $S_4$, $D_{6h}$, $D_6$, $C_{6h}$, $D_{3d}$, $D_{3h}$, $C_6$, $D_3$, $C_3$, $C_{3i}$, $C_{3v}$, $C_{3h}$ |
| | $\mathscr{E}_{\text{osc}} \parallel z$ | $S_4$, $D_{2d}$ |
| | $\mathscr{E}_{\text{osc}} \perp z$ | $D_{3h}$, $D_3$, $C_{3h}$, $C_{3v}$, $C_3$ |
| $\mathscr{E} \parallel z$ | $\mathscr{H}_{\text{osc}} \parallel z$ | $S_4$, $D_{2d}$ |
| | $\mathscr{E}_{\text{osc}} \parallel z$ | $S_4$, $D_{2d}$ |
| $\mathscr{E} \perp z$ | $\mathscr{H}_{\text{osc}} \parallel z$ | $D_{4h}$, $D_4$, $C_{4v}$, $D_{2d}$, $C_{4h}$, $C_4$, $S_4$, $D_{6h}$, $C_{6h}$, $C_{6v}$, $D_{3d}$, $D_{3h}$, $C_6$, $C_{3i}$, $C_{3v}$, $C_{3h}$ |
| | $\mathscr{E}_{\text{osc}} \parallel z$ | $S_4$, $D_{2d}$ |
| | $\mathscr{E}_{\text{osc}} \perp z$ | $D_{3h}$, $D_3$, $C_{3h}$, $C_{3v}$, $C_3$ |

## 12.6 EFFECTIVE HAMILTONIAN FOR THE SPIN–ORBIT MULTIPLET

Let us briefly examine the general case of the spin–orbit multiplet; that is, of the orbitally degenerate level $\Gamma$ and the nonzero spin $S$. The effective Hamiltonian method for the $S\Gamma$ multiplet [14, 46] is based on the same constructions of invariant expressions used in considering the pure spin level $^{2S+1}$A (Section 12.2) and the non-Kramers doublet $^1$E (Section 12.5). In summary:

(1) Find the symmetric $[\Gamma^2]$ and antisymmetric $\{\Gamma^2\}$ parts of the direct product $\Gamma \times \Gamma$.

(2) With the help of the Wigner–Eckart theorem, obtain the matrices $\mathbf{T}_{\gamma'}(\Gamma')$ of the imaginary operators ($\Gamma'$ from $\{\Gamma^2\}$) and of the real operators $V_{\gamma'}(\Gamma')$ ($\Gamma'$ from $[\Gamma^2]$).

(3) Using tables in [1], relate the components of $\mathbf{S}$ to the irreducible representations; that is, find $\hat{S}_{\Gamma\gamma}$;

(4) Find the sums of terms of the type $\Sigma_\Gamma c(\Gamma)\Sigma_\gamma \hat{\mathbf{T}}_\gamma(\Gamma)\hat{\mathbf{S}}_{\Gamma\gamma}$ (where $c(\Gamma)$ are parameters); that is, find the *zero-field splitting Hamiltonian*, linear with respect to $S$.

(5) The Zeeman Hamiltonian, linear in the field, comprises three types of scalar contributions. First, there are terms of the type $\Sigma_\Gamma g(\Gamma)\Sigma_\gamma \hat{\mathbf{S}}_{\Gamma\gamma}\mathscr{H}_{\Gamma\gamma}$; this part of the $S\Gamma$-term Zeeman Hamiltonian coincides exactly with the pure-spin Hamiltonian for the same point group. Secondly, the effective Hamiltonian includes an orbital Zeeman interaction of the type $\Sigma_\Gamma b(\Gamma)\Sigma_\gamma \hat{\mathbf{T}}_\gamma(\Gamma)\mathscr{H}_{\Gamma\gamma}$ (where $b(\Gamma)$ are parameters). Thirdly, it is necessary to construct terms including the products $\mathscr{H}_{\Gamma\gamma}\hat{\mathbf{S}}_{\Gamma'\gamma'}\hat{\mathbf{V}}_{\gamma''}(\Gamma'')$, which are $T$-even.

Higher-order (in $\hat{S}_i$ and $\mathscr{H}$) contributions, as well as the electric field terms of the effective Hamiltonian, are obtained in a similar way.

As an example of the procedure outlined above, let us make an application to the orbital doublet E of $\mathbf{D}_3$, assuming that the total spin is $S = \frac{1}{2}$ ($^2$E term). For this doublet the direct product $\mathrm{E} \times \mathrm{E} \doteq \mathrm{A}_1 + \mathrm{A}_2 + \mathrm{E}$, whence $[\mathrm{E}^2] \doteq \mathrm{A}_1 + \mathrm{E}$ and $\{\mathrm{E}^2\} \doteq \mathrm{A}_2$. In the E basis $x_\pm = \mp\sqrt{\frac{1}{2}}(x \pm \mathrm{i}y)$ (or $R_\pm$) and, using the Wigner–Eckart theorem and the tables in [1], we obtain

$$\mathbf{V}_+(\mathrm{E}) = \begin{bmatrix} 0 & 0 \\ 1 & 0 \end{bmatrix}, \quad \mathbf{V}_-(\mathrm{E}) = \begin{bmatrix} 0 & 1 \\ 0 & 0 \end{bmatrix}, \quad \mathbf{T}(\mathrm{A}_2) = \begin{bmatrix} 1 & 0 \\ 0 & -1 \end{bmatrix}. \tag{12.53}$$

The operators $\hat{S}_{\Gamma\gamma}$ are

$$\hat{S}_z = \hat{\mathbf{S}}_{\mathrm{A}_2}, \quad \hat{S}_+ \equiv \hat{\mathbf{S}}_{\mathrm{E}+} = -\sqrt{\tfrac{1}{2}}(\hat{S}_x + \mathrm{i}\hat{S}_y), \quad \hat{S}_- \equiv \hat{\mathbf{S}}_{\mathrm{E}-} \equiv \sqrt{\tfrac{1}{2}}(\hat{S}_x - \mathrm{i}\hat{S}_y).$$

Therefore the zero-field splitting Hamiltonian involves just one scalar and one parameter:

$$H_0 = \lambda \hat{\mathbf{T}}(\mathrm{A}_2)\hat{S}_z. \tag{12.54}$$

The spin component of the Zeeman Hamiltonian is expressed through (12.19), where the $z$ axis is parallel to $C_3$ (not to $C_4$, as it is in 12.19).

The orbital Zeeman interaction comprises just one term, $g'_{\parallel}\beta\hat{\mathbf{T}}(\mathrm{A}_2)\mathscr{H}_z$. Finally, the form of the terms of the type $\mathscr{H}_\alpha \hat{S}_\beta \hat{\mathbf{V}}_\gamma(\mathrm{E})$ is easy to establish from the following considerations. Since $\mathscr{H}_z = \mathscr{H}_{\mathrm{A}_2}$, the scalar in the group can be obtained by multiplying $\mathscr{H}_z$ by the $\mathrm{A}_2$ combination of $\hat{\mathbf{S}}_{\Gamma\gamma}$ and $\hat{\mathbf{V}}_\gamma(\mathrm{E})$:

$$\hat{\mathbf{w}}_{\mathrm{A}_2} = \sum_{\gamma_1\gamma_2} \hat{\mathbf{S}}_{\Gamma\gamma_1}\hat{\mathbf{V}}_{\gamma_2}(\mathrm{E})\langle \Gamma\gamma_1 \mathrm{E}\gamma_2|\mathrm{A}_2\rangle. \tag{12.55}$$

This expansion involves only $\hat{\mathbf{S}}_{\mathrm{E}\gamma_1}$ ($\mathrm{E}\times\mathrm{E}$ contains $\mathrm{A}_2$). Substituting the Clebsch–Gordan coefficients, we obtain

$$\hat{\mathbf{w}}_{\mathrm{A}_2} = -\hat{S}_-\hat{\mathbf{V}}_+(\mathrm{E}) + \hat{S}_+\hat{\mathbf{V}}_-(\mathrm{E}). \tag{12.56}$$

Similarly, it is possible to determine the two combinations that are multiplied by $\mathscr{H}_x$:

$$\sqrt{\tfrac{1}{2}}[-\hat{S}_+\hat{\mathbf{V}}_+(\mathrm{E}) - \hat{S}_-\hat{\mathbf{V}}_-(\mathrm{E})] \qquad -\hat{\mathbf{S}}_z[\hat{\mathbf{V}}_+(\mathrm{E}) + \hat{\mathbf{V}}_-(\mathrm{E})]. \tag{12.57}$$

Bringing together all these results, we write the effective Hamiltonian for the ${}^2\mathrm{E}$ doublet for $\mathscr{H}\parallel C_3(z)$ (the "parallel" field case):

$$\begin{aligned} H = {} & \lambda\hat{S}_z\hat{\mathbf{T}}(\mathrm{A}_2) + g_{\parallel}\beta\mathscr{H}_z\hat{S}_z + g'_{\parallel}\beta\mathscr{H}_z\hat{\mathbf{T}}(\mathrm{A}_2) \\ & + \sqrt{\tfrac{1}{2}}g''_{\parallel}\beta\mathscr{H}_z[-\hat{S}_-\hat{\mathbf{V}}_+(\mathrm{E}) + \hat{S}_+\hat{\mathbf{V}}_-(\mathrm{E})]. \end{aligned} \tag{12.58}$$

In the case of a "perpendicular" field $\mathscr{H}\parallel x$

$$\begin{aligned} H = {} & \lambda\hat{S}_z\hat{\mathbf{T}}(\mathrm{A}_2) + g_{\perp}\beta\mathscr{H}_x\hat{S}_x + \sqrt{\tfrac{1}{2}}g'_{\perp}\beta\mathscr{H}_x[-\hat{S}_+\hat{\mathbf{V}}_+(\mathrm{E}) + \hat{S}_-\hat{\mathbf{V}}_-(\mathrm{E})] \\ & + g''_{\perp}\beta\mathscr{H}_x\hat{S}_z[-\hat{\mathbf{V}}_+(\mathrm{E}) - \hat{\mathbf{V}}_-(\mathrm{E})]. \end{aligned} \tag{12.59}$$

It should be stressed again that the theory described in this chapter forms the basis of a *phenomenological* description. Group theory itself does not provide relationships between parameters of the effective Hamiltonian (although it may point to their physical character). The microscopic (quantum mechanical) description of the effective Hamiltonian parameters can be found in [14, 46].

## PROBLEMS

**12.1** Write the spin Hamiltonian for the ${}^4\mathrm{A}_2$ term in $\mathbf{D}_2$ and determine the energy levels.

**12.2** Prove that the quadratic terms $S_\alpha^2$ and $\hat{S}_\alpha \hat{S}_\beta$ do not appear in the spin Hamiltonian in cubic fields and show that the spin Hamiltonian for the cubic ($\mathbf{O}_h$) $^6A_1$ term ($s = \frac{5}{2}$) is of the form

$$H = D[\hat{S}_x^4 + \hat{S}_y^4 + \hat{S}_z^4 - \tfrac{1}{5}S(S+1)(3S^2 + 3S - 1)].$$

**12.3** Construct the spin Hamiltonian of the Zeeman interaction for Kramers doublets of $\mathbf{C}_{2v}$, $\mathbf{D}_3$ and $\mathbf{O}_h$.

**12.4** Construct the HFI Hamiltonian for the symmetry groups mentioned in Problem 12.3.

**12.5** Determine the linear electric field splitting of the $\mathbf{T}_d$ group $^3$A multiplet in an electric field oriented along the trigonal axis ($\mathscr{E}_x = \mathscr{E}_y = \mathscr{E}_z$).

**12.6** Using the tables in [1], construct the effective Hamiltonian (Section 12.5) for the $\mathbf{D}_{3h}$ group $^1\Gamma_5$ doublet and the $\mathbf{O}_h$ group $^1E_g$ doublet.

# 13 The Exchange Interaction in Polynuclear Coordination Compounds

## 13.1 THE HEISENBERG–DIRAC–VAN VLECK MODEL OF THE EXCHANGE INTERACTION†

*Polynuclear* or *cluster coordination compounds* are those that contain several metal ions bound by ligands. If the metal-ion shells are paramagnetic and unfilled, they are bound by the so-called *exchange interaction.* The main idea of the exchange Hamiltonian proposed by Heisenberg and Dirac [62, 79, 84] can be seen by considering the example of one-electron atoms in s states. We denote the one-electron part of the full Hamiltonian, including both kinetic energy and electron-nuclear attraction by $\hat{h}$ and interelectronic interaction by $\hat{g}$. The coordinate molecular functions $\Phi_{\pm}$ can be either even or odd under electron permutation:

$$\Phi_{\pm}(1, 2) = \frac{1}{[2(1 \pm \delta_{ab}^2)]^{1/2}} [\varphi_a(1)\varphi_b(2) \pm \varphi_a(2)\varphi_b(1)], \tag{13.1}$$

where $\delta_{ab} = \int \varphi_a(\mathbf{r})\varphi_b(\mathbf{r})\, d\tau$ is the overlap integral, $\varphi_a(1) \equiv \varphi_a(\mathbf{r}_1)$ and $\varphi_a(2) \equiv \varphi_a(\mathbf{r}_2)$. The total electron spin $S$ takes two values: 0 and 1, while the spin functions $|SM\rangle$ can be constructed using the coupling formula (see (5.27))

$$|SM\rangle = \sum_{m_1 m_2} |s_1 m_1\rangle |s_2 m_2\rangle \langle s_1 m_1 s_2 m_2 | SM\rangle. \tag{13.2}$$

In the present case $s_1 = s_2 = \frac{1}{2}$, $S = 0$ and 1, while the spin functions $|SM\rangle \equiv \chi(SM)$ are given by (5.31) and (5.32). The Pauli principle allows only those many-electron states that are antisymmetric under the exchange of any two electrons. Since the spin functions $|1M\rangle$ are symmetric while $|00\rangle$ are

† Some readers may leave this chapter for a 'second reading'.

antisymmetric, the wavefunctions of the allowed states can be written as

$$\left.\begin{aligned} \Psi(1M) &= \Phi_-(\mathbf{r}_1\mathbf{r}_2)|1\,M\rangle \quad (M=0,\pm 1), \\ \Psi(00) &= \Phi_+(\mathbf{r}_1\mathbf{r}_2)|00\rangle. \end{aligned}\right\} \tag{13.3}$$

Calculating the energy of the molecule using these wavefunctions we obtain the following result:

$$\begin{aligned} \epsilon(1) &= [(\epsilon_a + \epsilon_b + K_{ab}) - (2\epsilon_{ab}\delta_{ab} + J_{ab})]/(1-\delta_{ab}^2), \\ \epsilon(0) &= [(\epsilon_a + \epsilon_b + K_{ab}) + (2\epsilon_{ab}\delta_{ab} + J_{ab})]/(1+\delta_{ab}^2) \end{aligned} \tag{13.4}$$

with $\epsilon_a = \langle a|\hat{h}|a\rangle$, $\epsilon_b = \langle b|\hat{h}|b\rangle$, $\epsilon_{ab} = \langle a|\hat{h}|b\rangle$. In (13.4)

$$K_{ab} = \langle \varphi_a(1)\varphi_b(2)|\hat{g}|\varphi_a(1)\varphi_b(2)\rangle$$

is the Coulomb integral and

$$J_{ab} = \langle \varphi_a(1)\varphi_b(2)|\hat{g}|\varphi_b(1)\varphi_a(2)\rangle$$

is the exchange integral†.

The molecular level energies $\epsilon(S)$ with ($S = 0$ and 1) depend on the total spin; considering the case of identical atoms ($\epsilon_a = \epsilon_b = \epsilon_0$) we find

$$\epsilon(1) - \epsilon(0) = \frac{2(J_{ab} - K_{ab}\delta_{ab}^2) - 4\epsilon_0\delta_{ab}^2 + 4\delta_{ab}\epsilon_{ab}}{(1-\delta_{ab}^2)}. \tag{13.5}$$

The dependence of the energy on $S$ is given by the Hamiltonian in terms only of the spin operators:

$$\hat{H} = -2J\,\hat{\mathbf{S}}_1 \cdot \hat{\mathbf{S}}_2, \tag{13.6}$$

where $J$ is a parameter. Indeed, $\hat{\mathbf{S}}_1 \cdot \hat{\mathbf{S}}_2 = \frac{1}{2}(\hat{\mathbf{S}}^2 - \hat{s}_1^2 - \hat{s}_2^2)$; therefore

$$\epsilon(S) = -J[S(S+1) - s_1(s_1+1) - s_2(s_2+1)], \tag{13.7}$$

where $S = 0$ and 1, and $s_1 = s_2 = \frac{1}{2}$. Comparison of (13.5) and (13.7) shows that

$$J = \frac{K_{ab}\delta_{ab}^2 - J_{ab} + 2\epsilon_0\delta_{ab} - 2\epsilon_{ab}\delta_{ab}}{1-\delta_{ab}^4}. \tag{13.8}$$

The spin Hamiltonian (13.6) can be considered as an effective Hamiltonian, in which the quantity $J$ acts as a semiempirical parameter. In magnetochemical studies it is not usually calculated, but rather is determined so that optimal agreement is obtained between the theoretical and experimental temperature dependences of the cluster magnetic moment.

Van Vleck (see [62]) generalized the Heisenberg Hamiltonian with reference

† The notation $K$ for the Coulomb integral and $J$ for exchange integral are commonly used in the magnetochemistry of exchange systems. In quantum chemistry the opposite notations are usually used.

to many-electron atoms and wrote it as

$$\hat{H} = -2J\hat{\mathbf{s}}_a \cdot \hat{\mathbf{s}}_b, \tag{13.9}$$

where $s_a$ and $s_b$ are the total spins of the many-electron atoms:

$$\hat{\mathbf{s}}_a = \sum_i \hat{\mathbf{s}}_{ia}, \quad \hat{\mathbf{s}}_b = \sum_i \hat{\mathbf{s}}_{ib}.$$

The Hamiltonian described by (13.9) is called the *Heisenberg-Dirac–Van Vleck (HDVV) Hamiltonian*; it operates in the space of wavefunctions associated with total spins $s_a$ and $s_b$. Just as the Hamiltonian (13.6) is the spin equivalent of the Heitler–London scheme for two-electron atoms, the HDVV Hamiltonian (13.9) is the effective Hamiltonian used instead of the Heitler–London scheme for many-electron ions. The validity of this generalization is far from obvious. Indeed, the electronic Hamiltonian (13.6) has two eigenvalues for $S = 0$ and $S = 1$. Therefore the value of the only energy parameter $\epsilon(1) - \epsilon(0)$ can always be made to coincide with the calculated value, provided by (13.5) by a judicious choice of the semiempirical parameter $J$. Let us consider now the eigenvalues of the Hamiltonian (13.9):

$$\epsilon(S) = -J[S(S+1) - s_a(s_a+1) - s_b(s_b+1)], \tag{13.10}$$

where the value of the total spin $S$ is determined by the usual rule for addition of angular momenta:

$$S = s_a + s_b, \; s_a + s_b - 1, \; \ldots, \; |s_a - s_b|. \tag{13.11}$$

The total spin $S$ thus takes $2s_m + 1$ values, where $s_m$ is the lesser of $s_a$ and $s_b$. The energy spectrum contains $2s_m + 1$ levels and, therefore comprises $2s_m$ energy intervals, which satisfy the Landé rule:

$$\Delta\epsilon(S) = \epsilon(S) - \epsilon(S-1) = -2JS. \tag{13.12}$$

Meanwhile it is not a priori obvious that $2s_m + 1$ eigenvalues of the exact Schrödinger equation for a many-electron system are related by (13.12). It turns out that the Hamiltonian (13.9) is a reasonable approximation, whose framework of applicability is analysed in detail in [62, 63].

The HDVV model is generalized for polynuclear systems by introducing the following spin-Hamiltonian in place of (13.9):

$$\hat{H} = -2\sum_{ij} J_{ij}\hat{\mathbf{s}}_i \cdot \hat{\mathbf{s}}_j, \tag{13.13}$$

where $\hat{\mathbf{s}}_i$ are, as before, the total spins of many-electron ions, $J_{ij}$ are the exchange integrals, and the sum is over all ion pairs.

We emphasize that the HDVV model can describe interaction between ions only when their ground terms are *orbitally nondegenerate*; that is, when they are *spin multiplets*.

## 13.2 SPIN LEVELS OF SYMMETRIC TRIMERIC AND TETRAMERIC CLUSTERS

### 13.2.1 Trimeric clusters

The most thoroughly studied group of trimeric cluster coordination compounds includes many transition metal carboxylates, whose structure is shown in Fig. 13.1. The energy levels are characterized by the total spin $\acute{S}$ of the whole system. The allowed values of the total spin are given by the angular momentum addition rule. As mentioned above, for two ions having spins $s_1$ and $s_2$ the total spin $S$ takes values $s_1 + s_2, s_1 + s_2 - 1, \ldots, |s_1 - s_2|$. For three spins some of the total spin values $S$ may be repeated. Indeed, letting $s_1 = s_2 = s_3 = \frac{1}{2}$, it is easy to see that the total spin takes on values $S = \frac{1}{2}$ and $S = \frac{3}{2}$, and the state characterized by $S = \frac{1}{2}$ occurs twice. The rule for angular momentum coupling is symbolized by

$$\mathrm{D}^{(s_1)} \times \mathrm{D}^{(s_2)} \times \mathrm{D}^{(s_3)} = \sum_{S,\nu} \mathrm{D}^{(S)}_{(\nu)}, \tag{13.14}$$

where $\mathrm{D}^{(S)}$ is an irreducible representation of the rotation group. The subscript $\nu$ indicates states with identical total spin; these states differ in intermediate spin $S'$ in the coupling scheme for three angular momenta.† In the case under

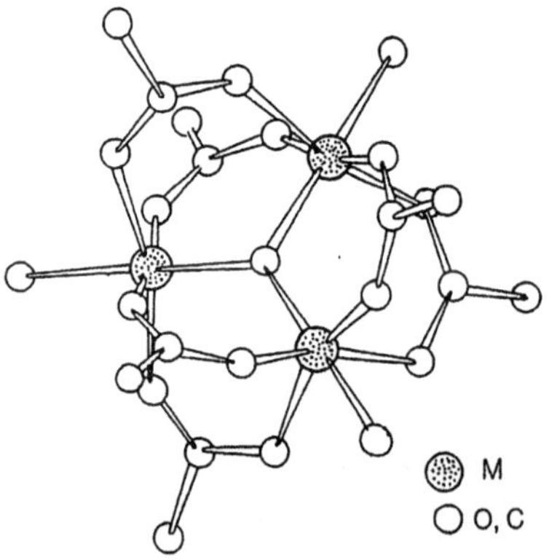


Fig. 13.1 Structure of a trimeric cation $[M_3O(CH_3COO)_6(H_2O)_3]^+$, with $M = Cr^{3+}$, $Fe^{3+}$, $Mn^{3+}$, $Rh^{3+}$, $Cr_2Fe$, $CrFe_2$ [62].

† Decomposition of the direct product of three irreducible representations of the rotation group $\mathrm{D}^{(s_1)}$, $\mathrm{D}^{(s_2)}$ and $\mathrm{D}^{(s_3)}$ can be presented as

$$\mathrm{D}^{(s_1)} \times \mathrm{D}^{(s_2)} \times \mathrm{D}^{(s_3)} = \sum_S n_S \mathrm{D}^{(S)}$$

where $n_S$ is the number of identical (possessing to the same $S$) irreducible representations $\mathrm{D}^{(S)}$. These repeated irreducible representations have different basis sets enumerated in symbolical expression (13.14) by $\nu \equiv S'$.

consideration, a trimeric cluster with $s_i = \frac{1}{2}$, the subscript $\nu$ enumerates states with identical $S = \frac{1}{2}$. These states differ in the values of the intermediate spin $S_{12} = s_1 + s_2$ ($S_{12} = 0, 1$).

The total spin of a trinuclear cluster with $s_i = \frac{3}{2}$ (e.g. $Cr^{3+}$) takes values $S = \frac{1}{2}, \frac{3}{2}, \frac{7}{2}, \frac{9}{2}$, according to the angular momentum addition scheme:

$$D^{(3/2)} \times D^{(3/2)} \times D^{(3/2)} = 2D^{(1/2)}_{(1,2)} + 4D^{(3/2)}_{(0,1,2,3)} + 2D^{(7/2)}_{(2,3)} + D^{(9/2)}_{(3)}. \quad (13.15)$$

It can be seen that the spin $S = \frac{1}{2}$ states are also repeated twice, the $S = \frac{3}{2}$ states are repeated four times, the $S = \frac{5}{2}$ states three times, etc. The subscripts in parentheses on the right-hand side of (13.15) indicate the values of the intermediate spins $S_{12}$ corresponding to the given total spin $S$.

The decomposition similar to (13.15) for a cluster with $s_i = \frac{5}{2}$ (e.g. $Fe^{3+}$) is of the form

$$D^{(5/2)} \times D^{(5/2)} \times D^{(5/2)} = 2D^{(1/2)}_{(2,3)} + 4D^{(3/2)}_{(1,2,3,4)} + 5D^{(7/2)}_{(1,2,3,4,5)} + 4D^{(9/2)}_{(2,3,4,5)}$$
$$+ 3D^{(11/2)}_{(3,4,5)} + 2D^{(13/2)}_{(4,5)} + D^{(15/2)}_{(5)}. \quad (13.16)$$

Let us consider a trigonal cluster with Hamiltonian

$$\hat{H} = -2J_0(\hat{\mathbf{s}}_1 \cdot \hat{\mathbf{s}}_2 + \hat{\mathbf{s}}_2 \cdot \hat{\mathbf{s}}_3 + \hat{\mathbf{s}}_3 \cdot \hat{\mathbf{s}}_1), \quad (13.17)$$

which can be expressed in terms of the total spin operator as

$$\hat{H} = -J_0\left(\hat{\mathbf{S}}^2 - \sum_i \hat{\mathbf{s}}_i^2\right), \quad (13.18)$$

so that its eigenvalues are of the form

$$\epsilon(s) = -J_0\left[S(S+1) - \sum_i^3 s_i(s_i+1)\right]. \quad (13.19)$$

It should be noted that these eigenvalues were found without constructing the actual wavefunctions $|SM\rangle$ from simple products of the type $\Pi_i|s_i m_i\rangle$. It can be seen that the energy levels of the exchange cluster depend on the total spin $S$. For a trinuclear cluster, as shown above, the state is also characterized by the intermediate spin $S_{12}$. However, from (13.19), the symmetric cluster energy is determined solely by the total spin and does not depend on the intermediate spin. In this sense one can talk about a *degeneracy over the intermediate spin*. Thus multiple degeneracy of the ground and excited states is characteristic of symmetric trimeric clusters. It should be noted that for antiferromagnetic clusters of this type, with half-integer spins, the ground state in the HDVV model is fourfold-degenerate ($2D^{(1/2)}$).

For heteronuclear clusters $M_2M'$ and distorted homonuclear clusters the HDVV Hamiltonian becomes

$$\hat{H} = -2[J_0(\hat{\mathbf{s}}_1 \cdot \hat{\mathbf{s}}_3 + \hat{\mathbf{s}}_2 \cdot \hat{\mathbf{s}}_3) + J_1 \hat{\mathbf{s}}_1 \cdot \hat{\mathbf{s}}_2]. \quad (13.20)$$

Its eigenvalues $\epsilon(S_{12}, S)$ depend on the intermediate spin $S_{12}$:

$$\epsilon(S_{12}, S) = -J_0\left[S(S+1) - \sum_{i=1}^{3} s_i(s_i+1)\right]$$
$$+\delta[S_{12}(S_{12}+1) - s_1(s_1+1) - s_2(s_2+1)], \qquad (13.21)$$

with $\delta = J_0 - J_1$.

## 13.2.2 Tetrameric clusters

Tetrameric clusters of practically all 3d and 4d elements have been synthesized and studied by X-ray analysis and magnetochemical methods (for reviews see [62–66]). Examples include the coordination compounds $Cu_4OX_6L_4$ (where X is a halogen and L an axial monodentate ligand) containing copper(II) tetrahedral clusters (Fig. 13.2) [61–63].

For a tetrahedral ($\mathbf{T_d}$) system

$$\hat{H} = -2J(\hat{s}_1 \cdot \hat{s}_2 + \hat{s}_1 \cdot \hat{s}_3 + \hat{s}_1 \cdot \hat{s}_4 + \hat{s}_2 \cdot \hat{s}_3 + \hat{s}_2 \cdot \hat{s}_4 + \hat{s}_3 \cdot \hat{s}_4). \qquad (13.22)$$

This Hamiltonian can be expressed in terms of the operator $\hat{S}^2$, its eigenvalues are

$$\epsilon(s) = -J\left[S(S+1) - \sum_{i=1}^{4} s_i(s_i+1)\right]. \qquad (13.23)$$

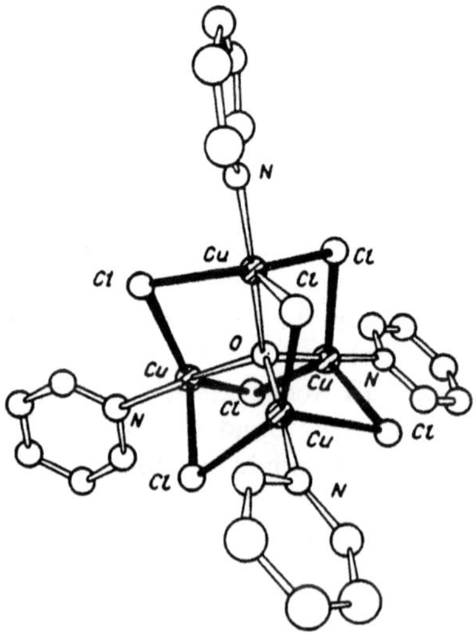


**Fig. 13.2** Structure of the complex $Cu_4OCl_6(C_6H_5N)_4$ (after [65]).

The wavefunctions associated with total spin $S$ can be chosen in one of the alternative schemes for addition of four angular momenta: $|s_1 s_2 (S_{12}) s_3 s_4 (S_{34}) SM\rangle$, $|s_1 s_2 (S_{12}) s_3 (S_{123}) s_4 SM\rangle$, etc. Each addition scheme is characterized by two intermediate spins $S'$ and $S''$ of $S_{ij}$ or $S_{ijk}$ type. The energy $\epsilon(S)$ does *not depend on the values of the intermediate spin*, which leads to multiple degeneracies. For example, for a tetrahedral cluster with $s_i = \frac{1}{2}$ the total spin is $S = 0$ for $S_{12} = 0$ and $S_{34} = 0$ as well as for $S_{12} = 1$ and $S_{34} = 1$. This leads to a twice-repeated representation $D^{(0)}$ $(2D^{(0)})$. The state with $S = 1$ is repeated three times for $S_{12} = 0$ and $S_{34} = 1$, for $S_{12} = 1$ and $S_{34} = 0$, and for $S_{12} = S_{34} = 1$. Finally, the state $D^{(2)}$ with the maximum spin $S = 2$ is nondegenerate with respect to the intermediate spin ($S_{12} = S_{34} = 1$).

For a tetrahedral cluster, distorted along the $C_2$ axis ($\mathbf{D}_{2d}$ symmetry), or a square-planar system ($\mathbf{D}_{4h}$),

$$\hat{H} = -2J(\hat{s}_1 \cdot \hat{s}_2 + \hat{s}_2 \cdot \hat{s}_3 + \hat{s}_3 \cdot \hat{s}_4 + \hat{s}_4 \cdot \hat{s}_1) - 2J_1(\hat{s}_2 \cdot \hat{s}_4 + \hat{s}_1 \cdot \hat{s}_3). \tag{13.24}$$

This Hamiltonian can be expressed in terms of the operators of intermediate spins $\hat{S}_{13}^2$ and $\hat{S}_{24}^2$, and its eigenvalues are found to be

$$\begin{aligned}\epsilon(S_{13}, S_{24}, S) = &-J[S(S+1) - S_{24}(S_{24}+1) - S_{13}(S_{13}+1)] \\ &- J_1\left[S_{13}(S_{13}+1) + S_{24}(S_{24}+1) - \sum_{i=1}^{4} s_i(s_i+1)\right].\end{aligned} \tag{13.25}$$

The energy levels here depend on the set of $S_{13}$ and $S_{24}$ values, so the degeneracies over the intermediate spins are partially removed.

## 13.3 GENERAL METHOD OF CALCULATION OF SPIN LEVELS IN THE HEISENBERG MODEL

### 13.3.1 Structure of the exchange Hamiltonian matrix

In the previous section we examined only clusters that were sufficiently symmetric for the HDVV Hamiltonian to be expressed in terms of the total and intermediate spin operators. In more complex systems this is impossible; hence there are no simple algebraic expressions for the spin-level energies.

Let us consider now the general form of a trimeric system with Hamiltonian

$$\hat{H} = -2(J_1 \hat{s}_2 \cdot \hat{s}_3 + J_2 \hat{s}_3 \cdot \hat{s}_1 + J_3 \hat{s}_1 \cdot \hat{s}_2), \tag{13.26}$$

where all three exchange integrals are different. The energy levels are enumerated by the total-spin values; however, the Hamiltonian (13.26) cannot be expressed in terms of $\hat{\mathbf{S}}^2$ or any of the operators $\hat{\mathbf{S}}_{ij}^2$.

Let us now consider the selection rules for the matrix elements of the Hamiltonian (13.26) from the point of view of group theory. This Hamiltonian is a scalar operator with respect to the rotation group. Wavefunctions with given total spin $S$ transform according to the $(2S+1)$-dimensional representation $D^{(S)}$ of the rotation group **K**. Wavefunctions with identical total spin $S$ correspond to repeating representations. For nonrepeating representations the wavefunctions $|SM\rangle$ are correct functions in the zeroth-order approximation with respect to the Hamiltonian (13.26), and therefore the energy of these states can be obtained as the diagonal matrix element

$$\epsilon(S) = \langle SM|\hat{H}|SM\rangle. \tag{13.27}$$

Being a scalar, the exchange Hamiltonian also has nonzero matrix elements between states that are transformed in the same way but belong to different bases for a repeated representation. In other words, the HDVV Hamiltonian has nonvanishing matrix elements of the form $\langle \nu SM|\hat{H}|\nu' SM\rangle$, where $\nu$ and $\nu'$ may differ. These matrix elements, which are diagonal with respect to the quantum number $M$, are indeed identical for all values of $M$ and the given $S$. It is obvious that if, instead of taking arbitrarily chosen simple products of the spin functions of all ions as the basis for the construction of the Hamiltonian (13.26), we use the wavefunctions associated with the given total spin $S$, then the $f$-dimensional matrix of the total Hamiltonian splits into blocks. Each block has the corresponding value of the given spin and its projection; the block dimension is equal to the number of repetitions of the total spin value in the decomposition described by (13.14). Thus, to obtain the value of the energy, it is sufficient to solve a secular equation of the form

$$|\langle \nu SM|\hat{H}|\nu' SM\rangle - \delta_{\nu\nu'}\epsilon| = 0. \tag{13.28}$$

The exchange-Hamiltonian eigenvalues for the case of repeating representations are expressed as superpositions of states with different intermediate spins:

$$|vSM\rangle = \sum_{\nu} c(\nu)|SM\rangle, \tag{13.29}$$

where the quantum number $v$ enumerates the eigenvalues of (13.28). The above considerations may considerably simplify the determination of exchange-energy levels. Let us consider for instance a trinuclear cluster with $s_i = \frac{1}{2}$ and having the general Hamiltonian (13.26). The $64 \times 64$-matrix can be decomposed, in accordance with (13.15) into two $2 \times 2$-matrices ($S = \frac{1}{2}$ and $S = \frac{7}{2}$), one rank-3 matrix ($S = \frac{5}{2}$) and one $4 \times 4$-matrix ($S = \frac{3}{2}$). The energy $\epsilon(\frac{9}{2})$ is calculated directly from (13.27).

### 13.3.2 Example of calculation of spin levels

The above considerations establish only the structure of the energy matrix. In order to find of its elements, it is necessary to construct wavefunctions of the type $|\nu SM\rangle$, where $\nu$ are various intermediate spin sets. For three ions with spins $S_1$, $S_2$ and $S_3$ we have to use a Wigner decomposition twice. First, in conformity with the general equation (5.27), we find the wavefunctions for two ions:

$$|S_{12}M_{12}\rangle = \sum_{m_1 m_2} |s_1 m_1\rangle |s_2 m_2\rangle \langle s_1 m_1 s_2 m_2 | SM\rangle. \tag{13.30}$$

The cluster wavefunctions $|S_{12}SM\rangle$ ($\nu = S_{12}$) are obtained by "attaching" the spin of the third ion:

$$|(S_{12})SM\rangle = \sum_{M_{12} m_3} |S_{12}M_{12}\rangle |s_3 m_3\rangle \langle S_{12}M_{12} s_3 m_3 | SM\rangle. \tag{13.31}$$

As an example we shall consider spin-level determination for the simple case $s_1 = s_2 = s_3 = \frac{1}{2}$. From (13.30) and (13.31) we obtain

$$\left.\begin{aligned} |(1)\tfrac{3}{2}\tfrac{3}{2}\rangle &= \alpha\alpha\alpha, \\ |(1)\tfrac{1}{2}\tfrac{1}{2}\rangle &= \sqrt{\tfrac{1}{6}}(2\beta\alpha\alpha - \alpha\beta\alpha - \alpha\alpha\beta), \\ |(0)\tfrac{1}{2}\tfrac{1}{2}\rangle &= \sqrt{\tfrac{1}{2}}(\alpha\beta\alpha - \alpha\alpha\beta), \end{aligned}\right\} \tag{13.32}$$

where $\alpha\beta\alpha = \alpha(1)\beta(2)\alpha(3)$ etc. Only states with maximum spin projection $M = S$ are given.

To calculate the Hamiltonian matrix elements, we shall use the following equations:

$$\left.\begin{aligned} \langle\alpha\alpha|\hat{\mathbf{s}}_1\cdot\hat{\mathbf{s}}_2|\alpha\alpha\rangle &= \langle\beta\beta|\hat{\mathbf{s}}_1\cdot\hat{\mathbf{s}}_2|\beta\beta\rangle = \tfrac{1}{4}, \\ \langle\alpha\beta|\hat{\mathbf{s}}_1\cdot\hat{\mathbf{s}}_2|\alpha\beta\rangle &= \langle\beta\alpha|\hat{\mathbf{s}}_1\cdot\hat{\mathbf{s}}_2|\beta\beta\rangle = -\tfrac{1}{4}, \\ \langle\alpha\beta|\hat{\mathbf{s}}_1\cdot\hat{\mathbf{s}}_2|\beta\alpha\rangle &= \langle\beta\alpha|\hat{\mathbf{s}}_1\cdot\hat{\mathbf{s}}_2|\alpha\beta\rangle = \tfrac{1}{2}. \end{aligned}\right\} \tag{13.33}$$

The energy $\epsilon(\frac{3}{2})$ is obtained directly from (13.27):

$$\epsilon(\tfrac{3}{2}) = -\tfrac{1}{2}(J_1 + J_2 + J_3). \tag{13.34}$$

For two repeating spin states with $S = \frac{1}{2}$ the secular equation (13.28) becomes

$$\begin{array}{cc} |(1)\tfrac{1}{2}\tfrac{1}{2}\rangle & |(0)\tfrac{1}{2}\tfrac{1}{2}\rangle \end{array}$$

$$\begin{vmatrix} J_1 + J_2 - \tfrac{1}{2}J_3 - \epsilon & \tfrac{1}{2}\sqrt{3}(J_2 - J_1) \\ \tfrac{1}{2}\sqrt{3}(J_2 - J_1) & \tfrac{3}{2}J_3 - \epsilon \end{vmatrix} = 0. \tag{13.35}$$

Solving the secular equation (13.35), we obtain

$$\epsilon_\pm(\tfrac{1}{2}) = \tfrac{1}{2}(J_1 + J_2 + J_3) \pm (J_1^2 + J_2^2 + J_3^2 - J_1 J_3 - J_2 J_3 - J_1 J_2)^{1/2}. \tag{13.36}$$

These energy levels correspond to the functions

$$\Psi_\pm = c_1^\pm |(1)\tfrac{1}{2}\tfrac{1}{2}\rangle + c_2^\pm |(0)\tfrac{1}{2}\tfrac{1}{2}\rangle, \tag{13.37}$$

where the coefficients $c_1^{\pm}$ and $c_2^{\pm}$ are of the forms

$$\left.\begin{aligned} c_1^+ &= \sqrt{\tfrac{1}{2}}(1-t)^{1/2}, \quad c_2^+ = \frac{|J_2 - J_1|}{(J_2 - J_1)\sqrt{2}}(1+t)^{1/2}, \\ c_1^- &= \sqrt{\tfrac{1}{2}}(1+t)^{1/2}, \quad c_2 = -\frac{|J_2 - J_1|}{(J_2 - J_1)\sqrt{2}}(1-t)^{1/2}, \end{aligned}\right\} \qquad (13.38)$$

with

$$t = \frac{2J_3 - J_1 - J_2}{2(J_1^2 + J_2^2 + J_3^2 - J_1J_2 - J_1J_3 - J_2J_3)^{1/2}}.$$

It can be seen that these correct zeroth-order wavefunctions are linear combinations of wavefunctions with $S' = 1$ and $S' = 0$. The index $v(=1,2)$, labelling the different bases, thus labels the two components in the eigenvalues [13.36], as well as the corresponding eigenfunctions (13.37).

### 13.3.3 The 6*j*- and 9*j*-symbols

Here we introduce the so-called 6*j* and 9*j*-symbols which are widely used in many problems of atomic and molecular physics†. In the next section we will employ these symbols in efficient calculations of spin-levels of multielectronic exchange systems. Let us consider three angular momenta ($j_1$,$j_2$ and $j_3$) of an arbitrary physical nature. We will deal with the spin angular momenta $s_1, s_2$ and $s_3$ bearing in mind their application to the exchange systems. Three spins, $s_1, s_2, s_3$, can be coupled to get full spin $S$ in three different ways

$$\begin{aligned} \text{I}: s_1 + s_2 &= S_{12}, \qquad S_{12} + s_3, - S; \\ \text{II}: s_2 + s_3 &= S_{23}, \qquad s_1 + S_{23} = S; \\ \text{III}: s_1 + s_3 &= S_{13}, \qquad S_{13} + s_2 = S. \end{aligned}$$

The corresponding wavefunctions will be denoted as

$$|s_1s_2(S_{12})s_3SM\rangle\ (\text{I}), \quad |s_1, s_2s_3(S_{23})SM\rangle\ (\text{II})$$

and

$$|s_1s_3(S_{13})s_2SM\rangle\ (\text{III}), \qquad (13.39)$$

which are the eigen-functions of $\hat{\mathbf{s}}_1^2, \hat{\mathbf{s}}_2^2, \hat{\mathbf{s}}_3^2, \hat{\mathbf{S}}_{ij}^2, \hat{\mathbf{S}}^2$ and $\hat{S}_z$. For example, for the scheme coupling I we have:

$$\left.\begin{aligned} \hat{\mathbf{s}}_i^2|s_1s_2(S_{12})s_3SM\rangle &= s_i(s_i+1)|s_1s_2(S_{12})s_3SM\rangle, \\ \mathbf{S}_{12}^2|s_1s_2(S_{12})s_3SM\rangle &= S_{12}(S_{12}+1)|s_1s_2(S_{12})s_3SM\rangle, \\ \mathbf{S}^2|s_1s_2(S_{12})s_3SM\rangle &= S(S+1)|s_1s_2(S_{12})s_3SM\rangle, \\ \hat{S}_z|s_1s_2(S_{12})s_3SM\rangle &= M|s_1s_2(S_{12})s_3SM\rangle. \end{aligned}\right\} \qquad (13.40)$$

† For full description see in [2, 41, 42, 61, 84].

The wavefunctions (13.39) can be obtained using the two-step coupling procedure (13.30), (13.31):

$$\left.\begin{aligned}
|s_1 s_2(S_{12})s_3 SM\rangle &= \sum_{m_1,m_2,m_3} \langle s_1 m_1|\langle s_2 m_2|\langle s_3 m_3| \\
&\times \langle s_1 m_1 s_2 m_2|S_{12} m_{12}\rangle\langle S_{12} m_{12} s_3 m_3|SM\rangle, \\
|s_1, s_2, s_3(S_{23})SM\rangle &= \sum_{m_1,m_2,m_3} \langle s_1 m_1|\langle s_2 m_2|\langle s_3 m_3| \\
&\times \langle s_2 m_2 s_3 m_3|S_{23} m_{23}\rangle\langle s_1 m_1 S_{23} m_{23}|SM\rangle, \\
|s_1 s_3(S_{13})s_2 SM\rangle &= \sum_{m_1,m_2,m_3} |s_1 m_1\rangle|s_2 m_2\rangle|s_3 m_3\rangle \\
&\times \langle s_1 m_1 s_3 m_3|S_{13} m_{13}\rangle\langle S_{13} m_{13} s_2 m_2|SM\rangle.
\end{aligned}\right\} \tag{13.41}$$

Three full sets of the wavefunctions (13.39) are connected by a unitary transformation; for example:

$$|s_1, s_2 s_3(S_{23})SM\rangle = \sum_{S_{12}} \langle s_1 s_2(S_{12})s_3 SM|s_1, s_2 s_3(S_{23})SM\rangle \times |s_1 s_2(S_{12})s_3 SM\rangle. \tag{13.42}$$

The coefficients of the unitary transformations connecting sets of wavefunctions belonging to the two different coupling schemes of three angular momenta are expressed in terms of the so-called Wigner 6*j*-symbols [2]:

$$\left.\begin{aligned}
&\langle s_1, s_2 s_3(S_{23})SM|s_1 s_2(S_{12})s_3 SM\rangle \\
&= (-1)^{s_1+s_2+s_3+S}[(2S_{12}+1)(2S_{23}+1)]^{1/2}\begin{Bmatrix} s_1 & s_2 & S_{12} \\ s_3 & S & S_{23} \end{Bmatrix}, \\
&\langle s_1 s_2(S_{12})s_3 SM|s_1 s_3(S_{13})s_2 SM\rangle \\
&= (-1)^{s_2+s_3+S_{12}+S_{13}}[(2S_{12}+1)(2S_{13}+1)]^{1/2}\begin{Bmatrix} s_2 & s_1 & S_{12} \\ s_3 & S & S_{13} \end{Bmatrix}, \\
&\langle s_1, s_2 s_3(S_{23})SM|s_1 s_3(S_{13})s_2 SM\rangle \\
&= (-1)^{s_2+s_3 S_{23}}[(2S_{13}+1)(2S_{23}+1)]^{1/2}\begin{Bmatrix} s_1 & s_3 & S_{13} \\ s_2 & S & S_{23} \end{Bmatrix}.
\end{aligned}\right\} \tag{13.43}$$

Comparing (13.41) and (13.42) one can express the 6*j*-symbols$\{:::\}$ in terms of

Clebsch–Gordan coefficients:

$$\begin{Bmatrix} s_1 & s_2 & S_{12} \\ s_3 & S & S_{23} \end{Bmatrix} = (-1)^{s_1+s_2+s_3+S}[(2S_{12}+1)(2S_{23}+1)]^{-1/2}$$

$$\times \sum_{m_1,m_2,m_3} \sum_{m_{12},m_{23}} \langle s_2 m_2 s_3 m_3 | S_{23} m_{23} \rangle$$

$$\times \langle s_1 m_1 S_{23} m_{23} | SM \rangle \langle s_1 m_1 s_2 m_2 | S_{12} m_{12} \rangle$$

$$\times \langle S_{12} m_{12} s_3 m_3 | SM \rangle. \tag{13.44}$$

In (13.44), the value of $M$ is fixed ($M = m_1 + m_2 + m_3$). Using (5.28) we can deduce the general expression for the $6j$-symbols in terms of $3j$-symbols:

$$\begin{Bmatrix} j_1 & j_2 & j_3 \\ j_4 & j_5 & j_6 \end{Bmatrix} = \sum_{\text{all } m} (-1)^{\sum_{i=1}^{6}(l_i - m_i)} \begin{pmatrix} j_1 & j_2 & j_3 \\ -m_1 & -m_2 & -m_3 \end{pmatrix}$$

$$\times \begin{pmatrix} j_1 & j_5 & j_6 \\ m_1 & -m_5 & m_6 \end{pmatrix} \begin{pmatrix} j_4 & j_2 & j_6 \\ m_4 & m_2 & -m_6 \end{pmatrix} \begin{pmatrix} j_4 & j_5 & j_3 \\ -m_4 & m_5 & m_3 \end{pmatrix}. \tag{13.45}$$

The $6j$-symbol differs from the Racah's $W$-coefficient in a simple phase factor:

$$\begin{Bmatrix} j_1 & j_2 & j_3 \\ j_4 & j_5 & j_6 \end{Bmatrix} = (-1)^{j_1+j_2+j_3+j_4+j_5} W(j_1 j_2 j_4 j_5; j_3 j_6). \tag{13.46}$$

For each $3j$-symbol in (13.45) the "triangle condition" (rule of angular momenta addition) must be fulfilled; for example, $\begin{pmatrix} j_1 & j_2 & j_3 \\ m_1 & m_2 & m_3 \end{pmatrix}$ vanish unless

$$j_1 j_2 \geq j_3 \geq |j_1 - j_2|, \quad j_1 + j_3 \geq j_2 \geq |j_1 - j_3|,$$

$$j_2 + j_3 \geq j_1 \geq |j_2 - j_3|, \quad \text{or briefly} \quad \Delta(j_1 j_2 j_3).$$

As is obvious from (13.45), four independent triangle conditions must be satisfied for the nonzero $6j$-symbol

$$\Delta(j_1 j_2 j_3), \quad \Delta(j_1 j_5 j_6), \quad \Delta(j_4 j_2 j_6) \quad \text{and} \quad \Delta(j_4 j_5 j_3),$$

or visually [84]

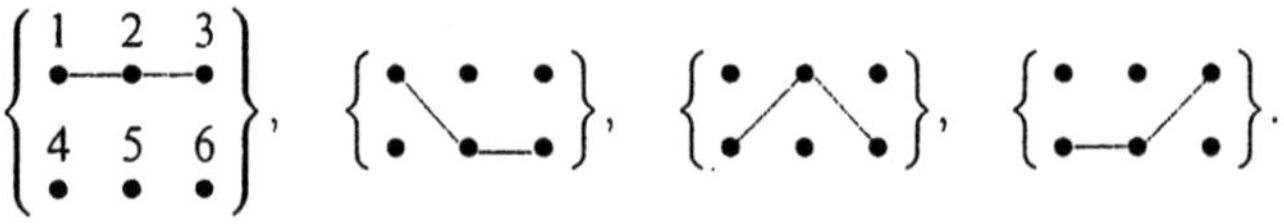


These triangle conditions are equivalent for those four faces of a tetrahedron as shown in Fig. 13.3.

The $6j$-symbol is symmetric irrespective of the interchange of any two columns or of switching the upper and lower members of any two rows; for

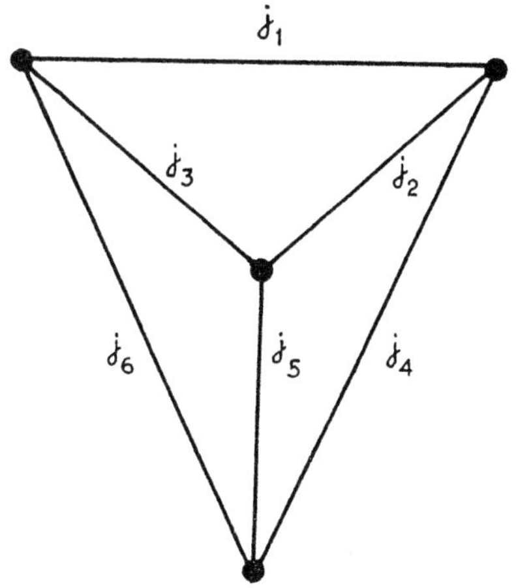


**Fig. 13.3** Geometrical interpretation for the triangle conditions for 6*j*-symbol.

example:

$$\begin{Bmatrix} j_1 & j_2 & j_3 \\ j_4 & j_5 & j_6 \end{Bmatrix} = \begin{Bmatrix} j_4 & j_5 & j_3 \\ j_1 & j_2 & j_6 \end{Bmatrix} = \begin{Bmatrix} j_2 & j_1 & j_3 \\ j_5 & j_4 & j_6 \end{Bmatrix}, \text{etc.} \tag{13.47}$$

The orthogonality and the normalization relationships arising from the unitary nature of the transformation (13.42) are valid:

$$\left.\begin{aligned} \sum_{j_3}(2j_3+1)(2j_6+1)\begin{Bmatrix} j_1 & j_2 & j_3 \\ j_4 & j_5 & j_6 \end{Bmatrix}\begin{Bmatrix} j_1 & j_2 & j_3 \\ j_4 & j_5 & j_6' \end{Bmatrix} &= \delta_{j_6 j_6'}, \\ \sum_{j_6}(2j_3+1)(2j_6+1)\begin{Bmatrix} j_1 & j_2 & j_3 \\ j_4 & j_5 & j_6 \end{Bmatrix}\begin{Bmatrix} j_1 & j_2 & j_3' \\ j_4 & j_5 & j_6 \end{Bmatrix} &= \delta_{j_3 j_3'}. \end{aligned}\right\} \tag{13.48}$$

A 6*j*-symbol with one zero element collapses to the simple algebraic form:

$$\begin{Bmatrix} j_1 & j_2 & 0 \\ j_4 & j_5 & j_6 \end{Bmatrix} = (-1)^{j_1+j_4+j_6}[(2j_1+1)(2j_4+1)]^{-1/2}\delta_{j_1 j_2}\delta_{j_4 j_5}. \tag{13.49}$$

There are algebraic expressions for many 6*j*-symbols, the most complete data are listed in [2]. Two important expressions are the following:

$$\begin{Bmatrix} j_1 & j_1 & 1 \\ j_2 & j_2 & j \end{Bmatrix} = (-1)^{j_1+j_2+j}\frac{j(j+1)-j_1(j_1+1)-j_2(j_2+1)}{[j_1(2j_1+1)(2j_2+2)j_2(2j_2+1)(2j_2+2)]^{1/2}}, \tag{13.50}$$

$$\begin{Bmatrix} j_1 & j_2 & j_3 \\ 1 & j_3 & j_2 \end{Bmatrix} = (-1)^{j_1+j_2+j_3+1}\frac{2[j_2(j_2+1)+j_3(j_3+1)-j_1(j_1+1)]}{[2j_2(2j_2+1)(2j_2+2)2j_3(2j_3+1)(2j_3+2)]^{1/2}}.$$

Finally, the Regge symmetry should be mentioned:

$$\begin{Bmatrix} j_1 & j_2 & j_3 \\ j_4 & j_5 & j_6 \end{Bmatrix} = \begin{Bmatrix} (j_1+j_2+j_4-j_5)/2 & (j_1+j_2+j_5-j_4)/2 & j_3 \\ (j_1+j_4+j_5-j_2)/2 & (j_4+j_2+j_5-j_1)/2 & j_6 \end{Bmatrix} \tag{13.51}$$

The tabulation of 6*j*-symbols is given by Rotenberg *et al.* [4], for symbols up to a maximum value of 8 for any argument, similar tables are listed in [2] for arguments up to 3.

Wigner's 9*j*-symbols of Fano's coefficients form a unitary matrix of full sets of wavefunctions belonging to three different schema of two by two coupling of four angular momenta [2]:

$$\begin{aligned}
\text{I:} &\quad s_1 + s_2 = S_{12}, \quad s_3 + s_4 = S_{34}, \quad S_{12} + S_{34} = S;\\
\text{II:} &\quad s_1 + s_3 = S_{13}, \quad s_2 + s_4 = S_{24}, \quad S_{13} + S_{24} = S;\\
\text{III:} &\quad s_1 + s_4 = S_{14}, \quad s_2 + s_3 = S_{23}, \quad S_{14} + S_{23} = S.
\end{aligned}$$

For scheme I the corresponding wavefunctions can be labelled as $|s_1 s_2(S_{12}) s_3 s_4(S_{34}) SM\rangle$ being the eigen-functions of all $\hat{s}_i^2, \hat{S}_{12}^2, \hat{S}_{34}^2, \hat{S}^2$ and $\hat{S}_z$. Applying a thrice-repeated Clebsch–Gordan expansion one can represent these wavefunctions as follows:

$$\begin{aligned}
&|s_1 s_2(S_{12}) s_3 s_4(S_{34}) SM\rangle\\
&= \sum_{m_1, m_2, m_3, m_4} \sum_{m_{12}, m_{34}} |s_1 m_1\rangle |s_2 m_2\rangle |s_3 m_3\rangle |s_4 m_4\rangle\\
&\times \langle s_1 m_1 s_2 m_2 | s_{12} m_{12}\rangle \langle s_3 m_3 s_4 m_4 | S_{34} m_{34}\rangle\\
&\times \langle S_{12} m_{12} S_{34} m_{34} | SM\rangle.
\end{aligned} \tag{13.52}$$

The element of the unitary matrix corresponding to the transformation between one of the schema of four angular momenta coupling one to another are connected with the 9*j*-symbols $\{:::\}$, which are independent on $M$ value:

$$\left.\begin{aligned}
&\langle s_1 s_2(S_{12}) s_3 s_4(S_{34}) SM | s_1 s_3(S_{13}) s_2 s_4(S_{24}) SM\rangle\\
&= [(2S_{12}+1)(2S_{13}+1)(2S_{24}+1)(2S_{34}+1)]^{1/2} \begin{Bmatrix} s_1 & s_2 & S_{12} \\ s_3 & s_4 & S_{34} \\ S_{13} & S_{24} & S \end{Bmatrix},\\
&\langle s_1 s_2(S_{12}) s_3 s_4(S_{34}) SM | s_1 s_4(S_{14}) s_2 s_3(S_{23}) SM\rangle\\
&= (-1)^{s_3+s_4-S_{34}} [(2S_{12}+1)(2S_{14}+1)(2S_{23}+1)(2S_{34}+1)]^{1/2}\\
&\times \begin{Bmatrix} s_1 & s_2 & S_{12} \\ s_4 & s_3 & S_{34} \\ S_{14} & S_{23} & S \end{Bmatrix},\\
&\langle s_1 s_3(S_{13}) s_2 s_4(S_{24}) SM | s_1 s_4(S_{14}) s_2 s_3(S_{23}) SM\rangle\\
&= (-1)^{s_3-s_4-S_{23}+S_{24}} [(2S_{13}+1)(2S_{24}+1)(2S_{14}+1)(2S_{23}+1)]^{1/2}\\
&\times \begin{Bmatrix} s_1 & s_3 & S_{13} \\ s_4 & s_2 & S_{24} \\ S_{14} & S_{23} & S \end{Bmatrix}.
\end{aligned}\right\} \tag{13.53}$$

The $9j$-symbols are expressed in terms of products of six Clebsch–Gordan coefficients:

$$\begin{Bmatrix} s_1 & s_2 & S_{12} \\ s_3 & s_4 & S_{34} \\ S_{13} & S_{24} & S \end{Bmatrix} = [(2S_{12}+1)(2S_{13}+1)(2S_{24}+1)(2S_{34}+1)]^{1/2}(2S+1)^{-1}$$

$$\times \sum_{\text{all } m_i} \sum_{\text{all } m_{ik}} \langle s_1 m_1 s_2 m_2 | S_{12} m_{12} \rangle \langle s_3 m_3 s_4 m_4 | S_{34} m_{34} \rangle$$

$$\times \langle S_{12} m_{12} S_{34} m_{34} | SM \rangle \langle s_1 m_1 s_3 m_3 | S_{13} m_{13} \rangle$$

$$\times \langle s_2 m_2 s_4 m_4 | S_{24} m_{24} \rangle \langle S_{13} m_{13} S_{24} m_{24} | SM \rangle. \tag{13.54}$$

Sometimes the expression of the $9j$-symbol through $3j$-symbols seems to be useful:

$$\begin{Bmatrix} s_1 & s_2 & s_3 \\ s_4 & s_5 & s_6 \\ s_7 & s_8 & s_9 \end{Bmatrix} = \sum_{\text{all } m} \begin{pmatrix} s_1 & s_2 & s_3 \\ m_1 & m_2 & m_3 \end{pmatrix} \begin{pmatrix} s_4 & s_5 & s_6 \\ m_4 & m_5 & m_6 \end{pmatrix}$$

$$\times \begin{pmatrix} s_7 & s_8 & s_9 \\ m_7 & m_8 & m_9 \end{pmatrix} \begin{pmatrix} s_1 & s_4 & s_5 \\ m_1 & m_4 & m_5 \end{pmatrix}$$

$$\times \begin{pmatrix} s_2 & s_5 & s_8 \\ m_2 & m_5 & m_8 \end{pmatrix} \begin{pmatrix} s_3 & s_6 & s_9 \\ m_3 & m_6 & m_9 \end{pmatrix}. \tag{13.55}$$

For the nonzero $9j$-symbol

$$\begin{Bmatrix} s_1 & s_2 & s_3 \\ s_4 & s_5 & s_6 \\ s_7 & s_8 & s_9 \end{Bmatrix}$$

the following triangle conditions must be held simultaneously:

$$\Delta(s_1\ s_2\ s_3), \quad \Delta(s_4\ s_5\ s_6), \quad \Delta(s_7\ s_8\ s_9),$$
$$\Delta(s_1\ s_4\ s_7), \quad \Delta(s_2\ s_5\ s_8), \quad \Delta(s_3\ s_6\ s_9).$$

Since $9j$-symbols are elements of a unitary matrix the orthogonality and normalization conditions must be satisfied:

$$\left.\begin{aligned} &\sum_{s_7, s_8}(2s_7+1)(2s_8+1) \begin{Bmatrix} s_1 & s_2 & s_3 \\ s_4 & s_5 & s_6 \\ s_7 & s_8 & s_9 \end{Bmatrix} \begin{Bmatrix} s_1 & s_2 & s_3' \\ s_4 & s_5 & s_6' \\ s_7 & s_8 & s_9 \end{Bmatrix} \\ &= [(2s_3+1)(2s_6+1)]^{-1} \delta_{s_3 s_3'} \delta_{s_6 s_6'}, \\ &\sum_{s_3, s_6}(2s_3+1)(2s_6 1) \begin{Bmatrix} s_1 & s_2 & s_3 \\ s_4 & s_5 & s_6 \\ s_7 & s_8 & s_9 \end{Bmatrix} \begin{Bmatrix} s_1 & s_2 & s_3 \\ s_4 & s_5 & s_6 \\ s_7' & s_8' & s_9 \end{Bmatrix} \\ &= [(2s_7+1)(2s_8+1)]^{-1} \delta_{s_7 s_7'} \delta_{s_8 s_8'}. \end{aligned}\right\} \tag{13.56}$$

The $9j$-symbol can be expressed in terms of $6j$-symbols:

$$\begin{Bmatrix} s_1 & s_2 & s_3 \\ s_4 & s_5 & s_6 \\ s_7 & s_8 & s_9 \end{Bmatrix} = \sum_j (-1)^{2j}(2j+1)$$

$$\times \begin{Bmatrix} s_1 & s_2 & s_3 \\ s_6 & s_9 & j \end{Bmatrix} \begin{Bmatrix} s_4 & s_5 & s_6 \\ s_2 & j & s_8 \end{Bmatrix} \begin{Bmatrix} s_7 & s_8 & s_9 \\ j & s_1 & s_4 \end{Bmatrix}. \qquad (13.57)$$

The $9j$-symbol possesses the following symmetry properties: (1) Transportation of any two columns or rows does not change the absolute value of the $9j$-symbol and results only in the phase factor $\kappa$; for example:

$$\begin{Bmatrix} s_1 & s_2 & s_3 \\ s_4 & s_5 & s_6 \\ s_7 & s_8 & s_9 \end{Bmatrix} = \kappa \begin{Bmatrix} s_2 & s_1 & s_3 \\ s_5 & s_4 & s_6 \\ s_8 & s_7 & s_9 \end{Bmatrix},$$

$$(13.58)$$

$$\begin{Bmatrix} s_1 & s_2 & s_3 \\ s_4 & s_5 & s_6 \\ s_7 & s_8 & s_9 \end{Bmatrix} = \kappa \begin{Bmatrix} s_4 & s_5 & s_6 \\ s_1 & s_2 & s_3 \\ s_7 & s_8 & s_9 \end{Bmatrix}.$$

The phase factor $\kappa = (-1)^{\Sigma_{i=1}^{9} s_i}$ and $\kappa = 1$ for odd and even numbers of transpositions respectively. (2) The $9j$-symbol is symmetric for each respective transportation

$$\begin{Bmatrix} s_1 & s_2 & s_3 \\ s_4 & s_5 & s_6 \\ s_7 & s_8 & s_9 \end{Bmatrix} = \begin{Bmatrix} s_1 & s_4 & s_7 \\ s_2 & s_5 & s_8 \\ s_3 & s_6 & s_9 \end{Bmatrix}. \qquad (13.59)$$

The $9j$-symbol with one zero collapses to give a $6j$-symbol [2]:

$$\begin{Bmatrix} s_1 & s_2 & s_3 \\ s_4 & s_5 & s_6 \\ s_7 & s_8 & 0 \end{Bmatrix} = (-1)^{s_3+s_2+s_4+s_7}[(2s_3+1)(2s_7+1)]^{-1/2}$$

$$\times \begin{Bmatrix} s_1 & s_2 & s_3 \\ s_5 & s_4 & s_7 \end{Bmatrix}. \qquad (13.60)$$

Using the symmetry property (1), one can use (13.60) for any zeroth $s_i$ in the $9j$-symbol.

If two indices equal zero we have:

$$\begin{Bmatrix} s_1 & s_2 & s_3 \\ s_4 & 0 & s_6 \\ s_7 & s_8 & 0 \end{Bmatrix} = \frac{(-1)^{s_1-s_2-s_3}}{(2s_2+1)(2s_3+1)} \delta_{s_4 s_6} \delta_{s_2 s_8} \delta_{s_3 s_6} \delta_{s_7 s_8}. \tag{13.61}$$

Finally, in the case of three zeroth elements, the $9j$-symbol reduces to simple expressions:

$$\left.\begin{aligned} \begin{Bmatrix} s_1 & s_2 & s_3 \\ s_4 & s_5 & s_6 \\ 0 & 0 & 0 \end{Bmatrix} &= \frac{\delta_{s_1 s_4} \delta_{s_2 s_5} \delta_{s_3 s_6}}{[(2s_1+1)(2s_2+1)(2s_3+1)]^{1/2}}, \\ \begin{Bmatrix} 0 & s_2 & s_3 \\ s_4 & 0 & s_6 \\ s_7 & s_8 & 0 \end{Bmatrix} &= \frac{(-1)^{2s_2} \delta_{s_2 s_3} \delta_{s_2 s_4} \delta_{s_2 s_6} \delta_{s_2 s_7} \delta_{s_2 s_8}}{(2s_2+1)^2}. \end{aligned}\right\} \tag{13.62}$$

Many algebraic expressions for $9j$-symbols plus tables for $(s_7, s_8, s_9) = (\frac{1}{2}, \frac{1}{2}, 0)$ or $(\frac{1}{2}, \frac{1}{2}, 1)$ and $0 \leq s_1, s_2, s_3, s_4, s_5, s_6 \leq 4$ are given in [2] (see also [85, 86]).

### 13.3.4 Application of irreducible tensor method, recoupling

The above example (Section 13.3.2) of spin-triad calculation with an arbitrary exchange parameter set illustrates a general point concerning Hamiltonian matrix structure within repeating spin states. Direct calculation for a polynuclear system, in particular the determination of exchange-Hamiltonian matrix elements, is an extremely cumbersome task. This calculation is already too elaborate for a triad with spins $s_i > \frac{1}{2}$. With the use of tensor algebra methods, developed mainly with reference to atomic shells [39, 40, 42], the calculation of matrices can be greatly simplified [62, 87, 89].

The spin operators $\hat{\mathbf{S}}$ are irreducible tensor operators $\hat{\mathbf{S}}_q^{(\kappa)}$ of rank 1 (the symbols $\kappa$ and $q$ indicate the rank and components of the tensor operator; $q = -\kappa, -\kappa+1, \ldots, \kappa$). The cyclic components of the operator $\hat{\mathbf{S}}$ can be considered as components of the irreducible tensor operator $\mathbf{S}_q^{(1)}$ $(q = 0, \pm 1)$:

$$\hat{\mathbf{S}}_0^{(1)} = \hat{S}_z, \quad \hat{\mathbf{S}}_{\pm 1}^{(1)} = \mp\sqrt{\tfrac{1}{2}}(\hat{S}_x \pm i\hat{S}_y). \tag{13.63}$$

The *tensor product* of the irreducible tensor operators $\hat{\mathbf{T}}^{(\kappa_1)}$ and $\mathbf{U}^{(\kappa_2)}$ is given by the Wigner decomposition

$$\{\hat{\mathbf{T}}^{(\kappa_1)} \otimes \hat{\mathbf{U}}^{(\kappa_2)}\}_Q^{(K)} = \sum_{q_1 q_2} \hat{\mathbf{T}}_{q_1}^{(\kappa_1)} \hat{\mathbf{U}}_{q_2}^{(\kappa_2)} \langle \kappa_1 q_1 \kappa_2 q_2 | KQ \rangle, \tag{13.64}$$

where the rank $K$ of the tensor product obeys the vector rule: $K = \kappa_1 + \kappa_2, \kappa_1 + \kappa_2 - 1, \ldots, |\kappa_1 - \kappa_2|$.

The exchange Hamiltonian of a trimeric compound with Hamiltonian (13.26) can be written in tensor form as

$$\hat{H} = 2\sqrt{\tfrac{1}{3}}[J_{12}\{\hat{\mathbf{S}}^{(1)}(1)\otimes\hat{\mathbf{S}}^{(1)}(2)\}^{(0)} + J_{13}\{\mathbf{S}^{(1)}(1)\otimes\mathbf{S}^{(1)}(3)\}^{(0)} + J_{23}\{\hat{\mathbf{S}}^{(1)}(2)\otimes\hat{\mathbf{S}}^{(1)}(3)\}^{(0)}] \quad (13.65)$$

where $\hat{\mathbf{S}}^{(1)}_{\mu}(i)$ $(i = 1, 2, 3)$ are the first rank irreducible tensors acting in the spin-spaces of individual ions. We have used the relations (arising from (13.63) and (13.64)) between the scalar product of vector operators $\mathbf{S}_i$ and the scalar convolution of first order irreducible tensors $\hat{\mathbf{S}}^{(1)}_{\mu}(i)$

$$\mathbf{s}_1\mathbf{s}_2 = -\sqrt{3}\left\{\hat{\mathbf{S}}^{(1)}(1)\otimes\hat{\mathbf{S}}^{(1)}(2)\right\}^{(0)}, \quad \text{etc.}$$

Calculations of the matrix elements of the Hamiltonian (13.65) (as well as the matrix elements of more general spin-coupling operators) can be performed by means of the so-called *recoupling procedure.*

Let us start with a dimeric system consisting of two ions $i$ and $j$ with spins $s_i$ and $s_j$ and consider the matrix elements of the two-particle tensor operator $\hat{\mathbf{Q}}^{\kappa_{ij}}_{\mu_{ij}}$ ("particles" in this context are the ions with the total spins $s_i$ and $s_j$):

$$\hat{\mathbf{Q}}^{(\kappa_{ij})}_{\mu_{ij}} = \left\{\hat{\mathbf{S}}^{(\kappa_i)}(i)\otimes\hat{\mathbf{S}}^{(\kappa_j)}(j)\right\}^{(\kappa_{ij})}_{\mu_{ij}} = \sum_{\mu_i\mu_j}\hat{\mathbf{S}}^{(\kappa_i)}_{\mu_i}(i)\hat{\mathbf{S}}^{(\kappa_j)}_{\mu_j}(j)\langle\kappa_i\mu_i\kappa_j\mu_j|\kappa_{ij}\mu_{ij}\rangle \quad (13.66)$$

In (13.66) $\hat{\mathbf{S}}^{(\kappa_i)}_{\mu_i}(i)$ and $\hat{\mathbf{S}}^{(\kappa_j)}_{\mu_j}(j)$ are the irreducible tensor operators acting in the spin spaces of ions $i$ and $j$ respectively. Matrix elements of these one-particle operators are expressed by means of the Wigner–Eckart theorem

$$\langle s_i m_i|\hat{\mathbf{S}}^{(\kappa_i)}_{\mu_i}|s_i m_i'\rangle = (-1)^{s_i - m_i}\langle s_i\|\hat{\mathbf{S}}^{(\kappa_i)}\|s_i\rangle\begin{pmatrix} s_i & \kappa_i & s_i \\ -m_i & \mu_i & m_i' \end{pmatrix}. \quad (13.67)$$

The reduced matrix elements in (13.67) for $\kappa_i = 0, 1$ and 2 are:

$$\langle s\|\hat{\mathbf{S}}^{(0)}\|s\rangle = (2s+1)^{1/2},$$
$$\langle s\|\hat{\mathbf{S}}^{(1)}\|s\rangle = [s(s+1)(2s+1)]^{1/2},$$
$$\langle s\|\mathbf{S}^{(2)}\|s\rangle = (1/\sqrt{6}[2s(2s+3)(2s+2)(2s+1)(2s-1)]^{1/2}. \quad (13.68)$$

While calculating the matrix elements of $\hat{\mathbf{Q}}^{(\kappa_{ij})}_{\mu_{ij}}$ let us use as the basis set wavefunctions of two interacting ions belonging to the full spins $S_{ij} = s_i + s_j, s_i + s_j - 1, \ldots |s_i - s_j|$ and to the projections $m_{ij}$:

$$|s_i s_j S_{ij} m_{ij}\rangle = \sum_{m_i, m_j}|s_i m_i\rangle|s_j m_j\rangle\langle s_i m_i s_j m_j|s_{ij} m_{ij}\rangle. \quad (13.69)$$

Applying the Wigner–Eckart theorem (6.4) to the irreducible tensor $\hat{\mathbf{Q}}_{\mu_{ij}}^{(\kappa_{ij})}$ we obtain:

$$\langle s_i s_j S_{ij} m_{ij} | \hat{\mathbf{Q}}_{\mu_{ij}}^{(\kappa_{ij})} | s_i s_j S'_{ij} m'_{ij} \rangle = \frac{(-1)^{2\kappa_{ij}}}{(2s_{ij}+1)^{1/2}} \langle s_i s_j S_{ij} \| \hat{\mathbf{Q}}^{(\kappa_{ij})} \| s_i s_j S'_{ij} \rangle$$

$$\times \langle S_{ij} m_{ij} | S'_{ij} m'_{ij} \kappa_{ij} \mu_{ij} \rangle \equiv (-1)^{s_{ij}-m_{ij}} \langle s_i s_j S_{ij} \| \hat{\mathbf{Q}}^{(\kappa_{ij})} \| s_i s_j S'_{ij} \rangle \begin{pmatrix} s_{ij} & \kappa_{ij} & S'_{ij} \\ -m_{ij} & \mu_{ij} & m'_{ij} \end{pmatrix}. \tag{13.70}$$

The quantity $\langle s_i s_j S_{ij} \| \hat{\mathbf{Q}}^{(\kappa_{ij})} \| s_i s_j S'_{ij} \rangle$ is the so-called two-particle reduced matrix element. Using the Clebsch–Gordan expansion for $\hat{\mathbf{Q}}_{\mu_{ij}}^{(\kappa_{ij})}$ (13.66) and for basis sets (13.69) one can express the two-particle reduced matrix element in terms of the one-particle ones [2, 62]:

$$\langle s_i s_j S_{ij} \| \{\hat{\mathbf{S}}^{(\kappa_i)}(i) \otimes \hat{\mathbf{S}}^{(\kappa_j)}(j)\}^{(\kappa_{ij})} \| s_i s_j S'_{ij} \rangle = [(2S_{ij}+1)(2\kappa_{ij}+1)(2S'_{ij}+1)]^{1/2}$$

$$\times \begin{Bmatrix} \kappa_{ij} & S_{ij} & S'_{ij} \\ \kappa_i & S_i & S_i \\ \kappa_j & S_j & S_j \end{Bmatrix} \langle s_i \| \hat{\mathbf{S}}^{(\kappa_i)}(i) \| s_i \rangle \langle s_j \| \hat{\mathbf{S}}^{(\kappa_j)}(j) \| s_j \rangle. \tag{13.71}$$

In (13.71) $\{\vdots \ \vdots \ \vdots\}$ is the $9j$-symbol defined in Section 13.3.3 and $\langle s_i \| \hat{\mathbf{S}}^{(\kappa_i)}(i) \| s_i \rangle$ is the reduced matrix elements of the irreducible tensor operators $\hat{\mathbf{S}}_{\mu}^{(\kappa_i)}(i)$ (equation (13.68). *Equation* (13.71) *allows us to express a matrix element of the two-particle operator involving spin variables of two interacting subsystems in terms of one-particle matrix elements relating to each system (recoupling).* It should be stressed that in contrast to the illustrative example of Section 13.3.2 no wavefunctions in the explicit form have been used.

Now we can proceed to the consideration of a more complicated case of a trimeric system. The Hamiltonian (13.63) can be considered as a particular case of the generalized spin-Hamiltonian of the interaction of three ions with spins $s_1, s_2, s_3$:

$$\hat{H}^{(3)} = \sum_{\kappa} (2\kappa+1)^{1/2} \{c^{(\kappa)}$$

$$(\kappa_1 \kappa_2 \kappa_{12} \kappa_3) \otimes \{\{\hat{\mathbf{S}}^{(\kappa_1)}(1) \otimes \hat{\mathbf{S}}^{(\kappa_2)}(2)\}^{(\kappa_{12})} \otimes \hat{\mathbf{S}}^{(\kappa_3)}(3)\}^{(\kappa)}\}^{(0)}, \tag{13.72}$$

As earlier in (13.72), $\hat{\mathbf{S}}^{(\kappa_i)}(i)$ is the irreducible tensor of rank $\kappa_i$ acting in the spin-space of $i$-th ion, $0 \le \kappa_i \le 2S_i, \kappa_{12} = \kappa_1 + \kappa_2, \kappa_1 + \kappa_2 - 1, \ldots, |\kappa_1 - \kappa_2|$, $c^{(\kappa)}(\ldots)$ are the numerical parameters. Hamiltonian (13.72) involves three-particle interactions and the original Hamiltonians (13.26) and (13.65) can be easily deduced from (13.72). Actually providing for example, $\kappa_1 = \kappa_2 = 1, \kappa_3 = 0, \kappa_{12} = 0$ and comparing (13.72) and (13.26) we obtain

$$c^{(0)}(1\,1\,0\,0)\{\hat{\mathbf{S}}^{(1)}(1) \otimes \hat{\mathbf{S}}^{(1)}(2)\}^{(0)} \equiv -2J_{12}\mathbf{s}_1\mathbf{s}_2, \tag{13.73}$$

so that

$$c^{(0)}(1\,1\,0\,0) = 2\sqrt{3}J_{12}.$$

Applying the recoupling procedure to the three particle operator

$$\hat{\mathbf{Q}}^{(\kappa)}_{\mu} = \{\{\hat{\mathbf{S}}^{(\kappa_1)}(1)\otimes\hat{\mathbf{S}}^{(\kappa_2)}(2)\}^{(\kappa_{12})}\otimes\hat{\mathbf{S}}^{(\kappa_3)}(3)\}^{(\kappa)}_{\mu}, \tag{13.74}$$

one can obtain the following result for the reduced matrix element in the basis set $|s_1s_2(S_{12})s_3SM\rangle$ of a trimeric cluster consisting of spins $s_1, s_2, s_3$:

$$\begin{aligned}
&\langle s_1s_2(S_{12})s_3S\|\hat{\mathbf{Q}}^{(\kappa)}\|s_1s_2(S'_{12})s_3S'\rangle \\
&= [(2S+1)(2S'+1)(2\kappa+1)(2S_{12}+1)(2S'_{12}+1)(2\kappa_{12}+1)]^{1/2} \\
&\quad\times \begin{Bmatrix} S_{12} & S'_{12} & \kappa_{12} \\ s_3 & s_3 & \kappa_3 \\ S & S' & \kappa \end{Bmatrix} \begin{Bmatrix} s_1 & s_1 & \kappa_1 \\ s_2 & s_2 & \kappa_2 \\ S_{12} & S'_{12} & \kappa_{12} \end{Bmatrix} \\
&\quad\times \langle s_1\|\hat{\mathbf{S}}^{(\kappa_1)}\|s_1\rangle\langle s_2\|\hat{\mathbf{S}}^{(\kappa_2)}\|s_2\rangle\langle s_3\|\hat{\mathbf{S}}^{(\kappa_3)}\|s_3\rangle.
\end{aligned} \tag{13.75}$$

In the case of isotropic Heisenberg exchange under consideration ($\kappa = 0$ and one of $\kappa_i = 0$ in (13.74)) the $9j$-symbols in (13.75) reduce to $6j$-symbols (Section 13.3.3). Using (13.75) one can easily obtain the following matrix elements of three operator terms of exchange Hamiltonian (13.26) [62, 87]:

$$\begin{aligned}
&\langle s_1s_2(S_{12})s_3SM|\mathbf{s}_1\mathbf{s}_2|s_1s_2(S'_{12})s_3S'M'\rangle \\
&\quad= \delta_{SS'}\delta_{MM'}\delta_{S_{12}S'_{12}}[S_{12}(S_{12}+1) - s_1(s_1+1) - s_2(s_2+1)]/2, \\
&\langle s_1s_2(S_{12})s_3SM|s_1s_3|\mathbf{s}_1\mathbf{s}_2(S'_{12})s_3S'M'\rangle \\
&\quad= \delta_{SS'}\delta_{MM'}(-1)^{s_1+s_2+s_3+S+1}[(2S_{12}+1)(2S'_{12}+1) \\
&\quad\times (2s_1+1)(s_1+1)s_1(2s_1+3)s_3(s_3+1)]^{1/2} \\
&\quad\times \begin{Bmatrix} S_{12} & S'_{12} & 1 \\ s_3 & s_3 & S \end{Bmatrix} \begin{Bmatrix} s_1 & s_1 & 1 \\ S'_{12} & S_{12} & s_2 \end{Bmatrix}, \\
&\langle s_1s_2(S_{12})s_3SM|\mathbf{s}_2\mathbf{s}_3|s_1s_2(S'_{12})s_3S'M\rangle \\
&\quad= \delta_{SS'}\delta_{MM'}(-1)^{S_{12}+S'_{12}+s_1+s_2+s_3+S+1}[(2S_{12}+1) \\
&\quad\times (2S'_{12}+1)s_1(s_1+1)(2s_1+1)s_3(s_3+1)(2s_3+1)]^{1/2} \\
&\quad\times \begin{Bmatrix} S_{12} & S'_{12} & 1 \\ s_3 & s_3 & S \end{Bmatrix} \begin{Bmatrix} S_{12} & S'_{12} & 1 \\ s_2 & s_2 & s_1 \end{Bmatrix}.
\end{aligned} \tag{13.76}$$

As pointed out in Section 13.3.1, isotropic exchange couples the states of the system with the same quantum numbers $S$ and $M$ in conformity with (13.76).

The generalized Hamiltonian for a tetrameric cluster can be presented in different coupling schemes. The coupling scheme

$$s_1 + s_2 = S_{12}, \quad s_3 + s_4 = S_{34}, \quad S_{12} + S_{34} = S$$

seems to be convenient. Using this scheme one can present the generalized exchange spin-Hamiltonian as follows:

$$\hat{H}^{(4)} = \sum_{\kappa} (2\kappa + 1)\{c^{(\kappa)}(\kappa_1\kappa_2\kappa_{12}\kappa_3\kappa_4\kappa_{34}) \otimes \hat{\mathbf{X}}^{(\kappa)}\}^{(0)}, \tag{13.77}$$

where the four-particle tensor operator $X_{\mu}^{(\kappa)}$ is

$$\hat{\mathbf{X}}_{\mu}^{(\kappa)} = \{\{\hat{\mathbf{S}}^{(\kappa_1)}(1) \otimes \hat{\mathbf{S}}^{(\kappa_2)}(2)\}^{(\kappa_{12})} \otimes \{\hat{\mathbf{S}}^{(\kappa_3)}(3) \otimes \hat{\mathbf{S}}^{(\kappa_4)}(4)\}^{(\kappa_{34})}\}_{\mu}^{(\kappa)} \tag{13.78}$$

and $c^{(\kappa)}(\kappa_1\kappa_2\kappa_{12}\kappa_3\kappa_4\kappa_{34}\}$ are the numerical parameters of the generalized Hamiltonian connected with the exchange parameters.

The three-step recoupling procedure can be performed to give the following result for the reduced matrix element of the four-particle operator $\hat{\mathbf{X}}_{\mu}^{(\kappa)}$ [62, 87]:

$$\begin{aligned}
&\langle s_1 s_2 (S_{12}) s_3 s_4 (S_{34}) S \| \hat{\mathbf{X}}^{(\kappa)} \| s_1 s_2 (S'_{12}) s_3 s_4 (S_{34}) S' \rangle \\
&= [(2S+1)(2S'+1)(2\kappa+1)(2S_{12}+1)(2S'_{12}+1)(2\kappa_{12}+1) \\
&\times (2\kappa_{34}+1)(2\kappa'_{34}+1)]^{1/2} \begin{Bmatrix} S_{12} & S'_{12} & \kappa_{12} \\ S_{34} & S'_{34} & \kappa_{34} \\ S & S' & \kappa \end{Bmatrix} \\
&\times \begin{Bmatrix} s_1 & s_1 & \kappa_1 \\ s_2 & s_2 & \kappa_2 \\ S_{12} & S'_{12} & \kappa_{12} \end{Bmatrix} \begin{Bmatrix} s_3 & s_3 & \kappa_3 \\ s_4 & s_4 & \kappa_4 \\ S_{34} & S'_{34} & \kappa_{34} \end{Bmatrix} \\
&\times \langle s_1 \| \hat{\mathbf{S}}^{(\kappa_1)} \| s_1 \rangle \langle s_2 \| \hat{\mathbf{S}}^{(\kappa_2)} \| s_2 \rangle \langle s_3 \| \hat{\mathbf{S}}^{(\kappa_3)} \| s_3 \rangle \langle s_4 \| \hat{\mathbf{S}}^{(\kappa_4)} \| s_4 \rangle.
\end{aligned} \tag{13.79}$$

As in the previous case of a trinuclear system one pass to the Heisenberg-type isotropic exchange interaction in a tetrameric system by putting (13.78) and (13.79) $\kappa = \kappa_i = \kappa_j = 0$ and $\kappa_k = \kappa_l = 1$, collapsing $9j$-symbols in (13.79) to $6j$-symbols. Matrix elements of the isotropic exchange Hamiltonian

$$\hat{H} = -2J_{12}\mathbf{s}_1\mathbf{s}_2 - 2J_{13}\mathbf{s}_1\mathbf{s}_3 - 2J_{14}\mathbf{s}_1\mathbf{s}_4 - 2J_{23}\mathbf{s}_2\mathbf{s}_3 - 2J_{14}\mathbf{s}_1\mathbf{s}_4 \tag{13.80}$$

are given in [62].

The recoupling procedure presented here can be generalized for more complicated high-nuclearity exchange systems (see [90]), for example for extended linear chains of spins [62, 87]. The matrix elements of the exchange Hamiltonian can be found be recurrence [62, 87, 89].

Thus the application of tensor methods makes it possible to obtain closed form expressions for matrix elements of exchange Hamiltonian without resorting to the cumbersome calculation of many particle wavefunctions of the system of interacting ions.

## 13.4 GROUP-THEORETICAL CLASSIFICATION OF EXCHANGE MULTIPLETS

### 13.4.1 "Accidental" degeneracy

The spin-state theory described in Sections 13.1–13.3 is phenomenological, since it is based on the HDVV model. Precisely because of this, it is attractive for those interested in the practical magnetochemistry of polynuclear complexes. Its range of applicability is determined by the approximations interent in analysis of the Heitler–London many-electron scheme.

On the basis of general group-theoretical considerations, the analysis of trigonal three-nuclear clusters has already led us to the conclusion that, since in $\mathbf{D}_3$ there are no representations (including double-valued ones) whose degree of degeneracy exceeds two, even an undistorted system with half-integer spins cannot give rise to a quadruplet ground state or to states with higher degree of degeneracy. Therefore in such a system (e.g. the trigonal cluster $Cr_3$) there are necessarily interactions that remove "unphysical" degeneracies. Likewise, in a tetrahedral tetrameric system ($s_i = \frac{1}{2}$) there must exist interactions that remove the degeneracy of the first excited state $3\mathrm{D}^{(1)}$.

### 13.4.2 Spin–orbit multiplets

The nature of "accidental" degeneracies relative to intermediate spin values is ascertained with the help of the group-theoretical classification of exchange multiplets. Note the obvious fact that the HDVV Hamiltonian operator in the spin space only and in the HDVV model the energy levels are enumerated by the values of the total spin of the system under consideration, while, according to the Wigner theorem, each eigenfunction must have the corresponding total spin as well as an irreducible representation of the cluster point group. *The group-theoretical classification procedure makes it possible to establish the correspondence between the states $nD^{(S)}$ belonging to n-fold repeated $D^{(S)}$ of the HDVV model and the spin–orbit multiplets $S\Gamma$ of a polynuclear many-electron cluster.*

A detailed description of the procedure (i.e. group-theoretical classification of exchange multiplets) is given in [62, 63]. The mathematical approach developed in [62, 63] is based on the use of the representations of the so-called unitary groups which are outside the scope of this book. Here we describe a more simple efficient procedure of the group-theoretical classification of exchange multiplets based on the Slater's elegant idea [91] employed firstly in the problem of mixed-valence clusters in [105]. Let us illustrate a new approach by examples considering step-by-step the mathematical procedure.

**Example 13.1** Consider in detail the most simple symmetric trimeric cluster with ion spins $s_i = \frac{1}{2} (i = \mathrm{a,b,c})$. The metal skeleton of the system belongs to the point group

$\mathbf{D}_{3h}$. Depending on the ligand environment, the full symmetry of the trigonal molecule belongs to one of the trigonal groups ($\mathbf{D}_{3h}$, $\mathbf{C}_{3v}$ etc.). The system under consideration is supposed to have $\mathbf{C}_{3v}$ symmetry. An example of such a type of system is investigated in detail in [63, 92, 93] and represents the trime clear cooper (II) cluster ($d^9, s_i = \frac{1}{2}$) in crystals of $\mu$-hydroxo-$\mu$-(pyridine-2-carbaldehyde oximato)-$\mu_3$-sufato-tricooper

$$Cu_3(C_6H_5N_2O)_3(OH)SO_3 \cdot 10.5H_2O$$

with a metal core of $\mathbf{D}_{3h}$ symmetry and a $\mathbf{C}_{3v}$ symmetry for the tri-cooper core with axial oxygen (Fig. 13.4).

The three spin coupling scheme (13.14) looks as follows:

$$D^{(1/2)} \times D^{(1/2)} \times D^{(1/2)} = (D^{(0)} + D^{(1)}) \times D^{(1/2)} = 2D^{(1/2)} + D^{(3/2)}$$

Thus the HDVV model provides three spin levels $S'(S)$:

$$(0)\tfrac{1}{2}, (1)\tfrac{1}{2} \quad \text{and} \quad (1)\tfrac{3}{2}, \quad \text{or} \quad 2D^{(1/2)} \quad \text{and} \quad D^{(3/2)}.$$

In conformity with (13.19) the energy levels $\epsilon(S)$ for the symmetric trigonal system are independent on intermediate spin $S'$ and depend on the total spin value $S$ only; thus two levels with $S = \frac{1}{2}$ are degenerate over the values of intermediate spin ($S' = 0$ and 1).

Let us denote three nondegenerate spin-orbitals localized on the a, b and c centers of a trimer as $a_{m_a}(s_a)$, $b_{m_b}(s_l)$ and $c_{m_c}(s_c)$ with $s_a = s_b = s_c = \frac{1}{2}$.

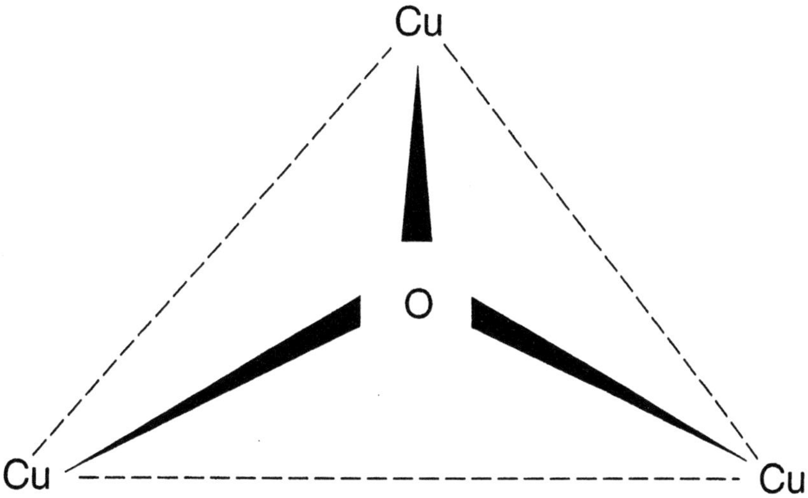


**Fig. 13.4** Structure of tri-cooper core with axial oxygen of the complex $Cu_3(C_6H_5N_2O)_3(OH)SO_4 \cdot 10.5H_2O$.

For example:

$$a_{m_a}(s_a) = a(\mathbf{r} - \mathbf{R}_a)\chi(s_a m_a)$$

where $a(\mathbf{r} - \mathbf{R}_a)$ is the coordinate wavefunction of the electron localized on the ion a with the position-vector $\mathbf{R}_a$; $\chi(s_a m_a)$ is the spin wavefunction of this electron. The Heitler–London wavefunctions of the system satisfying the Pauli principle and depending on both the electronic space coordinates and spin variables are expressed in terms of Slater determinants

$$|a_{m_a}(s_a)b_{m_b}(s_b)c_{m_c}(s_c)|$$

with fixed $s_a, s_b, s_c$ and different sets of $\{m_a, m_b, m_c\}$ as it is schematized in Fig. 13.5. Each determinant belongs to a certain value of full spin projection

$$M = m_a + m_b + m_c$$

but not to the definite full spin $S$. To construct the Heitler–London wavefunctions belonging to a certain intermediate spin (say, $S_{ab}$) and full spin $S$ quantum numbers we must use the three spin-coupling scheme (see (13.30) and (13.31)):

$$|(S_{ab})SM\rangle = \sum_{m_a, m_b} \sum_{m_{ab}, m_c} |a_{m_a}(s_a)b_{m_b}(s_b)c_{m_c}(s_c)| \times \langle s_a m_a s_b m_b | S_{ab} m_{ab}\rangle \langle S_{ab} m_{ab} s_c m_c | SM\rangle. \qquad (13.81)$$

It should be emphasized that the wavefunctions (13.81) are the Heitler–London eigenfunctions of the real ("physical") Hamiltonian of the triatomic system involving kinetic energy of electrons and intra- and interatomic interactions, but not the eigenfunctions of the effective HDVV Hamiltonian acting only in the spin-space of interacting ions. To construct the wavefunctions belonging to the irreducible representations of the point symmetry group $\mathbf{C}_{3v}$ of the system ($S\Gamma$-states) we must apply the projection operator $\hat{P}_{\Gamma\gamma}$ (see Section 6.3.2)) to the wavefunctions (13.81):

$$|\nu S\Gamma M\gamma\rangle = \hat{P}_{\Gamma\gamma}|(S_{ab})SM\rangle, \qquad (13.82)$$

where the symbol $\nu$ enumerates the identical (repeated) $S\Gamma$-states.

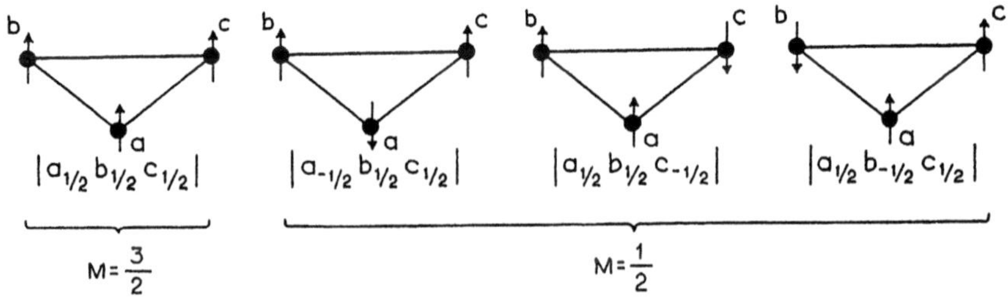


**Fig. 13.5** Pictorial representation of the Heitler–London basis set for the trimeric cluster with $s_a = s_b = s_c = \frac{1}{2}$.

The procedure for the construction of all $S\Gamma$-states is relatively simple for the trimer with $s_i = \frac{1}{2}$, but the calculations become cumbersome for more complex clusters involving many-electron ions. On the other hand, important information (see next sections) can be extracted only from the knowledge of the allowed $S\Gamma$-states and the correspondence between $n\mathrm{D}^{(S)}$-states of spin-coupling scheme and $S\Gamma$-states of the multi-electronic Heitler–London scheme. To determine the allowed $S\Gamma$-states and to establish the one-to-one correspondence between $n\mathrm{D}^{(S)}$ and $S\Gamma$ (group-theoretical classification of spin-states) one can avoid a complicated procedure of constructing the $S\Gamma$-wavefunctions according to (13.81) and (13.82).

Let us consider the determinant

$$|\mathrm{a}_{1/2}(\tfrac{1}{2})\mathrm{b}_{1/2}(\tfrac{1}{2})\mathrm{c}_{1/2}(\tfrac{1}{2})| \equiv |\mathrm{a}_{1/2}\mathrm{b}_{1/2}\mathrm{c}_{1/2}|, \tag{13.83}$$

with maximum spin projection $M = m_\mathrm{a} + m_\mathrm{b} + m_\mathrm{c} = \frac{3}{2}$ (Fig. 13.5). This is the only state with $M = \frac{3}{2}$ which obviously belongs to the maximum full spin $S = \frac{3}{2}$.

Operations of the group $\mathbf{C}_{3\mathrm{v}}$ induce permutations of the localized orbitals according to the general definition (3.58) of a "rotated function";
for example

$$\hat{C}_3|\mathrm{a}_{1/2}\mathrm{b}_{1/2}\mathrm{c}_{1/2}| = |\mathrm{b}_{1/2}\mathrm{c}_{1/2}\mathrm{a}_{1/2}|. \tag{13.84}$$

The initial order of one-electron orbitals in the resulting determinant can be restored by the two permutations, hence the sign of the result remains unchanged

$$\hat{C}_3|\mathrm{a}_{1/2}\mathrm{b}_{1/2}\mathrm{c}_{1/2}| = |\mathrm{a}_{1/2}\mathrm{b}_{1/2}\mathrm{c}_{1/2}|. \tag{13.85}$$

Therefore the character of $\hat{C}_3$ is $+1$. Operation $\hat{\sigma}_\mathrm{v}$ changes places by two orbitals, so the restoration of the initial order of orbitals changes the sign of the determinant; for example

$$\hat{\sigma}_\mathrm{v}(c)|\mathrm{a}_{1/2}\mathrm{b}_{1/2}\mathrm{c}_{1/2}| = |\mathrm{b}_{1/2}\mathrm{a}_{1/2}\mathrm{c}_{1/2}| = -|\mathrm{a}_{1/2}\mathrm{b}_{1/2}\mathrm{c}_{1/2}| \tag{13.86}$$

and we find that $\chi^{(M=\frac{3}{2})}(\hat{\sigma}_\mathrm{v}) = -1$ (the notation $\chi^{(M)}(\hat{R})$ for the characters indicates that the basis of the representation is formed by the set of determinant functions belonging to a given full spin projection $M$). Since three operations $\hat{\sigma}_\mathrm{v}$ belong to the same class we find that $\chi^{(M=\frac{3}{2})}(\hat{\sigma}_\mathrm{v}) = -1$ for all $\hat{\sigma}_\mathrm{v}$. The results are collected in Table 13.1 where the characters of $\mathbf{C}_{3\mathrm{v}}$ are also given for convenience.

**Table 13.1** Characters of transformations of the basis functions $|\mathrm{a}_{m_\mathrm{a}}(\frac{1}{2})\mathrm{b}_{m_\mathrm{b}}(\frac{1}{2})\mathrm{c}_{m_\mathrm{c}}(\frac{1}{2})|$ under the operations of $\mathbf{C}_{3\mathrm{v}}$ group.

| $\mathbf{C}_{3\mathrm{v}}$ | $\hat{E}$ | $2\hat{C}_3$ | $3\hat{\sigma}_\mathrm{v}$ | |
|---|---|---|---|---|
| $A_1$ | 1 | 1 | 1 | |
| $A_2$ | 1 | 1 | −1 | |
| E | 2 | −1 | 0 | |
| $A_2$ | 1 | 1 | −1 | $\chi^{(M=\frac{3}{2})}(\hat{R})$ |
| $A_2 + E$ | 3 | 0 | −1 | $\chi^{(M=\frac{1}{2})}(\hat{R})$ |

Let us now consider the states with $M = \frac{1}{2}$. There are three such type determinants containing one reversed spin (Fig. 13.5):

$$|a_{-1/2}b_{1/2}c_{1/2}|, \quad |a_{1/2}b_{-1/2}c_{1/2}|, \quad |a_{1/2}b_{1/2}c_{-1/2}|. \tag{13.87}$$

Operation $\hat{C}_3$ transforms each determinant into one other, for example:

$$\hat{C}_3|a_{-1/2}b_{1/2}c_{1/2}| = |b_{-1/2}c_{1/2}a_{1/2}| = |a_{1/2}b_{-1/2}c_{1/2}|, \text{etc.} \tag{13.88}$$

This transformation of three determinants (13.87) under rotation $\hat{C}_3$ can be presented in the matrix form according to the general expression (3.64):

$$\begin{aligned} &\hat{C}_3[|a_{-1/2}b_{1/2}c_{1/2}|\,|a_{1/2}b_{-1/2}c_{1/2}|\,|a_{1/2}b_{1/2}c_{-1/2}|] \\ &\quad = [|a_{-1/2}b_{1/2}c_{1/2}|\,|a_{1/2}b_{-1/2}c_{1/2}|\,|a_{1/2}b_{1/2}c_{-1/2}|]\mathbf{D}^{(M=1/2)}(\hat{C}_3) \end{aligned} \tag{13.89}$$

with

$$\mathbf{D}^{(M=1/2)}(\hat{C}_3) = \begin{vmatrix} 0 & 0 & 1 \\ 1 & 0 & 0 \\ 0 & 1 & 0 \end{vmatrix}.$$

Thus $\chi^{(M=\frac{1}{2})}(\hat{C}_3) = 0$, while for each $\hat{\sigma}_v$ operation $\chi^{(M=\frac{1}{2})}(\hat{\sigma}_v) = -1$. The characters $\chi^{(M=\frac{1}{2})}(\hat{R})$ of the three-dimensional representation $D^{(M=\frac{1}{2})}$ with the basis (13.87) are given in Table 13.1. From Table 13.1 we see that the representation $D^{(M=\frac{3}{2})}$ is the irreducible representation $A_2$ while the three-dimensional representation $D^{(M=\frac{1}{2})}$ with the $M = \frac{1}{2}$-basis proves to be reducible. Decomposing it into the irreducible parts (using (4.12)) we find

$$\Gamma^{(M=\frac{1}{2})} \doteq A_2 + E$$

as shown in the left column of Table 13.1.

Since the only determinant $|a_{1/2}b_{1/2}c_{1/2}|$ with $M = \frac{3}{2}$ is just the state with $S = \frac{3}{2}$ we conclude that the corresponding $S\Gamma$-term of the Heitler–London scheme for trimer is $^4A_2$. The set of determinants with $M = \frac{1}{2}$ belongs to the states with full spins $S = \frac{3}{2}$ and $S = \frac{1}{2}$. To find the characters of the representations relating to $S = \frac{1}{2}$ we must extract the $\chi^{(M=\frac{3}{2})}(\hat{R}) \equiv \chi^{(S=\frac{3}{2})}(\hat{R})$ from the $\chi^{(M=\frac{1}{2})}(\hat{R})$:

$$\chi^{(S=\frac{1}{2})}(\hat{R}) = \chi^{(M=\frac{1}{2})}(\hat{R}) - \chi^{(S=\frac{3}{2})}(\hat{R}).$$

In other words

$$\Gamma(S = \tfrac{1}{2}) = \Gamma(M = \tfrac{1}{2}) - \Gamma(S = \tfrac{3}{2}) = (A_2 + E) - A_2 = E$$

and we arrive at the conclusion that two states with $S = \frac{1}{2}$ ($S' = 0$ and 1) form the $^2E$ term of the Heitler–London scheme.

The result of the group-theoretical classification of the many-electron Heitler–London states for a $\mathbf{C}_{3v}$-trimer with $s_i = \frac{1}{2}$ is expressed symbolically as

$$2D_{HL}^{(1/2)} \doteq {}^2E, \quad D_{HL}^{(3/2)} \doteq {}^4A_2, \tag{13.90}$$

where the symbol HL indicates that the $D^{(S)}$ representations belong to the spin-coupling in the Heitler–London scheme (see (13.81)). The result (13.90) shows that the Heitler–London scheme for three one-electron atoms placed in the vertices of a symmetric triangle leads to an orbital doublet $^2E$ and an orbital single $^4A_2$ (this result is valid for all trigonal groups with adopted labels of irreducible representations). Comparing this result with the spin-levels of the symmetric triangle obtained within the scope of the HDVV model (equation (13.19)) we see that spin levels with $S = \frac{1}{2}$, twice repeating in the HDVV model correspond to $^2E$ term of Heitler–London scheme. It should be noted that spin-wavefunction (eigen-functions of the HDVV Hamiltonian) are expressed in terms of simple products, but not determinants, of spin-functions of interacting ions (equation (13.32)). In this case for example the application of the $\hat{\sigma}_v$ operation to the $\alpha(a)\alpha(b)\alpha(c)$ does not change the sign of this product (compare with (13.86)).

**Example 13.2** Consider a trimeric ($\mathbf{C}_{3v}$) cluster with $s_a = s_b = s_c = 1$. Each ion energy scheme is supposed to consist of two nondegenerate one-electron orbitals $\varphi_i$ and $\psi_i$ ($i = a, b$ and c), the ground state being the two-electron high spin state $s_i = 1$, according to the Hund's rule (Fig. 13.6). The wavefunctions of each ion (say, ion a) with a specific full spin $s_a$ and its projection $m_a$ can be expressed in terms of two-electron determinants

$$|\varphi_a(s_1 m_1)\,\psi_a(s_2 m_2)| \quad (\text{with } s_1 = s_2 = \tfrac{1}{2})$$

using the general rule (5.27) for two spin coupling. For example, for ion a we obtain the wavefunction

$$a_{m_a}(s_a) = \sum_{m_1, m_2} |\varphi_a(\tfrac{1}{2}m_1)\psi_a(\tfrac{1}{2}m_2)|\langle \tfrac{1}{2}m_1 \tfrac{1}{2}m_2 | s_a m_a\rangle \tag{13.91}$$

with $s_a = 1$ for the high-spin state. The basis set of Heitler–London wavefunctions belonging to the specific value of $M = m_a + m_b + m_c$ consists of six-order determinants formed by two-orbital wavefunctions:

$$|a_{m_a}(1)b_{m_b}(1)c_{m_c}(1)| \equiv |a_{m_a}b_{m_b}c_{m_c}| \quad (m_a, m_b, m_c = 0, \pm 1),$$

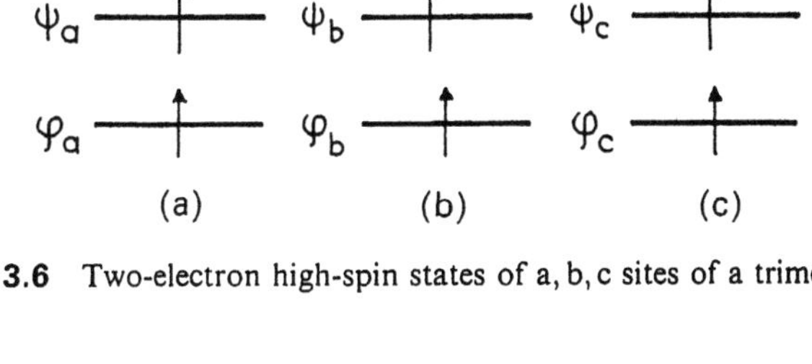


**Fig. 13.6** Two-electron high-spin states of a, b, c sites of a trimeric cluster.

two electrons in the six-electron system. The three-spin coupling scheme in the case under consideration looks as follows

$$\mathrm{D}^{(1)} \times \mathrm{D}^{(1)} \times \mathrm{D}^{(1)} = (\mathrm{D}^{(0)} + \mathrm{D}^{(1)} + \mathrm{D}^{(2)}) \times \mathrm{D}^{(1)} = \mathrm{D}^{(0)} + 3\mathrm{D}^{(1)}_{(0,1,2)} + 2\mathrm{D}^{(2)}_{(1,2)} + \mathrm{D}^{(3)}. \tag{13.92}$$

The only determinant with $M = 3$

$$|\mathrm{a}_1\, \mathrm{b}_1\, \mathrm{c}_1|$$

belongs to the state with maximum full spin $S = 3$. Applying $\hat{C}_3$ we find

$$\hat{C}_3|\mathrm{a}_1\mathrm{b}_1\mathrm{c}_1| = |\mathrm{b}_1\mathrm{c}_1\mathrm{a}_1| \equiv |\mathrm{a}_1\mathrm{b}_1\mathrm{c}_1|, \tag{13.93}$$

hence $\chi^{(M=3)}(\hat{C}_3) = 1$. For $\hat{\sigma}_v$ operation we obtain

$$\hat{\sigma}_v(\mathrm{c})|\mathrm{a}_1\mathrm{b}_1\mathrm{c}_1| = |\mathrm{b}_1\mathrm{a}_1\mathrm{c}_1|. \tag{13.94}$$

One can easily see that in contrast to the previous case of one-electron atoms (equation 13.86)), transposition of two-electron atomic states does not change the sign of determinants and we find that $\chi^{(M=3)}(\hat{\sigma}_v) = 1$. There are three determinants with $M = 2$:

$$|\mathrm{a}_0\mathrm{b}_1\mathrm{c}_1|, \quad |\mathrm{a}_1\mathrm{b}_0\mathrm{c}_1|, \quad |\mathrm{a}_1\mathrm{b}_1\mathrm{c}_0|. \tag{13.95}$$

Characters of a three-dimensional representation with the basis (13.95) are given in Table 13.2.

The six-dimensional basis set with $M = 1$ can be divided into two subsets:

$$\begin{aligned} &\mathrm{I:} \quad |\mathrm{a}_1\mathrm{b}_0\mathrm{c}_0|, \quad |\mathrm{a}_0\mathrm{b}_1\mathrm{c}_0|, \quad |\mathrm{a}_0\mathrm{b}_0\mathrm{c}_1|; \\ &\mathrm{II:} \quad |\mathrm{a}_{-1}\mathrm{b}_1\mathrm{c}_1|, \quad |\mathrm{a}_1\mathrm{b}_{-1}\mathrm{c}_1|, \quad |\mathrm{a}_1\mathrm{b}_1\mathrm{c}_1|. \end{aligned} \tag{13.96}$$

It is obvious that three determinants belonging to each subset transform independently

**Table 13.2** Characters of transformations of the basis functions $|\mathrm{a}_{m_a}(1)\mathrm{b}_{m_b}(1)\mathrm{c}_{m_c}(1)|$ under operations of the $\mathbf{C}_{3v}$ group.

| $\mathbf{C}_{3v}$ | $\hat{E}$ | $2\hat{C}_3$ | $3\hat{\sigma}_v$ | |
|---|---|---|---|---|
| $A_1$ | 1 | 1 | 1 | $\chi^{(M=3)}(\hat{R})$ |
| $A_1 + E$ | 3 | 0 | 1 | $\chi^{(M=2)}(\hat{R})$ |
| $A_1 + E$ | 3 | 0 | 1 | $\chi_{\mathrm{I}}^{(M=1)}(\hat{R})$ |
| $A_1 + E$ | 3 | 0 | 1 | $\chi_{\mathrm{II}}^{(M=1)}(\hat{R})$ |
| $A_1$ | 1 | 1 | 1 | $\chi_{\mathrm{I}}^{(M=0)}(\hat{R})$ |
| $A_1 + A_2 + E$ | 6 | 0 | 0 | $\chi_{\mathrm{II}}^{(M=0)}(\hat{R})$ |

under the operations of $\mathbf{C}_{3v}$. Finally we obtain two subsets for seven determinants belonging to $M = 0$:

$$\begin{aligned} \text{I:} \quad & |a_0 b_0 c_0|; \\ \text{II:} \quad & |a_0 b_1 c_{-1}|, \quad |a_0 b_{-1} c_1|, \quad |a_1 b_0 c_{-1}|, \\ & |a_{-1} b_0 c_1|, \quad |a_1 b_{-1} c_0|, \quad |a_{-1} b_1 c_0|. \end{aligned} \tag{13.97}$$

Characters are collected in Table 13.2.

We see that

$$\Gamma^{(M=3)} = \Gamma^{(S=3)} = A_1.$$

Decomposing $\Gamma^{(M=2)}$ into the irreducible components we find:

$$\Gamma^{(M=2)} = A_1 + E,$$

hence

$$\Gamma^{(S=2)} = \Gamma^{(M=2)} - \Gamma^{(S=3)} = E.$$

In a similar way one can obtain

$$\begin{aligned} \Gamma^{(M=1)} &= 2A_1 + 2E, \\ \Gamma^{(S=1)} &= \Gamma^{(M=1)} - \Gamma^{(S=3)} - \Gamma^{(S=2)} = A_1 + E; \\ \Gamma^{(M=0)} &= 2A_1 + A_2 + 2E, \\ \Gamma^{(S=0)} &= \Gamma^{(M=0)} - \Gamma^{(S=3)} - \Gamma^{(S=2)} - \Gamma^{(S=1)} = A_2. \end{aligned}$$

The Heitler–London spin-orbital $S\Gamma$ multiplets corresponding to $nD^{(S)}$ states (13.92) of HDVV model are the following:

$$\begin{aligned} & D^{(0)} \doteq {}^1A_2, \quad 3D^{(1)} \doteq {}^3A_1 + {}^3E, \\ & 2D^{(2)} \doteq {}^5E, \quad D^{(3)} \doteq {}^7A_1. \end{aligned} \tag{13.98}$$

We see that the spin-singlet ground state of an antiferromagnetic cluster with $s_i = 1$ is the orbital singlet ${}^1A_2$. The first excited state $3D^{(1)}$ of HDVV model involves two different $S\Gamma$ multiplets (orbital singlet ${}^3A_1$ and orbital doublet ${}^3E$), the next level $2D^{(2)}$ represents the orbital doublet ${}^5E$, and, finally, $D^{(3)}$ is an orbital singlet ${}^7A_1$. Thus in some cases the HDVV model leads to an "accidental" degeneracy of different multiplets $S\Gamma$, which, according to the Wigner theorem, must have different energies. This problem will be discussed in the next section.

**Example 13.3** Consider a trimeric ($\mathbf{C}_{3v}$) cluster with spins of ions $s_i = \frac{3}{2}$. Trimeric chromium (Cr (III),$d^3$) carboxilates $[Cr_3\ O(CH_3COO)_6(H_2O)_3]^+$ are exemplified in Fig.

13.1. The HDVV model spin states are determined by the decomposition (13.15). To form three-electron high-spin states ($s_i = \frac{3}{2}$) one must suppose that three orbitals (say, $\varphi_i, \psi_i$ and $\chi_i$) of each ion are involved. The three-electron wavefunction of each ion with total spin projection $m_i (i = \mathrm{a, b, c})$ can be represented as a third-order determinant

$$\mathrm{a}_{m_i}(s_i) = |\varphi_i(\tfrac{1}{2}m_1)\psi_i(\tfrac{1}{2}m_2)\chi_i(\tfrac{1}{2}m_3)|, \quad (m_i = m_1 + m_2 + m_3).$$

Using the three-spin coupling scheme (equations (13.30) and (13.31)) one can construct the $|(S_{12})s_i m_i\rangle$ states for each ion. For high spin ($s_i = \frac{3}{2}$) state $s_{12} = 1$ and we find for example ion a ($\mathrm{a}_{m_\mathrm{a}} \equiv \mathrm{a}_{m_\mathrm{a}}(\frac{3}{2})$)

$$\mathrm{a}_{m_\mathrm{a}} = \sum_{m_1, m_2} \sum_{m_{12}, m_3} |\varphi_\mathrm{a}(\tfrac{1}{2}m_1)\psi_\mathrm{a}(\tfrac{1}{2}m_2)\chi_\mathrm{a}(\tfrac{1}{2}m_3)| \\ \times \langle \tfrac{1}{2}m_1 \tfrac{1}{2}m_2 | 1 m_{12}\rangle \langle 1 m_{12} \tfrac{1}{2} m_3 | \tfrac{3}{2} m_3\rangle. \tag{13.99}$$

Starting with the state of cluster with the maximum value of $M = S = \frac{9}{2}$

$$|\mathrm{a}_{3/2}\mathrm{b}_{3/2}\mathrm{c}_{3/2}| \tag{13.100}$$

we find that $\chi^{(M=\frac{9}{2})}(\hat{C}_3) = 1$. Taking into account that transposition of three-electronic atomic states changes the sign of the determinant we find that

$$\chi^{(M=\frac{9}{2})}(\hat{\sigma}_\mathrm{v}) = -1 \qquad \text{(Table 13.3)}$$

There are three determinants with $M = 7/2$ belonging to $S = 9/2$ and $S = 7/2$:

$$|\mathrm{a}_{1/2}\mathrm{b}_{3/2}\mathrm{c}_{3/2}|, \quad |\mathrm{a}_{3/2}\mathrm{b}_{1/2}\mathrm{c}_{3/2}|, \quad |\mathrm{a}_{3/2}\mathrm{b}_{3/2}\mathrm{c}_{1/2}|. \tag{13.101}$$

Six determinants with $M = \frac{5}{2}(S = \frac{9}{2}, \frac{7}{2}, \frac{5}{2})$ form two subsets

$$\begin{aligned} &\text{I:} \quad |\mathrm{a}_{-1/2}\mathrm{b}_{3/2}\mathrm{c}_{3/2}|, \quad |\mathrm{a}_{3/2}\mathrm{b}_{-1/2}\mathrm{c}_{3/2}|, \quad |\mathrm{a}_{3/2}\mathrm{b}_{3/2}\mathrm{c}_{-1/2}|; \\ &\text{II:} \quad |\mathrm{a}_{1/2}\mathrm{b}_{1/2}\mathrm{c}_{3/2}|, \quad |\mathrm{a}_{1/2}\mathrm{b}_{3/2}\mathrm{c}_{1/2}|, \quad |\mathrm{a}_{3/2}\mathrm{b}_{1/2}\mathrm{c}_{1/2}|. \end{aligned} \tag{13.102}$$

For $M = \frac{3}{2}$ ($S = \frac{9}{2}, \frac{7}{2}, \frac{5}{2}, \frac{3}{2}$) we find:

$$\begin{aligned} &\text{I:} \quad |\mathrm{a}_{-3/2}\mathrm{b}_{3/2}\mathrm{c}_{3/2}|, \quad |\mathrm{a}_{3/2}\mathrm{b}_{-3/2}\mathrm{c}_{3/2}|, \quad |\mathrm{a}_{3/2}\mathrm{b}_{3/2}\mathrm{c}_{-3/2}|; \\ &\text{II:} \quad \begin{cases} |\mathrm{a}_{-1/2}\mathrm{b}_{1/2}\mathrm{c}_{3/2}|, \quad |\mathrm{a}_{1/2}\mathrm{b}_{-1/2}\mathrm{c}_{3/2}|, \\ |\mathrm{a}_{-1/2}\mathrm{b}_{3/2}\mathrm{c}_{1/2}|, \quad |\mathrm{a}_{1/2}\mathrm{b}_{3/2}\mathrm{c}_{-1/2}|, \\ |\mathrm{a}_{3/2}\mathrm{b}_{-1/2}\mathrm{c}_{1/2}|, \quad |\mathrm{a}_{3/2}\mathrm{b}_{1/2}\mathrm{c}_{-1/2}|. \end{cases} \end{aligned} \tag{13.103}$$

Finally, there are three subsets of determinants with $M = \frac{1}{2}(S = \frac{9}{2}, \frac{7}{2}, \frac{5}{2}, \frac{3}{2}, \frac{1}{2})$:

$$\begin{aligned} &\text{I:} \quad |\mathrm{a}_{-1/2}\mathrm{b}_{-1/2}\mathrm{c}_{3/2}|, \quad |\mathrm{a}_{-1/2}\mathrm{b}_{3/2}\mathrm{c}_{-1/2}|, \quad |\mathrm{a}_{3/2}\mathrm{b}_{-1/2}\mathrm{c}_{-1/2}|; \\ &\text{II:} \quad |\mathrm{a}_{1/2}\mathrm{b}_{1/2}\mathrm{c}_{-1/2}|, \quad |\mathrm{a}_{1/2}\mathrm{b}_{-1/2}\mathrm{c}_{1/2}|, \quad |\mathrm{a}_{-1/2}\mathrm{b}_{1/2}\mathrm{c}_{1/2}|; \\ &\text{III:} \quad \begin{cases} |\mathrm{a}_{1/2}\mathrm{b}_{3/2}\mathrm{c}_{-3/2}|, \quad |\mathrm{a}_{1/2}\mathrm{b}_{-3/2}\mathrm{c}_{3/2}|, \\ |\mathrm{a}_{3/2}\mathrm{b}_{1/2}\mathrm{c}_{-3/2}|, \quad |\mathrm{a}_{-3/2}\mathrm{b}_{1/2}\mathrm{c}_{3/2}|, \\ |\mathrm{a}_{3/2}\mathrm{b}_{-3/2}\mathrm{c}_{1/2}|, \quad |\mathrm{a}_{-3/2}\mathrm{b}_{3/2}\mathrm{c}_{1/2}|. \end{cases} \end{aligned} \tag{13.104}$$

The characters of operations of the $\mathbf{C}_{3v}$ group for all basis sets and the results of the reduction of the corresponding representations are collected in Table 13.3. We find that

$$\begin{aligned}
\Gamma^{(M=\frac{9}{2})} &= \Gamma^{(s=\frac{9}{2})} = A_2, \\
\Gamma^{(s=\frac{7}{2})} &= \Gamma^{(M=\frac{7}{2})} - \Gamma^{(S=\frac{9}{2})} = E, \\
\Gamma^{(S=\frac{5}{2})} &= \Gamma^{(M=\frac{5}{2})} - \Gamma^{(S=\frac{9}{2})} - \Gamma^{(S=\frac{7}{2})} = A_2 + E, \\
\Gamma^{(S=\frac{3}{2})} &= \Gamma^{(M=\frac{3}{2})} - \Gamma^{(S=\frac{9}{2})} - \Gamma^{(S=\frac{7}{2})} - \Gamma^{(S=\frac{5}{2})} = A_1 + A_2 + E, \\
\Gamma^{(S=\frac{1}{2})} &= \Gamma^{(M=\frac{1}{2})} - \Gamma^{(S=\frac{9}{2})} - \Gamma^{(S=\frac{7}{2})} - \Gamma^{(S=\frac{5}{2})} - \Gamma^{(S=\frac{3}{2})} = E.
\end{aligned}$$

The results of the group-theoretical classification of the Heitler–London states are the following [62]:

$$\begin{aligned}
&2D^{(1/2)} \doteq {}^2E, \quad 4D^{(3/2)} \doteq {}^4A_1 + {}^4A_2 + {}^4E, \\
&3D^{(5/2)} \doteq {}^6A_2 + {}^6E, \quad 2D^{(7/2)} \doteq {}^8E, \\
&D^{(9/2)} \doteq {}^{10}A_2.
\end{aligned} \tag{13.105}$$

As in the case of a trimeric cluster with $s_i = \frac{1}{2}$ (Example 13.1) the ground state of a trigonal trimer with $s_i = \frac{3}{2}$ is the orbital doublet ${}^2E$. The first excited state $4D^{(3/2)}$ of the HDVV model involves three different multiplets (${}^4A_1, {}^4A_2$ and ${}^4E$), the next level, $3D^{(5/2)}$, involves two multiplets (${}^6A_2$ and ${}^6E$), where the $2D^{(7/2)} \doteq {}^8E$ is orbitally degenerate, and, finally, $D^{(9/2)}$ is the orbital singlet ${}^{10}A_2$. We see that, as in previous

**Table 13.3** Characters of transformations of the basis functions $|a_{m_a}(\frac{3}{2}) b_{m_b}(\frac{3}{2}) c_{m_c}(\frac{3}{2})|$ under operations of the $\mathbf{C}_{3v}$ group.

| $\mathbf{C}_{3v}$ | $\hat{E}$ | $2\hat{C}_3$ | $3\hat{\sigma}_v$ | |
|---|---|---|---|---|
| $A_2$ | 1 | 1 | −1 | $\chi^{(M=\frac{9}{2})}(\hat{R})$ |
| $A_2 + E$ | 3 | 0 | −1 | $\chi^{(M=\frac{7}{2})}(\hat{R})$ |
| $A_2 + E$ | 3 | 0 | −1 | $\chi_{I}^{(M=\frac{5}{2})}(\hat{R})$ |
| $A_2 + E$ | 3 | 0 | −1 | $\chi_{II}^{(M=\frac{5}{2})}(\hat{R})$ |
| $A_2 + E$ | 3 | 0 | −1 | $\chi_{I}^{(M=\frac{3}{2})}(\hat{R})$ |
| $A_1 + A_2 + 2E$ | 6 | 0 | 0 | $\chi_{II}^{(M=\frac{3}{2})}(\hat{R})$ |
| $A_2$ | 1 | 1 | −1 | $\chi_{III}^{(M=\frac{3}{2})}(\hat{R})$ |
| $A_2 + E$ | 3 | 0 | −1 | $\chi_{I}^{(M=\frac{1}{2})}(\hat{R})$ |
| $A_2 + E$ | 3 | 0 | −1 | $\chi_{II}^{(M=\frac{1}{2})}(\hat{R})$ |
| $A_1 + A_2 + 2E$ | 6 | 0 | 0 | $\chi_{III}^{(M+\frac{1}{2})}(\hat{R})$ |

cases, the HDVV model leads to an "accidental" degeneracy of different $S\Gamma$ multiplets.

**Example 13.4** Consider tetrameric tetrahedral ($\mathbf{T}_d$) cluster with $s_i = \frac{1}{2}$. Eigenvalues (13.23) depend on total spin values $S$ and do not depend on the values of two intermediate spins in the four spins coupling scheme (Section 13.2.2):

$$D^{(1/2)} \times D^{(1/2)} \times D^{(1/2)} \times D^{(1/2)} = 2D^{(0)} + 3D^{(1)} + D^{(2)}. \tag{13.106}$$

The Heitler–London state with the maximum spin projection $M = S = 2$ is represented by the only determinant constructed from the four spin-orbitals $a_{m_a}(\frac{1}{2})$, $b_{m_b}(\frac{1}{2})$, $c_{m_c}(\frac{1}{2})$ and $d_{m_d}(\frac{1}{2})$ with $m_i = \frac{1}{2}$:

$$|a_{1/2}b_{1/2}c_{1/2}d_{1/2}|. \tag{13.107}$$

Symmetry operations of $\mathbf{T}_d$ (Fig. 2.33a) result in permutations of orbitals, for example

$$\hat{C}_2(z)|a_{1/2}b_{1/2}c_{1/2}d_{1/2}| = |b_{1/2}a_{1/2}d_{1/2}c_{1/2}| = |a_{1/2}b_{1/2}c_{1/2}d_{1/2}|.$$

We find that $\chi^{(M=2)}(\hat{C}_2) = 1, \chi^{(M=2)}(\hat{C}_3) = 1, \chi^{(M=2)}(\hat{S}_4) = -1, \chi^{(M=2)}(\hat{\sigma}_d) = -1$. The basis set with $M = 1$ includes four determinants

$$\left.\begin{array}{ll} |a_{-1/2}b_{1/2}c_{1/2}d_{1/2}|, & |a_{1/2}b_{-1/2}c_{1/2}d_{1/2}|, \\ |a_{1/2}b_{1/2}c_{-1/2}d_{1/2}|, & |a_{1/2}b_{1/2}c_{1/2}d_{-1/2}|. \end{array}\right\} \tag{13.108}$$

Finally, for $M = 0$ we have a six-dimensional representation with the basis:

$$\left.\begin{array}{ll} |a_{-1/2}b_{-1/2}c_{1/2}d_{1/2}|, & |a_{-1/2}b_{1/2}c_{-1/2}d_{1/2}|, \\ |a_{-1/2}b_{1/2}c_{1/2}d_{1/2}|, & |a_{1/2}b_{-1/2}c_{-1/2}d_{1/2}|, \\ |a_{1/2}b_{-1/2}c_{1/2}d_{-1/2}|, & |a_{1/2}b_{1/2}c_{-1/2}d_{-1/2}|. \end{array}\right\} \tag{13.109}$$

Applying symmetry operations to basis sets (13.107)–(13.109) we find the characters $\chi^{(M)}(\hat{R})$ and the irreducible parts of $\Gamma^{(M)}$ (Table 13.4). We see that $\Gamma^{(S=2)} + A_2$, $\Gamma^{(S=1)} = T_1$, $\Gamma^{(S=0)} = E$ and the correspondence between Heitler–London's $S\Gamma$ multiplets and $nD^{(S)}$ states is the following

$$2D^{(0)} \doteq {}^1E, \quad 3D^{(1)} \doteq {}^3T_1, \quad D^{(2)} \doteq {}^5A_2. \tag{13.110}$$

Results of the group-theoretical classification show that the two-fold "accidental" degeneracy of the spin-singlet level $2D^{(0)}$ ($S_{ab} = 0$, $S_{cd} = 0$; $S_{ab} = 1$, $S_{cd} = 1$) corresponds to the orbital doublet $^1E$. Three-fold degenerate level $3D^{(1)}$ with $S = 1(S_{ab} = 0, S_{cd} = 1; S_{ab} = 1, S_{cd} = 0; S_{ab} = 1, S_{cd} = 1)$ corresponds to the orbital triplet $^3T_1$ of the Heitler–London scheme.

**Example 13.5** Consider a square-planar metalic core with $s_i = \frac{1}{2}$. The cluster including ligand surrounding is supposed to belong to the point group $\mathbf{D}_4$. As in the

**Table 13.4** Characters of transformations of the basis functions $|a_{m_a}(\frac{1}{2})b_{m_b}(\frac{1}{2})c_{m_c}(\frac{1}{2})d_{m_d}(\frac{1}{2})|$ under the operations of $\mathbf{T}_d$.

| $\mathbf{T}_d$ | $\hat{E}$ | $3\hat{C}_2$ | $8\hat{C}_3$ | $6\hat{S}_4$ | $6\hat{\sigma}_d$ | |
|---|---|---|---|---|---|---|
| $A_1$ | 1 | 1 | 1 | 1 | 1 | |
| $A_2$ | 1 | 1 | 1 | −1 | −1 | |
| E | 2 | 2 | −1 | 0 | 0 | |
| $T_1$ | 3 | −1 | 0 | 1 | −1 | |
| $T_2$ | 3 | −1 | 0 | −1 | 1 | |
| $A_2$ | 1 | 1 | 1 | −1 | −1 | $\chi^{(M=2)}(\hat{R})$ |
| $A_2 + T_1$ | 4 | 0 | 1 | 0 | −2 | $\chi^{(M=1)}(\hat{R})$ |
| $A_2 + E + T_1$ | 6 | 2 | 0 | 0 | −2 | $\chi^{(M=0)}(\hat{R})$ |

previous case, the spin-coupling scheme is given by Equation (13.106). As distinguished from the $\mathbf{T}_d$ cluster the eigenvalues of the HDVV Hamiltonian (13.25) for tetragonal clusters depends not only on the full spin $S$ but also on two intermediate spins (for example $S_{ac}$ and $S_{bd}$). The basis set of the Heitler–London states is given by equations (13.107)–(13.109). The results of the calculation of characters $\chi^{(M)}(\hat{R})$ and the corresponding irreducible components of $\Gamma^{(M)}$ are collected in Table 13.5. Using the data of Table 13.5 one can express the results of the group-theoretical classification as follows [62]:

$$2D^{(0)} \doteq {}^1A_1 + {}^1B_1, \quad 3D^{(1)} \doteq {}^3A_2 + {}^3E, \quad D^{(2)} \doteq {}^5B_1. \qquad (13.111)$$

**Table 13.5** Characters of transformations of the basis functions $|a_{m_a}(\frac{1}{2})b_{m_b}(\frac{1}{2})c_{M_c}(\frac{1}{2})d_{m_d}(\frac{1}{2})|$ under the operations of $\mathbf{D}_4$.

| $\mathbf{D}_4$ | $\hat{E}$ | $2\hat{C}_4$ | $\hat{C}_4^2$ | $2\hat{C}_2$ | $2\hat{C}_2'$ | |
|---|---|---|---|---|---|---|
| $A_1$ | 1 | 1 | 1 | 1 | 1 | |
| $A_2$ | 1 | 1 | 1 | −1 | −1 | |
| $B_1$ | 1 | −1 | 1 | 1 | −1 | |
| $B_2$ | 1 | −1 | 1 | −1 | 1 | |
| E | 2 | 0 | −2 | 0 | 0 | |
| $B_1$ | 1 | −1 | 1 | 1 | −1 | $\chi^{(M=2)}(\hat{R})$ |
| $A_2 + B_1 + E$ | 4 | 0 | 0 | 0 | −2 | $\chi^{(M=1)}(\hat{R})$ |
| $A_1 + A_2 + 2B_1 + E$ | 6 | 0 | 2 | 2 | −2 | $\chi^{(M=0)}(\hat{R})$ |

In a similar way one can obtain the $S\Gamma$ multiplets for the overall $\mathbf{D}_{4h}$ symmetry:

$$2D^{(0)} \doteq {}^1A_{1g} + {}^1B_{1g}, \quad 3D^{(1)} \doteq {}^3A_{2g} + {}^3E_u, \quad D^{(2)} \doteq {}^5B_{1g}. \qquad (13.112)$$

From Equations (13.111) and (13.112) one can see that the degeneracy of $2D^{(0)}$ existing in the tetrahedral system is removed in tetragonal clusters, while the degeneracy of $3D^{(1)}({}^3T_1$ in $T_d)$ is removed only partially $({}^3A_2 + {}^3E)$ in $\mathbf{D}_4$ and ${}^3A_{2g} + {}^3E_u$ in $\mathbf{D}_{4h}$). Rewriting (13.25) as

$$\epsilon(S_{ac}, S_{bd}) = -JS(S+1) - (J_1 - J)[S_{ac}(S_{ac}+1) - S_{bd}(S_{bd}+1)] - 3J_1, \qquad (13.113)$$

we see that for $S = 1$ two levels $\epsilon(1,0,1)$ and $\epsilon(0,1,1)$ have the same energy (Table 13.6) in conformity with the fact of the two-fold orbital degeneracy of ${}^3E({}^3E_u)$. It should be stressed that the group-theoretical classification procedure gives the one-to-one correspondence between $nD^{(S)}$ and $S\Gamma$-multiplets but does not provide information concerning the interrelation between the quantum numbers of intermediate spins and the irreducible representations $\Gamma$. For example, in the case under consideration, each of the two spin-singlet states of HDVV Hamiltonian $(S_{ab}, S_{cd}, S) = (0,0,0), (1,1,0)$ can not be related to a certain Heitler–London multiplet ${}^1A_1$ or ${}^1B_1$. To solve this problem one must construct the Heitler–London wavefunctions using the four spin coupling scheme

$$|(S_{ac})(S_{bd})SM\rangle = \sum_{m_a, m_c, m_a} \sum_{m_{ac}, m_{bd}} |a_{m_a} b_{m_b} c_{m_c} d_{m_d}|$$
$$\times \langle \tfrac{1}{2}m_a \tfrac{1}{2}m_c | S_{ac}m_{ac}\rangle \langle \tfrac{1}{2}m_b \tfrac{1}{2}m_d | S_{bd}m_{bd}\rangle \langle S_{ac}m_{ac}S_{bd}m_{bd}|SM\rangle \qquad (13.114)$$

and the projection operator $\hat{P}_{\Gamma\gamma}$ with $\Gamma = A_1(A_{1g}), B_1(B_{1g})$ for $S = 0, A_2(A_{2g})$, $E(E_u)(S=1)$ and $B_1\ (B_{1g})(S=2)$ in $\mathbf{D}_4(\mathbf{D}_{4h})$.
For example

$$|0,0,0\rangle = \tfrac{1}{2}(|a\bar{c}b\bar{d}| + |\bar{a}c\bar{b}d| - |a\bar{c}\bar{b}d| - |\bar{a}cb\bar{d}|), \qquad (13.115)$$

**Table 13.6** Spin states for an exchange tetrameric cluster $(s_i = \frac{1}{2})$ of tetragonal symmetry.

| $(S_{ac}, S_{bd}, S)$ | $S\Gamma(D_4)$ | $S\Gamma(D_{4h})$ | $\epsilon(S_{ac}, S_{bd}, S)$ |
|---|---|---|---|
| $(1,1,2)$ | ${}^5B_1$ | ${}^5B_{1g}$ | $-2J - 7J_1$ |
| $(1,0,1)$ | ${}^3Ex$ | ${}^3E_u x$ | $-5J_1$ |
| $(0,1,1)$ | ${}^3Ey$ | ${}^3E_u y$ | $-5J_1$ |
| $(1,1,1)$ | ${}^3A_2$ | ${}^3A_{2g}$ | $2J - 7J_1$ |
| $(0,0,0)$ | ${}^1A_1$ | ${}^1A_{1g}$ | $-3J_1$ |
| $(1,1,0)$ | ${}^1B_1$ | ${}^1B_{1g}$ | $4J - 7J_1$ |

where $a \equiv a(m_a = \frac{1}{2}), \bar{a} \equiv a(m_a = -\frac{1}{2})$, etc. Applying the projection operator $\hat{P}_{\Gamma\gamma}$ we obtain

$$\hat{P}_{A_{1g}}|0,0,0\rangle = |0,0,0\rangle,$$

$$\hat{P}_{\Gamma\gamma}|0,0,0\rangle = 0 \quad \text{for} \quad \Gamma \neq A_{1g}$$

and, hence

$$|0,0,0\rangle = |{}^1A_{1g}\rangle.$$

In the same way one can obtain the wavefunctions of all states (only functions with $M = S$) [94]:

$$|{}^1B_{1g}\rangle = \tfrac{1}{2}(|a\bar{c}b\bar{d}| + |\bar{a}c\bar{b}d| + |\bar{a}cb\bar{d}| + |a\bar{c}\bar{b}d|),$$

$$|{}^1A_{1g}\rangle = \tfrac{1}{2}(|a\bar{c}b\bar{d}| - |a\bar{c}\bar{b}d| - |\bar{a}cb\bar{d}| + |\bar{a}c\bar{b}d|),$$

$$|{}^3A_{2g}\rangle = \tfrac{1}{2}(|acb\bar{d}| + |ac\bar{b}d| - |a\bar{c}bd| - |\bar{a}cbd|),$$

$$|{}^3E_u x\rangle = \tfrac{1}{2}(|acb\bar{d}| - |acb\bar{d}| + |\bar{a}cbd| - |a\bar{c}bd|),$$

$$|{}^3E_u y\rangle = \tfrac{1}{2}(|acb\bar{d}| - |ac\bar{b}d| - |\bar{a}cbd| + |a\bar{c}bd|),$$

$$|{}^5B_{1g}\rangle = |acbd|. \tag{13.116}$$

The correspondence between quantum numbers $(S_{ac}, S_{bd}, S)$ and $S\Gamma$ is given in Table 13.6.

**Example 13.6** Consider a hexameric cluster with $s_i = \frac{1}{2}$, possessing full molecular symmetry $\mathbf{C}_{6v}$ (Fig. 13.7). As an example of such a type (but more complex) of high nuclearity system, the mixed metal ion-organic radical compound containing six manganese ions ($s_i = \frac{5}{2}$) and six radicals ($s_i = \frac{1}{2}$) can be mentioned [90]. Fig. 13.8 shows the structure [95] of $[Mn(hfac)_2(NITPh)]_6$(hfac = hexafluoroacetilacetonate, NITPh = 2-phenyl-4,4,5,5,-tetrametil-4,5-dihydro-1H-imidazolyl-1-oxyl-3-oxide). For

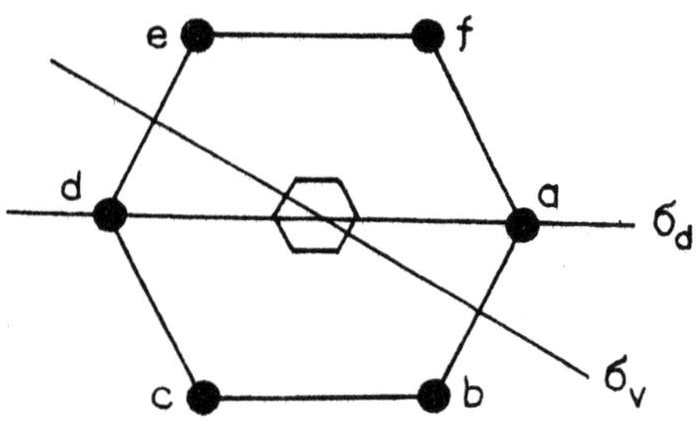


**Fig. 13.7** Metallic skeleton of a hexameric $\mathbf{C}_{6v}$ cluster.

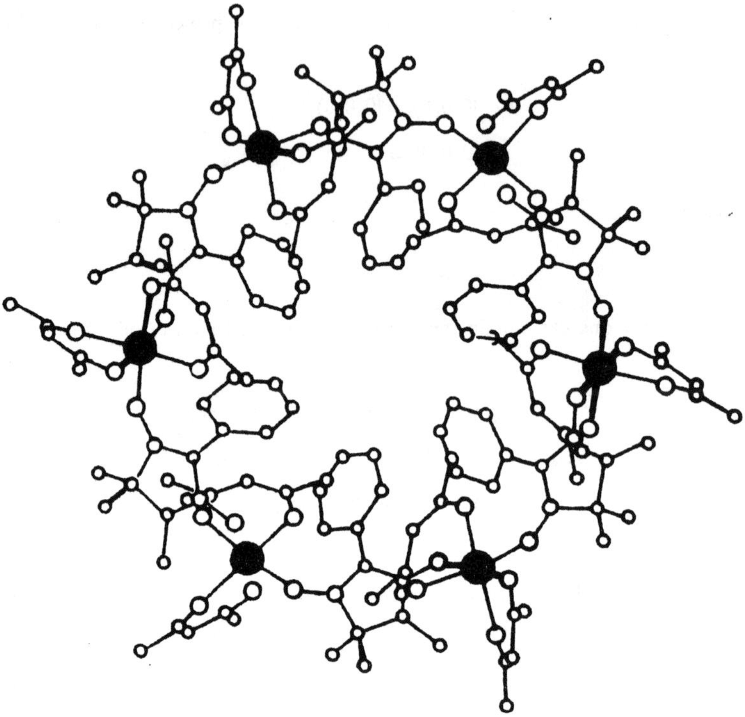

**Fig. 13.8** Crystal structure of $[Mn(hfac)_2\ (NITPh)]$ [94].

six spins $S_i = \frac{1}{2}$ (six-membered organic radicals ring) we have the following coupling scheme

$$\underbrace{D^{(1/2)} \times D^{(1/2)} \dots D^{(1/2)}}_{6\ \text{times}} = D^{(3)} + 5D^{(2)} + 9D^{(1)} + 5D^{(0)}, \qquad (13.117)$$

classifying the spin states of HDVV Hamiltonian. For $M = S = 3$ we have the only determinant $|a_{1/2}\, b_{1/2}\, c_{1/2}\, d_{1/2}\, f_{1/2}|$. Rotations $\hat{C}_6, \hat{C}_6^3 \equiv \hat{C}_2$ and reflection $\hat{\sigma}_v$ induce the odd numbers of transpositions of a, b, ... f orbitals

$$\hat{C}_6|a_{1/2}\, b_{1/2}\, c_{1/2}\, d_{1/2}\, e_{1/2}\, f_{1/2}| = |b_{1/2}\, c_{1/2}\, d_{1/2}\, e_{1/2}\, f_{1/2}\, a_{1/2}|$$
$$= -|a_{1/2}\, b_{1/2}\, c_{1/2}\, d_{1/2}\, e_{1/2}\, f_{1/2}|, \text{etc.}, \qquad (13.118)$$

while the rotation $\hat{C}_3 \equiv \hat{C}_6^2$ and the reflections $\hat{\sigma}_d$ result in the even numbers of transpositions, and hence $\chi^{(M=3)}(\hat{C}_6) = -1, \chi^{(M=3)}(\hat{C}_6^2) = 1$, etc. (Table 13.7). We find that $D^{(3)} \doteq {}^7B_2$.

For $M = 2$ we have six determinants with one reversed spin ($|a_{-1/2}\, b_{1/2}\, c_{1/2}\, d_{1/2}\, e_{1/2}\, f_{1/2}|$ etc.) and $\Gamma^{(M=2)} = A_1 + B_2 + E_1 + E_2$. Fifteen determinants with $M = 1$ form three subsets as illustrated in Fig. 13.9 (determinants belonging

**Table 13.7.** Characters of transformations of the basis functions $|a_{m_a} b_{m_b} c_{m_c} d_{m_d} e_{m_e} f_{m_f}|$ under operations of $\mathbf{C}_{6v}$ group.

| $\mathbf{C}_{6v}$ | $\hat{E}$ | $2\hat{C}_6$ | $2\hat{C}_3$ | $\hat{C}_2$ | $3\hat{\sigma}_v$ | $3\hat{\sigma}_d$ | |
|---|---|---|---|---|---|---|---|
| $A_1$ | 1 | 1 | 1 | 1 | 1 | 1 | |
| $A_2$ | 1 | 1 | 1 | 1 | −1 | −1 | |
| $B_1$ | 1 | −1 | 1 | −1 | 1 | −1 | |
| $B_2$ | 1 | −1 | 1 | −1 | −1 | 1 | |
| $E_1$ | 2 | 1 | −1 | −2 | 0 | 0 | |
| $E_2$ | 2 | −1 | −1 | 2 | 0 | 0 | |
| $B_2$ | 1 | −1 | 1 | −1 | −1 | 1 | $\chi^{(M=3)}(\hat{R})$ |
| $A_1 + B_2 + E_1 + E_2$ | 6 | 0 | 0 | 0 | 0 | 2 | $\chi^{(M=2)}(\hat{R})$ |
| $A_2 + B_2 + E_1 + E_2$ | 6 | 0 | 0 | 0 | −2 | 0 | $\chi_{I}^{(M=1)}(\hat{R})$ |
| $A_1 + B_2 + E_1 + E_2$ | 6 | 0 | 0 | 0 | 0 | 2 | $\chi_{II}^{(M=1)}(\hat{R})$ |
| $B_2 + E_1$ | 3 | 0 | 0 | −3 | −1 | 1 | $\chi_{III}^{(M=1)}(\hat{R})$ |
| $A_1 + B_2 + E_1 + E_2$ | 6 | 0 | 0 | 0 | 0 | 2 | $\chi_{I}^{(M=0)}(\hat{R})$ |
| $A_1 + A_2 + B_1 + B_2 + 2E_1 + 2E_2$ | 12 | 0 | 0 | 0 | 0 | 0 | $\chi_{II}^{(M=0)}(\hat{R})$ |
| $A_1 + B_2$ | 2 | 0 | 2 | 0 | 0 | 2 | $\chi_{III}^{(M=0)}(\hat{R})$ |

to different subsets are unmixed by the symmetry operations). Finally, twenty determinants with $M = 0$ can be also distributed over three independent subsets (Fig. 13.9). Representations $\Gamma^{(M)}$ are collected in Table 13.7. Corresponding irreducible representations $\Gamma^{(S)}$ can be obtained in the same way as in previous examples. The spin-orbital multiplets are

$$\begin{aligned} &D^{(3)} \doteq {}^7B_2, \quad 5D^{(2)} \doteq {}^5A_1 + {}^5E_1 + {}^5E_2, \\ &9D^{(1)} \doteq {}^3A_2 + 2\,{}^3B_2 + 2\,{}^3E_1 + {}^3E_2, \\ &5D^{(0)} \doteq 2\,{}^1A_1 + {}^1B_1 + {}^1E_1. \end{aligned} \tag{13.119}$$

We see that each repeated spin multiplet $nD^{(S)}$ contains several $S\Gamma$- states. Let us notice that in the case under consideration some $S\Gamma$-terms prove to be repeating ($2\,{}^3B_2$ and $2\,{}^3E_1$ in $9D^{(1)}$, $2\,{}^1A_1$ in $5D^{(0)}$. For these repeated $S\Gamma$-terms we have to diagonalize the HDVV Hamiltonian as it was stated in Section 5.62, the order of each mixing-matrix being equal to the number of repeated $S\Gamma$-terms. As to the six-membered $s_i = \frac{1}{2}$-ring this means that in the symmetry adopted basis ($S\Gamma$-states) the $5D^{(2)}$-block of full energy $64 \times 64$ matrix proves to be diagonal, the $9D^{(1)}$-submatrix is blocked and consists of two one-dimensional and two $2 \times 2$-matrices; finally, the $5D^{(0)}$ block gives two one-dimensional and one $2 \times 2$-matrix. Hence the problem of the energy levels of the

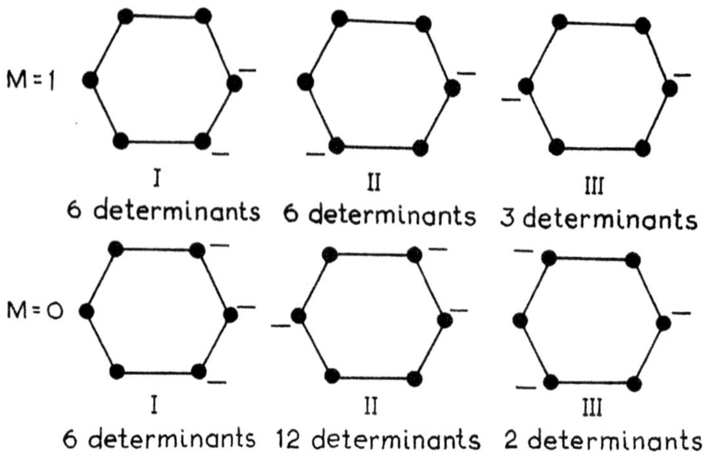


**Fig. 13.9** Pictorial representation for the states with $M = 1$ and $M = 0$ for a six-membered ring of spins $s_i = \frac{1}{2}$.

six-membered $s_i = \frac{1}{2}$-ring can be solved "by hand". Of course, for high-spin and high-nuclearity clusters (for instance, for the mixed manganese ion-organic radical cluster presented in Fig. 13.8) possessing extended basis sets the group-theoretical classification procedure and diagonalization of HDVV-Hamiltonian can be performed only by the use of computer.

An impressive example of such type calculations is presented by Gatteschi *et al.* [90]. Table 13.8 shows the total spin states and the corresponding irreducible representations of $\mathbf{D}_6$ for the six $s_i = 1/2$ and six $s_2 = 5/2$ spins† (mixed manganese-organic radical ring, Fig. 13.8). From Table 13.8 we see that by using the results of the group-theoretical classification and the symmetry adopted basis one can simplify essentially the problem of the eigen-values of the exchange system.

### 13.4.3 Conclusions from the group-theoretical classification

The group-theoretical classification of exchange multiplets enables us to understand the origin of multiple degeneracies and to identify the splitting mechanism [62, 63]. Let us consider for example some conclusions drawn from the classification of the symmetric ($\mathbf{D}_3$) trimeric cluster with half-integer spins. In the case of antiferromagnetic exchange the ground state $2D^{(1/2)}$ of such a system corresponds to the doublet $^2E$. We can see that the degeneracy of the $2D^{(1/2)}$ multiplet over the intermediate spin in the HDVV model is not accidental. This degeneracy of the HDVV model corresponds to the twofold orbital degeneracy of Heitler–London states with total spin $S = \frac{1}{2}$. Therefore

† Spin-functions (but not orbital functions) are classified in [90].

**Table 13.8** Total spin states and irreducible representations of $\mathbf{D}_6$ for six $s_1 = 1/2$ and six $s_2 = 5/2$ spins ($n$ is the number of $S$-states) (after D. Gatteschi *et al.* [90]).

| S | $A_1$ | $A_2$ | $B_1$ | $B_2$ | $E_1$ | $E_2$ | $n$ |
|---|---|---|---|---|---|---|---|
| 18 | 1 | 0 | 0 | 0 | 0 | 0 | 1 |
| 17 | 1 | 0 | 2 | 0 | 2 | 2 | 11 |
| 16 | 9 | 2 | 5 | 4 | 9 | 11 | 60 |
| 15 | 21 | 16 | 25 | 14 | 37 | 35 | 220 |
| 14 | 64 | 41 | 52 | 47 | 101 | 107 | 620 |
| 13 | 126 | 113 | 138 | 107 | 245 | 239 | 1452 |
| 12 | 277 | 227 | 251 | 239 | 487 | 499 | 2966 |
| 11 | 459 | 437 | 485 | 425 | 913 | 901 | 5434 |
| 10 | 802 | 719 | 758 | 741 | 1499 | 1521 | 9060 |
| 9 | 1164 | 1139 | 1208 | 1117 | 2319 | 2297 | 113860 |
| 8 | 1695 | 1580 | 1627 | 1614 | 3247 | 3281 | 19562 |
| 7 | 2135 | 2118 | 2203 | 2084 | 4287 | 4253 | 25620 |
| 6 | 2679 | 2538 | 2583 | 2585 | 5161 | 5208 | 31123 |
| 5 | 2899 | 2902 | 2995 | 2855 | 5857 | 5810 | 34985 |
| 4 | 3102 | 2948 | 2982 | 3008 | 5990 | 6050 | 36120 |
| 3 | 2784 | 2818 | 2904 | 2758 | 5652 | 5592 | 33752 |
| 2 | 2390 | 2244 | 2254 | 2312 | 4576 | 4644 | 27640 |
| 1 | 1460 | 1534 | 1596 | 1466 | 3062 | 2994 | 18168 |
| 0 | 608 | 492 | 466 | 562 | 1018 | 1088 | 6340 |

we arrive at the conclusion that Coulomb interactions between cluster electrons and nuclei cannot remove this degeneracy. Taking account of Coulomb interactions at any order of perturbation theory can only lead to a shift of the ground level.

Orbital symmetry analysis makes it possible to understand the physical nature of the splitting. Spin functions with $S = \frac{1}{2}$ provide the representation $D^{(1/2)} \doteq \bar{E}$ of the trigonal group ($\bar{E}$ being a double-valued representation). To classify the fine structure of energy levels, we set up the direct product of the representations carried by orbital and spin components belonging to the total wavefunction of the ground quadruplet, and decompose it into its irreducible parts: $E \times D^{(1/2)} = E \times \bar{E} = \bar{A}_1 + \bar{A}_2 + \bar{E}$. Thus, the ground state includes two

Kramers doublets: $\bar{A}_1 + \bar{A}_2 \equiv 2\bar{A}$ (complex-conjugate representation of the double group $\mathbf{D}'_3$) and $\bar{E}$. We see that for trigonal clusters with half-integer spins the ground state in the case of antiferromagnetic exchange is split into two doublets $2\bar{A}$ and $\bar{E}$. The splitting is due to spin–orbit interaction. The unphysical degeneracy in the HDVV model is related here to the fact that this model does not take spin–orbit interaction into account.

It should be mentioned that the reasoning applied above provides only the number and the classification of double-valued representations related to the fine-structure levels. It is important to know the order of perturbation theory in which the multiplet ${}^{2S+1}\Gamma$ is split by spin–orbit interaction. We use the rules stated in Section 11.3.3. In trigonal groups the orbital angular momentum $\mathbf{L}$ transforms according to two irreducible representations: $A_2$ ($\hat{L}_z$ component) and E ($\hat{L}_x$, $\hat{L}_y$); $\hat{\mathbf{L}}$ is a purely imaginary operator, and the matrix in the basis of the orbital doublet E is nonzero if $\{E^2\}$ (the antisymmetric part of the product $E \times E$) involves representations according to which the orbital angular momentum components are transformed. Since $\{E^2\} = A_2$, spin–orbit interaction splits the $^2E$ level in first order of perturbation theory. The corresponding spin–orbit interaction operator is of the form

$$\hat{H}_{so} = \lambda \hat{L}_z \hat{S}_z, \tag{13.120}$$

which coincides with (12.54).

Let us now consider the excited states of antiferromagnetic trimers with $s_i > \frac{1}{2}$. To the $4D^{(3/2)}$ level of the HDVV model there correspond three orbital states $^4A_1$, $^4A_2$ and $^4E$. We thus obtain a new result, qualitatively distinguishing the excited and ground states. If in the ground state, without taking spin–orbit interaction into account, the degeneracy is exact then the excited-state degeneracy is the consequence of the limited scope of the HDVV model, even in the absence of spin–orbit interaction. The fact that the orbital part of the wavefunctions for spin-quadruplet states has three irreducible representations $A_1$, $A_2$ and E shows that in a $\mathbf{D}_3$ symmetry system there are Coulomb interactions that remove the 16-fold degeneracy of the $4D^{(3/2)}$ state. Thus the HDVV model does not account fully for Coulomb exchange interaction between many-electron metal ions. This conclusion is exact, since it is based on group-theoretical considerations only. The latter, of course, do not reveal the nature of the interactions part (not taken into account) or of the corresponding splitting mechanism of "accidentally" degenerate levels in the HDVV model. The multiplets $^4A_1$, $^4A_2$ and $^4E$ are subjected to further splittings into Kramers doublets. Since $D^{(3/2)} \doteq 2\bar{A} + \bar{E}$, we obtain

$$A_1 \times D^{(3/2)} = A_2 \times D^{(3/2)} = 2\bar{A} + \bar{E}, \qquad E \times D^{(3/2)} = 2\bar{A} + 3\bar{E}.$$

The splitting is again due to spin–orbit interaction. Since the mean value of the orbital angular momentum for nondegenerate states is zero, orbital singlets with $S \geq 1$ ($^4A_1$ and $^4A_2$) are split in second (or higher) order of perturbation theory. The doublet $^4E$ is split at first order; however, the degeneracy is not

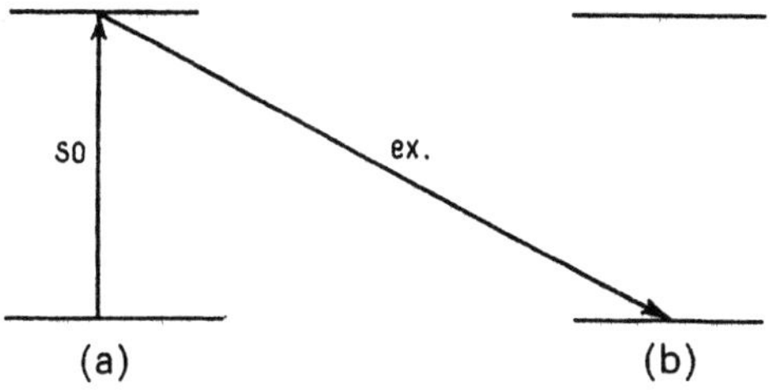


**Fig. 13.10** The origin of AC exchange between ions a and b. Arrows show the mixing between spin–orbit (so) and exchange (ex) interactions.

completely removed. The final splitting is achieved when second-order spin–orbit interaction is taken into account. Other excited states of trimeric clusters are analysed in a similar way.

The conclusions mentioned above can describe the qualitative aspects of the fine structure of exchange multiplets. In a number of cases the Heisenberg model does not provide a complete description of the energy spectrum. The more detailed qualitative theory is given in [62, 93, 96].

### 13.4.4 Non-Heisenberg exchange interactions

The *fine* structure of exchange levels is the result of interactions between ions not taken into account in the HDVV model: these are conventionally known as *non-Heisenberg interactions* [62, 63]. We mention here the three most important ones.

(1) *Dzyaloshinskii–Moria antisymmetric (AS) exchange* leads to weak ferromagnetism; that is, to a slight relative inclination of sublattices in antiferromagnetic substances. AS exchange occurs in second order of perturbation theory owing to spin–orbit interaction and Heisenberg exchange, as shown in Fig. 13.10. It is described by the operator

$$\hat{H}_{AS} = \sum_{ij} \mathbf{G}_{ij}(\hat{\mathbf{s}}_i \times \hat{\mathbf{s}}_j), \tag{13.121}$$

which involves the vector product of spin operators and $\mathbf{G}_{ij} = -\mathbf{G}_{ji}$ is the vector constant of AS exchange.

(2) *The biquadratic exchange (BE) interaction* describes the interelectron exchange interaction, not taken into consideration by the HDVV model. For dimeric clusters the BQ operator is

$$\hat{H}_{BQ} = j(\hat{\mathbf{s}}_a \cdot \hat{\mathbf{s}}_b)^2. \tag{13.122}$$

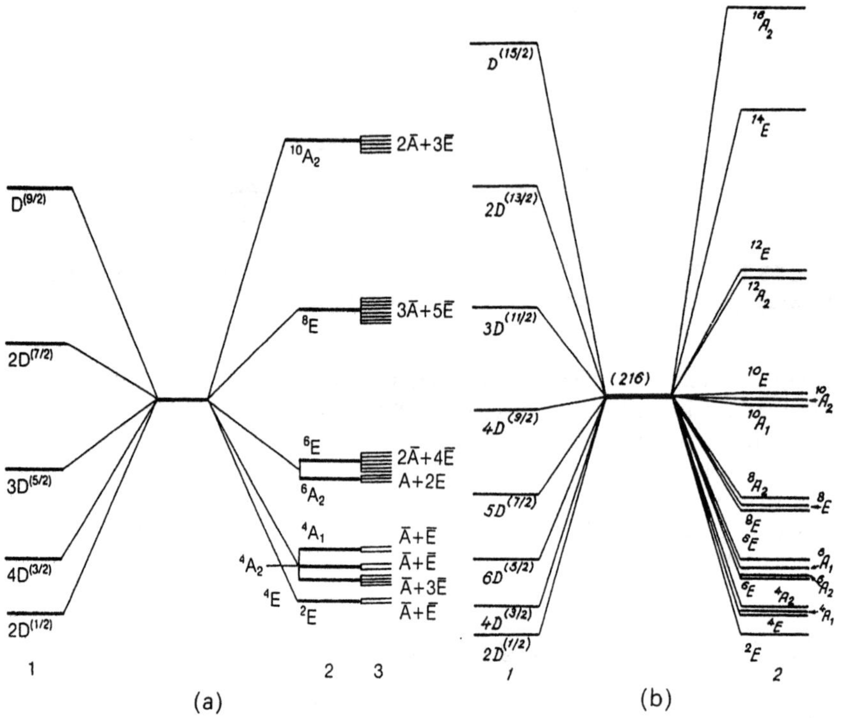


**Fig. 13.11** Level diagram of a trigonal trimeric cluster, taking account of non-Heisenberg exchange interactions for $s_i = \frac{3}{2}$ (a) and $\frac{5}{2}$ (b): 1, HDVV model; 2, splitting of exchange levels due to biquadratic exchange; 3, spin–orbit splitting (antisymmetric exchange) [62, 63].

For trimers the BE operator contains two independent parameters $j'$ and $j''$:

$$\hat{H}_{\text{BQ}} = j' \sum_{ij} (\hat{\mathbf{s}}_i \cdot \hat{\mathbf{s}}_j)^2 + j'' \left( \sum_{ij} \hat{\mathbf{s}}_i \cdot \hat{\mathbf{s}}_j \right)^2. \qquad (13.123)$$

Microscopic evaluations show that $j'$ and $j''$ are usually far less than the exchange parameter $J$ of the HDVV model.

(3) *The anisotropic* ($AN$) *exchange* Hamiltonian for a dimeric cluster is

$$\hat{H}_{\text{AN}} = \text{D}(3\hat{S}_{\text{a}x}\hat{S}_{\text{b}x} - \hat{\mathbf{s}}_{\text{a}} \cdot \hat{\mathbf{s}}_{\text{b}}) + \text{E}(\hat{S}_{\text{a}x}\hat{S}_{\text{b}x} - \hat{S}_{\text{a}y}\hat{S}_{\text{b}y}). \qquad (13.124)$$

A more detailed account is given in [62]. Here we mention only that exchange Hamiltonian terms appear at second order in the spin–orbit interaction.

From the above it is obvious that the AS exchange splits the quadruplet $2D^{(1/2)} \doteq {}^2E$, while BE exchange leads to $4D^{(3/2)}$ splitting into the terms ${}^4A_1$, ${}^4A_2$ and ${}^4E$. Therefore, summarizing the results of group-theoretical classification of exchange multiplets and those of microscopic analysis, allow one to determine all types of non-Heisenberg interactions that split the HDVV model multiplets (Table 13.9). The influence of non-Heisenberg exchange on the spectra of trimeric clusters with $s_i = \frac{3}{2}$ and $s_i = \frac{5}{2}$ is illustrated in Fig. 13.11.

**Table 13.9** Principles of non-Heisenberg exchange interaction selection [62].

| Results of HDVV model | Group-theoretical classification | Active non-Heisenberg exchange forms |
|---|---|---|
| (1) Accidental degeneracy $S(\ldots S'\ldots)$ | Several multiplets ${}^{2S+1}\Gamma$, $nD^{(S)} \doteq \Sigma_i {}^{2S+1}\Gamma_i$ | Biquadratic exchange and other isotropic forms of higher-rank exchange |
| (2) Accidental degeneracy $S(\ldots S'\ldots)$, $S=0$ | One multiplet ${}^1\Gamma$ | Degeneracy is exact and not removed by non-Heisenberg exchange; it can be removed only by an external field |
| (3) Accidental degeneracy $S \neq 0$. | One multiplet ${}^{2S+1}\Gamma$ (a) $S=\frac{1}{2}$ $\{\Gamma^2\}$ contains $\Gamma(\hat{\mathbf{L}})$ | Antisymmetric exchange |
| | (b) $S>\frac{1}{2}$, $\Gamma \neq A$, $\{\Gamma^2\}$ contains $\Gamma(\hat{\mathbf{L}})$ | Antisymmetric exchange plus anisotropic exchange |
| (4) Nondegenerate multiplet (orbital singlet) ${}^{2S+1}A$ | (a) $S=\frac{1}{2}$ | Splitting only by a magnetic field |
| | (b) $S>\frac{1}{2}$, (noncubic groups) | Anisotropic exchange |
| | (c) $S>\frac{3}{2}$, (cubic groups) | Higher-order anisotropic exchange |
| | (d) $S\leq\frac{3}{2}$, (cubic groups) | Degeneracy is removed by a magnetic field, electric field or strain |

## 13.5 PARAMAGNETIC RESONANCE AND HYPERFINE INTERACTIONS

The group-theoretical classification of exchange multiplets makes it possible to develop a phenomenological theory of EPR spectra. Indeed, in order to construct an effective Hamiltonian, we must know the point symmetry of the system, the total spin $S$ and the irreducible representation $\Gamma$ for the state under consideration (Section 12.6). These data, necessary for calculation of EPR spectra, are precisely those provided by group-theoretical classification. Therefore it is possible to apply the effective Hamiltonian method (Section 12.6) directly to cluster multiplets $S\Gamma$.

**Example 13.7** Consider a trimeric ($\mathbf{D}_3$) cluster ground state. For antiferromagnetic systems of this type the ground state is $^2E$. The term $^2E$ was analysed in Section 12.6; its zero-field splitting is due to spin–orbit interaction, and the Zeeman Hamiltonian in parallel and perpendicular fields is given by (12.58) and (12.59) respectively. The calculation of energy levels and EPR spectrum is given in [63].

It should be stressed here that the HDVV model does not give the EPR spectrum—first because of the "accidental" degeneracies and, secondly, because of Zeeman interaction isotropy.

*Hyperfine electron–nuclear interaction* can be described in a similar way to that outlined above [62, 63]. There are also a number of other physicochemical problems whose solution is facilitated by the use of group-theoretical classifications; these include

(1) the conditions under which binding appears between the quadrupole moment of the cluster electron shell and the ion nuclei;

(2) construction of the effective Hamiltonian of the electron–nuclear quadrupole interaction, hybrid states, and nuclear quadrupole resonance in tetrameric tetrahedral clusters;

(3) the Mössbauer effect in tetrameric cluster hybrid states;

(4) Jahn–Teller interaction in tetrameric and trimeric clusters.

## 13.6 GROUP-THEORETICAL CLASSIFICATION OF MULTIPLETS OF MIXED-VALENCE CLUSTERS AND CLUSTERS WITH PARTIAL DELOCALIZATION

Metal-ion valence is determined by the number of electrons in the unfilled shell. If two ions with different valences, one of which has an "extra" electron, occupy equivalent positions, this electron can be localized on either of them, with equal probability. Electron delocalization leads to the concept of *"mixed" valence*, the delocalization mechanism being considered as a "double exchange". A semiclassical double-exchange theory, based on model concepts, has been developed in [97, 98]. Within the scope of the many-electron microscopic model [63, 99, 100], it is possible to take account of both

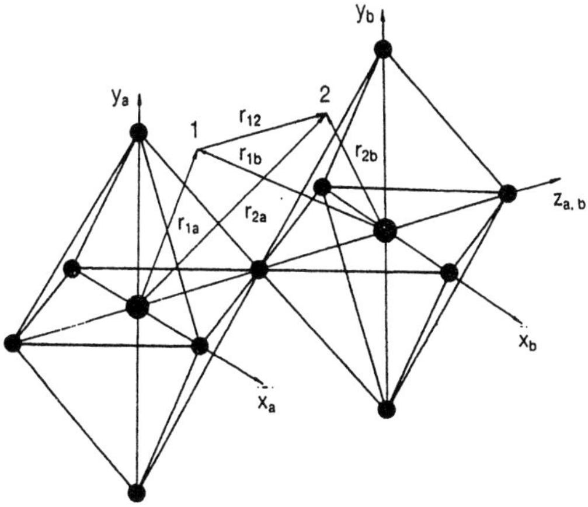


**Fig. 13.12** Model of a bi-octahedral cluster.

double exchange and Heisenberg exchange, as well as the crystal field and the degeneracy of the cluster ion–electron terms. In order to classify and construct many-electron states of mixed-valence clusters, group-theoretical methods are of considerable use. We shall illustrate this, taking the example of a bi-octahedral cluster of transition metal ions $d^1$–$d^2$ (Fig. 13.12).

Exchange interaction and transfer is due to the interaction of electrons from different centres with each other and with "alien" nuclei. The ground state of a $d^1$ ion in a cubic field is $^2T_2(t_2)$ and that of a $d^2$ ion is $^3T_1(t_2^2)$. The $t_2^2$ configuration also provides the terms $^1A_1$, $^1E$ and $^1T_2$. Assuming a weak interaction between centres, the cluster spectrum can be divided into three multiplet groups:

$$[^2T_2(t_2)]_a\,[^3T_1(t_2^2)]_b, \quad [^2T_2(t_2)]_a\,[^1E, {}^1T_2(t_2^2)]_b, \quad [^2T_2(t_2)]_a\,[^1A_1(t_2^2)]_b.$$

A problem that can be solved by group-theoretical classification is the determination of the allowed cluster $S\Gamma$ terms with a localized, but tunnelling, extra electron. These terms are better named *exchange-resonance terms*, since states of the system with different localizations of the extra electron are equivalent from the energetic point of view.

The group-theoretical classification procedure [99] comprises the following stages:

(1) The centres a and b are brought into coincidence; the hypothetical system thus obtained has $\mathbf{O}_h$ symmetry.

(2) The direct products of b-centre orbital irreducible representations ($A_{1g}$, $E_g$, $T_{1g}$, $T_{2g}$) and the a-centre representation $T_{2g}$ in $\mathbf{O}_h$ are taken.

(3) These direct products are decomposed into irreducible parts $\Gamma_i$ in $\mathbf{O}_h$.

(4) The a and b centres are separated so that they occupy their initial

positions. Assuming $\mathbf{C}_{4v}$ symmetry for the cluster with localized electrons, $\Gamma_i$ is reduced in $\mathbf{C}_{4v}$. The resulting $\mathbf{C}_{4v}$ group irreducible representations enumerate the orbital components of the wavefunctions of interacting $d^1$ and $d^2$ ions with localized electrons.

(5) The spin states are obtained according to the usual scheme of angular momentum addition: $S = s_a + s_b, \ldots, |s_a - s_b|$ for $s_a = \frac{1}{2}, s_b = 1 : S = \frac{3}{2}$ and $\frac{1}{2}$.

It should be noted that the Pauli principle does not impose in our case any restrictions on the resulting states (in contrast with the states of one many-electron ion in a crystal field; Sections 5.5 and 5.7), so all terms $^2\Gamma_i$ and $^4\Gamma_i$ are allowed.

**Example 13.8** Consider the ground $[^2T_{2g}]_a\ [^3T_{1g}]_b$ multiplet group of a bioctahedral $d^1$–$d^2$ cluster with localized electrons [99]. In the $\mathbf{O}_h$ group of "superposed" centres we take the direct product $T_{1g} \times T_{2g}$ and reduce the representations thus obtained in $\mathbf{C}_{4v}$:

$$\begin{array}{ccccccccc} T_{2g} \times T_{1g} \doteq & A_{2g} & + & E_g & + & T_{1g} & + & T_{2g} & (\mathbf{O}_h) \\ & \downarrow & & \downarrow & & \downarrow & & \downarrow & \\ & B_1 & & A_1 + B_1 & & A_2 + E & & B_2 + E & (\mathbf{C}_{4v}) \end{array}$$

Since $S = \frac{1}{2}$ and $\frac{3}{2}$, we obtain the terms of the $\mathbf{C}_{4v}$ system:

$$[^2T_{2g}]_a\,[^3T_{1g}]_b : {}^{4,2}A_1,\ {}^{4,2}A_2,\ 2\,{}^{4,2}B_1,\ {}^{4,2}B_2,\ 2\,{}^{4,2}E.$$

In a similar way we obtain

$$[^2T_2]_a\,[^1E, {}^1T_2]_b : 2\,{}^2A_1,\ 2\,{}^2A_2,\ 2B_1,\ 2\,{}^2B_2,\ 4\,{}^2E;$$

$$[^2T_2]_a\,[^1A_1]_b : {}^2B_2,\ {}^2E.$$

Stationary (resonance) states belong to $\mathbf{D}_{4h}$; that is, to the full symmetry group of the cluster, and are characterized by a certain parity. If $S\Gamma(\mathbf{C}_{4v})$ is a term of the cluster with a localized electron then, when the reduction $\mathbf{D}_{4h} \rightarrow \mathbf{C}_{4v}$ takes place, the resonant states $\Gamma_u$ and $\Gamma_g$ yield the irreducible representation $\Gamma$. From the tables in [1] we find the scheme of localized and resonant states

$$\begin{array}{lcccccccccc} \mathbf{C}_{4v} : & \multicolumn{2}{c}{A_1} & \multicolumn{2}{c}{A_2} & \multicolumn{2}{c}{B_1} & \multicolumn{2}{c}{B_2} & \multicolumn{2}{c}{E} \\ & \swarrow & \searrow & \swarrow & \searrow & \swarrow & \searrow & \swarrow & \searrow & \swarrow & \searrow \\ \mathbf{D}_{4h} : & A_{1g} & A_{2u} & A_{2g} & A_{1u} & B_{1g} & B_{2u} & B_{2g} & B_{1u} & E_g & E_u \end{array}$$

The allowed $S\Gamma$-multiplets of polynuclear mixed-valence clusters consisting of ions with orbitally nondegenerate ground states can be performed using the approach employed for exchange clusters in Section 13.4.3. Let us illustrate the receipt of the group-theoretical classification of exchange-tunnel multiplets by examples based on the approach developed in [105].

**Example 13.9** Consider trimeric mixed-valence $d^1 - d^1 - d^0$-cluster with $\mathbf{C}_{3v}$ symmetry for the full molecule. The site-symmetry of each transition metal ion belongs to the $\mathbf{C}_s$ group, so that the local crystal field removes degeneracy of d-states completely. The system under consideration involves two electrons shared among three sites (a, b and c) and migrating over three nondegenerate orbitals, which will be denoted as a, b and c (the corresponding spin-orbitals are $a_{m_a}(s_a)$, $b_{m_b}(s_b)$, $c_{m_c}(s_c)$ with $s_a = s_b = s_c = \frac{1}{2}$). For the sake of simplicity, the $a, b, c$ orbitals are supposed to be spherically symmetric. Full basis sets consist of twelve two-electron determinants

$$|a_{m_a}(\tfrac{1}{2})b_{m_c}(\tfrac{1}{2})|,\ |a_{m_a}(\tfrac{1}{2})c_{m_c}(\tfrac{1}{2})|,\ |b_{m_b}(\tfrac{1}{2})c_{m_c}(\tfrac{1}{2})|$$

which are related to the c,b and a-localizations of the "hole" (free center) respectively. For $M = S = 1$ we have three determinants:

$$|a_{1/2}(\tfrac{1}{2})b_{1/2}(\tfrac{1}{2})|,\ |a_{1/2}(\tfrac{1}{2})c_{1/2}(\tfrac{1}{2})|,\ |b_{1/2}(\tfrac{1}{2})c_{1/2}(\tfrac{1}{2})| \qquad (13.125)$$

The characters of the three-dimensional reducible representation of $\mathbf{C}_{3v}$ with basis set (13.125) are given in Table 13.10. Decomposing this representation into irreducible ones we find that

$$\Gamma^{(M=1)} = \Gamma^{(S=1)} = A_2 + E$$

For $M = 0$ we have six determinants

$$\underbrace{|a_{-1/2}(\tfrac{1}{2})b_{1/2}(\tfrac{1}{2})|,\ |a_{1/2}(\tfrac{1}{2})b_{-1/2}(\tfrac{1}{2})|,}_{\text{c - localized hole}}$$
$$\underbrace{|a_{-1/2}(\tfrac{1}{2})c_{1/2}(\tfrac{1}{2})|,\ |a_{1/2}(\tfrac{1}{2})c_{-1/2}(\tfrac{1}{2})|,}_{\text{b - localized hole}} \qquad (13.126)$$
$$\underbrace{|b_{-1/2}(\tfrac{1}{2})c_{1/2}(\tfrac{1}{2})|,\ |b_{1/2}(\tfrac{1}{2})c_{-1/2}(\tfrac{1}{2})|}_{\text{a - localized hole}}$$

for which we obtain (Table 13.10)

$$\Gamma^{(M=0)} = A_1 + A_2 + 2E.$$

**Table 13.10** Characters of transformations of the basis functions of $d^1 - d^1 - d^0$-mixed-valence trimer.

| $\mathbf{C}_{3v}$ | $\hat{E}$ | $2\hat{C}_3$ | $3\hat{\sigma}_v$ | |
|---|---|---|---|---|
| $A_2 + E$ | 3 | 0 | $-1$ | $\chi^{(M=1)}(\hat{R})$ |
| $A_1 + A_2 + 2E$ | 6 | 0 | 0 | $\chi^{(M=0)}(R)$ |

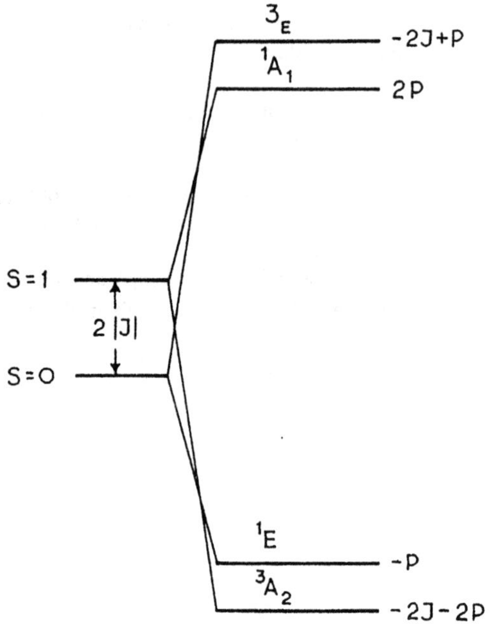


**Fig. 13.13** Energy levels of a $d^1-d^1-d^0$ (mixed-valence trimeric cluster ($J<0$, $P>0$).

Extracting $\Gamma^{(S=1)}$ from $\Gamma^{(M=0)}$ we find $\Gamma^{(S=0)}$. Hence for $d^1-d^1-d^0$- mixed-valence trimer we find two spin-triplets, $^3A_2$ and $^3E$, and two spin-singlets, $^1A_1$ and $^1E$.

It should once more be noted that the allowed $S\Gamma$ terms have been derived without constructing wavefunctions out of the system, as the last can usually be found using the projection operator. The energy two exchange levels split by the electron transfer (resonance splitting) [100] (Fig. 13.13):

$$\epsilon(^3A_2) = -2J - 2P, \quad \epsilon(^3E) = -2J + P,$$
$$\epsilon(^1A_1) = 2P, \qquad \epsilon(^1E) = -P. \tag{13.127}$$

In (13.127) $J$ is the exchange parameter for the $d^1-d^1$ interaction between two $d^1$ ions (Section 13.1); $P$ is the transfer parameter. The absolute value of resonance splitting for the $d^1-d^1-d^0$-system is independent on the full spin, but the orders of the A and E states are different for exchange levels with $S=1$ and $S=0$.

**Example 13.10** Consider the $d^1-d^1-d^2$ cluster with full $\mathbf{C}_{3v}$ symmetry. The one-electron spectrum of each ion is supposed to consist of two nondegenerate levels (orbitals $\varphi_i$ and $\psi_i$, Fig. 13.14) with the high-spin (Hund's) ground state ($S=1$) of the $d^2$ ion. The full basis set involves four-electron determinants

$$|a_{m_a}(s_a)b_{m_b}(s_b)c_{m_c}(s_c)|$$

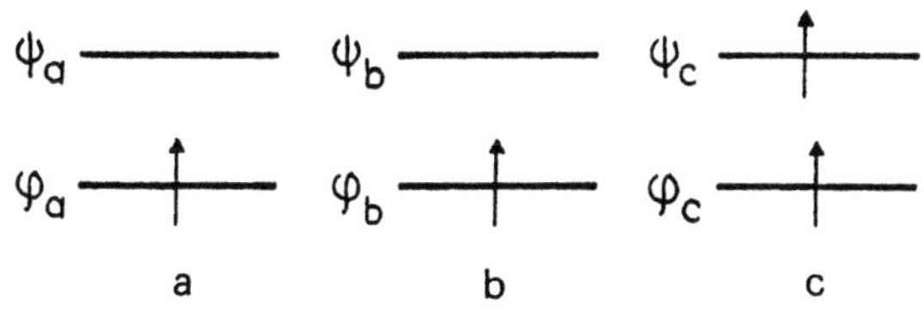


**Fig. 13.14** One-electron levels of $d^1 - d^1 - d^2$ mixed-valence cluster (c-localization).

relating to the three possible localizations of the "extra" electron. For example, for the c-localization $s_a = s_b = \frac{1}{2}, s_c = 1$ and $a_{m_a}(s_a)$, $b_{m_b}(s_b)$ are the one-electron spin-orbitals (with orbital parts $\varphi_a(\mathbf{r}), \varphi_b(\mathbf{r})$, while $c_{m_c}(s_c)$ is the two electron wavefunction of the high-spin state of ion c with the occupied $\varphi_c$ and $\psi_c$ orbitals (see Equation (13.91)). Each localized state of a mixed-valence trimer can be described by the scheme of coupling of three spins $s_a, s_b, s_c$. The spin-coupling scheme for a mixed-valence trimer $d^1 - d^2 - d^2$ can be symbolically presented as

$$3(D^{(1/2)} \times D^{(1/2)} \times D^{(1)}) = 3D^{(0)} + 6D^{(1)} + 3D^{(2)}, \tag{13.128}$$

where the direct product of three $D^{(s_i)}$ relates to the spin-coupling within each localized scheme; factor 3 takes into account three possibilities for localization. Let us derive a one-to-one correspondence between $nD^{(s)}$ states $S\Gamma$-terms of stationary (delocalized) states of a mixed-valence trimer (group-theoretical classification).

For $M = S = 2$ we have the only determinant for each localization, and find a three-dimensional representation with basis set

$$|a_{1/2}(\tfrac{1}{2})b_{1/2}(\tfrac{1}{2})c_1(1)|, \ |a_{1/2}(\tfrac{1}{2})b_1(1)c_{1/2}(\tfrac{1}{2})|, \ |a_1(1)b_{1/2}(\tfrac{1}{2})c_{1/2}(\tfrac{1}{2})|. \tag{13.129}$$

This reducible representation one can be decomposed as shown in Table 13.11. Basis functions for $M = 1$ form two independent subsets–three-dimensional (I) and six-dimensional (II). For the c-localized "extra" electron, the functions belonging to these

**Table 13.11** Characters of transformations for the basis functions of $d^1 - d^1 - d^2$-mixed valence trimer.

| $C_{3v}$ | $\hat{E}$ | $2\hat{C}_3$ | $3\hat{\sigma}_v$ | |
|---|---|---|---|---|
| $A_2 + E$ | 3 | 0 | −1 | $\chi^{(M=2)}(\hat{R})$ |
| $A_2 + E$ | 3 | 0 | −1 | $\chi_I^{(M=1)}(\hat{R})$ |
| $A_1 + A_2 + 2E$ | 6 | 0 | 0 | $\chi_{II}^{(M=1)}(\hat{R})$ |
| $A_1 + A_2 + 2E$ | 6 | 0 | 0 | $\chi_I^{(M=0)}(\hat{R})$ |
| $A_2 + E$ | 3 | 0 | −1 | $\chi_{II}^{(M=0)}(\hat{R})$ |
| $A_2 + E$ | 3 | 0 | −1 | $\chi_{III}^{(M=0)}(\hat{R})$ |

subsets are the following

$$\begin{aligned}
&\text{I:} \quad |a_{1/2}(\tfrac{1}{2})b_{1/2}(\tfrac{1}{2})c_0(1); \\
&\text{II:} \quad |a_{-1/2}(\tfrac{1}{2})b_{1/2}(\tfrac{1}{2})c_1(1)|, |a_{1/2}(\tfrac{1}{2})b_{-1/2}(\tfrac{1}{2})c_1(1)|.
\end{aligned} \tag{13.130}$$

Decomposing the nine-dimensional representation $\Gamma^{(M=1)}$ we find

$$\Gamma^{(M=1)} = A_1 + 2A_2 + 3E,$$

so that

$$\Gamma^{(S=1)} = \Gamma^{(M=1)} - \Gamma^{(S=2)} = A_1 + A_2 + 2E.$$

Finally, for $M = 0$ we have three independent subsets with c-localized states:

$$\begin{aligned}
&\text{I:} \quad |a_{-1/2}(\tfrac{1}{2})b_{1/2}(\tfrac{1}{2})c_0(1)|, \quad |a_{1/2}(\tfrac{1}{2})b_{-1/2}(\tfrac{1}{2})c_0(1)|; \\
&\text{II:} \quad |a_{1/2}(\tfrac{1}{2})b_{1/2}(\tfrac{1}{2})c_{-1}(1); \\
&\text{III:} \quad |a_{-1/2}(\tfrac{1}{2})b_{-1/2}(\tfrac{1}{2})c_1(1).
\end{aligned} \tag{13.131}$$

The corresponding characters are given in Table 13.11 and we find

$$\begin{aligned}
\Gamma^{(M=0)} &= A_1 + 3A_2 + 4E, \\
\Gamma^{(S=0)} &= \Gamma^{(M=0)} - \Gamma^{(S=1)} - \Gamma^{(S=2)} = A_1 + E.
\end{aligned}$$

The results of group-theoretical classification for $d^1 - d^1 - d^2$ trimer can be presented as follows:

$$3D^{(2)} \doteq {}^5A_2 + {}^5E, \quad 6D^{(1)} \doteq {}^3A_1 + {}^3A_2 + 2{}^3E, \quad 3D^{(0)} \doteq {}^1A_1 + {}^1E. \tag{13.132}$$

We can conclude that the $S\Gamma$-basis diagonalizes the full Hamiltonian, twice-repeated ${}^2E$ terms are mixed by double exchange.

**Example 13.11** Consider a tetrameric $d^1 - d^1 - d^1 - d^2$ tetrahedral ($\mathbf{T}_d$) cluster. This system plays the role of a model of mixed-valence four-iron ferredoxines consisting of three Fe(II) and one Fe(III) in oxidized form $[Fe_4S_4]^{n+}$ ($n = 1, 2, 3$) [101]. The spin coupling scheme for four spins $s_1 = s_2 = s_3 = \frac{1}{2}$, $s_4 = 1$ with allowance for the existence of four localizations (a, b, c and d) of an "extra" electron can be presented as follows:

$$\begin{aligned}
4(D^{(1/2)} \times D^{(1/2)} \times D^{(1/2)} \times D^{(1)}) &= 4[(2D^{(1/2)} + D^{(3/2)}) \times D^{(1)}] \\
&= 12D^{(1/2)} + 12D^{(3/2)} + 4D^{(5/2)}.
\end{aligned} \tag{13.133}$$

For maximum projection of full spin $M = S = \frac{5}{2}$ we have a four-dimensional representation with the Heitler–London basis set formed by the following fifth-order determinants:

$$|a_{1/2}\, b_{1/2}\, c_{1/2}\, d_1|, \quad |a_{1/2}\, b_{1/2}\, c_1\, d_{1/2}|,$$
$$|a_{1/2}\, b_1\, c_{1/2}\, d_{1/2}|, \quad |a_1\, b_{1/2}\, c_{1/2}\, d_{1/2}|. \qquad (13.134)$$

The basis of the twelve-dimensional representation $D^{(M=3/2)}$ ($S = \frac{5}{2}$ and $\frac{3}{2}$) consists of two independent subsets:

$$\text{I:} \quad |a_{1/2}\, b_{1/2}\, c_{1/2}\, d_0|;$$
$$\text{II:} \quad |a_{-1/2}\, b_{1/2}\, c_{1/2}\, d_1|, \quad |a_{1/2}\, b_{-1/2}\, c_{1/2}\, d_1|, \quad |a_{1/2}\, b_{1/2}\, c_{-1/2}\, d_1|, \qquad (13.135)$$

where only d-localized states are presented. Finally, d-localized functions belonging to the basis of the twelve-dimensional representation $D^{(M=1/2)}$ are the following:

$$\text{I:} \quad |a_{-1/2}\, b_{1/2}\, c_{1/2}\, d_0|, \quad |a_{1/2}\, b_{-1/2}\, c_{1/2}\, d_0|, \quad |a_{1/2}\, b_{1/2}\, c_{-1/2}\, d_0|;$$
$$\text{II:} \quad |a_{1/2}\, b_{1/2}\, c_{1/2}\, d_{-1}|;$$
$$\text{III:} \quad |a_{-1/2}\, b_{-1/2}\, c_{1/2}\, d_1|, \quad |a_{-1/2}\, b_{1/2}\, c_{-1/2}\, d_1|, \quad |a_{1/2}\, b_{-1/2}\, c_{-1/2}\, d_1|.$$

The characters of the reducible representations $D^{(M)}$ are given in Table 13.12. From Table 13.12 we find that

$$\Gamma^{(S=\frac{5}{2})} = A_2 + T_1,$$
$$\Gamma^{(S=\frac{3}{2})} = \Gamma^{(M=\frac{3}{2})} - \Gamma^{(S=\frac{5}{2})} = A_2 + E + 2T_1 + T_2,$$
$$\Gamma^{(S=\frac{1}{2})} = \Gamma^{(M=\frac{1}{2})} - \Gamma^{(S=\frac{5}{2})} - \Gamma^{(S=\frac{3}{2})} - A_2 + E + 2T_1 + T_2.$$

The results of the group-theoretical classification for the mixed-valence $d^1 - d^1 - d^1 - d^2$-tetrameric cluster can be expressed as follows:

$$4D^{(5/2)} \doteq {}^6A_2 + {}^6T_1,$$
$$12D^{(3/2)} \doteq {}^4A_2 + {}^4E + 2\,{}^4T_1 + {}^4T_2,$$
$$12D^{(1/2)} \doteq {}^2A_2 + {}^2E + 2\,{}^2T_1 + {}^2T_2. \qquad (12.137)$$

**Table 13.12** Characters of transformations for the basis functions of $d^1 - d^1 - d^1 - d^2$-mixed-valence tetramer.

| $T_d$ | $\hat{E}$ | $3\hat{C}_2$ | $8\hat{C}_3$ | $6\hat{S}_4$ | $6\hat{\sigma}_d$ | |
|---|---|---|---|---|---|---|
| $A_2 + T_1$ | 4 | 0 | 1 | 0 | −2 | $\chi^{(M=\frac{5}{2})}(\hat{R})$ |
| $A_2 + T_1$ | 4 | 0 | 1 | 0 | −2 | $\chi_{I}^{(M=\frac{3}{2})}(\hat{R})$ |
| $A_2 + E + 2T_1 + T_2$ | 12 | 0 | 0 | 0 | −2 | $\chi_{II}^{(M=\frac{3}{2})}(\hat{R})$ |
| $A_2 + E + 2T_1 + T_2$ | 12 | 0 | 0 | 0 | −2 | $\chi_{I}^{(M=\frac{1}{2})}(\hat{R})$ |
| $A_2 + T_1$ | 4 | 0 | 1 | 0 | −2 | $\chi_{II}^{(M=\frac{1}{2})}(\hat{R})$ |
| $A_2 + E + 2T_1 + T_2$ | 12 | 0 | 0 | 0 | −2 | $\chi_{III}^{(M=\frac{1}{2})}(\hat{R})$ |

The energy â levels of the tetrahedral $d^1 - d^1 - d^1 - d^2$-system are found in [102], including the $2 \times 2$-matrices of double exchange for twice-repeated $2\,{}^4T_1$ and $2\,{}^2T_1$ states.

The $S\Gamma$-terms for the case of a distorted tetrahedron can be obtained from (12.137) by reducing of the irreducible representations. The cases of rhombic $\mathbf{D}_{2d}$ and trigonal $\mathbf{C}_{3v}$ symmetries are considered in [103, 104]. Group $\mathbf{D}_{2d}$ is on the one side a subgroup of the $\mathbf{D}_{4h}$ group and, on the other side, a subgroup of the $\mathbf{T}_d$ group. This makes it possible to treat the irreducible representations of $\mathbf{D}_{2d}$ as a result of the reduction of irreducible representations of $\mathbf{T}_d$ and $\mathbf{D}_{4h}$ groups as follows:

$$\begin{array}{lccccccccc}
\mathbf{T}_d & : & A_1 & A_2 & \multicolumn{2}{c}{E} & \multicolumn{2}{c}{T_1} & \multicolumn{2}{c}{T_2} \\
 & & \downarrow & \downarrow & \swarrow & \searrow & \swarrow & \searrow & \swarrow & \searrow \\
\mathbf{D}_{2d} & : & A_1 & B_1 & A_1 & B_1 & A_2 & E & B_2 & E \\
 & & \uparrow & \uparrow & \uparrow & \uparrow & \uparrow & \uparrow & \uparrow & \uparrow \\
\mathbf{D}_{4h} & : & A_{1g} & B_{1g} & A_{1g} & B_{1g} & A_{2g} & E_u & B_{2g} & E_u
\end{array}$$

Taking into account the result (13.137) one can derive the following $S\Gamma$-states for a square-planar $d^1 - d^1 - d^1 - d^2$ mixed-valence cluster [102]:

$$\begin{aligned}
4D^{(5/2)} &\doteq {}^6A_{1g} + {}^6B_{1g} + {}^6E_u, \\
12D^{(3/2)} &\doteq 2\,{}^4A_{2g} + 2\,{}^4B_{1g} + {}^4A_{1g} + {}^4B_{2g} + 3\,{}^4E_u, \\
12D^{(1/2)} &\doteq 2\,{}^2A_{2g} + 2\,{}^2B_{1g} + {}^2A_{1g} + {}^2B_{2g} + 3\,{}^2E_u.
\end{aligned} \tag{13.138}$$

**Example 13.12** Consider $d^2 - d^2 - d^2 - d^1$-terameric tetrahedral ($\mathbf{T}_d$) mixed-valence cluster [105]. As in the previous case we consider high-spin states of $d^2$-ions ($s_1 = s_2 = s_3 = 1$, $s_4 = \frac{1}{2}$) and hence for the "hole"-type $d^2 - d^2 - d^2 - d^1$ tetramer we find the following spin-states:

$$4(D^{(1)} \times D^{(1)} \times D^{(1)} \times D^{(1/2)}) = 4D^{(7/2)} + 12D^{(5/2)} + 20D^{(3/2)} + 16D^{(1/2)}. \tag{13.139}$$

Four seventh-order determinants

$$|a_{1/2} b_1 c_1 d_1|, \quad |a_1 b_{1/2} c_1 d_1|, \quad |a_1 b_1 c_{1/2} d_1|, \quad |a_1 b_1 c_1 d_{1/2}| \tag{13.140}$$

form the basis of the $D^{(M=\frac{7}{2})}$ representation. 16.determinants with $M = \frac{5}{2}$ belong to states with $S = \frac{7}{2}$ and $\frac{5}{2}$ and form two subsets. For d-localization of the "extra" hole we obtain:

$$\begin{aligned}
&\text{I:} \quad |a_1 b_1 c_1 d_{-1/2}|; \\
&\text{II:} \quad |a_1 b_1 c_0 d_{1/2}|, \quad |a_1 b_0 c_1 d_{1/2}|, \quad |a_0 b_1 c_1 d_{1/2}|.
\end{aligned} \tag{13.141}$$

36 basis functions of the $D^{(M=\frac{3}{2})}$ representation are formed by three subsets with the

d-localized representatives:

$$
\begin{aligned}
&\text{I:} && |a_1 b_1 c_{-1} d_{1/2}|, && |a_1 b_{-1} c_1 d_{1/2}|, && |a_{-1} b_1 c_1 d_{1/2}|; \\
&\text{II:} && |a_1 b_0 c_0 d_{1/2}|, && |a_0 b_1 c_0 d_{1/2}|, && |a_0 b_0 c_1 d_{1/2}|; \\
&\text{III:} && |a_1 b_1 c_0 d_{-1/2}|, && |a_1 b_0 c_1 d_{-1/2}|, && |a_0 b_1 c_1 d_{-1/2}|;
\end{aligned}
\tag{13.142}
$$

Finally, 52 $M = \frac{1}{2}$-determinants can be divided into four subsets, the d-localized states are the following:

$$
\begin{aligned}
&\text{I:} && |a_0 b_0 c_0 d_{1/2}|; \\
&\text{II:} && |a_1 b_1 c_{-1} d_{-1/2}|, && |a_1 b_{-1} c_1 d_{-1/2}|, && |a_{-1} b_1 c_1 d_{-1/2}|; \\
&\text{III:} && |a_1 b_0 c_0 d_{-1/2}|, && |a_0 b_1 c_0 d_{-1/2}|, && |a_0 b_0 c_1 d_{-1/2}|; \\
&\text{IV:} && |a_1 b_{-1} c_0 d_{1/2}|, && |a_1 b_0 c_{-1} d_{1/2}|, && |a_0 b_{-1} c_1 d_{1/2}|, \\
& && |a_{-1} b_1 c_0 d_{1/2}|, && |a_{-1} b_0 c_1 d_{1/2}|, && |a_0 b_1 c_{-1} d_{1/2}|.
\end{aligned}
\tag{13.143}
$$

While calculating the characters of $\mathbf{T}_d$ group operations one must take into account the fact that any transpositions of the two-electron states $a_{m_a}(1)$, $b_{m_b}(1)$ and $c_{m_c}(1)$ do not change the sign of the determinant. The result of the group-theoretical classification is the following:

$$
\begin{aligned}
&4D^{(7/2)} \doteq {}^8A_1 + {}^8T_2, \\
&12D^{(5/2)} \doteq {}^6A_1 + {}^6E + {}^6T_1 + 2\,{}^6T_2, \\
&20D^{(3/2)} \doteq {}^4A_1 + 2\,{}^4E + 2\,{}^4T_1 + 3\,{}^4T_2, \\
&16D^{(1/2)} \doteq {}^2A_1 + {}^2A_2 + {}^2E + 2\,{}^2T_1 + 2\,{}^2T_2.
\end{aligned}
\tag{13.144}
$$

Matrices of the double exchange for the repeated $S\Gamma$-terms are given in [105].

The group-theoretical approach presented here for the exchange and mixed valence clusters can be easily applied to the systems with the partially delocalized electron. These systems consist of the exchange moiety involving localized electrons (ions with the fixed oxidation degrees) and mixed valence moiety, the last involves an "extra" electron sharing among several ions with localized electrons. In order to derive the $S\Gamma$-multiplets of such type systems one can combine the results obtained for the exchange and mixed-valence clusters.

**Example 13.13.** Polyoxovanadate $K_6\ [H_3KV_{12}As_3O_{39}(AsO_4)]\cdot 8H_2O$ contains a localized trimeric $V_3$(IV) unit and the mixed valence unit consisting of one electron moving among three V(V) ions (Fig. 13.15) [106]. This cluster can be referred to as the system with the partial delocalized electron. Exchange $d^1 - d^1 - d^1$-trimer (a, b, c-sites) and mixed-valence $d^1 - d^0 - d^0$-cluster ($\alpha, \beta, \gamma$-sites) are placed in parallel planes, so

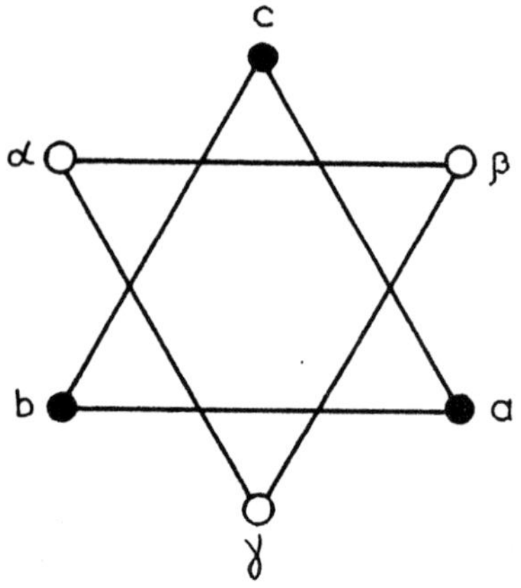


**Fig. 13.15** Scheme of magnetic sites of $[H_3KV_{12}As_3O_{39}(AsD_4)]^{6+}$ (•, localized), (o, delocalized) [106].

that the symmetry group of each moiety is $\mathbf{D}_{3h}$ ("site-symmetry") and the overall symmetry is $\mathbf{C}_{3v}$.

The local crystal field acting on each site is supposed to remove the degeneracy of d-orbitals. The basis set of the system is formed by $3 \cdot 2^4 = 48$ determinants $|a_{m_a} b_{m_b} c_{m_c} \alpha_{m_\alpha}|$, etc. Applying the operation of $\mathbf{C}_{3v}$ to this basis set one can find $S\Gamma$-multiplets of the system with partial delocalization. For example, for maximum $M = S = 2$ we have the three-dimensional basis

$$|a_{1/2} b_{1/2} c_{1/2} \alpha_{1/2}|, \quad |a_{1/2} b_{1/2} c_{1/2} \beta_{1/2}|, \quad |a_{1/2} b_{1/2} c_{1/2} \gamma_{1/2}|, \tag{13.145}$$

with $\Gamma^{(M=2)} = \Gamma^{(S=2)} = A_2 + E$ and, hence, we find ${}^5A_2$ and ${}^5E$ multiplets. In the case under consideration the procedure can be simplified. In fact, $\mathbf{C}_{3v}$ (full symmetry group) is a subgroup of $\mathbf{D}_{3h}$ (site-symmetries of both moieties) and $C_3$ axis and $\sigma_v$ planes are common for the exchange and mixed-valence trimers. In this case $S\Gamma$-multiplets of the whole system can be found from the decomposition of the direct products

$$D^{(S_{ex})} \times D^{(S_{mv})} \quad \text{and} \quad \Gamma_{ex} \times \Gamma_{mv}$$

where $S_{ex}\Gamma_{ex}$ and $S_{mv}\Gamma_{mv}$ are the $S\Gamma$-multiplets of the exchange and mixed valence units. As was found (Example 13.1) $S_{ex}\Gamma_{ex} = {}^2E + {}^4A_2$, while for one electron shared among three sites of a triangle $S_{mv}\Gamma_{mv} = {}^2A_1 + {}^2E$. The direct product can be symbolically expressed as

$$({}^2E + {}^4A_2) \times ({}^2A_1 + {}^2E) = {}^1A_1 + {}^1A_2 + {}^1E + {}^3A_1 + 2\,{}^3A_2 + 2\,{}^3E + {}^5A_2 + {}^5E. \tag{13.146}$$

The decomposition (13.146) classifies the $S\Gamma$-states of the four-electron system with partial delocalization. More complicated high-nuclearity vanadium clusters with partially delocalized electrons are considered in [94]. Energy levels of distorted trimeric

**Table 13.13** Group-theoretical classification and energies for exchange-resonance multiplets of a mixed-valence $d^4$–$d^4$–$d^5$ cluster [100] ($J$ is the Heisenberg exchange integral and $P$ is the transfer parameter).

| $S$ | $\Gamma(\mathbf{D}_3)$ | $\epsilon$ |
|---|---|---|
| $\frac{13}{2}$ | $A_1$ | $-48J + 2P$ |
| | E | $-48J - P$ |
| $\frac{11}{2}$ | E | $-35J + \frac{9}{5}P$ |
| | $A_1$ | $-35J - \frac{3}{5}P$ |
| | $A_2$ E | $-35J - P$ |
| $\frac{9}{2}$ | $A_1$, E | $-24J + \frac{8}{5}P$ |
| | E | $-24J - \frac{2}{5}P$ |
| | $A_1$, $A_2$ E | $-24J - P$ |
| $\frac{11}{2}$ | $A_1$, $A_2$, E | $-15J + \frac{7}{5}P$ |
| | $A_1$, E | $-15J - \frac{1}{5}P$ |
| | $A_2$, E, E | $-15J - P$ |
| $\frac{5}{2}$ | $A_1$, E, E | $-8J + \frac{6}{5}P$ |
| | $A_1$, $A_2$, E | $-8J$ |
| | $A_1$, $A_2$, E, E | $-8J - P$ |
| $\frac{3}{2}$ | $A_2$, E | $-3J + P$ |
| | $A_1$, E, E | $-3J + \frac{1}{5}P$ |
| | $A_1$, $A_2$, E | $-3J - P$ |
| $\frac{1}{2}$ | $A_1$ | $\frac{4}{5}P$ |
| | $A_2$, E | $\frac{2}{5}P$ |
| | E | $-P$ |

and tetrameric manganese and iron mixed-valence clusters and a special role for topology are considered in [107, 108]. The most complete consideration of the symmetric $d^n - d^n - d^{n\pm 1}$ is given in [100]; some results for $d^4 - d^4 - d^5$ ($\mathbf{D}_3$) are shown in Table 13.13.

## PROBLEMS

**13.1** Express the HDVV Hamiltonian of a tetrahedral cluster distorted along the $C_3$ axis ($\mathbf{C}_{3v}$) in terms of the total and intermediate spin operators and determine the eigenvalues.

**13.2** Construct the exchange Hamiltonian matrix of a trimeric cluster with $s_1 = s_2 = \frac{1}{2}$ and $s_3 = 1$ with three different exchange integrals.

**13.3** Using Clebsch–Gordan coefficients, find the spin wavefunctions of three-particle systems with $s_1 = s_2 = s_3 \equiv s = \frac{1}{2}$, (13.32), $s = 1$ and $s = \frac{3}{2}$.

**13.4** Express the two-particle matrix element in (13.71) in terms of one-particle ones.

**13.5** Prove that $\mathrm{E} \times \bar{\mathrm{E}} = \bar{\mathrm{A}}_1 + \bar{\mathrm{A}}_2 + \bar{\mathrm{E}}$ for the double group $\mathbf{D}_3'$.

**13.6** Construct the matrix of Heisenberg's Hamiltonian for tetrameric cluster with $s_i = \frac{1}{2}$ and $s_i = 1$.

**13.7** Classify the $S\Gamma$- multiplets for a tetrahedral cluster with $s_i = 1$.

**13.8** Prove that external magnetic and electric fields do not split the ground state $2\mathrm{D}^{(0)} = {}^1\mathrm{E}$ (Section 13.4.2) of tetrahedral ($\mathbf{T}_d$) exchange cluster with $s_i = \frac{1}{2}$ at first order of perturbation theory. Consider the selection rules in second order of perturbation theory.

# 14 Vibrational Spectra and Electron–Vibrational Interactions

## 14.1 NORMAL VIBRATIONS

### 14.1.1 Degrees of freedom. Normal coordinates

Atoms in molecules are bound by elastic forces and therefore undergo complex motions, which at any time determine the instantaneous configuration of the molecule. Any instantaneous displacement of an atom from its equilibrium position can be described by three Cartesian projections $\Delta X$, $\Delta Y$ and $\Delta Z$ in a coordinate system fixed to this atom; this system is called *local*. Each atom can be said to have three *degrees of freedom*. A nonlinear molecule, comprising $N$ atoms has $3N$ degrees of freedom. Among these only $3N - 6$ are related to vibrational motion, since 3 describe the translational motion of the molecule as a whole and 3 its rotation. For a linear molecule it is not possible to talk about rotational motion about its axis; therefore the number of degrees of freedom in this case is $3N - 5$.

The positions of all the atoms in a nonlinear molecule relative to their equilibrium points are thus described by $3N - 6$ variables. Instead of Cartesian displacements of the atoms $\Delta R_{\alpha i}$ (where $i$ is the atom number and $\alpha = x, y, z$ indicate the displacement projections), we introduce the linear combinations

$$Q_{\bar{\Gamma}\bar{\gamma}} = \sum_{\alpha i} U_{\alpha i} \Delta R_{\alpha i}. \tag{14.1}$$

Such linear combinations are called *symmetrized coordinates* belonging to the irreducible vibrational representation $\bar{\Gamma}$. The $Q_{\bar{\Gamma}\bar{\gamma}}$ are the basis functions of the irreducible representations $\bar{\Gamma}$ and describe the atomic displacements. For the $M_3$ complex (metal core of a trimeric cluster) $3N - 6 = 3$; the symmetrized displacements of such a molecule are illustrated in Fig. 14.1. Three

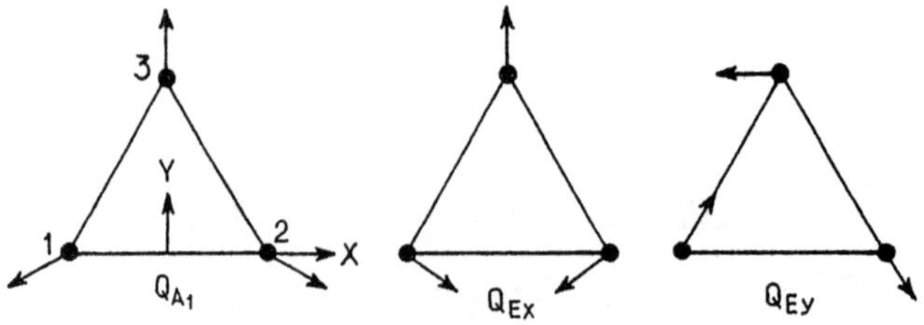


**Fig. 14.1** Symmetrized coordinates of $M_3$ complex (metal core of a trimeric cluster).

symmetrized coordinates form the basis of the reducible three-dimensional representation of $\mathbf{D}_{3h}$; this representation comprises two irreducible representations—the totally symmetric $A_1'$ ($Q_{A_1'}$) and the two-dimensional E ($Q_{Ex}$ and $Q_{Ey}$). The corresponding symmetrized displacements (in the molecular coordinate system) are of the forms

$$\left.\begin{aligned} A_1': \quad & Q_{A_1'} = \sqrt{\tfrac{1}{3}}[-\tfrac{1}{2}(\sqrt{3}X_1 + Y_1) + \tfrac{1}{2}(\sqrt{3}X_2 - Y_2) + Y_3], \\ & Q_{Ex} = \sqrt{\tfrac{1}{3}}[\tfrac{1}{2}(\sqrt{3}X_1 - Y_1) - \tfrac{1}{2}(\sqrt{3}X_2 + Y_2) + Y_3], \\ E: \quad & \\ & Q_{Ey} = \sqrt{\tfrac{1}{3}}[\tfrac{1}{2}(X_1 + \sqrt{3}Y_1) + \tfrac{1}{2}(X_2 - \sqrt{3}Y_2) - X_3], \end{aligned}\right\} \quad (14.2)$$

where $X_i$, $Y_i$ and $Z_i$ are the displacements of the $i$th ion (Fig. 14.1). In molecules and structures with a large number of atoms some irreducible vibrational representations are repeating. Thus in the molecule $BF_3$ and the ion $CO_3^{2-}$ (a regular triangle with a central atom) there are two type-E′ vibrations, to which there correspond two pairs of symmetrized displacements: $Q_{ax}$, $Q_{ay}$ and $Q_{bx}$, $Q_{by}$.

In the case of nonrepeating vibrational representations the symmetrized displacements are the normal coordinates of the molecule. The normal coordinates for repeating representations are obtained by "mixing" of the symmetrized displacements, that transform like the same basis vector. Thus for different E representations $Q_x^{(1)} = C_a^{(1)} Q_{ax} + C_b^{(1)} Q_{bx}$, $Q_x^{(2)} = C_a^{(2)} Q_{ax} + C_b^{(2)} Q_{bx}$ etc. Normal coordinates describe so-called *normal vibrations* (*normal modes*). The vibrational movement of the molecule can be represented as a set of normal vibrations; that is, of *independent harmonic oscillators*. Each normal vibration is characterized by its frequency $\omega = 2\pi\nu$ (with $\nu = 1/T$, where $T$ is the period), which is a spectroscopic characteristic of the molecule and depends on the magnitude and nature of the interatomic elastic interactions, determined by the chemical bonding.

If $\bar{\Gamma}$ is a degenerate (i.e. not unidimensional) representation then to the set of normal coordinates $Q_{\bar{\Gamma}\bar{\gamma}}$ there corresponds a single frequency $\omega_{\bar{\Gamma}}$, and the mode $\bar{\Gamma}$ is called a *degenerate mode*. The Schrödinger equation for vibrational motion is decomposed into a sum of independent equations for harmonic oscillators:

$$\left(-\frac{\hbar^2}{2M_{\bar\Gamma}}\frac{\partial^2}{\partial Q^2_{\bar\Gamma\bar\gamma}}+\frac{M_{\bar\Gamma}\omega^2_{\bar\Gamma}}{2}Q^2_{\bar\Gamma\bar\gamma}\right)\Phi_{n_{\bar\Gamma\bar\gamma}}(Q_{\bar\Gamma\bar\gamma})=\epsilon_{n_{\bar\Gamma\bar\gamma}}\Phi_{n_{\bar\Gamma\bar\gamma}}(Q_{\bar\Gamma\bar\gamma}), \qquad (14.3)$$

where $M_{\bar\Gamma}$ is the reduced mass of the vibration, $\Phi_{n_{\bar\Gamma\bar\gamma}}(Q_{\bar\Gamma\bar\gamma})$ is a harmonic-oscillator wavefunction, $\epsilon_{n_{\bar\Gamma\bar\gamma}}$ is the energy of the vibration. From quantum mechanics it is known that the energy spectrum of a harmonic oscillator depends on the quantum number $n_{\bar\Gamma\bar\gamma}=0, 1, 2, \ldots$:

$$\epsilon_{n_{\bar\Gamma\bar\gamma}}=\hbar\omega_{\bar\Gamma}(n_{\bar\Gamma\bar\gamma}+\tfrac{1}{2}). \qquad (14.4)$$

The wavefunction $\Phi_{n_{\bar\Gamma\bar\gamma}}$ is given by

$$\Phi_{n_{\bar\Gamma\bar\gamma}}=N(n_{\bar\Gamma\bar\gamma})\exp\left(-\frac{M_{\bar\Gamma}\omega_{\bar\Gamma}Q^2_{\bar\Gamma\bar\gamma}}{2\hbar}\right)H_{n_{\bar\Gamma\bar\gamma}}\left(Q_{\bar\Gamma\bar\gamma}\left(\frac{M_{\bar\Gamma}\omega_{\bar\Gamma}}{\hbar}\right)^{1/2}\right), \qquad (14.5)$$

where $N(n_{\bar\Gamma\bar\gamma})$ is a normalization factor, and $H_n(\ldots)$ are Hermite polynomials:

$$H_n(x)=(-1)^n\frac{\mathrm{d}^n}{\mathrm{d}x^n}\mathrm{e}^{-x^2},$$

$$H_0=1,\qquad H_1=2x,\qquad H_2=4x^2-2,\qquad H_3=8x^3-12x.$$

The theory of molecular vibrations is described in detail in [109, 110]. We shall discuss below only some of its aspects, related to the application of group theory.

### 14.1.2 Classification of normal vibrations

The set of $3N$ displacements of atoms in an $N$-atom molecule, whose atoms are in an equilibrium configuration, forms the basis of a $3N$-dimensional representation of the molecular point group. This reducible representation $\Gamma$ can be reduced into irreducible representations $\bar\Gamma$. After eliminating the translational and rotational motions, we obtain various types of vibrational modes. This procedure for obtaining the irreducible representations $\bar\Gamma$ corresponding to the molecular vibrations is called *classification of normal vibrations*.

Let us illustrate the classification method using the simple example of the water molecule. Three atoms have nine degrees of freedom, so that nine displacements $X_1$, $Y_1$, $Z_1$, $X_2$, $Y_2$, $Z_2$, $X_3$, $Y_3$ and $Z_3$ (Fig. 14.2) transform into each other by $\mathbf{C}_{2\mathrm{v}}$ group operations. It is easy to write down the nine-dimensional representation. In so doing, it should be borne in mind that the group operations interchange the places of the hydrogen atoms, leaving the position of the oxygen atom unchanged and transforming the local coordinate systems; for example, $\hat C_2 Z_2=Z_3$, $\hat C_2 Y_2=-Y_3$, $\hat\sigma_{\mathrm{h}}X_1=-X_1$ etc. Taking account of these simple relations, we write down the operation $\hat C_2$ in matrix

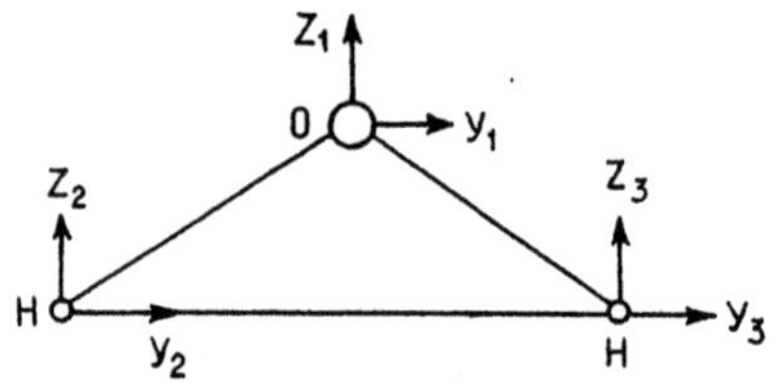


**Fig. 14.2** Local coordinate systems of a water molecule (the $X_1$, $X_2$ and $X_3$ axes are perpendicular to the molecular plane and directed upwards out of the page).

form as

$$\hat{C}_2 \begin{bmatrix} X_1 \\ Y_1 \\ Z_1 \\ X_2 \\ Y_2 \\ Z_2 \\ X_3 \\ Y_3 \\ Z_3 \end{bmatrix} = \begin{bmatrix} -1 & 0 & 0 & 0 & 0 & 0 & 0 & 0 & 0 \\ 0 & -1 & 0 & 0 & 0 & 0 & 0 & 0 & 0 \\ 0 & 0 & 1 & 0 & 0 & 0 & 0 & 0 & 0 \\ 0 & 0 & 0 & 0 & 0 & 0 & -1 & 0 & 0 \\ 0 & 0 & 0 & 0 & 0 & 0 & 0 & -1 & 0 \\ 0 & 0 & 0 & 0 & 0 & 0 & 0 & 0 & 1 \\ 0 & 0 & 0 & -1 & 0 & 0 & 0 & 0 & 0 \\ 0 & 0 & 0 & 0 & -1 & 0 & 0 & 0 & 0 \\ 0 & 0 & 0 & 0 & 0 & 1 & 0 & 0 & 0 \end{bmatrix} \begin{bmatrix} X_1 \\ Y_1 \\ Z_1 \\ X_2 \\ Y_2 \\ Z_2 \\ X_3 \\ Y_3 \\ Z_3 \end{bmatrix}. \tag{14.6}$$

In a similar way it is possible to find the matrices of all operations of the group—in other words, a nine-dimensional reducible representation, whose basis is formed by the displacement components of the three atoms. The characters of this representation are determined directly; for instance,

$$\chi(\hat{C}_2) = -1.$$

Knowing these characters, we can decompose the representation into irreducible representations.

However, it is more convenient to use the method based on site symmetry, already employed for molecular orbital classification (Sections 7.3 and 8.2). The classification of normal vibrations involves the following stages:

(1) determination of the site group $\mathbf{G}_S(i)$ of each atom $A_i$ in the molecule;

(2) calculation of the matrix characters for each atom's $X_i$, $Y_i$, $Z_i$ displacement transformations due to the atomic site-group operations;

(3) summation of the contributions of each atom to the total character;

(4) calculation of the characters from the full group of the molecule, taking into account that the characters of operations $\hat{R}$ not included in site groups are zero;

**Table 14.1** Characters of H- and O-atom displacements in site groups.

| $\mathbf{C}_s$ | $\hat{E}$ | $\hat{\sigma}_v(yz)$ |
|---|---|---|
| $\chi_{2H}$ | 6 | 2 |

| $\mathbf{C}_{2v}$ | $\hat{E}$ | $\hat{C}_2$ | $\hat{\sigma}_v(yz)$ | $\hat{\sigma}_v(xz)$ |
|---|---|---|---|---|
| $\chi_O$ | 3 | $-1$ | 1 | 1 |

(5) determination of characters related to the translational and rotational motions of the molecule as a whole;

(6) subtraction from the character of the complete representation thus obtained the characters of the translational and rotational motions (note that displacements of the whole molecule along the $x$, $y$ and $z$ axes are transformed in translational motion as components of a polar vector, while in rotational motion they are transformed as components of an axial vector);

(7) reduction of the irreducible representation thus obtained gives the types $\bar{\Gamma}$ of normal vibrational modes.

**Example 14.1** Consider the water molecule: the site symmetry of the hydrogen atoms is $\mathbf{C}_s$ and that of the oxygen atom is $\mathbf{C}_{2v}$ (the full group of the molecule). The operation $\hat{\sigma}_v \equiv \hat{\sigma}(yz)$ (Fig. 14.2) reverses the sign of $X_2$, leaving $Y_2$ and $Z_2$ unchanged; therefore in the site group of one H atom the matrix of this operation is

$$\mathbf{D}(\hat{\sigma}_v) = \begin{bmatrix} -1 & 0 & 0 \\ 0 & 1 & 0 \\ 0 & 0 & 1 \end{bmatrix}.$$

The character $\chi_H(\hat{\sigma}_v) = 1$. Table 14.1 gives the site-group characters for the two H atoms and the O atom. The passage from $\mathbf{C}_s$ to $\mathbf{C}_{2v}$ for the hydrogen atoms provides the characters for the whole molecule (see Table 14.2).

Let us now use the rule giving the characters of translational and rotational motions. From the character table of $\mathbf{C}_{2v}$ we find that the polar vector components form the basis of three representations $A_1(z)$, $B_1(x)$ and $B_2(y)$, while the axial vector components form the basis of $A_2(R_z)$, $B_1(R_y)$ and $B_2(R_x)$. Subtracting from the total character $\chi_{H_2O}$ the characters of these irreducible representations, we find the character of the vibrational

**Table 14.2** Displacement characters of the $H_2O$ molecule.

| $\mathbf{C}_{2v}$ | $\hat{E}$ | $\hat{C}_2$ | $\hat{\sigma}_v(yz)$ | $\hat{\sigma}_v(xz)$ |
|---|---|---|---|---|
| $\chi_O$ | 3 | $-1$ | 1 | 1 |
| $\chi_{2H}$ | 6 | 0 | 2 | 0 |
| $\chi_{H_2O}$ | 9 | $-1$ | 3 | 0 |

**Table 14.3** Characters of the vibrational representation of the $H_2O$ molecule.

| $\mathbf{C}_{2v}$ | $\hat{E}$ | $\hat{C}_2$ | $\hat{\sigma}_v(yz)$ | $\hat{\sigma}_v(xz)$ |
|---|---|---|---|---|
| $\chi_{H_2O}$ | 9 | $-1$ | 3 | 0 |
| $A_1(z)$ | 1 | 1 | 1 | 1 |
| $B_1(x)$ | 1 | $-1$ | $-1$ | 1 |
| $B_2(y)$ | 1 | $-1$ | 1 | $-1$ |
| $A_2(R_z)$ | 1 | 1 | $-1$ | $-1$ |
| $B_1(R_y)$ | 1 | $-1$ | $-1$ | 1 |
| $B_2(R_z)$ | 1 | $-1$ | 1 | $-1$ |
| $\chi^{(vib)}_{H_2O}$ | 3 | 1 | 1 | 3 |

representation $\chi^{(vib)}_{H_2O}$ (see Table 14.3). The reducible representation $\chi^{(vib)}_{H_2O}$ is reduced as usual according to (4.12):

$$\Gamma^{(vib)}_{H_2O} \equiv \bar{\Gamma} = 2A_1 + B_2.$$

This means that the water molecule has two totally symmetric ($A_1$) normal modes and one of $B_2$ type.

### 14.1.3 Construction of normal coordinates

The method described in Section 14.1.2 makes it possible to classify the normal vibrations without knowing the normal coordinates themselves. Since the symmetrized displacements and normal coordinates are the basis functions of the irreducible representations of the point group, they may be constructed using the projection operators. In the present case we may choose any displacement (e.g. $X_1$) of any (say the $i$th) atom, applying to it the projection operator $\hat{P}^{\bar{\Gamma}}_{\bar{\gamma}}$ (Section 6.3.2). The operator $\hat{P}^{\bar{\Gamma}}_{\bar{\gamma}}$ projects the space of all displacements $X_i$, $Y_i$, $Z_i$ onto the symmetrized displacement $Q_{\bar{\Gamma}\bar{\gamma}}$:

$$\hat{P}^{\bar{\Gamma}}_{\bar{\gamma}} X_i = Q_{\bar{\Gamma}\bar{\gamma}}. \tag{14.7}$$

If the application of $\hat{P}^{\bar{\Gamma}}_{\bar{\gamma}}$ to $X_i$ gives zero, this displacement is not contained in the symmetrized coordinate $Q_{\bar{\Gamma}\bar{\gamma}}$; to obtain this coordinate, it is necessary to take another displacement. Such calculations are explained in detail in many textbooks; therefore we restrict ourselves here to the above remarks. As an example, we take the normal coordinates of an octahedral ($\mathbf{O}_h$) complex. Classification of normal vibrations leads to the result

$$\bar{\Gamma} = A_{1g} + E_g + T_{2g} + 2T_{1u} + T_{2u}.$$

Thus the octahedral complex is characterized by three types of even vibrations and three types of odd. Translational and rotational coordinates form the bases of $T_{1u}$ (polar vector) and $T_{1g}$ (axial vector) respectively. Figure 14.3 shows the displacements. The atomic displacements corresponding to normal coordinates of even type are

$$Q_{A_{1g}} = \sqrt{\tfrac{1}{6}}(Y_2 - Y_5 + Z_3 - Z_6 + X_1 - X_4) \quad (A_{1g}),$$

$$\left.\begin{aligned} Q_{E_g u} &= \tfrac{1}{2}\sqrt{\tfrac{1}{3}}(2Z_3 - 2Z_6 - X_1 + X_4 - Y_2 + Y_5) \\ Q_{E_g v} &= \tfrac{1}{2}(X_1 - X_4 - Y_2 + Y_5) \end{aligned}\right\} \quad (E_g),$$

$$\left.\begin{aligned} Q_{T_2\xi} &= \tfrac{1}{2}(Z_2 - Z_5 + Y_3 - Y_6), \\ Q_{T_2\eta} &= \tfrac{1}{2}(X_3 - X_6 - Z_1 - Z_4), \\ Q_{T_2\zeta} &= \tfrac{1}{2}(Y_1 - Y_4 + X_2 - X_5) \end{aligned}\right\} \quad (T_{2g}).$$

It should be stressed that the projection operator makes it possible to construct symmetrized displacements, which are the normal coordinates only for nonrepeating vibrational representations. The normal coordinates for

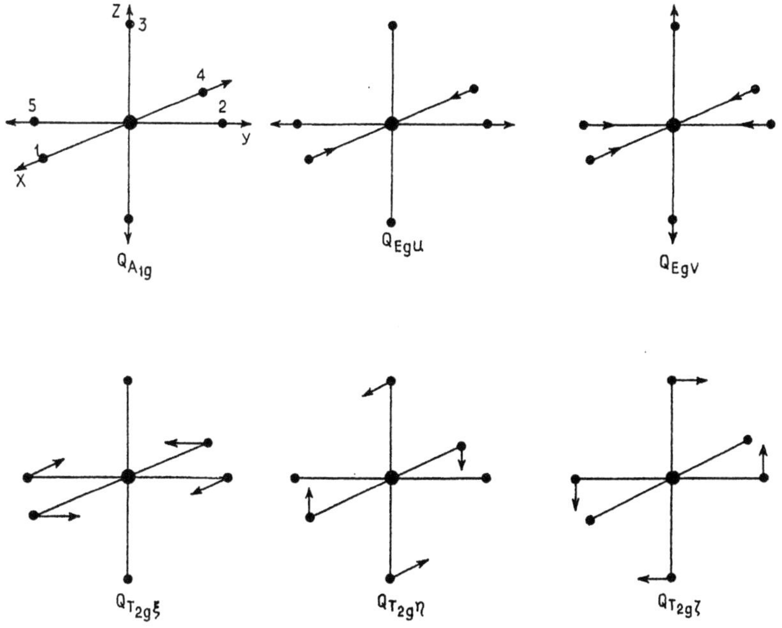


**Fig. 14.3** Even normal displacements of an octahedral complex.

repeating $\bar{\Gamma}$ ($2T_{1u}$ in the above examples) are determined by the solution of the dynamical vibrational problem.

## 14.2 SELECTION RULES FOR INFRARED ABSORPTION AND COMBINATION LIGHT SCATTERING

The intervals between the electron energy levels of complexes and molecules are usually far greater than the energy $\hbar\omega_{\bar{\Gamma}}$ of vibrational quanta. Therefore if the external radiation frequency comes into resonance with one of the frequencies $\omega_{\bar{\Gamma}}$, the molecule does not change its ground electronic state—it changes only the energy of the vibration. Since vibrational energies correspond to the infrared spectral region, there is infrared (IR) light absorption. The intensity of IR absorption by a type-$\bar{\Gamma}$ vibration is given by matrix elements of the form

$$M_\alpha(n_{\bar{\Gamma}\bar{\gamma}} \rightarrow n'_{\bar{\Gamma}\bar{\gamma}}) = \int_{-\infty}^{\infty} \Phi_{n'_{\bar{\Gamma}\bar{\gamma}}}(Q_{\bar{\Gamma}\bar{\gamma}})\hat{P}_\alpha \Phi_{n_{\bar{\Gamma}\bar{\gamma}}}(Q_{\bar{\Gamma}\bar{\gamma}})\,\mathrm{d}Q_{\bar{\Gamma}\bar{\gamma}}, \tag{14.8}$$

where $\hat{P}_\alpha$ ($\alpha = x, y, z$) are the components of the dipole moment related to atom displacements in the molecular coordinate system. At low temperatures $k_B T < \hbar\omega$ nearly all molecules are in the ground vibrational state $n_{\bar{\Gamma}\bar{\gamma}} = 0$. IR absorption is therefore the result of the transition to the first vibrational level $(0 \rightarrow 1)$. To analyse the selection rules for IR absorption, it is necessary to know the transformation properties of the harmonic-oscillator wavefunctions $\Phi_{n_{\bar{\Gamma}\bar{\gamma}}}(Q_{\bar{\Gamma}\bar{\gamma}})$ for $n_{\bar{\Gamma}\bar{\gamma}} = 0$ and 1. Consider (14.4) and (14.5), yielding for a nondegenerate mode

$$\left.\begin{aligned} \Phi_0(Q) &= N_0 \exp\left(-\frac{M\omega}{2\hbar}Q^2\right), \\ \Phi_1(Q) &= N_1 2Q \exp\left(-\frac{M\omega}{2\hbar}Q^2\right). \end{aligned}\right\} \tag{14.9}$$

The exponential function here is totally symmetric, since it is unchanged by the substitution $Q \rightarrow -Q$. Therefore the wavefunction $\Phi_0 (n = 0)$ describing the ground state of the harmonic oscillator for nondegenerate vibrations is transformed according to the totally symmetric representation. The wavefunction $\Phi_1 (n = 1)$ is transformed according to the same representation as the normal coordinate $Q$.

Let us now consider a degenerate vibration; for example $T_{1u}$, i.e. the vibration of an octahedral complex with normal coordinates $Q_x$, $Q_y$, $Q_z$ (the basis of $T_{1u}$) and frequency $\omega_{T_{1u}} = \omega$. The energy of such a vibration,

$$\epsilon(n_x, n_y, n_z) = \hbar\omega(n_x + n_y + n_z + \tfrac{3}{2}) \tag{14.10}$$

is determined by three quantum numbers $n_x$, $n_y$ and $n_z$, while the wavefunctions, by virtue of the independence of the normal vibrations, are written as

$$\Phi_{n_x n_y n_z}(Q_x, Q_y, Q_z) = N(n_x)N(n_y)N(n_z)\Phi_{n_x}(Q_x)\Phi_{n_y}(Q_y)\Phi_{n_z}(Q_z). \quad (14.11)$$

For the ground state

$$\Phi_{000} = N_0^3 \exp\left[-\frac{M\omega}{2\hbar}(Q_x^2 + Q_y^2 + Q_z^2)\right]. \quad (14.12)$$

The first excited state, with energy $\hbar\omega$, is threefold-degenerate, since it is realized by populating the first excited state of one of the three $T_{1u}$ vibrations

$$n_x = 1,\ n_y = n_z = 0; \qquad n_x = n_z = 0,\ n_y = 1; \qquad n_x = n_y = 0,\ n_z = 1.$$

The three wavefunctions of this degenerate vibration are

$$\left.\begin{aligned}\Phi_{100} &= N_1 N_0^2 2Q_x \exp\left[-\frac{M\omega}{2\hbar}(Q_x^2 + Q_y^2 + Q_z^2)\right],\\ \Phi_{010} &= N_1 N_0^2 2Q_y \exp\left[-\frac{M\omega}{2\hbar}(Q_x^2 + Q_y^2 + Q_z^2)\right],\\ \Phi_{001} &= N_1 N_0^2 2Q_z \exp\left[-\frac{M\omega}{2\hbar}(Q_x^2 + Q_y^2 + Q_z^2)\right].\end{aligned}\right\} \quad (14.13)$$

The expression $Q_x^2 + Q_y^2 + Q_z^2$ in the exponent is a scalar product:

$$\begin{aligned}\varphi_{A_{1g}} &= \sum_{\bar{\gamma}} Q_{T_{1u}\bar{\gamma}} Q_{T_{1u}\bar{\gamma}} \langle T_{1u}\bar{\gamma}\, T_{1u}\bar{\gamma} \mid A_{1g}\rangle\\ &\equiv \sqrt{\tfrac{1}{3}}(Q_x^2 + Q_y^2 + Q_z^2),\end{aligned} \quad (14.14)$$

and hence is invariant under all point-group operations. Therefore the wavefunction of the ground vibrational level is also transformed according the totally symmetric representation. In the case of a degenerate vibration, the three functions $\Phi_{100}$, $\Phi_{010}$ and $\Phi_{001}$ are transformed as three coordinates; that is, according to the irreducible representation $T_{1u}$. In the general case the wavefunctions of the first excited state of the mode $\bar{\Gamma}$ form the basis of the irreducible representation $\bar{\Gamma}$.

We can now return to the selection rules for IR absorption; these result from the form of the matrix element (14.8) and the conclusions concerning the characteristics of the harmonic-oscillator wavefunctions. Since $\hat{P}_\alpha$ are vector components, *an IR transition is allowed if the direct product $\bar{\Gamma} \times \Gamma(\mathbf{P})$ includes the totally symmetric representation* $A_{1g}$. In other words, *IR absorption is possible for the mode $\bar{\Gamma}$ if $\bar{\Gamma}$ is a representation according to which vector components are transformed.* In particular, this means that in systems having an inversion centre IR absorption can occur only for odd vibrations. In the present case of a $T_{1u}$ mode in an $\mathbf{O}_h$ system the transition $0 \rightarrow 1$ is allowed. It is then said that the $T_{1u}$ mode is *active in IR absorption*. The mode $T_{2u}$ (when odd) is not active in IR absorption.

In the example considered in Section 14.1.2 of the $H_2O$ molecule all vibrational modes $2A_1$ and $B_2$ are active in IR absorption. Indeed, the components of the dipole moment $\hat{P}_z$ and $\hat{P}_y$ are transformed according to the irreducible representations $A_1$ and $B_2$ respectively (Table 14.3), which is just what the selection rules require. It also becomes obvious that for $z$ polarization (Fig. 14.2) the $A_1$ modes are active, while for $y$ polarization the $B_2$ mode is active.

*Raman scattering* of light is a process during which a photon scattered by a molecule increases or reduces its energy by an amount equal to the energy $\hbar\omega$ of the molecular vibrational quanta. During this scattering the electronic state of the molecule does not change, and in the vibrational subsystem the transition $0 \to 1$ or $1 \to 0$, (14.4), takes place. The selection rules for Raman scattering are easy to obtain if in (14.8) the vector components $\hat{P}_\alpha$ are replaced by the components of the polarizability tensor $\hat{\mathscr{P}}_{\alpha\beta} \sim \hat{P}_\alpha \hat{P}_\beta$. The six independent quantities $\hat{\mathscr{P}}_{xx}$, $\hat{\mathscr{P}}_{yy}$, $\hat{\mathscr{P}}_{zz}$, $\hat{\mathscr{P}}_{xy}$, $\hat{\mathscr{P}}_{xz}$ and $\hat{\mathscr{P}}_{yz}$ are transformed according to the $\bar{\Gamma}$ representations contained in the symmetric square $[\Gamma^2(\hat{\mathbf{P}})]$ (Section 11.3.2) of the vector representation $\bar{\Gamma}$ (the antisymmetric part is zero since $\hat{\mathscr{P}}_{\alpha\beta} = \hat{\mathscr{P}}_{\beta\alpha}$). This gives the following selection rule: *a type-$\bar{\Gamma}$ vibration is active in Raman scattering if the direct product* $[\Gamma^2(\hat{\mathbf{P}})] \times \bar{\Gamma}$ *includes a totally symmetric representation; that is, if the vibrational representation $\bar{\Gamma}$ is contained in* $[\Gamma^2(\hat{\mathbf{P}})]$. From this it follows, in particular, that in centrosymmetric systems the odd vibrations are not active in Raman scattering—only even vibrations can be active. For example, for an octahedral complex $\Gamma(\hat{\mathbf{P}}) = T_{1u}$ and $[T_{1u}^2] = A_{1g} + E_g + T_{2g}$. Hence the $A_{1g}$, $E_g$ and $T_{2g}$ modes, i.e. all types of even vibrations, are active in Raman scattering.

In systems without an inversion centre the vibrational states are not characterized by a particular parity, and a mode may be active in both IR absorption and Raman scattering.

## 14.3 ELECTRON–VIBRATIONAL INTERACTIONS

Electron–vibrational interactions determine a variety of spectral properties of molecules and crystals [111–113], and play important roles in structural phase transitions [114] and many other effects in chemistry [115]. Here we shall touch on only a few problems that are related to the application of symmetry theory.

Let $W(\mathbf{r}, \mathbf{R})$ be the potential energy of interaction between valence electrons and atomic cores. As before, $\mathbf{r}$ is the set of electronic coordinates and $\mathbf{R}$ the set of nuclear coordinates, describing an arbitrary, not equilibrium, configuration. Let us expand $W(\mathbf{r}, \mathbf{R})$ in terms of the symmetrized displacements $Q_{\bar{\Gamma}\bar{\gamma}}$, measured from the symmetric configuration coordinates $\mathbf{R}_0$:

$$W(\mathbf{r}, \mathbf{R}) = W(\mathbf{r}, \mathbf{R}_0) + \sum_{\bar{\Gamma}\bar{\gamma}} \left( \frac{\partial W(\mathbf{r}, \mathbf{R})}{\partial Q_{\bar{\Gamma}\bar{\gamma}}} \right)_{Q_{\bar{\Gamma}\bar{\gamma}}=0} Q_{\bar{\Gamma}\bar{\gamma}}. \tag{14.15}$$

Here $W(\mathbf{r}, \mathbf{R}_0)$ is the static field of the atom cores, while the second term is the energy of interaction of electrons with nuclear displacements—the linear (relative to atomic displacements) *electron–vibrational (vibronic) interaction.* The quantity

$$V_{\bar{\Gamma}\bar{\gamma}}(\mathbf{r}) = \left( \frac{\partial W(\mathbf{r}, \mathbf{R})}{\partial Q_{\bar{\Gamma}\bar{\gamma}}} \right)_{Q_{\bar{\Gamma}\bar{\gamma}}=0} \tag{14.16}$$

transforms like the basis function $\bar{\gamma}$ of the irreducible representation $\bar{\Gamma}$; that is, like the normal coordinate $Q_{\bar{\Gamma}\bar{\gamma}}$. Thus the vibronic interaction operator

$$H_{\mathrm{eV}} = \sum_{\bar{\Gamma}\bar{\gamma}} V_{\bar{\Gamma}\bar{\gamma}}(\mathbf{r})\, Q_{\bar{\Gamma}\bar{\gamma}} \tag{14.17}$$

is a point-group scalar, i.e. it is invariant under simultaneous transformations applied to electronic variables ($\mathbf{r}$) and nuclear variables ($Q$), resulting from the group operations.

The function $V_{\bar{\Gamma}\bar{\gamma}}(\mathbf{r})$ is the derivative of the potential energy with respect to the displacement $Q_{\bar{\Gamma}\bar{\gamma}}$; in other words, it is a *generalized force*. This force acts on the atoms located at the points $\mathbf{R}_0$ ($Q_{\bar{\Gamma}\bar{\gamma}} = 0$), and its sources are the electrons with coordinates $\mathbf{r}$. The matrix element

$$\langle \Gamma\gamma | V_{\bar{\Gamma}\bar{\gamma}} | \Gamma\gamma \rangle = \int \varphi^*_{\Gamma\gamma}(\mathbf{r})\, V_{\bar{\Gamma}\bar{\gamma}}(\mathbf{r}) \varphi_{\Gamma\gamma}(\mathbf{r})\, \mathrm{d}\tau, \tag{14.18}$$

where $\varphi_{\Gamma\gamma}(\mathbf{r})$ the electronic wavefunction is calculated for the atomic configuration $\mathbf{R}_0$, gives the average force acting on the atoms of this configuration.

Continuing the expansion (14.15), we obtain the *quadratic vibronic interaction*:

$$H'_{\mathrm{eL}} = \sum_{\bar{\Gamma}_1\bar{\gamma}_1} \sum_{\bar{\Gamma}_2\bar{\gamma}_2} W_{\bar{\Gamma}_1\bar{\gamma}_1\bar{\Gamma}_2\bar{\gamma}_2}(\mathbf{r})\, Q_{\bar{\Gamma}_1\bar{\gamma}_1} Q_{\bar{\Gamma}_2\bar{\gamma}_2}. \tag{14.19}$$

The quantities

$$W_{\bar{\Gamma}_1\bar{\gamma}_1\bar{\Gamma}_2\bar{\gamma}_2}(\mathbf{r}) = \left( \frac{\partial^2 W(\mathbf{r}, \mathbf{R})}{\partial Q_{\bar{\Gamma}_1\bar{\gamma}_1} Q_{\bar{\Gamma}_2\bar{\gamma}_2}} \right)_{Q=0} \tag{14.20}$$

will transform according to some reducible representation $\bar{\Gamma}$. If $\bar{\Gamma}_1$ and $\bar{\Gamma}_2$ are modes related to different vibrational representations ($\bar{\Gamma}_1 \neq \bar{\Gamma}_2$) then $\bar{\Gamma}$ is determined from the decomposition of the direct product $\bar{\Gamma}_1 \times \bar{\Gamma}_2$. If $\bar{\Gamma}_1 = \bar{\Gamma}_2 \equiv \bar{\Gamma}$ (one vibrational representation), $\bar{\Gamma}$ is contained in the symmetric square $[\bar{\Gamma}^2]$.

Using the Clebsch–Gordan inverse expansion (5.20), the products of the type $Q_{\bar{\Gamma}_1\bar{\gamma}_1}Q_{\bar{\Gamma}_2\bar{\gamma}_2}$ and the quantities $W_{\bar{\Gamma}_1\bar{\gamma}_1\bar{\Gamma}_2\bar{\gamma}_2}(\mathbf{r})$ can be written as

$$\left.\begin{aligned} Q_{\bar{\Gamma}_1\bar{\gamma}_1}Q_{\bar{\Gamma}_2\bar{\gamma}_2} &= \sum_{\bar{\Gamma}\bar{\gamma}} \Theta_{\bar{\Gamma}\bar{\gamma}}(\bar{\Gamma}_1\bar{\Gamma}_2)\langle \bar{\Gamma}\bar{\gamma}\,|\,\bar{\Gamma}_1\bar{\gamma}_1\bar{\Gamma}_2\bar{\gamma}_2\rangle, \\ W_{\bar{\Gamma}_1\bar{\gamma}_1\bar{\Gamma}_2\bar{\gamma}_2} &= \sum_{\bar{\Gamma}\bar{\gamma}} W_{\bar{\Gamma}\bar{\gamma}}(\bar{\Gamma}_1\bar{\Gamma}_2)\langle \bar{\Gamma}\bar{\gamma}\,|\,\bar{\Gamma}_1\bar{\gamma}_1\bar{\Gamma}_2\bar{\gamma}_2\rangle. \end{aligned}\right\} \tag{14.21}$$

Substituting (14.21) into (14.19) and using the orthogonality relation (5.18) for Clebsch–Gordan coefficients, we obtain the *quadratic vibronic interaction* [116]:

$$H'_{\mathrm{eL}} = \sum_{\bar{\Gamma}\bar{\gamma}}\sum_{\bar{\Gamma}_1\bar{\Gamma}_2} W_{\bar{\Gamma}\bar{\gamma}}(\bar{\Gamma}_1\bar{\Gamma}_2)\,\Theta_{\bar{\Gamma}\bar{\gamma}}(\bar{\Gamma}_1\bar{\Gamma}_2). \tag{14.22}$$

The use of (14.22) rather than (14.19) is convenient since the operators $W_{\bar{\Gamma}\bar{\gamma}}$ and $\Theta_{\bar{\Gamma}\bar{\gamma}}$ are transformed according to the irreducible representations of the point group.

**Example 14.2** Consider the octahedral complex $ML_6$, with $\bar{\Gamma} = A_{1g}, \bar{E}_g, T_{2g}, 2T_{1u}, T_{2u}$. The quadratic interaction with the totally symmetric vibration takes the simple form

$$H'_{\mathrm{eL}} = W_{A_{1g}}(A_{1g}A_{1g})Q^2_{A_{1g}}. \tag{14.23}$$

For the $E_g$ mode $[E_g^2] \doteq A_{1g} + E_g \equiv \bar{\Gamma}$, and

$$\Theta_{E_g\bar{\gamma}} = \sum_{\bar{\gamma}_1,\bar{\gamma}_2 = u,v} Q_{E_g\bar{\gamma}_1}Q_{E_g\bar{\gamma}_2}\langle E_g\bar{\gamma}_1 E_g\bar{\gamma}_2\,|\,E_g\bar{\gamma}\rangle. \tag{14.24}$$

Using the tables in [1], we obtain

$$H'_{\mathrm{eL}} = \sqrt{\tfrac{1}{2}}W_{A_{1g}}(Q_u^2 + Q_v^2) + \sqrt{\tfrac{1}{2}}W_{E_g u}(Q_v^2 - Q_u^2) + \sqrt{2}\,W_{E_g v}Q_uQ_v. \tag{14.25}$$

Likewise it is easy to obtain the other terms of the Hamiltonian, that are quadratic in $Q$.

## 14.4 JAHN–TELLER EFFECT

### 14.4.1 Jahn–Teller theorem

Let $\epsilon_0(\Gamma)$ be an electronic term, degenerate for a certain symmetric atomic configuration $\mathbf{R}_0$, for which all displacements $Q_{\bar{\Gamma}\bar{\gamma}} = 0$. We shall find the correction to the energy $\epsilon_\Gamma(Q)$ of this term for small displacements characterized by the set of coordinates $Q_{\bar{\Gamma}\bar{\gamma}}$. To do this, we write down the secular equation for the operator of linear interaction of electrons with displacements:

$$\left| \sum_{\bar{\Gamma}\bar{\gamma}} \langle \Gamma\gamma_1 \,|\, V_{\bar{\Gamma}\bar{\gamma}} \,|\, \Gamma_2\gamma_2 \rangle Q_{\bar{\Gamma}\bar{\gamma}} - \epsilon(Q)\,\delta_{\gamma_1\gamma_2} \right| = 0, \qquad (14.26)$$

where $|\Gamma\gamma\rangle \equiv \varphi_{\Gamma\gamma}(\mathbf{r})$ is the wavefunction of the electrons in the symmetric atomic configurations (i.e. for $Q = 0$). The energy correction $\epsilon(Q)$ is nonzero if some of the matrix elements $\langle \Gamma\gamma_1 \,|\, V_{\bar{\Gamma}\bar{\gamma}} \,|\, \Gamma\gamma_2 \rangle$ are nonzero. Since it is calculated using one set of functions $\varphi_{\Gamma\gamma}(\mathbf{r})$, the matrix of the operator $V_{\bar{\Gamma}\bar{\gamma}}(\mathbf{r})$ is nonzero if $\bar{\Gamma}$ is contained in the symmetric square $[\Gamma^2]$ (not in the direct product $\Gamma \times \Gamma$). On the other hand, the fact that there are energy corrections $\epsilon(Q)$ means that there is a generalized force acting on atoms at the point $Q = 0$. Owing to the action of this force, the symmetric configuration of a many-atom system corresponding to the degenerate term $\Gamma$ is unstable. Therefore, from the point of view of group-theoretical selection rules, the determination of the stability of a symmetric atomic configuration is reduced to finding whether there is an active vibrational mode $\bar{\Gamma}$. Jahn and Teller (see [73]) showed that *a symmetric configuration of any many-atom system with a degenerate electron term is unstable.* This was done by an exhaustive examination of all crystal point groups, and calculating symmetric squares $[\Gamma^2]$. For double-valued representations $\Gamma$ the antisymmetric product $\{\Gamma^2\}$ must contain $\bar{\Gamma}$ (see Section 11.3.3 and Appendix III). A general proof of the Jahn–Teller theorem has been obtained (for details see [113]).

### 14.4.2 Adiabatic potentials

To obtain the system energy for *slow* atomic displacements, i.e. in the adiabatic approximation, it is necessary to add to the atom energy $\epsilon(Q)$ in the average electronic field the elastic (potential) energy of atomic interaction $\Sigma_{\bar{\Gamma}\bar{\gamma}} \frac{1}{2}(M_{\bar{\Gamma}}\omega_{\bar{\Gamma}}^2)\, Q_{\bar{\Gamma}\bar{\gamma}}^2$. The function of the coordinates $Q$ thus obtained

$$U_\Gamma(Q) = \sum_{\bar{\Gamma}\bar{\gamma}} \tfrac{1}{2} M_{\bar{\Gamma}} \omega_{\bar{\Gamma}}^2 Q_{\bar{\Gamma}\bar{\gamma}}^2 + \epsilon_\Gamma(Q), \qquad (14.27)$$

is called the *adiabatic potential* of the $\Gamma$ term. The matrix elements in (14.26) are found from the Wigner–Eckart theorem.

**Example 14.3** For the orbital doublet $E_g$ of the cubic group $\mathbf{O}_h$ ($\bar{\Gamma} = A_{1g}$ and $E_g$)

$$H_{eL} = V_1(\mathbf{r})\, Q_1 + V_u(\mathbf{r})\, Q_u + V_v(\mathbf{r})\, Q_v, \qquad (14.28)$$

where $V_1 \equiv V_{A_{1g}}$, $V_u \equiv V_{E_g u}$ etc. With the help of the Wigner–Eckart theorem, we find the matrix representation of $H_{eL}$ in the basis of two functions $\varphi_{E_g u}(\mathbf{r})$ and $\varphi_{E_g v}(\mathbf{r})$ of the doublet $E_g$:

$$\mathbf{H}_{eL} = V_A \begin{bmatrix} 1 & 0 \\ 0 & 1 \end{bmatrix} Q_1 + V_E \begin{bmatrix} -1 & 0 \\ 0 & 1 \end{bmatrix} Q_u + V_E \begin{bmatrix} 0 & 1 \\ 1 & 0 \end{bmatrix} Q_v, \qquad (14.29)$$

where $V_A$ and $V_E$ are the reduced matrix elements

$$V_A = \sqrt{\tfrac{1}{2}} \langle E_g \| V_{A_{1g}} \| E_g \rangle, \qquad V_E = \tfrac{1}{2} \langle E_g \| V_{E_g} \| E_g \rangle. \qquad (14.30)$$

These quantities are called *coupling constants*, and describe the interactions with the totally symmetric and E vibrations. They are independent from the point of view of symmetry, the relations between the elements of matrix (14.29) being determined by the Wigner–Eckart theorem. Solving the secular equation (14.26), we obtain

$$\epsilon(Q) = V_A Q_1 \pm V_E (Q_u^2 + Q_v^2)^{1/2}.$$

The adiabatic potential as a function of $Q_1$ is

$$U_E(Q_1) = \tfrac{1}{2} M_1 \omega_1^2 Q_1^2 + V_A Q_1, \qquad (14.31)$$

and is easily rewritten in the form

$$U_E(Q_1) = \tfrac{1}{2} M_1 \omega_1^2 (Q_1 - Q_0)^2 - \frac{V_A^2}{M_1 \omega_1^2}, \qquad (14.32)$$

where $Q_0 = -V_A / M_1 \omega_1^2$ is the displaced equilibrium value of the normal coordinate, while $-V_A^2 / M_1 \omega_1^2 = \Delta\epsilon$ is the correction to the energy. In the space of the degenerate vibration E active in the Jahn–Teller effect (the E(e) problem) the expression is more complicated:

$$U_E(Q_u, Q_v) = \tfrac{1}{2} M_E \omega_E^2 (Q_u^2 + Q_v^2) \pm V_E (Q_u^2 + Q_v^2)^{1/2}. \qquad (14.33)$$

The adiabatic potential is an axially symmetric surface of the "Mexican hat" (or "sombrero") type (Fig. 14.4). A more detailed treatment taking account of quadratic terms of the type described by (14.15), shows that the latter lead to "corrugation" of the surface and formation of minima corresponding to tetragonal distortions of the octahedron.

**Example 14.4** Consider the orbital triplet $T_{2g}$ of $\mathbf{O}_h$. The symmetric square $[T_{2g}^2] \doteq A_{1g}^2 + E_g + T_{2g}$ determines the active modes. The adiabatic potential is a complicated surface in the space of five normal coordinates of the $E_g$ and $T_{2g}$ vibrations active in the Jahn–Teller effect. Since the parameters $V_E$ and $V_{T_2}$ are independent, it is reasonable to consider the particular cases $V_{T_2} = 0$ (the T(e) problem) and $V_E = 0$ (the $T(t_2)$ problem). For the former, using the Wigner–Eckart theorem in the $T_2$ basis $\varphi_\xi$, $\varphi_\eta$, and $\varphi_\zeta$ we obtain

$$\mathbf{H}_{eL} = V_E \left( \begin{bmatrix} -\tfrac{1}{2} & 0 & 0 \\ 0 & -\tfrac{1}{2} & 0 \\ 0 & 0 & 1 \end{bmatrix} Q_u + \begin{bmatrix} \tfrac{1}{2}\sqrt{3} & 0 & 0 \\ 0 & -\tfrac{1}{2}\sqrt{3} & 0 \\ 0 & 0 & 0 \end{bmatrix} Q_v \right), \qquad (14.34)$$

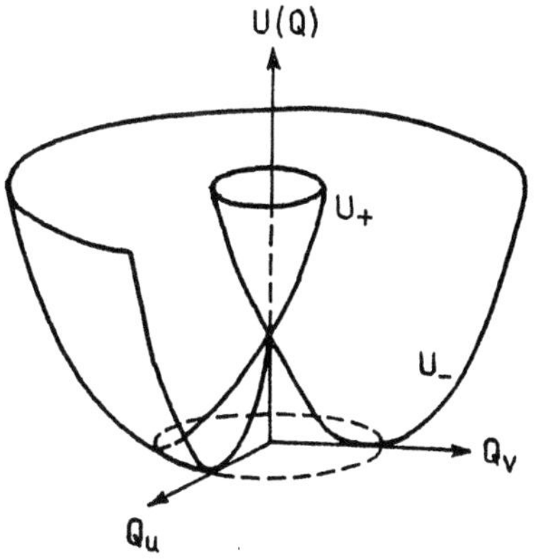


**Fig. 14.4** Adiabatic potential of the E(e) problem ("Mexican hat" or "Sombrero").

where $V_E = \sqrt{\frac{1}{3}} \langle T_{2g} \| V_{E_g} \| T_{2g} \rangle$. Equation (14.16) gives the three sheets of the adiabatic potential:

$$\left.\begin{aligned} U_\xi &= \tfrac{1}{2} M_E \omega_E^2 (Q_u^2 + Q_v^2) + V_E(-\tfrac{1}{2}Q_u + \tfrac{1}{2}\sqrt{3}Q_v), \\ U_\eta &= \tfrac{1}{2} M_E \omega_E^2 (Q_u^2 + Q_v^2) + V_E(-\tfrac{1}{2}Q_u - \tfrac{1}{2}\sqrt{3}Q_v), \\ U_\zeta &= \tfrac{1}{2} M_E \omega_E^2 (Q_u^2 + Q_v^2) + V_E Q_u. \end{aligned}\right\} \tag{14.35}$$

In contrast with the E(e) problem, the adiabatic potential conserves its parabolic form (Fig. 14.5(a)). To the three minima of the adiabatic potential (Fig. 14.5(b)) there correspond three tetragonally distorted ($\mathbf{D}_{4h}$) configurations of the octahedral complex. The case under consideration is known as the Jahn–Teller static effect when the distorted configurations are stable and the system does not jump between them. In the case of the so-called dynamic Jahn–Teller effect, the system moves between distorted configurations (for a more detailed explanation see [112, 116]). The quadratic interaction causes the frequencies $\omega_E$ of atomic vibrations to vary in the vicinity of equilibrium configurations and leads to redefinition of the normal coordinates [113].

If the $T(t_2)$ interaction predominates, the octahedral complex has four minima with trigonal distortions ($\mathbf{D}_{3d}$). A more detailed description can be found in [116]. Unlike the result given by (14.35), in the $T(t_2)$ problem the adiabatic surface is complicated and approaches a parabolic form only near deep minima for a strong $T(t_2)$ interaction. Generalization of the above approach [116] makes it possible to find the extremal points of the adiabatic potential in the case of a multimode Jahn–Teller effect; that is, when interaction with crystal acoustic and optical modes is taken into account.

The above aspects of the Jahn–Teller effect are relevant only in the adiabatic description; that is, in the determination of the form of the potential surface on the basis of symmetry considerations. A complete theory of Jahn–Teller energy must take account of the nonadiabatic effects [111, 112, 113, 116].

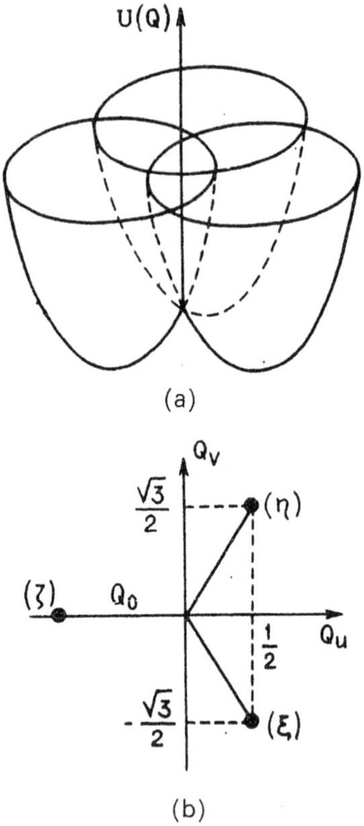


**Fig. 14.5** Adiabatic potential of the T(e) problem: (a) surface shape; (b) coordinates of the minima in the $(Q_u, Q_v)$ plane.

## 14.5 OPTICAL-BAND SPLITTING IN THE STATIC JAHN–TELLER EFFECT

When vibronic interactions in the molecular and impurity-ion spectra of crystals are strong, wide bands of light absorption and luminescence appear [116, 117]. The width of these bands is the result of generation or absorption of vibrational quanta accompanying electronic transitions. In the adiabatic approximation such processes are described by configuration curves or adiabatic potentials. Adiabatic potential minima are shifted owing to the unequal influence of the average field of electrons in states A and $\Gamma$ on atomic configurations (Fig. 14.6). Absorption takes place when atoms oscillate near positions $Q_1$ when the configuration is unchanged (the Franck–Condon principle). Emission is preceded by the establishment of a new equilibrium configuration $Q_2$. Therefore the absorption band maxima $\Omega_m^{(a)}$ and

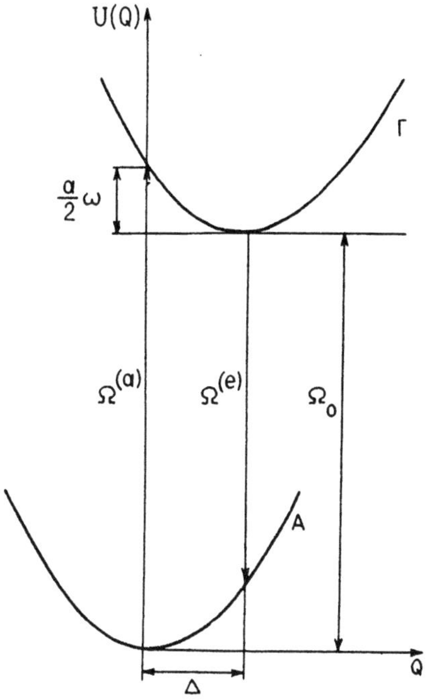


**Fig. 14.6** Adiabatic potentials in the model of two orbitally nondegenerate levels: $\Omega_m^{(a)}$ and $\Omega_m^{(e)}$ are the frequencies of the absorption band and luminescence band maxima, and $\Omega_0$ is the frequency of the purely electronic transition.

luminescence band maxima $\Omega_m^{(e)}$ do not coincide. The quantity $\Delta^2 = (Q_1 - Q_2)^2$ is called the *Stokes loss parameter* (alternatively the *heat-release constant* or *Pekar–Huang–Kun* parameter).

The above description holds only for orbitally nondegenerate electronic terms. The Jahn–Teller effect qualitatively modifies this simple situation, owing to a decrease of symmetry in equilibrium configurations. Let us consider this behaviour, taking examples of the static Jahn–Teller effect [116].

**Example 14.5** Consider the singlet–doublet (A → E) transition in a tetragonal ($\mathbf{D}_{4h}$) complex. Since $[E^2] = A_1 + B_1 + B_2$, the two nondegenerate modes $B_1$ and $B_2$ are active in the Jahn–Teller effect (Section 14.4.1). A pictorial representation of the atomic displacements in these vibrations is shown in Fig. 14.7(a). In the basis $\varphi_x$, $\varphi_y$ of the orbital E doublet the matrix representation of the vibronic Hamiltonian is

$$\mathbf{H}_{eL} = V_1 \begin{bmatrix} 0 & 1 \\ 1 & 0 \end{bmatrix} Q_1 + V_2 \begin{bmatrix} 1 & 0 \\ 0 & -1 \end{bmatrix} Q_2, \tag{14.36}$$

where $Q_1 = Q_{B_1}$ and $Q_2 = Q_{B_2}$ are the corresponding vibronic coupling parameters. It

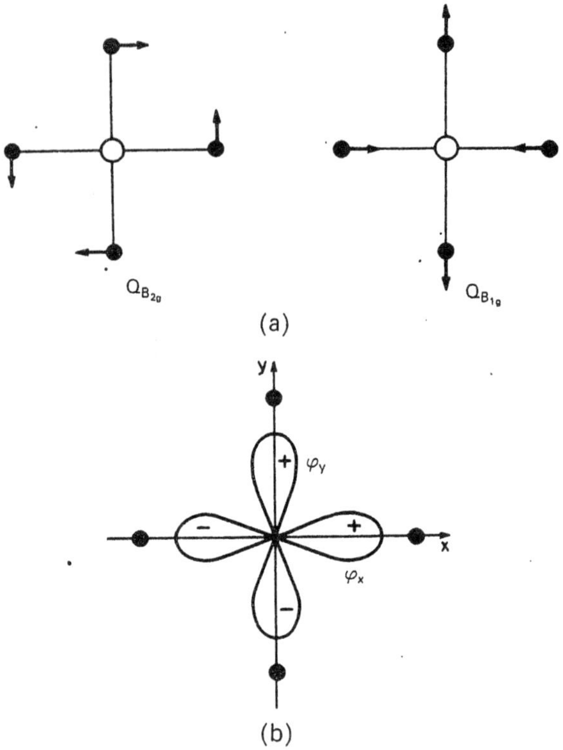


**Fig. 14.7** Tetragonal centre: (a) atomic displacements for $B_1$ and $B_2$ vibrations; (b) spatial distributions of $p_x$ and $p_y$ electronic states.

is clear from physical consideration that $|V_1| \gg |V_2|$, at least for those metals that are coupled to the ligands mainly by ionic bonding. Figure 14.7(b) shows the electron density distribution for the two p states. It can be seen that displacements of $B_1$ type interact more strongly with the unfilled electron shell than those of $B_2$ type. Putting $V_2 = 0$, we obtain the static Jahn–Teller effect in $Q_1$ space. In this case we arrive at an adiabatic potential consisting of two wells:

$$U_E(Q_1) = \tfrac{1}{2}M_1\omega_1^2(Q_1 \pm Q_0)^2 - \frac{V_1^2}{M_1\omega_1^2}, \tag{14.37}$$

where $Q_0 = V_E/M_1\omega_1^2$ is the shift in the coordinate $Q_1$ due to the static Jahn–Teller effect. Figure 14.8 shows that two elementary optical bands arising from transitions in two adiabatic surfaces coincide. Thus in the case under consideration the static Jahn–Teller effect is not manifested in the optical band shape.

**Example 14.6** Consider a transition $A \to T_2$ in a cubic ($\mathbf{O}_h$) complex. In the ground orbital singlet state A there is no Jahn–Teller interaction, and the minimum is situated

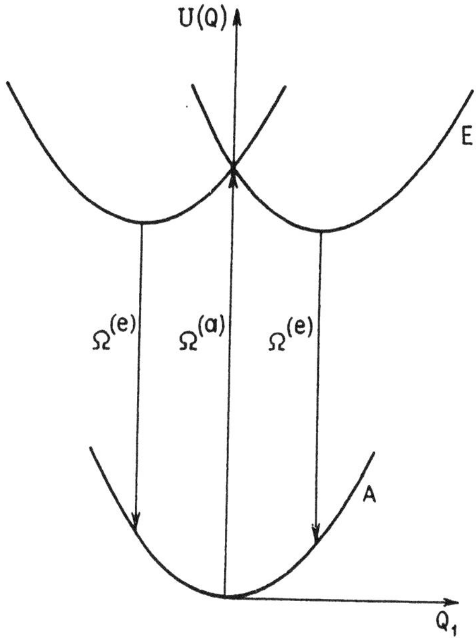


**Fig. 14.8** Adiabatic potential scheme for the transition $A \rightarrow E(b_1)$ in a tetragonal complex.

at the point $Q_u = Q_v = 0$ in the plane of normal coordinates $(Q_u, Q_v)$ of the E mode. In the orbital triplet $T_1\,(T_2)$ the adiabatic potential (14.35) has the form of three paraboloids, whose minima are located at the vertices of a regular triangle (Fig. 14.9). From Fig. 14.9 it can be seen that to the three transitions $A \rightarrow T_2\gamma$ $(\gamma = \xi, \eta, \zeta)$ there correspond Stokes loss parameters of identical magnitude. Hence to all three transitions there corresponds one absorption band and zero-phonon line. Of course, the same result is obtained for emission. the absorption band is not split, owing to the fact that the equilibrium atomic configuration in the initial state possesses cubic $\mathbf{O}_h$ symmetry. In this configuration the excited state is not split, and so there is just one band.

**Example 14.7** Consider the multiplet–multiplet $E(b_1) \rightarrow E'(b_1)$ transition in a tetragonal ($\mathbf{D}_4$) complex. Both initial and final electronic states are orbital doublets, the unfilled shell interacting with the same vibrational mode $B_1$ in these states. Figure 14.10 shows adiabatic curves and optical absorption and emission bands for the doublet–doublet transitions. Since the equilibrium atomic configurations possess a distorted symmetry ($\mathbf{D}_2$) optical bands are split owing to splitting of both the initial and final adiabatic electronic states.

**Example 14.8** Consider the multiplet–multiplet transition $T_2(e) \rightarrow T_2'(e)$, where $T_2$ and $T_2'$ are two orbital triplets. The coordinates of the minima are shown in Fig. 14.11. It can be seen that to the nine transitions $T_2\gamma \rightarrow T_2'\gamma'$ $(\gamma, \gamma' = \xi, \eta, \zeta)$ there correspond two Stokes loss parameters. The complex band is split into two sub-bands owing to the

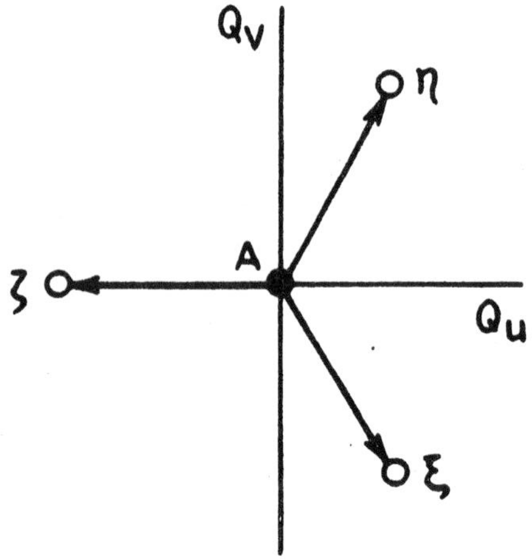


**Fig. 14.9** Displacements of equilibrium coordinates for singlet–multiplet transitions $A \to T_2\gamma$.

fact that each of the equilibrium configurations is tetragonal ($\mathbf{D}_{4h}$). This decrease in symmetry leads to splitting of both the initial and final states into a singlet and a doublet. Figure 14.11 illustrates the splittings of the $T_2(e) \to T_2'(e)$ band when the signs of $T_2(e)$ and $T_2'(e)$ coupling parameters are the same (a) and when they are opposite (b).

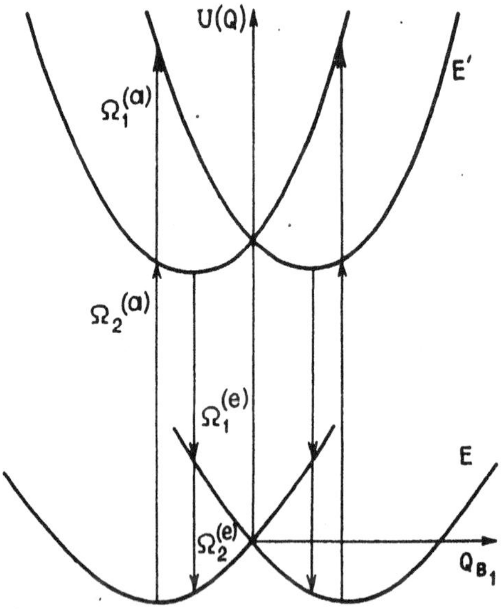


**Fig. 14.10** Optical-band splitting of $E(b_1) \to E(b_1)$ doublet–doublet transitions. $\Omega^{(a)}_{1(2)}$ are the absorption band maxima and $\Omega^{(e)}_{1(2)}$ the luminescence band maxima.

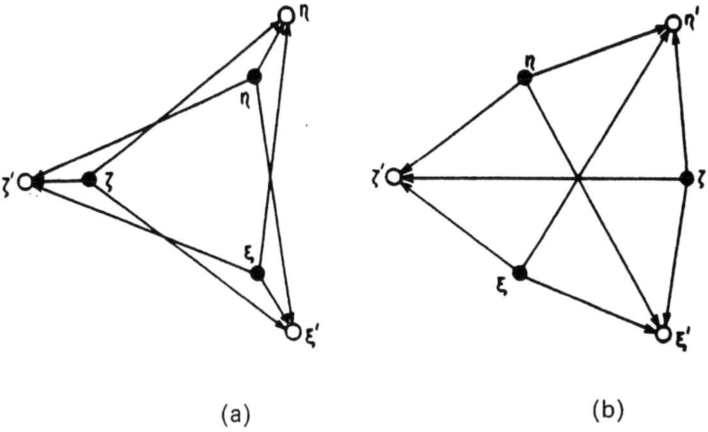


**Fig. 14.11** Displacements of equilibrium coordinates for multiplet–multiplet transitions $T_2\gamma \rightarrow T_2'\gamma'$.

**Example 14.9** Consider the multiplet–multiplet $E(b_2) \rightarrow E'(b_1)$ transition in a tetragonal ($\mathbf{D}_4$) complex. In contrast with the $E(b_1) \rightarrow E'(b_1)$ transition, we now consider the case of vibronic interaction with different Jahn–Teller modes in the initial and final electronic states. As is obvious from Fig. 14.12, in the equilibrium atomic configuration corresponding to the initial state there is no splitting of the final state. Because of this, the absence of a common Jahn–Teller mode for both initial and final electronic states results in an unsplit optical band.

These examples illustrate the following general rules of electron–vibrational band splitting in the static Jahn–Teller effect:

(1) singlet–multiplet transitions result in unsplit vibronic optical bands ($A \rightarrow T(e)$ transitions in cubic centres, and $A \rightarrow E(b_1)$, $A \rightarrow E(b_2)$ transitions in tetragonal centres);

(2) multiplet–multiplet transitions lead to split optical bands in the case of common Jahn–Teller active modes of the electronic states involved in the optical transition ($T(e) \rightarrow T'(e)$ transitions in cubic centres, and $E(b_1) \rightarrow E'(b_1)$, $E(b_2) \rightarrow E'(b_2)$ transitions in tetragonal centres);

(3) there is no optical band splitting in the case of multiplet–multiplet transitions if the initial and final states interact with different Jahn–Teller modes ($E(b_1) \rightarrow E'(b_2)$ transitions in tetragonal centres).

The bands of light absorbtion and luminescence are split if in the equilibrium atomic configurations corresponding to the initial state $\Gamma_1$ the excited configuration $\Gamma_2$ is also split. The case of the dynamic Jahn–Teller effect is more complicated, and is discussed in detail in [116, 117].

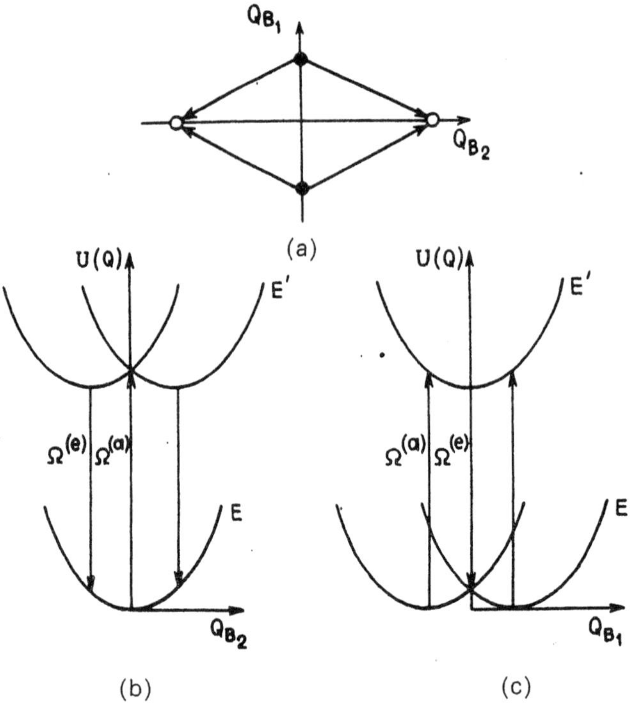


**Fig. 14.12** Multiplet—multiplet transition $E(b_1) \rightarrow E(b_2)$ in a tetragonal centre. (a) $(Q_1, Q_2)$ plane—minima are shifted along different axes. (b, c) Sections of potential surfaces by vertical planes along coordinates $Q_1$ and $Q_2$; the transitions are indicated by arrows. Absorption and emission bands are not split.

## 14.6 VIBRONIC SATELLITES OF ELECTRONIC LINES. VIBRONICALLY INDUCED FORBIDDEN TRANSITIONS

In the case of relatively weak vibronic interaction, along with purely electronic (zero-phonon) lines in the spectrum there appear *vibronic satellites*, i.e. one-quantum electron–vibrational lines. Photon absorption with simultaneous vibrational quantum absorption or emission is determined by a second-order matrix element. If the energy denominators in the equation describing the second-order matrix element of $\Gamma \rightarrow \Gamma_2$ transition as having a certain average value $\Delta$, we obtain

$$A(\Gamma_1 \rightarrow \Gamma_2, n_{\bar{\Gamma}} \rightarrow n_{\bar{\Gamma}} \pm 1) = \frac{2}{\Delta} \langle \Gamma_2 \gamma_2, n_{\bar{\Gamma}} \pm 1 | \mathbf{u} \cdot \hat{\mathbf{d}} \sum_{\bar{\gamma}} V_{\bar{\Gamma}\bar{\gamma}} Q_{\bar{\Gamma}\bar{\gamma}} | \Gamma_1 \gamma_1, n_{\bar{\Gamma}} \rangle, \tag{14.38}$$

where

$$|\Gamma_i \gamma_i n_{\bar{\Gamma}}\rangle = \varphi_{\Gamma_i \gamma_i}(\mathbf{r}) \prod_{\bar{\gamma}} \Phi_{n_{\bar{\Gamma}\bar{\gamma}}}(Q_{\bar{\Gamma}\bar{\gamma}}) \tag{14.39}$$

is the product of the electron wavefunction and the product of harmonic-oscillator wavefunctions for all vibrational modes. In writing down the quantum numbers $n$, it is recognized that the operator $\hat{H}_{eL}$ may change its quantum number by the addition of unity. The integral over the electronic coordinates involves products of vector components $\hat{\mathbf{d}}$ and functions $V_{\bar{\Gamma}\bar{\gamma}}(\mathbf{r})$ of the form $\hat{d}_\alpha V_{\bar{\Gamma}\bar{\gamma}}$. The set of these quantities forms the basis of the reducible representation $\Gamma(\hat{\mathbf{d}}) \times \bar{\Gamma}$, which can be decomposed into irreducible representations $\Gamma(\hat{\mathbf{d}}) \times \bar{\Gamma} \doteq \Sigma_i \bar{\Gamma}_i$. This leads to a selection rule for vibronic satellites of the zero-phonon line: *the one-quantum (with respect to the vibration $\bar{\Gamma}$) vibronic transition $\Gamma_1 \rightleftarrows \Gamma_2$ is allowed if the direct product $\Gamma_1 \times \Gamma_2$ contains at least one of the irreducible representations $\bar{\Gamma}_i$ contained in the direct product $\Gamma(\hat{\mathbf{d}}) \times \Gamma$.*

**Example 14.10** Consider, in the case of **O**$_h$ symmetry, the transition $A_{1g} \to T_{1u}$, with $\Gamma(\hat{\mathbf{d}}) = T_{1u}$. $\Gamma_1 \times \Gamma_2 = A_{1g} \times T_{1u} = T_{1u}$. Taking direct products $\Gamma(\hat{\mathbf{d}}) \times \bar{\Gamma}$, we have

$$\begin{aligned} T_{1u} \times A_{1g} &= \textcircled{T_{1u}}, \quad T_{1u} \times E_g = \textcircled{T_{1u}} + T_{2u}, \\ T_{1u} \times T_{2g} &= A_{2u} + E_u + \textcircled{T_{1u}} + T_{2u}, \\ T_{1u} \times T_{1u} &= A_{1g} + E_g + T_{1g} + T_{2g}, \\ T_{1u} \times T_{2u} &= A_{2g} + E_g + T_{1g} + T_{2g}. \end{aligned}$$

It can be seen that $\bar{\Gamma} = T_{1u}$ (encircled) is contained in $\Gamma(\hat{\mathbf{d}}) \times \bar{\Gamma}$ for all even $\bar{\Gamma}$ modes. Therefore for the electronic transition $A_{1g} \to T_{1u}$ satellites involving $A_{1g}$, $E_g$ and $T_{2g}$ vibrations are allowed.

In some cases purely electronic transitions are forbidden by the selection rules (Section 9.1) although these are satisfied for one or more vibronic satellites. If this is so then there is no zero-phonon line in the absorption or emission spectrum, all of the spectral intensity being contained in the vibronic lines. The vibrations lift the prohibition from the electronic transitions, which are therefore called *vibronically induced forbidden transitions.*

**Example 14.11** Consider the transition $A_{1g} \to E_g$ in an octahedral complex. Since $A_{1g} \times E_u = E_g$ (i.e. it does not contain $\Gamma(\hat{\mathbf{d}}) = T_{1u}$), the transition is forbidden as an electric dipole transition. Vibronic satellites are allowed if $\Gamma(\hat{\mathbf{d}}) \times \bar{\Gamma}$ contains $E_u$. From the decomposition obtained in Example 14.10, it can be seen that the selection rule is satisfied for $\bar{\Gamma} = T_{2g}$. Therefore the transition $A_{1g} \to E_g$ is allowed when $T_{2g}$ vibration is involved.

In centrosymmetric complexes parity considerations forbid certain transitions, such as the d–d transitions between terms resulting from the d shell. As well as weak odd crystal fields, odd vibrations can lead to lifting of the prohibition from such transitions.

**Example 14.12** Consider for a one-electron d ion in an octahedral complex the transition $T_{2g} \rightarrow E_g$ (alternatively, a $d^9$ ion and the "hole" transition $E_g \rightarrow T_{2g}$). The direct product $T_{2g} \times E_g \doteq T_{1g} + T_{2g}$ shows that in the electric dipole approximation the prohibition is lifted from the transition by $T_{1u}$ and $T_{2u}$ vibrations.

It is important to bear in mind two essential differences between the effects of odd static fields and those of odd vibrations, seen in forbidden transition spectra:

(1) odd crystal fields allow purely electronic transitions, while for the vibronic mechanism there are no purely electronic transitions, which remain forbidden;

(2) low-symmetry fields result in optical-line dichroism (Section 9.1), while vibronically induced lines in cubic crystals are isotropic, like purely electronic lines (Section 9.1.3), the reason being that time-averaged atomic vibrational motion does not reduce molecular symmetry, and thus does not lead to polarization dichroism.

## 14.7 POLARIZATION DEPENDENCE OF THE VIBRONIC SATELLITE INTENSITY

Polarization (angular) dependences of vibronic satellite intensity are determined using the method given in Section 17 of [116], which is similar to that described in Section 9.5.2 above with reference to two-photon transitions. We present here only some results.

(1) In cubic groups, for the forbidden transition $A_{1g} \rightarrow E_u$ (which is allowed when $T_{2g}$ vibration is involved), we obtain the angular dependence $F(\bar{\Gamma})$ as

$$F(T_{2g}) \sim l^2 + m^2 + n^2 = 1. \tag{14.40}$$

Single-photon line isotropy is a common characteristic of cubic groups; this characteristic refers to single-quantum satellites of both forbidden and allowed transitions.

(2) In the groups $\mathbf{C}_{nv}$ $(n > 2)$ a vector is transformed according to two representations, one of which is doubly degenerate (E). In $\mathbf{C}_{3v}$ the transition $A_1 \rightarrow E$ is allowed for $\sigma$ polarization, and the single-quantum satellite for the E vibrations shows the angular dependence

$$F(E) \sim l^2 + m^2 + \alpha n^2, \tag{14.41}$$

where $\alpha$ is a parameter. Thus the E satellite is allowed for $\sigma$ and $\pi$ polarizations, and its angular dependence is the most general among those possible in $\mathbf{C}_{3v}$.

(3) In triaxial groups $\hat{d}_x$, $\hat{d}_y$ and $\hat{d}_z$ are transformed according to three irreducible representations. For $\mathbf{C}_{2v}$ we have $\Gamma(\hat{d}_x) = B_1$, $\Gamma(\hat{d}_y) = B_2$ and $\Gamma(\hat{d}_z) = A_1$. The single-quantum satellite of the transition $A_1 \rightarrow B_1$ for $B_2$ vibrations gives

$$F(B_2) \sim n^2, \tag{14.42}$$

which distinguishes it from the zero-phonon transition, for which

$$F \sim m^2,$$

The selection rules and polarization dependences for vibronic satellites of two-phonon spectra are analysed in [72]. A transition $\Gamma_1 \rightarrow \Gamma_2$ of this type is allowed if $\Gamma_1 \times \Gamma_2$ includes irreducible representations from the decomposition $\Gamma(\hat{\mathbf{d}}) \times \Gamma(\hat{\mathbf{d}}) \times \bar{\Gamma}$.

A method of analysing the angular dependences of component intensities for vibronic satellites split by external fields was developed in [118], in which there are tables giving the polarization dependences of component intensities for single-quantum lines in the case of cubic symmetry groups in differently oriented electric, magnetic and distortion fields.

## 14.8 ELECTRON–VIBRATIONAL INTERACTION IN MIXED-VALENCE CLUSTERS

Since an "extra" electron produces local deformation of the ligand environment the vibronic interaction in mixed valence compound is usually strong enough and plays a significant role in spectroscopic and magnetic manifestations of the phenomenon of mixed valency. The vibronic interaction operator for a mixed-valence compound can be deduced using the general procedure described in Section 14.3. Since most of the mixed-valence compounds present polyatomic systems with a complicated molecular structure and a great many degrees of freedom, the rigorous deduction of the vibronic Hamiltonian proves to be cumbersome. We describe in this section the vibronic model for mixed-valence systems proposed by Piepho, Krausz and Schatz (PKS model [119]) (see review article [120] and references therein). The PKS model is an excellent example of a simple and efficient approach to a very complicated problem. Let us consider a dimeric mixed-valence cluster consisting of two identical moieties a and b and denote the full-symmetric ("breathing") coordinates of these moieties as $Q_a$ and $Q_b$.

In the framework of the PKS model these two coordinates $Q_a$ and $Q_b$ can be approximately considered as normal ones for weakly interacting moieties a and b and, hence, the vibrational Hamiltonian can be written as follows

$$\hat{H}_v = \hat{H}_{av} + \hat{H}_{bv}, \qquad (14.44)$$

where $\hat{H}_{a(b)v}$ is the harmonic oscillator Hamiltonian for the a(b) unit in the absence of the "extra" electron:

$$\hat{H}_{a(b)v} = -\frac{\hbar^2}{2M}\frac{\partial^2}{\partial Q_{a(b)}^2} + \frac{M\omega^2}{2}Q_{a(b)}^2. \qquad (14.45)$$

The "extra" electron is supposed to occupy nondegenerate orbital $\varphi_a(\mathbf{r})$ or $\varphi_b(\mathbf{r})$ and hence interacts only with the full-symmetric motion of the near-neighbour environment. The vibronic interaction operator (14.17) linear in atomic displacements may be written as follows:

$$\hat{H}_{ev} = V_a(\mathbf{r})Q_a + V_b(\mathbf{r})Q_b, \qquad (14.46)$$

where

$$V_{a(b)}(\mathbf{r}) = \left(\frac{\partial W(\mathbf{r},\mathbf{R})}{\partial Q_{a(b)}}\right)_{Q_{a(b)}=0} \qquad (14.47)$$

$W(\mathbf{r},\mathbf{R})$ is the potential energy of the interaction of the "extra" electron with the atomic cores (Equation (14.15)).

Because of the inversion symmetry of the dimeric cluster

$$\langle \varphi_a|V_a(\mathbf{r})|\varphi_a\rangle = \langle \varphi_b|V_b(\mathbf{r})|\varphi_b\rangle \equiv V.$$

Neglecting the matrix elements $\langle \varphi_a|V_a(\mathbf{r})|\varphi_b\rangle = \langle \varphi_a|V_b(\mathbf{r})|\varphi_b\rangle$ and using as the basis $\varphi_a$ and $\varphi_B$ one can present the $\hat{H}_{ev}$ operator in the matrix form as follows:

$$\hat{H}_{ev} = V\begin{matrix}\varphi_a & \varphi_b \\ \begin{vmatrix}1 & 0 \\ 0 & 0\end{vmatrix}\end{matrix}Q_a + V\begin{matrix}\varphi_a & \varphi_b \\ \begin{vmatrix}0 & 0 \\ 0 & 1\end{vmatrix}\end{matrix}Q_b. \qquad (14.48)$$

Assuming that the "extra" electron does not change the vibrational frequency $\omega$ while introducing the dimensionless vibrational coordinates

$$q_{a(b)} = \left(\frac{M\omega}{\hbar}\right)^{1/2}Q_{a(b)},$$

one can rewrite the total Hamiltonian of the system in the following form

$$\hat{H} = \hat{H}_v + \hat{H}_{ev} + \hat{H}_{tr}. \qquad (14.49)$$

In (14.49) $\hat{H}_{tr}$ is the electron transfer operator connecting two localized states

$\varphi_a$ and $\varphi_b$, and in the matrix form looks like the following:

$$\hat{\mathbf{H}}_{tr} = P\begin{vmatrix} 0 & 1 \\ 1 & 0 \end{vmatrix},$$

(with the columns labelled $\varphi_a$, $\varphi_b$), where

$$P = \int \varphi_a(\mathbf{r})\hat{h}\varphi_b(\mathbf{r})\,d\tau \tag{14.50}$$

is the transfer integral.

In dimensionless coordinates $H_v$ and $H_{ev}$ are the following:

$$H_v = \frac{\hbar\omega}{2}\left(q_a^2 + q_b^2 - \frac{\partial^2}{\partial^2 q_a^2} - \frac{\partial^2}{\partial q_b^2}\right), \tag{14.51}$$

$$\mathbf{H}_{ev} = v\begin{vmatrix} 1 & 0 \\ 0 & 0 \end{vmatrix} q_a + v\begin{vmatrix} 0 & 0 \\ 0 & 1 \end{vmatrix} q_b, \tag{14.52}$$

(with the columns of each matrix labelled $\varphi_a$, $\varphi_b$), with the vibronic coupling parameter

$$v = \left(\frac{\hbar}{M\omega}\right)^{1/2} V. \tag{14.53}$$

Let us move on to consider the symmetry adopted basis in the electronic and vibrational subsystem introducing odd and even electronic wavefunctions and vibrational coordinates

$$\varphi_\pm(\mathbf{r}) = \frac{1}{\sqrt{2}}[\varphi_a(\mathbf{r}) \pm \varphi_b(\mathbf{r})], \tag{14.54}$$

$$q_\pm = \frac{1}{\sqrt{2}}(q_a \pm q_b). \tag{14.55}$$

The $\varphi_\pm$ basis diagonalizes the $\mathbf{H}_{tr}$ giving the two electronic levels

$$\epsilon_\pm = \pm P. \tag{14.56}$$

The symmetric (even) coordinate $q_+$ describes the synchronous vibrational motion of both a and b moieties, meanwhile the antisymmetric (odd) coordinate $q_-$ is related to the out-of-phase "breathing" motion of the two subunits. This vibration is sketched in Fig. 14.13 for the octahedral environment of a and b ions.

In the symmetry adopted basis (14.54) and (14.55) the components of the total Hamiltonian take the following form ($v/\sqrt{2} \to v$):

$$\hat{H}_v = \frac{\hbar\omega}{2}\left(q_+^2 + q_-^2 - \frac{\partial^2}{\partial q_+^2} - \frac{\partial^2}{\partial q_-^2}\right), \tag{14.57}$$

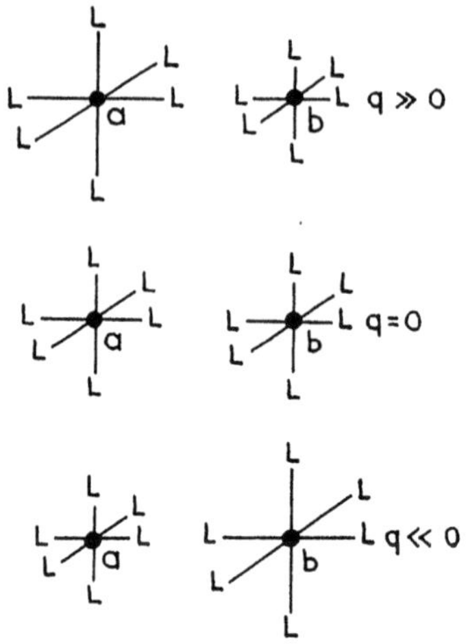


**Fig. 14.13** Out-of-phase of two octahedral subunits of a mixed-valence dimeric cluster.

$$\hat{\mathbf{H}}_{\text{ev}} = v \begin{array}{c} \begin{matrix} \varphi_+ & \varphi_- \end{matrix} \\ \begin{vmatrix} 1 & 0 \\ 0 & 1 \end{vmatrix} \end{array} q_+ + v \begin{array}{c} \begin{matrix} \varphi_+ & \varphi_- \end{matrix} \\ \begin{vmatrix} 0 & 1 \\ 1 & 0 \end{vmatrix} \end{array} q_-, \tag{14.58}$$

$$\mathbf{H}_{\text{tr}} = P \begin{array}{c} \begin{matrix} \varphi_+ & \varphi_- \end{matrix} \\ \begin{vmatrix} 1 & 0 \\ 0 & -1 \end{vmatrix} \end{array}. \tag{14.59}$$

As is clear from (14.57) and (14.58), the interaction with the symmetric mode can be eliminated by the linear shift of $q_+$ to arrive at the Hamiltonian

$$\hat{\mathbf{H}} = \frac{\hbar\omega}{2}\left(q^2 - \frac{\partial^2}{\partial q^2}\right) + v \begin{array}{c} \begin{matrix} \varphi_+ & \varphi_- \end{matrix} \\ \begin{vmatrix} 0 & 1 \\ 1 & 0 \end{vmatrix} \end{array} q + P \begin{array}{c} \begin{matrix} \varphi_+ & \varphi_- \end{matrix} \\ \begin{vmatrix} 1 & 0 \\ 0 & -1 \end{vmatrix} \end{array}, \tag{14.60}$$

where $q \equiv q_-$. This Hamiltonian describes the mixing of the electronic states of two closely spaced levels $\epsilon_+, \epsilon_-$ via out-of-phase vibration. This vibration proves to be active in the pseudo Jahn–Teller effect. The corresponding adiabatic potentials are the following:

$$U_{\pm}(q) = \frac{\hbar\omega}{2} q^2 \pm \sqrt{P^2 + v^2 q^2}. \tag{14.61}$$

Fig. 14.14 illustrates the adiabatic potentials for a symmetric dimer in the cases of completely localized electronic states ($P = 0$), strong ($v^2 > |P|\hbar\omega$) and weak ($v^2 < |P|\hbar\omega$) vibronic coupling. In the first case the adiabatic potentials represent two intersecting parabolas corresponding to the independent vibrations of the a and b moieties with the localized "extra" electron. In the case of strong vibronic coupling we find the double-well adiabatic potential, where the minima of the lower branch corresponds to the localization of the "extra" electron at one of two ions. The lower branch has the only minimum in the point $q = 0$ of full symmetry and provides the weak vibronic coupling. In this case the vibronic interaction proves to be suppressed by the electron transfer. In the more complicated case of the different subunits a and b (unsymmetric dimer) the PKS model can be generalized by introducing in the electronic Hamiltonian the perturbation $\hat{W}$ leading to the difference in the electron energies of the a and b sites:

$$\begin{matrix} & \varphi_a & \varphi_b \end{matrix}$$
$$\hat{\mathrm{W}} = W \begin{vmatrix} 1 & 0 \\ 0 & -1 \end{vmatrix}. \qquad (14.62)$$

In the case of the non-equivalent moieties of the PKS model the adiabatic potential becomes unsymmetric:

$$U_{\pm}(q) = \frac{\hbar\omega}{2} q^2 \pm \sqrt{P^2 + (vq + W)^2}. \qquad (14.63)$$

Fig. 14.15a illustrates the case of uncoupled states of an "extra" electron on two moieties with different energies. In the case of weak transfer (Fig. 14.15b), lower branch $U_-(q)$ possesses two minima with different energies. Provided that the transfer is strong enough (Fig. 14.15c) the shallow minimum disappears; that means the full localization of the system on the one center of a dimer.

One can see that the PKS model provides a very clear physical picture of vibrationally-assisted electron-transfer in a mixed-valence dimer. Symmetry consideration can also be used in solving the dynamic (quantum-mechanical) problem. To solve the quantum mechanical problem one must choose the basis set for diagonalization of the total Hamiltonian. The "exact" vibronic functions can be expressed in terms of products of electronic wavefunctions $\varphi_{\pm}(r)$ and the harmonic oscillator functions

$$\varphi_+(\mathbf{r})\Phi_n(q) \quad \text{and} \quad \varphi_-(\mathbf{r})\Phi_n(q) \quad (n = 0, 1, \ldots \infty) \qquad (14.64)$$

The parity of $\Phi_n(q)$ is determined by factor $(-1)^n$ (see (14.15)), therefore we can easily derive parities of different products (14.64):

$$\left.\begin{matrix} \varphi_+(\mathbf{r})\Phi_{2n}(q) \\ \varphi_-(\mathbf{r})\Phi_{2n+1}(q) \end{matrix}\right\} \text{even}, \qquad \left.\begin{matrix} \varphi_-(\mathbf{r})\Phi_{2n}(q) \\ \varphi_+(\mathbf{r})\Phi_{2n+1}(q) \end{matrix}\right\} \text{odd}. \qquad (14.65)$$

As the Hamiltonian of the system is usually invariant, in the case under consideration the Hamiltonian commutes with the operation of inversion.

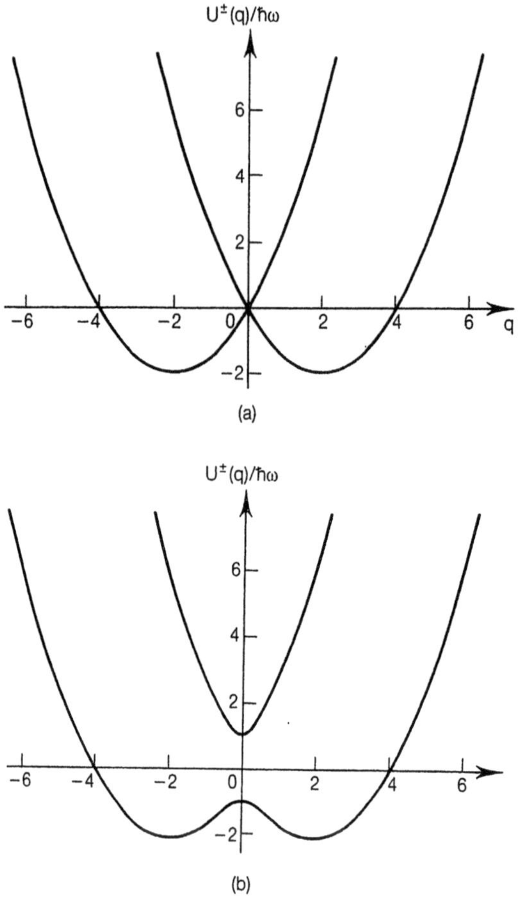


**Fig. 14.14** Adiabatic potentials of a symmetric mixed-valence dimer ($v = 2\hbar\omega$): (a) uncoupled moieties ($p = 0$); (b) strong vibronic coupling ($p - 2\hbar\omega$), (c) weak vibronic coupling, or strong transfer ($p = g\hbar\omega$).

Hence the matrix of the Hamiltonian in the basis (14.65) can be blocked into two submatrices corresponding to the even and odd basis.

The electron-vibrational wavefunctions $\Phi_\nu^{\pm}(\mathbf{r}, q)$ must be either odd or even:

$$\begin{aligned} \Phi_\nu^+(\mathbf{r}, q) &= \sum_{n=0}^{\infty}[c_{2n}^{\nu}\varphi_+(\mathbf{r})\Phi_{2n}(q) + c_{2n+1}^{\nu}\varphi_-(\mathbf{r})\Phi_{2n+1}(q)], \\ \Phi_\nu^-(\mathbf{r}, q) &= \sum_{n=0}^{\infty}[b_{2n}^{\nu}\varphi_-(\mathbf{r})\Phi_{2n}(q) + b_{2n+1}^{\nu}\varphi_+(\mathbf{r})\Phi_{2n+1}(q)], \end{aligned} \tag{14.66}$$

where $\nu$ enumerates the vibronic levels.

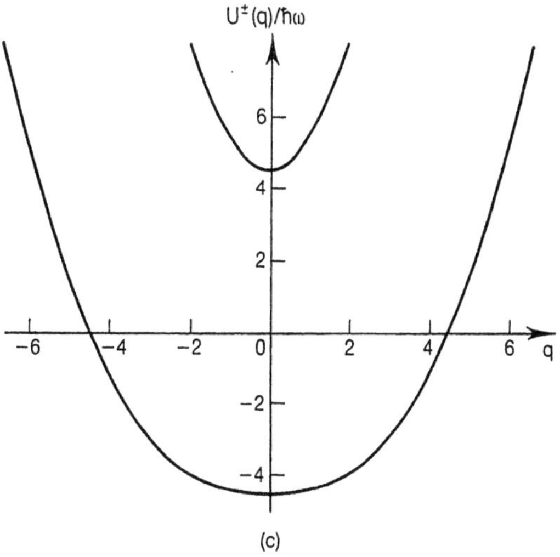


**Fig. 14.14** (*Continued*).

The real order of the matrix to be diagonalized depends on the vibronic coupling parameter. In the case of strong and moderate vibronic coupling the so-called intermediate coupling basis seems to be efficient [121, 122]. In the framework of the named approach the basis set for a symmetric case consists of the products of electronic-vibrational wavefunctions:

$$\varphi_{\mathrm{a}}(\mathbf{r})\Phi_n(q - q_{\mathrm{a}}), \quad \varphi_{\mathrm{b}}(\mathbf{r})\Phi_n(q - q_{\mathrm{b}}) \tag{14.67}$$

where $q_{\mathrm{a}}$ and $q_{\mathrm{b}}$ ($q_{\mathrm{a}} = -q_{\mathrm{b}} \equiv q_0$) are the equilibrium positions of the power branch of the double-well adiabatic potential, while $\Phi_n(q - q_{\mathrm{a(b)}})$ are the harmonic oscillator functions with the shifted equilibrium position of out-of-phase vibration. Since $q_0$ depends on both the vibronic parameters $v$ and transfer integral $P$ the basis set (14.167) provides a good convergence for the computational procedure of the total Hamiltonian diagonalization. The symmetry argument can be employed as well. Actually applying the operation of inversion $\hat{I}$ to the vibrational components we find:

$$\begin{aligned}\hat{I}\Phi_n(q - q_{\mathrm{a}}) &= \Phi_n(-q - q_{\mathrm{a}}) = (-1)^n\Phi_n(q + q_{\mathrm{a}})\\ &= (-1)^n\Phi_n(q - q_{\mathrm{b}})\end{aligned}$$

and therefore one can construct the even $(F_n^+(q))$ and odd $(F_n^-(q))$ superposition of vibrational functions

$$F_n^{\pm}(q) = \Phi_n(q - q_{\mathrm{a}}) \pm (-1)^n\Phi_n(q - q_{\mathrm{b}}). \tag{14.68}$$

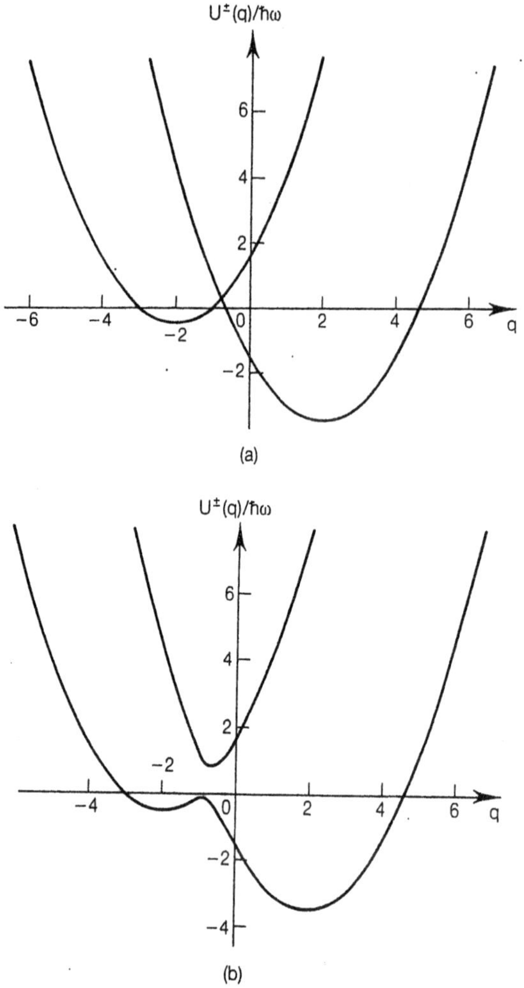


**Fig. 14.15** Adiabatic potentials of an unsymmetric mixed-valence dimer ($v = 2\hbar\omega$, $W = 1.5\hbar\omega$): (a) uncoupled electronic states ($P = 0$); (b) weak transfer ($P = \hbar\omega$); (c) strong transfer ($P = 3\hbar\omega$).

The symmetry adopted basis set is the following:

$$\left.\begin{matrix} \varphi_+(\mathbf{r})F_n^+(q) \\ \varphi_-(\mathbf{r})F_n^-(q) \end{matrix}\right\}\text{even}, \qquad \left.\begin{matrix} \varphi_+(\mathbf{r})F_n^-(q) \\ \varphi_-(\mathbf{r})F_n^+(q) \end{matrix}\right\}\text{odd}.$$

The solutions of the dynamic problem are given in [121, 122]. To generalize the PKS model for a trimeric mixed-valence cluster [123] one must take into account three full symmetric modes with coordinates $q_a, q_b, q_c$ belonging to the

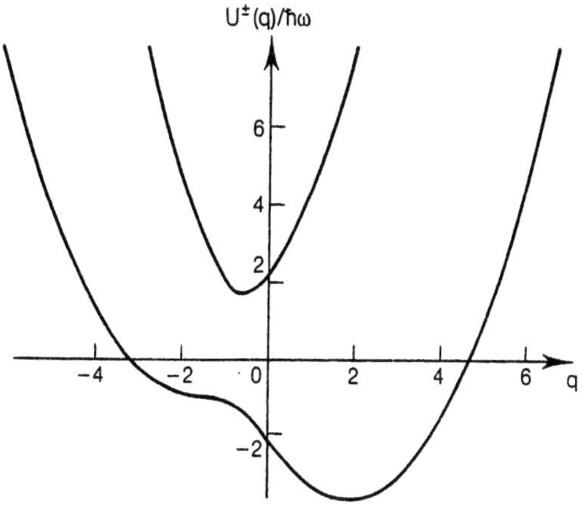


(c)

**Fig. 14.15** *(Continued).*

three sites a, b and c and the three possibilities of an "extra" electron localization. Assuming trigonal ($\mathbf{C}_{3v}$) molecular symmetry and employing the projection procedure one can find the symmetrized vibrational coordinates of the PKS model belonging to the $A_1$ and E irreducible representation of $\mathbf{C}_{3v}$

$$q_{A_1} = \frac{1}{\sqrt{3}}(q_a + q_b + q_c);$$

$$q_{Ex} = \frac{1}{\sqrt{2}}(q_b - q_a), \tag{14.69}$$

$$q_{Ey} = -\frac{1}{\sqrt{6}}(2q_c - q_a - q_b).$$

Pictorial representations of the vibrational motion are presented in Fig. 14.16. For the only electron sharing between three spinless atomic cores ($d^1 - d^0 - d^0$) we can construct the symmetry adopted electronic basis

$$\varphi_{A_1}(\mathbf{r}) = \frac{1}{\sqrt{3}}[\varphi_a(\mathbf{r}) + \varphi_b(\mathbf{r}) + \varphi_c(\mathbf{r})],$$

$$\varphi_{Ex}(\mathbf{r}) = \frac{1}{\sqrt{2}}[\varphi(\mathbf{r}) - \varphi_a(\mathbf{r})], \tag{14.70}$$

$$\varphi_{Ey}(\mathbf{r}) = -\frac{1}{\sqrt{6}}[2\varphi_c(\mathbf{r}) - \varphi_a(\mathbf{r}) - \varphi_b(\mathbf{r}).$$

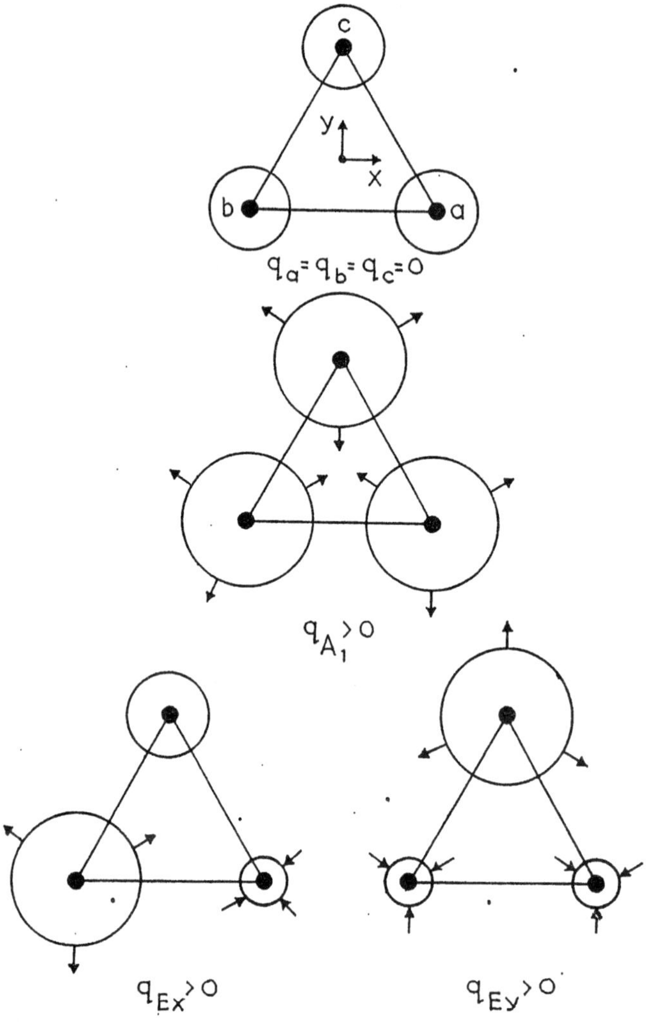


**Fig. 14.16** Pictorial representation of the vibrational motion of the trimeric cluster in the PKS model.

The vibronic interaction operator now includes three terms

$$\hat{H}_{ev} = V_a(\mathbf{r})q_a + V_b(\mathbf{r})q_b + V_c(\mathbf{r})q_c, \tag{14.71}$$

and the matrix of $\hat{H}_{ev}$ in symmetrized basis looks as follows (the interaction with $A_1$ mode proportional to the unit matrix is omitted):

$$
\mathbf{H}_{ev} = -\frac{v}{\sqrt{3}} \begin{vmatrix} 0 & q_{Ex} & q_{Ey} \\ q_{Ex} & \frac{1}{\sqrt{2}} q_{Ey} & \frac{1}{\sqrt{2}} q_{Ex} \\ q_{Ey} & \frac{1}{\sqrt{2}} q_{Ex} & -\frac{1}{\sqrt{2}} q_{Ey} \end{vmatrix} . \tag{14.72}
$$

The two-mode pseudo-Jahn–Teller problem leads to a complicated adiabatic potential and the complicated distribution of electronic density in a three-center system (see review article [123] dealing with the vibronic problem and references therein).

Using the symmetry arguments one can apply the PKS model to more complex systems. Symmetry adopted vibrational coordinates of the PKS model

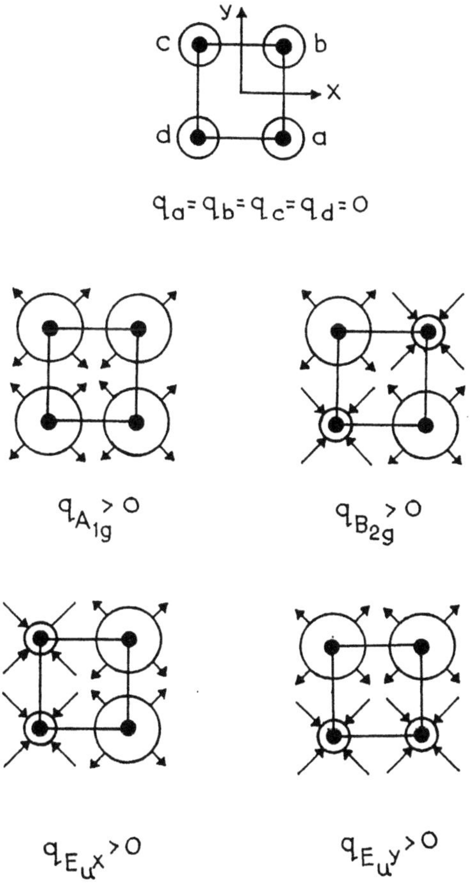


**Fig. 14.17** Pictorial representation of vibrational coordinates of the PKS model for a square planar mixed-valence cluster [94].

for a square planar $\mathbf{D}_{4h}$ cluster belong to $A_{1g}$, $B_{2g}$ and $E_u$ irreducible representations of $\mathbf{D}_{4h}$ [94]:

$$
\begin{aligned}
q_{A_{1g}} &= \tfrac{1}{2}(q_a + q_b + q_c + q_d);\\
q_{B_{2g}} &= \tfrac{1}{2}(q_a - q_b + q_c - q_d);\\
q_{E_u x} &= \tfrac{1}{2}(q_a + q_b - q_c - q_d),\\
q_{E_u y} &= \tfrac{1}{2}(-q_a + q_b + q_c - q_d).
\end{aligned}
\tag{14.73}
$$

The corresponding vibrational motion is shown in Fig. 14.17. Pseudo-Jahn–Teller problems for one and two electrons delocalized over four centers have been considered in [94].

Finally the PKS vibrational coordinates for mixed-valence tetrahedral ($\mathbf{T}_d$) clusters ($A_1$ and $T_2$ irreducible representations) looks as follows:

$$
\begin{aligned}
q_{A_1} &= \tfrac{1}{2}(q_a + q_b + q_c + q_d);\\
q_{T_2\xi} &= \tfrac{1}{2}(q_a - q_b - q_c - q_d),\\
q_{T_2\eta} &= \tfrac{1}{2}(-q_a + q_b - q_c + q_d),\\
q_{T_2\zeta} &= \tfrac{1}{2}(-q_a - q_b + q_c + q_d).
\end{aligned}
\tag{14.74}
$$

and are sketched in Fig. 14.18. The adiabatic potentials for the one-electron problem ($d^1 - d^0 - .d^0 - d^0$ cluster) are considered in [124, 125].

The PKS model in its conventional form deals with the nondegenerate electronic orbitals belonging to each site. In the cases of a high-symmetric (say, octahedral) local environment of ions with unfilled shells, this simplified assumption proves to be invalid and we arrive at a far more complicated multimode problem. Consider for example a bioctahedral cluster (Fig. 13.12) of the $d^1 - d^2$-type. The cubic field ground state of $d^1$-ion is $^2T_2$, while for the $d^2$-ion the ground state is $^3T_1$ (Chapter 5) and we have the complicated Jahn–Teller problems for each of the interacting submits. In both the $T_1$ and $T_2$ states the $A_1$, E and $T_2$ vibrations of the local environment are active. The corresponding atomic displacements of a bioctahedral cluster with allowance $T_1(a_1 + e)$ and $T_2(a_1 + e)$ local Jahn–Teller interaction are presented in Fig. 14.19 [126], where the symmetry groups for each type of distorted system are pointed out. One can see that one $\mathbf{C}_{4v}$ and four $\mathbf{C}_{2v}$ configurations are possible, corresponding to five energies of distortions.

Fig. 14.19 schematizes the distribution of electronic densities belonging to five Jahn–Teller type distortions (orbitals $\xi, \eta, \zeta$ for $T_2$ and $\alpha, \beta, \gamma$ for $T_1$ (the last are similar to three *P*-orbitals). This picture illustrates the origin of five Jahn–Teller energies arising from the electronic density distributions on each center and their mutual dispositions as shown in Fig. 14.20.

An important additional effect of elastic interaction between centers via a

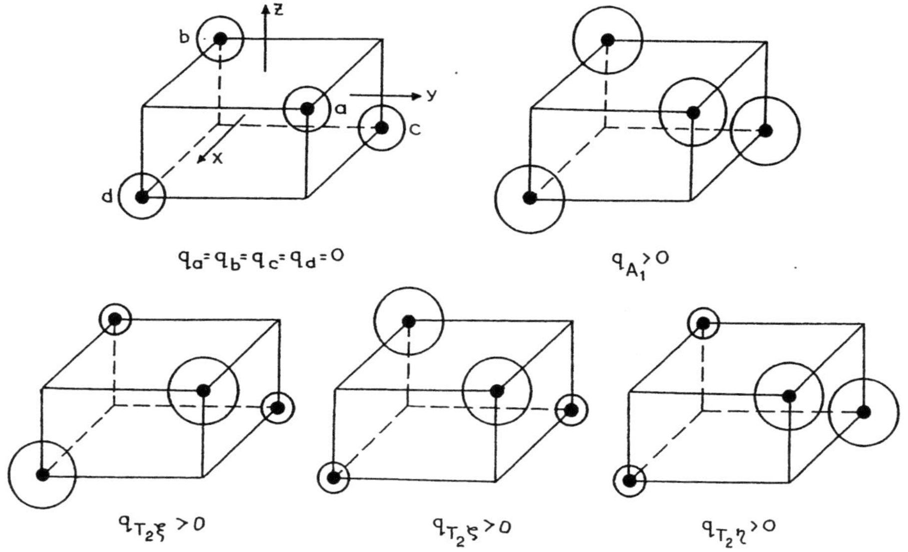


**Fig. 14.18** Vibrational coordinates of the PKS model for a tetrahedral ($\mathbf{T}_d$) cluster.

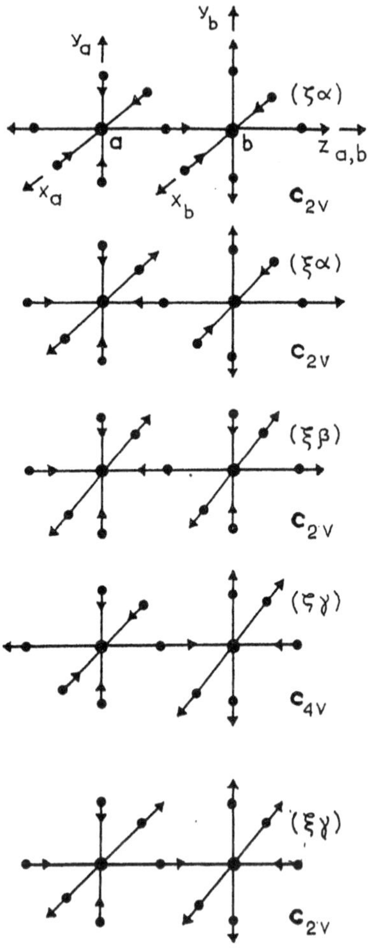


**Fig. 14.19** Displacements of atoms of the $d^1 - d^2(^2T_2 - ^3T_1)$ cluster ($D_{\infty h}$) due to the local Jahn–Teller effect.

phonon field [126] provides a contribution to the energies of the Jahn–Teller minima of the dimeric cluster.

Consideration of the vibronic effect in multielectronic mixed-valence systems proves to be far more complicated because of the existence of repeated $S\Gamma$-terms which are mixed by both the transfer and vibronic interactions. The combined action of transfer, Jahn–Teller and pseudo-Jahn–Teller vibronic interactions for trimeric systems $d^1 - d^1 - d^0$ and $d^1 - d^1 - d^2$ have been considered in [127]. For example, for the $d^1 - d^1 - d^2$-mixed valence trimer the

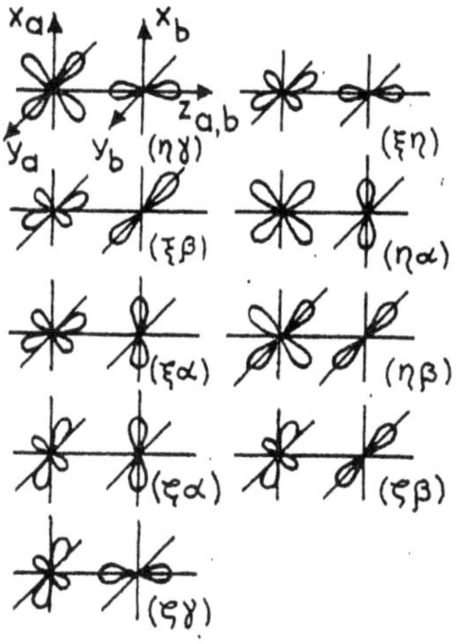


**Fig. 14.20** Mutual disposition of $T_2(\xi, \eta, \zeta)$ and $T_1(\alpha, \beta, \gamma)$ orbitals of $d^1 - d^2$-bioctahedral cluster [126].

complicated problem of vibronic mixing of $^3A_1, ^3A_2, 2^3E$ through E-type vibrations must be solved $((A_1 + A_2 + E + E) \otimes e$-problem).

## PROBLEMS

**14.1** Classify the normal vibrations of a square-planar molecule $ML_4$, an octahedral complex $ML_6$ and a tetrahedral complex $ML_4$.

**14.2** For the molecules in Problem 14.1 determine the normal modes active in IR absorption and Raman scattering.

**14.3** Determine the selection rules for vibronic satellites of the d–d transitions with identical spin multiplicity in an octahedral complex with $d^2$ and $d^3$ transition metal ions (the allowed terms are given in Sections 5.5 and 5.7).

**14.4** Using the projection operator to construct the vibrational coordinates of a trimeric ($\mathbf{C}_{3v}$) and tetrameric ($\mathbf{T}_d$) clusters in the framework of the PKS model.

# Appendix I

## Characters of Point Groups

1. The $C_S$ and $C_i$ groups

| $C_s$ | E | $\sigma_h$ | | |
|---|---|---|---|---|
| A′ | 1 | 1 | $x, y, R_z$ | $x^2, y^2, z^2, xy$ |
| A″ | 1 | −1 | $z, R_x, R_y$ | yz, xz |

| $C_i$ | E | I | | |
|---|---|---|---|---|
| $A_g$ | 1 | 1 | $R_x, R_y, R_z$ | $x^2, y^2, z^2$ xy,xz,yz |
| $A_u$ | 1 | -1 | x,y,z | |

2. The $C_n$ group

| $C_1$ | E |
|---|---|
| A | 1 |

| $C_2$ | E | $C_2$ | | |
|---|---|---|---|---|
| A | 1 | 1 | $z, R_z$ | $x^2, y^2, z^2, xy$ |
| B | 1 | -1 | $x, y, R_x, R_y$ | yz, xz |

| $C_3$ | E | $C_3$ | $C_3^2$ | | $\omega = \exp(2\pi i/3)$ |
|---|---|---|---|---|---|
| A | 1 | 1 | 1 | $z, R_z$ | $x^2 + y^2, z^2$ |
| E | $\begin{Bmatrix} 1 \\ 1 \end{Bmatrix}$ | $\begin{matrix} \omega \\ \omega^* \end{matrix}$ | $\begin{matrix} \omega^* \\ \omega \end{matrix}$ | $(x, y)(R_x, R_y)$ | $(x^2 - y^2, xy)(yz, xz)$ |

| $C_4$ | E | $C_4$ | $C_2$ | $C_4^3$ | | |
|---|---|---|---|---|---|---|
| A | 1 | 1 | 1 | 1 | $z, R_z$ | $x^2 + y^2, z^2$ |
| B | 1 | -1 | 1 | -1 | | $x^2 - y^2, xy$ |
| E | $\begin{Bmatrix} 1 \\ 1 \end{Bmatrix}$ | $\begin{matrix} i \\ -i \end{matrix}$ | $\begin{matrix} -1 \\ -1 \end{matrix}$ | $\begin{matrix} -i \\ i \end{matrix}$ | $(x, y)(R_x, R_y)$ | (yz, xz) |

| $\mathbf{C_5}$ | E | $C_5$ | $C_5^2$ | $C_5^3$ | $C_5^4$ | | $\omega=\exp(2\pi i/5)$ |
|---|---|---|---|---|---|---|---|
| A | 1 | 1 | 1 | 1 | 1 | $z, R_z$ | $x^2+y^2, z^2$ |
| $E_1$ | 1 | $\omega$ | $\omega^2$ | $\omega^{2*}$ | $\omega^*$ | $(x,y)(R_x,R_y)$ | $(yz, xz)$ |
| | 1 | $\omega^*$ | $\omega^{2*}$ | $\omega^2$ | $\omega$ | | |
| $E_2$ | 1 | $\omega^2$ | $\omega^*$ | $\omega$ | $\omega^{2*}$ | | $(x^2-y^2, xy)$ |
| | 1 | $\omega^{2*}$ | $\omega$ | $\omega^*$ | $\omega^2$ | | |

| $\mathbf{C_6}$ | E | $C_6$ | $C_3$ | $C_2$ | $C_3^2$ | $C_6^5$ | | $\omega=\exp(2\pi i/6)$ |
|---|---|---|---|---|---|---|---|---|
| A | 1 | 1 | 1 | 1 | 1 | 1 | $z, R_z$ | $x^2+y^2, z^2$ |
| B | 1 | -1 | 1 | -1 | 1 | -1 | | |
| $E_1$ | 1 | $\omega$ | $-\omega^*$ | -1 | $-\omega$ | $\omega^*$ | $(x, y)$ | $(yz, xz)$ |
| | 1 | $\omega^*$ | $-\omega$ | -1 | $-\omega^*$ | $\omega$ | $(R_x, R_y)$ | |
| $E_2$ | 1 | $-\omega^*$ | $-\omega$ | 1 | $-\omega^*$ | $-\omega$ | | $(x^2-y^2, xy)$ |
| | 1 | $-\omega$ | $-\omega^*$ | 1 | $-\omega$ | $-\omega^*$ | | |

| $\mathbf{C_7}$ | E | $C_7$ | $C_7^2$ | $C_7^3$ | $C_7^4$ | $C_7^5$ | $C_7^6$ | | $\omega=\exp(2\pi i/7)$ |
|---|---|---|---|---|---|---|---|---|---|
| A | 1 | 1 | 1 | 1 | 1 | 1 | 1 | $z, R_z$ | $x^2+y^2, z^2$ |
| $E_1$ | 1 | $\omega$ | $\omega^2$ | $\omega^3$ | $\omega^{3*}$ | $\omega^{2*}$ | $\omega^*$ | $(x,y)$ | $(xz, yz)$ |
| | 1 | $\omega^*$ | $\omega^{2*}$ | $\omega^{3*}$ | $\omega^3$ | $\omega^2$ | $\omega$ | $(R_x, R_y)$ | |
| $E_2$ | 1 | $\omega^2$ | $\omega^{3*}$ | $\omega^*$ | $\omega$ | $\omega^3$ | $\omega^{2*}$ | | $(x^2-y^2, xy)$ |
| | 1 | $\omega^{2*}$ | $\omega^3$ | $\omega$ | $\omega^*$ | $\omega^{3*}$ | $\omega^2$ | | |
| $E_3$ | 1 | $\omega^3$ | $\omega^*$ | $\omega^2$ | $\omega^{2*}$ | $\omega$ | $\omega^{3*}$ | | |
| | 1 | $\omega^{3*}$ | $\omega$ | $\omega^{2*}$ | $\omega^2$ | $\omega^*$ | $\omega^3$ | | |

| $\mathbf{C_8}$ | E | $C_8$ | $C_4$ | $C_8^2$ | $C_2$ | $C_8^5$ | $C_4^3$ | $C_4^7$ | | $\omega = \exp(2\pi i/8)$ |
|---|---|---|---|---|---|---|---|---|---|---|
| A | 1 | 1 | 1 | 1 | 1 | 1 | 1 | 1 | $z, R_z$ | $x^2+y^2, z^2$ |
| B | 1 | -1 | 1 | -1 | 1 | -1 | 1 | -1 | | |
| $E_1$ | $\{1$ | $\omega$ | i | $-\omega^*$ | -1 | $-\omega$ | -i | $\omega^*\}$ | $(x, y)$ | $(xz, yz)$ |
| | $\{1$ | $\omega^*$ | -i | $-\omega$ | -1 | $-\omega^*$ | i | $\omega\}$ | $(R_x, R_y)$ | |
| $E_2$ | $\{1$ | i | -1 | i | 1 | i | -1 | -i$\}$ | | $(x^2-y^2, xy)$ |
| | $\{1$ | -i | -1 | -i | 1 | -i | -1 | i$\}$ | | |

3. The $D_n$ Groups

| $\mathbf{D_2}$ | E | $C_2(z)$ | $C_2(y)$ | $C_2(x)$ | | |
|---|---|---|---|---|---|---|
| A | 1 | 1 | 1 | 1 | | $x^2, y^2, z^2$ |
| $B_1$ | 1 | 1 | -1 | -1 | $z, R_z$ | xy |
| $B_2$ | 1 | -1 | 1 | -1 | $y, R_y$ | xz |
| $B_3$ | 1 | -1 | -1 | 1 | $x, R_x$ | yz |

| $\mathbf{D_3}$ | E | $2C_3$ | $3C_2$ | | |
|---|---|---|---|---|---|
| $A_1$ | 1 | 1 | 1 | | $x^2+y^2, z^2$ |
| $A_2$ | 1 | 1 | -1 | $z, R_z$ | |
| E | 2 | -1 | 0 | $(x, y)(R_x, R_y)$ | $(x^2-y^2, xy)(xz, yz)$ |

| $\mathbf{D_4}$ | E | $2C_4$ | $C_2(=C_4^2)$ | $2C_2'$ | $2C_2''$ | | |
|---|---|---|---|---|---|---|---|
| $A_1$ | 1 | 1 | 1 | 1 | 1 | | $x^2+y^2, z^2$ |
| $A_2$ | 1 | 1 | 1 | -1 | -1 | $z, R_z$ | |
| $B_1$ | 1 | -1 | 1 | 1 | -1 | | $x^2-y^2$ |
| $B_2$ | 1 | -1 | 1 | -1 | 1 | | xy |
| E | 2 | 0 | -2 | 0 | 0 | $(x, y)(R_x, R_y)$ | $(xz, yz)$ |

| $\mathbf{D_4}$ | E | $2C_4$ | $C_2(=C_4^2)$ | $2C_2'$ | $2C_2''$ | | |
|---|---|---|---|---|---|---|---|
| $A_1$ | 1 | 1 | 1 | 1 | 1 | | $x^2+y^2, z^2$ |
| $A_2$ | 1 | 1 | 1 | -1 | -1 | $z, R_z$ | |
| $B_1$ | 1 | -1 | 1 | 1 | -1 | | $x^2-y^2$ |
| $B_2$ | 1 | -1 | 1 | -1 | 1 | | $xy$ |
| E | 2 | 0 | -2 | 0 | 0 | $(x, y)(R_x, R_y)$ | $(xz, yz)$ |

| $\mathbf{D_6}$ | E | $2C_6$ | $2C_3$ | $C_2$ | $3C_2'$ | $3C_2'$ | | |
|---|---|---|---|---|---|---|---|---|
| $A_1$ | 1 | 1 | 1 | 1 | 1 | 1 | | $x^2+y^2, z^2$ |
| $A_2$ | 1 | 1 | 1 | 1 | -1 | -1 | $z, R_z$ | |
| $B_1$ | 1 | -1 | 1 | -1 | 1 | -1 | | |
| $B_2$ | 1 | -1 | 1 | -1 | -1 | 1 | | |
| $E_1$ | 2 | 1 | -1 | -2 | 0 | 0 | $(x, y)(R_x, R_y)$ | $(xz, yz)$ |
| $E_2$ | 2 | -1 | -1 | 2 | 0 | 0 | | $(x^2-y^2, xy)$ |

4. The $C_{nv}$ groups

| $\mathbf{C_{2v}}$ | E | $C_2$ | $\sigma_v(xz)$ | $\sigma_v'(yz)$ | | |
|---|---|---|---|---|---|---|
| $A_1$ | 1 | 1 | 1 | 1 | z | $x^2, y^2, z^2$ |
| $A_2$ | 1 | 1 | -1 | -1 | $R_z$ | xy |
| $B_1$ | 1 | -1 | 1 | -1 | $x, R_y$ | xz |
| $B_2$ | 1 | -1 | -1 | 1 | $y, R_x$ | yz |

| $\mathbf{C_{3v}}$ | E | $2C_3$ | $3\sigma_v$ | | |
|---|---|---|---|---|---|
| $A_1$ | 1 | 1 | 1 | z | $x^2+y^2, z^2$ |
| $A_2$ | 1 | 1 | -1 | $R_z$ | |
| E | 2 | -1 | 0 | $(x, y)(R_x, R_y)$ | $(x^2-y^2, xy)(xz, yz)$ |

| $C_{4v}$ | $E$ | $2C_4$ | $C_2$ | $2\sigma_v$ | $2\sigma_d$ | | |
|---|---|---|---|---|---|---|---|
| $A_1$ | 1 | 1 | 1 | 1 | 1 | $z$ | $x^2+y^2, z^2$ |
| $A_2$ | 1 | 1 | 1 | -1 | -1 | $R_z$ | |
| $B_1$ | 1 | -1 | 1 | 1 | -1 | | $x^2-y^2$ |
| $B_2$ | 1 | -1 | 1 | -1 | 1 | | $xy$ |
| $E$ | 2 | 0 | -2 | 0 | 0 | $(x, y)(R_x, R_y)$ | $(xz, yz)$ |

| $C_{5v}$ | $E$ | $2C_5$ | $2C_5^2$ | $5\sigma_v$ | | |
|---|---|---|---|---|---|---|
| $A_1$ | 1 | 1 | 1 | 1 | $z$ | $x^2+y^2, z^2$ |
| $A_2$ | 1 | 1 | 1 | -1 | $R_z$ | |
| $E_1$ | 2 | $2\cos 72^\circ$ | $2\cos 144^\circ$ | 0 | $(x, y)(R_x, R_y)$ | $(xz, yz)$ |
| $E_2$ | 2 | $2\cos 144^\circ$ | $2\cos 72^\circ$ | 0 | | $(x^2-y^2, xy)$ |

| $C_{6v}$ | $E$ | $2C_6$ | $2C_3$ | $C_2$ | $3\sigma_v$ | $3\sigma_d$ | | |
|---|---|---|---|---|---|---|---|---|
| $A_1$ | 1 | 1 | 1 | 1 | 1 | 1 | $z$ | $x^2+y^2, z^2$ |
| $A_2$ | 1 | 1 | 1 | 1 | -1 | -1 | $R_z$ | |
| $B_1$ | 1 | -1 | 1 | -1 | 1 | -1 | | |
| $B_2$ | 1 | -1 | 1 | -1 | -1 | 1 | | |
| $E_1$ | 2 | 1 | -1 | -2 | 0 | 0 | $(x, y)(R_x, R_y)$ | $(xz, yz)$ |
| $E_2$ | 2 | -1 | -1 | 2 | 0 | 0 | | $(x^2-y^2, xy)$ |

5. The $C_{nh}$ Groups

| $C_{2h}$ | $E$ | $C_2$ | $I$ | $\sigma_h$ | | |
|---|---|---|---|---|---|---|
| $A_g$ | 1 | 1 | 1 | 1 | $R_z$ | $x^2, y^2, z^2, xy$ |
| $B_g$ | 1 | -1 | 1 | -1 | $R_x, R_y$ | $xz, yz$ |
| $A_u$ | 1 | 1 | -1 | -1 | $z$ | |
| $B_u$ | 1 | -1 | -1 | 1 | $x, y$ | |

| $\mathbf{C_{3h}}$ | $E$ | $C_3$ | $C_3^2$ | $\sigma_h$ | $S_3$ | $S_3^5$ | | $\omega = \exp(2\pi i/3)$ |
|---|---|---|---|---|---|---|---|---|
| $A'$ | 1 | 1 | 1 | 1 | 1 | 1 | $R_z$ | |
| $E'$ | 1 | $\omega$ | $\omega^*$ | 1 | $\omega$ | $\omega^*$ | $(x, y)$ | |
| | 1 | $\omega^*$ | $\omega$ | 1 | $\omega^*$ | $\omega$ | | |
| $A''$ | 1 | 1 | 1 | −1 | −1 | −1 | $z$ | |
| $E''$ | 1 | $\omega$ | $\omega^*$ | −1 | $-\omega$ | $-\omega^*$ | $(R_x, R_y)$ | |
| | 1 | $\omega^*$ | $\omega$ | −1 | $-\omega^*$ | $-\omega$ | | |

| $\mathbf{C_{4h}}$ | $E$ | $C_4$ | $C_2$ | $C_4^3$ | $I$ | $S_4^3$ | $\sigma_h$ | $S_4$ | | |
|---|---|---|---|---|---|---|---|---|---|---|
| $A_g$ | 1 | 1 | 1 | 1 | 1 | 1 | 1 | 1 | $R_z$ | $x^2 + y^2, z^2$ |
| $B_g$ | 1 | -1 | 1 | -1 | 1 | -1 | 1 | -1 | | $x^2 - y^2, xy$ |
| $E_g$ | 1 | i | -1 | -i | 1 | i | -1 | -i | $(R_x, R_y)$ | $(xz, yz)$ |
| | 1 | -i | -1 | i | 1 | -i | -1 | i | | |
| $A_u$ | 1 | 1 | 1 | 1 | -1 | -1 | -1 | -1 | $z$ | |
| $B_u$ | 1 | -1 | 1 | -1 | -1 | 1 | -1 | 1 | | |
| $E_u$ | 1 | i | -1 | -i | -1 | -i | 1 | i | $(x, y)$ | |
| | 1 | -i | -1 | i | -1 | i | 1 | -i | | |

| $\mathbf{C_{5h}}$ | $E$ | $C_5$ | $C_5^2$ | $C_5^3$ | $C_5^4$ | $\sigma_h$ | $S_5$ | $S_5^7$ | $S_5^3$ | $S_5^9$ | | $\omega = \exp(2\pi i/5)$ |
|---|---|---|---|---|---|---|---|---|---|---|---|---|
| $A'$ | 1 | 1 | 1 | 1 | 1 | 1 | 1 | 1 | 1 | 1 | $R_z$ | $x^2 + y^2, z^2$ |
| $E_1'$ | 1 | $\omega$ | $\omega^2$ | $\omega^{2*}$ | $\omega^*$ | 1 | $\omega$ | $\omega^2$ | $\omega^{2*}$ | $\omega^*$ | $(x, y)$ | |
| | 1 | $\omega^*$ | $\omega^{2*}$ | $\omega^2$ | $\omega$ | 1 | $\omega^*$ | $\omega^{2*}$ | $\omega^2$ | $\omega$ | | |
| $E_2'$ | 1 | $\omega^2$ | $\omega^*$ | $\omega$ | $\omega^{2*}$ | 1 | $\omega^2$ | $\omega^*$ | $\omega$ | $\omega^{2*}$ | | $(x^2 - y^2, xy)$ |
| | 1 | $\omega^{2*}$ | $\omega$ | $\omega^*$ | $\omega^2$ | 1 | $\omega^{2*}$ | $\omega$ | $\omega^*$ | $\omega^2$ | | |
| $A''$ | 1 | 1 | 1 | 1 | 1 | −1 | −1 | −1 | −1 | −1 | $z$ | |
| $E_1''$ | 1 | $\omega$ | $\omega^2$ | $\omega^{2*}$ | $\omega^*$ | −1 | $-\omega$ | $-\omega^2$ | $-\omega^{2*}$ | $-\omega^*$ | $(R_x, R_y)$ | $(xz, yz)$ |
| | 1 | $\omega^*$ | $\omega^{2*}$ | $\omega^2$ | $\omega$ | −1 | $-\omega^*$ | $-\omega^{2*}$ | $-\omega^2$ | $-\omega$ | | |
| $E_2''$ | 1 | $\omega^2$ | $\omega^*$ | $\omega$ | $\omega^{2*}$ | −1 | $-\omega^2$ | $-\omega^*$ | $-\omega$ | $-\omega^{2*}$ | | |
| | 1 | $\omega^{2*}$ | $\omega$ | $\omega^*$ | $\omega^2$ | −1 | $-\omega^{2*}$ | $-\omega$ | $-\omega^*$ | $-\omega^2$ | | |

| $\mathbf{C_{6h}}$ | E | $C_6$ | $C_3$ | $C_2$ | $C_3^2$ | $C_6^5$ | I | $S_3^5$ | $S_6^5$ | $\sigma_h$ | $S_6$ | $S_3$ | | $\omega = \exp(2\pi i/6)$ |
|---|---|---|---|---|---|---|---|---|---|---|---|---|---|---|
| $A_g$ | 1 | 1 | 1 | 1 | 1 | 1 | 1 | 1 | 1 | 1 | 1 | 1 | $R_z$ | |
| $B_g$ | 1 | -1 | 1 | -1 | 1 | -1 | 1 | -1 | 1 | -1 | 1 | -1 | | |
| $E_{1g}$ | 1 | $\omega$ | $-\omega^*$ | −1 | $-\omega$ | $\omega^*$ | 1 | $\omega$ | $-\omega^*$ | −1 | $-\omega$ | $\omega^*$ | $(R_x, R_y)$ | (xz, yz) |
| | 1 | $\omega^*$ | $-\omega$ | −1 | $-\omega^*$ | $\omega$ | 1 | $\omega^*$ | $-\omega$ | −1 | $-\omega^*$ | $\omega$ | | |
| $E_{2g}$ | 1 | $-\omega^*$ | $-\omega$ | 1 | $-\omega^*$ | $-\omega$ | 1 | $-\omega^*$ | $-\omega$ | 1 | $-\omega^*$ | $-\omega$ | | $(x^2 - y^2, xy)$ |
| | 1 | $-\omega$ | $-\omega^*$ | 1 | $-\omega$ | $-\omega^*$ | 1 | $-\omega$ | $-\omega^*$ | 1 | $-\omega$ | $-\omega^*$ | | |
| $A_u$ | 1 | 1 | 1 | 1 | 1 | 1 | −1 | −1 | −1 | −1 | −1 | −1 | z | |
| $B_u$ | 1 | −1 | 1 | −1 | 1 | −1 | −1 | 1 | −1 | 1 | −1 | 1 | | |
| $E_{1u}$ | 1 | $\omega$ | $-\omega^*$ | −1 | $-\omega$ | $\omega^*$ | −1 | $-\omega$ | $\omega^*$ | 1 | $\omega$ | $-\omega^*$ | (x, y) | |
| | 1 | $\omega^*$ | $-\omega$ | −1 | $-\omega^*$ | $\omega$ | −1 | $-\omega^*$ | $\omega$ | 1 | $\omega^*$ | $-\omega$ | | |
| $E_{2u}$ | 1 | $-\omega^*$ | $-\omega$ | 1 | $-\omega^*$ | $-\omega$ | −1 | $\omega^*$ | $\omega$ | −1 | $\omega^*$ | $\omega$ | | |
| | 1 | $-\omega$ | $-\omega$ | 1 | $-\omega$ | $-\omega^*$ | −1 | $\omega$ | $\omega^*$ | −1 | $\omega$ | $\omega^*$ | | |

6.The $D_{nh}$ groups

| $\mathbf{D_{2h}}$ | E | $C_2(z)$ | $C_2(y)$ | $C_2(x)$ | I | σ(xy) | σ(xz) | σ(yz) | | |
|---|---|---|---|---|---|---|---|---|---|---|
| $A_g$ | 1 | 1 | 1 | 1 | 1 | 1 | 1 | 1 | | $x^2, y^2, z^2$ |
| $B_{1g}$ | 1 | 1 | -1 | -1 | 1 | 1 | -1 | -1 | $R_z$ | xy |
| $B_{2g}$ | 1 | -1 | 1 | -1 | 1 | -1 | 1 | -1 | $R_y$ | xz |
| $B_{3g}$ | 1 | -1 | -1 | 1 | 1 | -1 | -1 | 1 | $R_x$ | yz |
| $A_u$ | 1 | 1 | 1 | 1 | -1 | -1 | -1 | -1 | | |
| $B_{1u}$ | 1 | 1 | -1 | -1 | -1 | -1 | 1 | 1 | z | |
| $B_{2u}$ | 1 | -1 | 1 | -1 | -1 | 1 | -1 | 1 | y | |
| $B_{3u}$ | 1 | -1 | -1 | 1 | -1 | 1 | 1 | -1 | x | |

| $D_{3h}$ | $E$ | $2C_3$ | $3C_2$ | $\sigma_h$ | $2S_3$ | $3\sigma_v$ | | |
|---|---|---|---|---|---|---|---|---|
| $A'_1$ | 1 | 1 | 1 | 1 | 1 | 1 | | $x^2+y^2, z^2$ |
| $A'_2$ | 1 | 1 | -1 | 1 | 1 | -1 | $R_z$ | |
| $E'$ | 2 | -1 | 0 | 2 | -1 | 0 | $(x, y)$ | $(x^2-y^2, xy)$ |
| $A''_1$ | 1 | 1 | 1 | -1 | -1 | -1 | | |
| $A''_2$ | 1 | 1 | -1 | -1 | -1 | 1 | $z$ | |
| $E''$ | 2 | -1 | 0 | -2 | 1 | 0 | $(R_x, R_y)$ | $(xz, yz)$ |

| $D_{4h}$ | $E$ | $2C_4$ | $C_2$ | $2C'_2$ | $2C''_2$ | $I$ | $2S_4$ | $\sigma_h$ | $2\sigma_v$ | $2\sigma_d$ | | |
|---|---|---|---|---|---|---|---|---|---|---|---|---|
| $A_{1g}$ | 1 | 1 | 1 | 1 | 1 | 1 | 1 | 1 | 1 | 1 | | $x^2+y^2, z^2$ |
| $A_{2g}$ | 1 | 1 | 1 | -1 | -1 | 1 | 1 | 1 | -1 | -1 | $R_z$ | |
| $B_{1g}$ | 1 | -1 | 1 | -1 | 1 | 1 | -1 | 1 | -1 | 1 | | $x^2-y^2$ |
| $B_{2g}$ | 1 | -1 | 1 | -1 | 1 | 1 | -1 | 1 | -1 | 1 | | $xy$ |
| $E_g$ | 2 | 0 | -2 | 0 | 0 | 2 | 0 | -2 | 0 | 0 | $(R_x, R_y)$ | $(xz, yz)$ |
| $A_{1u}$ | 1 | 1 | 1 | 1 | 1 | -1 | -1 | -1 | -1 | -1 | | |
| $A_{2u}$ | 1 | 1 | 1 | -1 | -1 | -1 | -1 | -1 | 1 | 1 | $z$ | |
| $B_{1u}$ | 1 | -1 | 1 | 1 | -1 | -1 | 1 | -1 | -1 | 1 | | |
| $B_{2u}$ | 1 | -1 | 1 | -1 | 1 | -1 | 1 | -1 | 1 | -1 | | |
| $E_u$ | 2 | 0 | -2 | 0 | 0 | -2 | 0 | 2 | 0 | 0 | $(x, y)$ | |

| $D_{5h}$ | $E$ | $2C_5$ | $2C_5^2$ | $5C_2$ | $\sigma_h$ | $2S_5$ | $2S_5^3$ | $5\sigma_v$ | | |
|---|---|---|---|---|---|---|---|---|---|---|
| $A'_1$ | 1 | 1 | 1 | 1 | 1 | 1 | 1 | 1 | | $x^2+y^2, z^2$ |
| $A'_2$ | 1 | 1 | 1 | -1 | 1 | 1 | 1 | -1 | $R_z$ | |
| $E'_1$ | 2 | 2cos72° | 2cos144° | 0 | 2 | 2cos72° | 2cos144° | 0 | $(x, y)$ | |
| $E'_2$ | 2 | 2cos144° | 2cos72° | 0 | 1 | 2cos144° | 2cos72° | 0 | | $(x^2-y^2, xy)$ |
| $A''_1$ | 1 | 1 | 1 | 1 | -1 | -1 | -1 | -1 | | |
| $A''_2$ | 1 | 1 | 1 | -1 | -1 | -1 | -1 | 1 | $z$ | |
| $E''_1$ | 2 | 2cos72° | 2cos144° | 0 | -2 | -2cos72° | -2cos144° | 0 | $(R_x, R_y)$ | $(xz, yz)$ |
| $E''_2$ | 2 | 2cos144° | 2cos72° | 0 | -2 | -2cos144° | -2cos72° | 0 | | |

| $D_{6h}$ | E | $2C_6$ | $2C_3$ | $C_2$ | $3C_2'$ | $3C_2''$ | I | $2S_3$ | $2S_6$ | $\sigma_h$ | $3\sigma_d$ | $3\sigma_v$ | | |
|---|---|---|---|---|---|---|---|---|---|---|---|---|---|---|
| $A_{1g}$ | 1 | 1 | 1 | 1 | 1 | 1 | 1 | 1 | 1 | 1 | 1 | 1 | | $x^2+y^2, z^2$ |
| $A_{2g}$ | 1 | 1 | 1 | 1 | −1 | −1 | 1 | 1 | 1 | 1 | −1 | −1 | $R_z$ | |
| $B_{1g}$ | 1 | −1 | 1 | −1 | 1 | −1 | 1 | −1 | 1 | −1 | 1 | −1 | | |
| $B_{2g}$ | 1 | −1 | 1 | −1 | −1 | 1 | 1 | −1 | 1 | −1 | −1 | 1 | | |
| $E_{1g}$ | 2 | 1 | −1 | −2 | 0 | 0 | 2 | 1 | −1 | −2 | 0 | 0 | $(R_x, R_y)$ | (xz, yz) |
| $E_{2g}$ | 2 | −1 | −1 | 2 | 0 | 0 | 2 | −1 | −1 | 2 | 0 | 0 | | $(x^2-y^2, xy)$ |
| $A_{1u}$ | 1 | 1 | 1 | 1 | 1 | 1 | −1 | −1 | −1 | −1 | −1 | −1 | | |
| $A_{2u}$ | 1 | 1 | 1 | 1 | −1 | −1 | −1 | −1 | −1 | −1 | 1 | 1 | z | |
| $B_{1u}$ | 1 | −1 | 1 | −1 | 1 | −1 | −1 | 1 | −1 | 1 | −1 | 1 | | |
| $B_{2u}$ | 1 | −1 | 1 | −1 | −1 | 1 | −1 | 1 | −1 | 1 | 1 | −1 | | |
| $E_{1u}$ | 2 | 1 | −1 | −2 | 0 | 0 | −2 | −1 | 1 | 2 | 0 | 0 | (x, y) | |
| $E_{2u}$ | 2 | −1 | −1 | 2 | 0 | 0 | −2 | 1 | 1 | −2 | 0 | 0 | | |

| $D_{8h}$ | E | $2C_8$ | $2C_8^3$ | $2C_4$ | $C_2$ | $4C_2'$ | $4C_2''$ | I | $2S_8$ | $2S_8^3$ | $2S_4$ | $\sigma_h$ | $4\sigma_d$ | $4\sigma_v$ | | |
|---|---|---|---|---|---|---|---|---|---|---|---|---|---|---|---|---|
| $A_{1g}$ | 1 | 1 | 1 | 1 | 1 | 1 | 1 | 1 | 1 | 1 | 1 | 1 | 1 | 1 | | $x^2+y^2, z^2$ |
| $A_{2g}$ | 1 | 1 | 1 | 1 | 1 | -1 | -1 | 1 | 1 | 1 | 1 | 1 | -1 | -1 | $R_z$ | |
| $B_{1g}$ | 1 | -1 | -1 | 1 | 1 | 1 | -1 | 1 | -1 | -1 | 1 | 1 | 1 | -1 | | |
| $B_{2g}$ | 1 | -1 | -1 | 1 | 1 | -1 | 1 | 1 | -1 | -1 | 1 | 1 | -1 | 1 | | |
| $E_{1g}$ | 2 | $\sqrt{2}$ | $-\sqrt{2}$ | 0 | -2 | 0 | 0 | 2 | $\sqrt{2}$ | $-\sqrt{2}$ | 0 | -2 | 0 | 0 | $(R_x, R_y)$ | (xz, yz) |
| $E_{2g}$ | 2 | 0 | 0 | -2 | 2 | 0 | 0 | 2 | 0 | 0 | -2 | 2 | 0 | 0 | | $(x^2-y^2, xy)$ |
| $E_{3g}$ | 2 | $-\sqrt{2}$ | $\sqrt{2}$ | 0 | -2 | 0 | 0 | 2 | $-\sqrt{2}$ | $\sqrt{2}$ | 0 | -2 | 0 | 0 | | |
| $A_{1u}$ | 1 | 1 | 1 | 1 | 1 | 1 | 1 | -1 | -1 | -1 | -1 | -1 | -1 | -1 | | |
| $A_{2u}$ | 1 | 1 | 1 | 1 | 1 | -1 | -1 | -1 | -1 | -1 | -1 | -1 | 1 | 1 | z | |
| $B_{1u}$ | 1 | -1 | -1 | 1 | 1 | 1 | -1 | -1 | 1 | 1 | -1 | -1 | -1 | 1 | | |
| $B_{2u}$ | 1 | -1 | -1 | 1 | 1 | -1 | 1 | -1 | 1 | 1 | -1 | -1 | 1 | -1 | | |
| $E_{1u}$ | 2 | $\sqrt{2}$ | $-\sqrt{2}$ | 0 | -2 | 0 | 0 | -2 | $-\sqrt{2}$ | $\sqrt{2}$ | 0 | 2 | 0 | 0 | (x, y) | |
| $E_{2u}$ | 2 | 0 | 0 | -2 | 2 | 0 | 0 | -2 | 0 | 0 | 2 | -2 | 0 | 0 | | |
| $E_{3u}$ | 2 | $-\sqrt{2}$ | $\sqrt{2}$ | 0 | -2 | 0 | 0 | -2 | $\sqrt{2}$ | $-\sqrt{2}$ | 0 | 2 | 0 | 0 | | |

7. The $D_{nd}$ groups

| $D_{2d}$ | $E$ | $2S_4$ | $C_2$ | $2C_2'$ | $2\sigma_d$ | | |
|---|---|---|---|---|---|---|---|
| $A_1$ | 1 | 1 | 1 | 1 | 1 | | $x^2+y^2, z^2$ |
| $A_2$ | 1 | 1 | 1 | -1 | -1 | $R_z$ | |
| $B_1$ | 1 | -1 | 1 | 1 | -1 | | $x^2-y^2$ |
| $B_2$ | 1 | -1 | 1 | -1 | 1 | $z$ | $xy$ |
| $E$ | 2 | 0 | -2 | 0 | 0 | $(x, y); (R_x, R_y)$ | $(xz, yz)$ |

| $D_{3d}$ | $E$ | $2C_3$ | $3C_2$ | $I$ | $2S_6$ | $3\sigma_d$ | | |
|---|---|---|---|---|---|---|---|---|
| $A_{1g}$ | 1 | 1 | 1 | 1 | 1 | 1 | | $x^2+y^2, z^2$ |
| $A_{2g}$ | 1 | 1 | -1 | 1 | 1 | -1 | $R_z$ | |
| $E_g$ | 2 | -1 | 0 | 2 | -1 | 0 | $(R_x, R_y)$ | $(x^2-y^2, xy)$, $(xz, yz)$ |
| $A_{1u}$ | 1 | 1 | 1 | -1 | -1 | -1 | | |
| $A_{2u}$ | 1 | 1 | -1 | -1 | -1 | 1 | $z$ | |
| $E_u$ | 2 | -1 | 0 | -2 | 1 | 0 | $(x, y)$ | |

| $D_{4d}$ | $E$ | $2S_8$ | $2C_4$ | $2S_8^3$ | $C_2$ | $4C_2'$ | $4\sigma_d$ | | |
|---|---|---|---|---|---|---|---|---|---|
| $A_1$ | 1 | 1 | 1 | 1 | 1 | 1 | 1 | | $x^2+y^2, z^2$ |
| $A_2$ | 1 | 1 | 1 | 1 | 1 | -1 | -1 | $R_z$ | |
| $B_1$ | 1 | -1 | 1 | -1 | 1 | 1 | -1 | | |
| $B_2$ | 1 | -1 | 1 | -1 | 1 | -1 | 1 | $z$ | |
| $E_1$ | 2 | $\sqrt{2}$ | 0 | $-\sqrt{2}$ | -2 | 0 | 0 | $(x, y)$ | |
| $E_2$ | 2 | 0 | -2 | 0 | 2 | 0 | 0 | | $(x^2-y^2, xy)$ |
| $E_3$ | 2 | $-\sqrt{2}$ | 0 | $\sqrt{2}$ | -2 | 0 | 0 | $(R_x, R_y)$ | $(yz, xz)$ |

| $D_{5d}$ | E | $2C_5$ | $2C_5^2$ | $5C_2$ | I | $2S_{10}^3$ | $2S_{10}$ | $5\sigma_d$ | | |
|---|---|---|---|---|---|---|---|---|---|---|
| $A_{1g}$ | 1 | 1 | 1 | 1 | 1 | 1 | 1 | 1 | | $x^2+y^2, z^2$ |
| $A_{2g}$ | 1 | 1 | 1 | -1 | 1 | 1 | 1 | -1 | $R_z$ | |
| $E_{1g}$ | 2 | 2cos72° | 2cos144° | 0 | 2 | 2cos72° | 2cos144° | 0 | $(R_x, R_y)$ | (xz,yz) |
| $E_{2g}$ | 2 | 2cos144° | 2cos72° | 0 | 2 | 2cos144° | 2cos72° | 0 | | $(x^2-y^2, xy)$ |
| $A_{1u}$ | 1 | 1 | 1 | 1 | -1 | -1 | -1 | -1 | | |
| $A_{2u}$ | 1 | 1 | 1 | -1 | -1 | -1 | -1 | 1 | $z$ | |
| $E_{1u}$ | 2 | 2cos72° | 2cos144° | 0 | -2 | -2cos72° | -2cos144° | 0 | (x,y) | |
| $E_{2u}$ | 2 | 2cos144° | 2cos72° | 0 | -2 | -2cos144° | -2cos72° | 0 | | |

| $D_{6d}$ | E | $2S_{12}$ | $2C_6$ | $2S_4$ | $2C_3$ | $2S_{12}^5$ | $C_2$ | $6C_2'$ | $6\sigma_d$ | | |
|---|---|---|---|---|---|---|---|---|---|---|---|
| $A_1$ | 1 | 1 | 1 | 1 | 1 | 1 | 1 | 1 | 1 | | $x^2+y^2, z^2$ |
| $A_2$ | 1 | 1 | 1 | 1 | 1 | 1 | 1 | -1 | -1 | $R_z$ | |
| $B_1$ | 1 | -1 | 1 | -1 | 1 | -1 | 1 | 1 | -1 | | |
| $B_2$ | 1 | -1 | 1 | -1 | 1 | -1 | 1 | -1 | 1 | $z$ | |
| $E_1$ | 2 | $\sqrt{3}$ | 1 | 0 | -1 | $-\sqrt{3}$ | -2 | 0 | 0 | (x,y) | |
| $E_2$ | 2 | 1 | -1 | -2 | -1 | 1 | 2 | 0 | 0 | | $(x^2-y^2, xy)$ |
| $E_3$ | 2 | 0 | -2 | 0 | 2 | 0 | -2 | 0 | 0 | | |
| $E_4$ | 2 | -1 | -1 | 2 | -1 | -1 | 2 | 0 | 0 | | |
| $E_5$ | 2 | $-\sqrt{3}$ | 1 | 0 | -1 | $\sqrt{3}$ | -2 | 0 | 0 | $(R_x, R_y)$ | (xz,yz) |

8. The $S_n$ groups

| $S_4$ | E | $S_4$ | $C_2$ | $S_4^3$ | | |
|---|---|---|---|---|---|---|
| A | 1 | 1 | 1 | 1 | $R_z$ | $x^2+y^2, z^2$ |
| B | 1 | -1 | 1 | -1 | z | $x^2-y^2, xy$ |
| E | $\begin{Bmatrix} 1 \\ 1 \end{Bmatrix}$ | $\begin{matrix} i \\ -i \end{matrix}$ | $\begin{matrix} -1 \\ -1 \end{matrix}$ | $\begin{matrix} -i \\ -i \end{matrix}$ | $(x,y); (R_x, R_y)$ | (xz,yz) |

| $S_6$ | $E$ | $C_3$ | $C_3^2$ | $I$ | $S_6^5$ | $S_6$ | | $\omega = \exp(2\pi i/3)$ |
|---|---|---|---|---|---|---|---|---|
| $A_g$ | 1 | 1 | 1 | 1 | 1 | 1 | $R_z$ | $x^2+y^2, z^2$ |
| $E_g$ | $\left\{\begin{matrix} 1 \\ 1 \end{matrix}\right.$ | $\begin{matrix} \omega \\ \omega^* \end{matrix}$ | $\begin{matrix} \omega^* \\ \omega \end{matrix}$ | $\begin{matrix} 1 \\ 1 \end{matrix}$ | $\begin{matrix} \omega \\ \omega^* \end{matrix}$ | $\left.\begin{matrix} \omega^* \\ \omega \end{matrix}\right\}$ | $(R_x, R_y)$ | $(x^2-y^2, xy); (xz, yz)$ |
| $A_u$ | 1 | 1 | 1 | -1 | -1 | -1 | $z$ | |
| $E_u$ | $\left\{\begin{matrix} 1 \\ 1 \end{matrix}\right.$ | $\begin{matrix} \omega \\ \omega^* \end{matrix}$ | $\begin{matrix} \omega^* \\ \omega \end{matrix}$ | $\begin{matrix} -1 \\ -1 \end{matrix}$ | $\begin{matrix} -\omega \\ -\omega^* \end{matrix}$ | $\left.\begin{matrix} -\omega^* \\ -\omega \end{matrix}\right\}$ | $(x, y)$ | |

| $S_8$ | $E$ | $S_8$ | $C_4$ | $S_8^3$ | $C_2$ | $S_8^5$ | $C_4^3$ | $S_8^7$ | | |
|---|---|---|---|---|---|---|---|---|---|---|
| $A$ | 1 | 1 | 1 | 1 | 1 | 1 | 1 | 1 | $R_z$ | $x^2+y^2, z^2$ |
| $B$ | 1 | -1 | 1 | -1 | 1 | -1 | 1 | -1 | $z$ | |
| $E_1$ | $\left\{\begin{matrix} 1 \\ 1 \end{matrix}\right.$ | $\begin{matrix} \omega \\ \omega \end{matrix}$ | $\begin{matrix} i \\ -i \end{matrix}$ | $\begin{matrix} -\omega^* \\ -\omega \end{matrix}$ | $\begin{matrix} -1 \\ -1 \end{matrix}$ | $\begin{matrix} -\omega \\ -\omega \end{matrix}$ | $\begin{matrix} -i \\ i \end{matrix}$ | $\left.\begin{matrix} \omega \\ \omega \end{matrix}\right\}$ | $(x, y); (R_x, R_y)$ | |
| $E_2$ | $\left\{\begin{matrix} 1 \\ 1 \end{matrix}\right.$ | $\begin{matrix} i \\ -i \end{matrix}$ | $\begin{matrix} -1 \\ -1 \end{matrix}$ | $\begin{matrix} -i \\ i \end{matrix}$ | $\begin{matrix} 1 \\ 1 \end{matrix}$ | $\begin{matrix} i \\ -i \end{matrix}$ | $\begin{matrix} -1 \\ -1 \end{matrix}$ | $\left.\begin{matrix} -i \\ i \end{matrix}\right\}$ | | $(x^2-y^2, xy)$ |
| $E_3$ | $\left\{\begin{matrix} 1 \\ 1 \end{matrix}\right.$ | $\begin{matrix} -\omega \\ -\omega \end{matrix}$ | $\begin{matrix} -i \\ i \end{matrix}$ | $\begin{matrix} \omega \\ \omega \end{matrix}$ | $\begin{matrix} -1 \\ -1 \end{matrix}$ | $\begin{matrix} \omega \\ \omega \end{matrix}$ | $\begin{matrix} i \\ -i \end{matrix}$ | $\left.\begin{matrix} -\omega \\ -\omega \end{matrix}\right\}$ | | $(xz, yz)$ |

9. The cubic groups

| **T** | E | $4C_3$ | $4C_3^2$ | $3C_2$ | | $\omega = \exp(2\pi i/3)$ |
|---|---|---|---|---|---|---|
| A | 1 | 1 | 1 | 1 | | $x^2+y^2+z^2$ |
| E | 1 | $\omega$ | $\omega$ | 1 | | $(2z^2-x^2-y^2, x^2-y^2)$ |
| | 1 | $\omega$ | $\omega$ | 1 | | |
| T | 3 | 0 | 0 | -1 | $(R_x,R_y,R_z)$ | (xy,xz,yz) |

| $\mathbf{T_h}$ | E | $4C_3$ | $4C_3^2$ | $3C_2$ | I | $4S_6$ | $4S_6^5$ | $3\sigma_h$ | | $\omega = \exp(2\pi i/3)$ |
|---|---|---|---|---|---|---|---|---|---|---|
| $A_g$ | 1 | 1 | 1 | 1 | 1 | 1 | 1 | 1 | | $x^2+y^2+z^2$ |
| $E_g$ | 1 | $\omega$ | $\omega^*$ | 1 | 1 | $\omega$ | $\omega^*$ | 1 | | $(2z^2-x^2-y^2, x^2-y^2)$ |
| | 1 | $\omega^*$ | $\omega$ | 1 | 1 | $\omega^*$ | $\omega$ | 1 | | |
| $T_g$ | 3 | 0 | 0 | -1 | 1 | 0 | 0 | -1 | $(R_x,R_y,R_z)$ | (xy,xz,yz) |
| $A_u$ | 1 | 1 | 1 | 1 | -1 | -1 | -1 | -1 | | |
| $E_u$ | 1 | $\omega$ | $\omega^*$ | 1 | -1 | $-\omega$ | $-\omega^*$ | -1 | | |
| | 1 | $\omega^*$ | $\omega$ | 1 | -1 | $-\omega^*$ | $-\omega$ | -1 | | |
| $T_u$ | 3 | 0 | 0 | -1 | -1 | 0 | 0 | 1 | (x,y,z) | |

| $T_d$ | E | $8C_3$ | $3C_2$ | $6S_4$ | $6\sigma_d$ | | |
|---|---|---|---|---|---|---|---|
| $A_1$ | 1 | 1 | 1 | 1 | 1 | | $x^2+y^2+z^2$ |
| $A_2$ | 1 | 1 | 1 | -1 | -1 | | |
| E | 2 | -1 | 2 | 0 | 0 | | $(2z^2-x^2-y^2, x^2-y^2)$ |
| $T_1$ | 3 | 0 | -1 | 1 | -1 | $(R_x, R_y, R_z)$ | |
| $T_2$ | 3 | 0 | -1 | -1 | 1 | $(x, y, z)$ | $(xy, xz, yz)$ |

| O | E | $8C_3$ | $3C_2(=C_2^4)$ | $6C_4$ | $6C_2$ | | |
|---|---|---|---|---|---|---|---|
| $A_1$ | 1 | 1 | 1 | 1 | 1 | | $x^2+y^2+z^2$ |
| $A_2$ | 1 | 1 | 1 | -1 | -1 | | |
| E | 2 | -1 | 2 | 0 | 0 | | $(2z^2-x^2-y^2, x^2-y^2)$ |
| $T_1$ | 3 | 0 | -1 | 1 | -1 | $(R_x, R_y, R_z); (x, y, z)$ | |
| $T_2$ | 3 | 0 | -1 | -1 | 1 | | $(xy, xz, yz)$ |

| $O_h$ | E | $8C_3$ | $6C_2$ | $6C_4$ | $3C_2(=C_4^2)$ | I | $6S_4$ | $8S_6$ | $3\sigma_h$ | $6\sigma_d$ | | |
|---|---|---|---|---|---|---|---|---|---|---|---|---|
| $A_{1g}$ | 1 | 1 | 1 | 1 | 1 | 1 | 1 | 1 | 1 | 1 | | $x^2+y^2+z^2$ |
| $A_{2g}$ | 1 | 1 | -1 | -1 | 1 | 1 | -1 | 1 | 1 | -1 | | |
| $E_g$ | 2 | -1 | 0 | 0 | 2 | 2 | 0 | -1 | 2 | 0 | | $(2z^2-x^2-y^2, x^2-y^2)$ |
| $T_{1g}$ | 3 | 0 | 1 | -1 | -1 | 3 | -1 | 0 | -1 | -1 | $(R_x, R_y, R_z)$ | |
| $T_{2g}$ | 3 | 0 | 1 | -1 | -1 | 3 | -1 | 0 | -1 | 1 | | $(xy, xz, yz)$ |
| $A_{1u}$ | 1 | 1 | 1 | 1 | 1 | -1 | -1 | -1 | -1 | -1 | | |
| $A_{2u}$ | 1 | 1 | -1 | -1 | 1 | -1 | 1 | -1 | -1 | 1 | $(x, y, z)$ | |
| $E_u$ | 2 | -1 | 0 | 0 | 2 | -2 | 0 | 1 | -2 | 0 | | |
| $T_{1u}$ | 3 | 0 | -1 | 1 | -1 | -3 | -1 | 0 | 1 | 1 | | |
| $T_{2u}$ | 3 | 0 | 1 | -1 | -1 | -3 | 1 | 0 | 1 | -1 | | |

10. The groups $C_{\infty v}$ and $D_{\infty h}$ for linear molecules

| $C_{\infty v}$ | E | $2C_\infty^\Phi$ | L | $\infty\sigma_v$ | | |
|---|---|---|---|---|---|---|
| $A_1 \equiv \Sigma^+$ | 1 | 1 | L | 1 | $z$ | $x^2+y^2, z^2$ |
| $A_2 \equiv \Sigma^-$ | 1 | 1 | L | -1 | $R_z$ | |
| $E_1 \equiv \Pi$ | 2 | $2\cos\Phi$ | L | 0 | $(x,y);(R_x,R_y)$ | $(xz, yz)$ |
| $E_2 \equiv \Delta$ | 2 | $2\cos 2\Phi$ | L | 0 | | $(x^2-y^2, xy)$ |
| $E_3 \equiv \Phi$ | 2 | $2\cos 3\Phi$ | L | 0 | | |
| L | L | L | L | L | | |

| $\mathbf{D}_{\infty h}$ | E | $2C_{\infty}^{\Phi}$ | L | $\infty\sigma_v$ | I | $2S_{\infty}^{\Phi}$ | L | $C_2$ | | |
|---|---|---|---|---|---|---|---|---|---|---|
| $\Sigma_g^+$ | 1 | 1 | L | 1 | 1 | 1 | L | 1 | | $x^2+y^2, z^2$ |
| $\Sigma_g^-$ | 1 | 1 | L | -1 | 1 | 1 | L | -1 | $R_z$ | |
| $\Pi_g$ | 2 | $2\cos\Phi$ | L | 0 | 2 | $-2\cos\Phi$ | L | 0 | $(R_x, R_y)$ | (xz,yz) |
| $\Delta_g$ | 2 | $2\cos 2\Phi$ | L | 0 | 2 | $2\cos\Phi$ | L | 0 | | $(x^2-y^2, xy)$ |
| L | L | L | L | L | L | L | L | L | | |
| $\Sigma_u^+$ | 1 | 1 | L | 1 | -1 | -1 | L | -1 | $z$ | |
| $\Sigma_u^-$ | 1 | 1 | L | -1 | -1 | -1 | L | -1 | | |
| $\Pi_u^+$ | 2 | $2\cos\Phi$ | L | 0 | -2 | $2\cos\Phi$ | L | 0 | (x,y) | |
| $\Pi_u^-$ | 2 | $2\cos 2\Phi$ | L | 0 | -2 | $2\cos 2\Phi$ | L | 0 | | |
| L | L | L | L | L | L | L | L | L | | |

# APPENDIX II: Matrices of the Irreducible Representations of Selected Point Groups†

## NOTATION

$$x_+ = x + \mathrm{i}y, \qquad x_- = x - \mathrm{i}y, \qquad \epsilon = \mathrm{e}^{2\pi\mathrm{i}/3}, \qquad \bar{\epsilon} = \mathrm{e}^{-2\pi\mathrm{i}/3}.$$

## AII.1 O

The bases of the irreducible representations are as follows:

$$\mathrm{A}_1:\ \tfrac{1}{6}\sqrt{21}\,[Y_{40} + \sqrt{\tfrac{5}{14}}(Y_{44} + Y_{4,-4})], \qquad x^4 + y^4 + z^4 - \tfrac{3}{5}r^4;$$

$$\mathrm{A}_2:\ \sqrt{\tfrac{1}{2}}\,\mathrm{i}(Y_{32} - Y_{3,-2}), \qquad \mathrm{i}xyz;$$

$$\mathrm{E}:\ \begin{cases} (1) & Y_{20}; \quad -(3z^2 - r^2); \\ (2) & \sqrt{\tfrac{1}{2}}(Y_{22} + Y_{2,-2}), \quad -\sqrt{3}(x^2 - y^2); \end{cases}$$

$$\mathrm{T}_1:\ \begin{cases} (1) & Y_{11}, \quad -\mathrm{i}x_+; \\ (2) & Y_{10}, \quad \mathrm{i}\sqrt{2}z; \\ (3) & Y_{1,-1}, \quad \mathrm{i}x_-; \end{cases}$$

$$\mathrm{T}_2:\ \begin{cases} (1) & Y_{21}, \quad zx_+; \\ (2) & \sqrt{\tfrac{1}{2}}\,\mathrm{i}(Y_{22} - Y_{2,-2}), \quad \sqrt{2}xy; \\ (3) & Y_{2,-1}, \quad -zx_-. \end{cases}$$

The matrices are shown in Table AI.1.

† After [7].

**Table A1.1** Matrices for irreducible representations of **O**.

| **O** | $\hat{E}$ | $\hat{C}_4^2(z)$ | $\hat{C}_4^2(x)$ | $\hat{C}_4^2(y)$ |
|---|---|---|---|---|
| $A_1$ | 1 | 1 | 1 | 1 |
| $A_2$ | 1 | 1 | 1 | 1 |
| E | $\begin{bmatrix} 1 & 0 \\ 0 & 1 \end{bmatrix}$ | $\begin{bmatrix} 1 & 0 \\ 0 & 1 \end{bmatrix}$ | $\begin{bmatrix} 1 & 0 \\ 0 & 1 \end{bmatrix}$ | $\begin{bmatrix} 1 & 0 \\ 0 & 1 \end{bmatrix}$ |
| $T_1$ | $\begin{bmatrix} 1 & 0 & 0 \\ 0 & 1 & 0 \\ 0 & 0 & 1 \end{bmatrix}$ | $\begin{bmatrix} -1 & 0 & 0 \\ 0 & 1 & 0 \\ 0 & 0 & -1 \end{bmatrix}$ | $\begin{bmatrix} 0 & 0 & -1 \\ 0 & -1 & 0 \\ -1 & 0 & 0 \end{bmatrix}$ | $\begin{bmatrix} 0 & 0 & 1 \\ 0 & -1 & 0 \\ 1 & 0 & 0 \end{bmatrix}$ |
| $T_2$ | $\begin{bmatrix} 1 & 0 & 0 \\ 0 & 1 & 0 \\ 0 & 0 & 1 \end{bmatrix}$ | $\begin{bmatrix} -1 & 0 & 0 \\ 0 & 1 & 0 \\ 0 & 0 & -1 \end{bmatrix}$ | $\begin{bmatrix} 0 & 0 & 1 \\ 0 & -1 & 0 \\ 1 & 0 & 0 \end{bmatrix}$ | $\begin{bmatrix} 0 & 0 & -1 \\ 0 & -1 & 0 \\ -1 & 0 & 0 \end{bmatrix}$ |

**Table A1.1 continued** Matrices for irreducible representations of **O**.

| **O** | $\hat{C}_4^3(z)$ | $\hat{C}_4(z)$ | $\hat{C}_4(x)$ | $\hat{C}_4(y)$ |
|---|---|---|---|---|
| $A_1$ | 1 | 1 | 1 | 1 |
| $A_2$ | −1 | −1 | −1 | −1 |
| E | $\begin{bmatrix} 1 & 0 \\ 0 & -1 \end{bmatrix}$ | $\begin{bmatrix} 1 & 0 \\ 0 & -1 \end{bmatrix}$ | $-\frac{1}{2}\begin{bmatrix} 1 & \sqrt{3} \\ \sqrt{3} & -1 \end{bmatrix}$ | $-\frac{1}{2}\begin{bmatrix} 1 & \sqrt{3} \\ \sqrt{3} & -1 \end{bmatrix}$ |
| $T_1$ | $\begin{bmatrix} i & 0 & 0 \\ 0 & 1 & 0 \\ 0 & 0 & -i \end{bmatrix}$ | $\begin{bmatrix} -i & 0 & 0 \\ 0 & 1 & 0 \\ 0 & 0 & i \end{bmatrix}$ | $-\frac{1}{2}\begin{bmatrix} -1 & i\sqrt{2} & 1 \\ i\sqrt{2} & 0 & i\sqrt{2} \\ 1 & i\sqrt{2} & -1 \end{bmatrix}$ | $-\frac{1}{2}\begin{bmatrix} 1 & i\sqrt{2} & 1 \\ i\sqrt{2} & 0 & i\sqrt{2} \\ -1 & i\sqrt{2} & 1 \end{bmatrix}$ |
| $T_2$ | $\begin{bmatrix} i & 0 & 0 \\ 0 & -1 & 0 \\ 0 & 0 & -i \end{bmatrix}$ | $\begin{bmatrix} -i & 0 & 0 \\ 0 & -1 & 0 \\ 0 & 0 & i \end{bmatrix}$ | $-\frac{1}{2}\begin{bmatrix} -1 & -\sqrt{2} & 1 \\ \sqrt{2} & 0 & -\sqrt{2} \\ 1 & \sqrt{2} & 1 \end{bmatrix}$ | $-\frac{1}{2}\begin{bmatrix} 1 & \sqrt{2} & 1 \\ -\sqrt{2} & 0 & \sqrt{2} \\ 1 & -\sqrt{2} & 1 \end{bmatrix}$ |

**Table A1.1 Continued** Matrices for irreducible representations of **O**.

| **O** | $\hat{C}_4^3(y)$ | $\hat{C}_4(y)$ | $\hat{C}_3(x\bar{y}\bar{z})$ | $\hat{C}_3^2(x\bar{y}\bar{z})$ |
|---|---|---|---|---|
| $A_1$ | 1 | 1 | 1 | 1 |
| $A_2$ | −1 | −1 | 1 | 1 |
| E | $\frac{1}{2}\begin{bmatrix}-1 & \sqrt{3}\\ \sqrt{3} & 1\end{bmatrix}$ | $\frac{1}{2}\begin{bmatrix}-1 & \sqrt{3}\\ \sqrt{3} & 1\end{bmatrix}$ | $-\frac{1}{2}\begin{bmatrix}1 & \sqrt{3}\\ -\sqrt{3} & 1\end{bmatrix}$ | $-\frac{1}{2}\begin{bmatrix}1 & -\sqrt{3}\\ \sqrt{3} & 1\end{bmatrix}$ |
| $T_1$ | $\frac{1}{2}\begin{bmatrix}1 & \sqrt{2} & 1\\ -\sqrt{2} & 0 & \sqrt{2}\\ 1 & -\sqrt{2} & 1\end{bmatrix}$ | $\frac{1}{2}\begin{bmatrix}1 & -\sqrt{2} & 1\\ \sqrt{2} & 0 & -\sqrt{2}\\ 1 & \sqrt{2} & 1\end{bmatrix}$ | $-\frac{1}{2}\begin{bmatrix}-i & -\sqrt{2} & i\\ i\sqrt{2} & 0 & i\sqrt{2}\\ -i & \sqrt{2} & i\end{bmatrix}$ | $\frac{1}{2}\begin{bmatrix}-i & i\sqrt{2} & -i\\ \sqrt{2} & 0 & -\sqrt{2}\\ i & i\sqrt{2} & i\end{bmatrix}$ |
| $T_2$ | $-\frac{1}{2}\begin{bmatrix}1 & i\sqrt{2} & -1\\ i\sqrt{2} & 0 & i\sqrt{2}\\ -1 & i\sqrt{2} & 1\end{bmatrix}$ | $\frac{1}{2}\begin{bmatrix}-1 & i\sqrt{2} & 1\\ i\sqrt{2} & 0 & i\sqrt{2}\\ 1 & i\sqrt{2} & -1\end{bmatrix}$ | $\frac{1}{2}\begin{bmatrix}-i & i\sqrt{2} & -i\\ \sqrt{2} & 0 & -\sqrt{2}\\ i & i\sqrt{2} & i\end{bmatrix}$ | $-\frac{1}{2}\begin{bmatrix}-i & -\sqrt{2} & i\\ i\sqrt{2} & 0 & i\sqrt{2}\\ -i & \sqrt{2} & i\end{bmatrix}$ |

**Table A1.1 Continued** Matrices for irreducible representations of **O**.

| **O** | $\hat{C}_3(\bar{x}y\bar{z})$ | $\hat{C}_3^2(\bar{x}y\bar{z})$ | $\hat{C}_3(\bar{x}\bar{y}z)$ | $\hat{C}_3^2(\bar{x}\bar{y}z)$ |
|---|---|---|---|---|
| $A_1$ | 1 | 1 | 1 | 1 |
| $A_2$ | 1 | 1 | 1 | 1 |
| E | $-\frac{1}{2}\begin{bmatrix} 1 & \sqrt{3} \\ -\sqrt{3} & 1 \end{bmatrix}$ | $-\frac{1}{2}\begin{bmatrix} 1 & -\sqrt{3} \\ \sqrt{3} & 1 \end{bmatrix}$ | $-\frac{1}{2}\begin{bmatrix} 1 & \sqrt{3} \\ -\sqrt{3} & 1 \end{bmatrix}$ | $-\frac{1}{2}\begin{bmatrix} 1 & -\sqrt{3} \\ \sqrt{3} & 1 \end{bmatrix}$ |
| $T_1$ | $\frac{1}{2}\begin{bmatrix} i & -\sqrt{2} & -i \\ i\sqrt{2} & 0 & i\sqrt{2} \\ i & \sqrt{2} & -i \end{bmatrix}$ | $-\frac{1}{2}\begin{bmatrix} i & i\sqrt{2} & i \\ \sqrt{2} & 0 & -\sqrt{2} \\ -i & i\sqrt{2} & -i \end{bmatrix}$ | $\frac{1}{2}\begin{bmatrix} -i & \sqrt{2} & i \\ i\sqrt{2} & 0 & i\sqrt{2} \\ -i & -\sqrt{2} & i \end{bmatrix}$ | $-\frac{1}{2}\begin{bmatrix} -i & i\sqrt{2} & -i \\ -\sqrt{2} & 0 & \sqrt{2} \\ i & i\sqrt{2} & i \end{bmatrix}$ |
| $T_2$ | $-\frac{1}{2}\begin{bmatrix} i & i\sqrt{2} & i \\ \sqrt{2} & 0 & -\sqrt{2} \\ -i & i\sqrt{2} & -i \end{bmatrix}$ | $\frac{1}{2}\begin{bmatrix} i & -\sqrt{2} & i \\ i\sqrt{2} & 0 & i\sqrt{2} \\ i & \sqrt{2} & -i \end{bmatrix}$ | $\frac{1}{2}\begin{bmatrix} i & i\sqrt{2} & i \\ -\sqrt{2} & 0 & \sqrt{2} \\ -i & i\sqrt{2} & -i \end{bmatrix}$ | $-\frac{1}{2}\begin{bmatrix} i & \sqrt{2} & -i \\ i\sqrt{2} & 0 & i\sqrt{2} \\ i & -\sqrt{2} & -i \end{bmatrix}$ |

**Table A1.1 Continued** Matrices for irreducible representations of **O**.

| **O** | $\hat{C}_3(xyz)$ | $\hat{C}_3^2(xyz)$ | $\hat{C}_2(\bar{x}y)$ | $\hat{C}_2(xy)$ |
|---|---|---|---|---|
| $A_1$ | 1 | 1 | 1 | 1 |
| $A_2$ | 1 | 1 | −1 | −1 |
| E | $-\frac{1}{2}\begin{bmatrix} 1 & \sqrt{3} \\ -\sqrt{3} & 1 \end{bmatrix}$ | $-\frac{1}{2}\begin{bmatrix} 1 & -\sqrt{3} \\ \sqrt{3} & 1 \end{bmatrix}$ | $\begin{bmatrix} 1 & 0 \\ 0 & -1 \end{bmatrix}$ | $\begin{bmatrix} 1 & 0 \\ 0 & -1 \end{bmatrix}$ |
| $T_1$ | $-\frac{1}{2}\begin{bmatrix} i & \sqrt{2} & -i \\ i\sqrt{2} & 0 & i\sqrt{2} \\ i & -\sqrt{2} & -i \end{bmatrix}$ | $\frac{1}{2}\begin{bmatrix} i & i\sqrt{2} & i \\ -\sqrt{2} & 0 & \sqrt{2} \\ -i & i\sqrt{2} & -i \end{bmatrix}$ | $\begin{bmatrix} 0 & 0 & -i \\ 0 & -1 & 0 \\ i & 0 & 0 \end{bmatrix}$ | $\begin{bmatrix} 0 & 0 & i \\ 0 & -1 & 0 \\ -i & 0 & 0 \end{bmatrix}$ |
| $T_2$ | $-\frac{1}{2}\begin{bmatrix} -i & i\sqrt{2} & -i \\ -\sqrt{2} & 0 & \sqrt{2} \\ i & i\sqrt{2} & i \end{bmatrix}$ | $\frac{1}{2}\begin{bmatrix} -i & \sqrt{2} & i \\ i\sqrt{2} & 0 & i\sqrt{2} \\ -i & -\sqrt{2} & i \end{bmatrix}$ | $\begin{bmatrix} 0 & 0 & i \\ 0 & 1 & 0 \\ -i & 0 & 0 \end{bmatrix}$ | $\begin{bmatrix} 0 & 0 & -i \\ 0 & 1 & 0 \\ i & 0 & 0 \end{bmatrix}$ |

**Table A1.1 Continued** Matrices for irreducible representations of **O**.

| **O** | $\hat{C}_2(yz)$ | $\hat{C}_2(\bar{y}z)$ | $\hat{C}_2(x\bar{z})$ | $\hat{C}_2(xz)$ |
|---|---|---|---|---|
| $A_1$ | 1 | 1 | 1 | 1 |
| $A_2$ | $-1$ | $-1$ | $-1$ | $-1$ |
| E | $-\frac{1}{2}\begin{bmatrix} 1 & \sqrt{3} \\ \sqrt{3} & -1 \end{bmatrix}$ | $-\frac{1}{2}\begin{bmatrix} 1 & \sqrt{3} \\ \sqrt{3} & -1 \end{bmatrix}$ | $\frac{1}{2}\begin{bmatrix} -1 & \sqrt{3} \\ \sqrt{3} & 1 \end{bmatrix}$ | $\frac{1}{2}\begin{bmatrix} -1 & \sqrt{3} \\ \sqrt{3} & 1 \end{bmatrix}$ |
| $T_1$ | $\frac{1}{2}\begin{bmatrix} -1 & i\sqrt{2} & 1 \\ -i\sqrt{2} & 0 & -i\sqrt{2} \\ 1 & i\sqrt{2} & -1 \end{bmatrix}$ | $\frac{1}{2}\begin{bmatrix} -1 & -i\sqrt{2} & 1 \\ i\sqrt{2} & 0 & i\sqrt{2} \\ 1 & -i\sqrt{2} & -1 \end{bmatrix}$ | $-\frac{1}{2}\begin{bmatrix} 1 & -\sqrt{2} & 1 \\ -\sqrt{2} & 0 & \sqrt{2} \\ 1 & \sqrt{2} & 1 \end{bmatrix}$ | $-\frac{1}{2}\begin{bmatrix} 1 & \sqrt{2} & -1 \\ \sqrt{2} & 0 & -\sqrt{2} \\ 1 & -\sqrt{2} & 1 \end{bmatrix}$ |
| $T_2$ | $\frac{1}{2}\begin{bmatrix} 1 & -\sqrt{2} & 1 \\ -\sqrt{2} & 0 & \sqrt{2} \\ 1 & \sqrt{2} & 1 \end{bmatrix}$ | $\frac{1}{2}\begin{bmatrix} 1 & \sqrt{2} & 1 \\ \sqrt{2} & 0 & -\sqrt{2} \\ 1 & -\sqrt{2} & 1 \end{bmatrix}$ | $\frac{1}{2}\begin{bmatrix} 1 & -i\sqrt{2} & -1 \\ i\sqrt{2} & 0 & i\sqrt{2} \\ 1 & -i\sqrt{2} & 1 \end{bmatrix}$ | $\frac{1}{2}\begin{bmatrix} 1 & i\sqrt{2} & -1 \\ -i\sqrt{2} & 0 & -i\sqrt{2} \\ -1 & i\sqrt{2} & 1 \end{bmatrix}$ |

## AII.2 $D_{4h}$

The bases of irreducible representations (Fig. 6.2) are as follows:

*even representations*

$A_{1g}$: $Y_{20}$, $-(3z^2 - r^2)$;

$A_{2g}$: $Y_{10}$, $-ixy(x^2 - y^2)$;

$B_{1g}$: $\sqrt{\tfrac{1}{2}}(Y_{22} + Y_{2,-2})$, $-(x^2 - y^2)$;

$B_{2g}$: $\sqrt{\tfrac{1}{2}}\,i(Y_{22} - Y_{2,-2})$, $xy$;

$E_g$: $\begin{cases} (1) & Y_{11}, \quad -zx_+; \\ (2) & Y_{1,-1}, \quad -zx_-; \end{cases}$

*odd representations*

$A_{1u}$: $Y_{20}$, $-xyz(x^2 - y^2)$;

$A_{2u}$: $Y_{10}$, $iz$;

$B_{1u}$: $\sqrt{\tfrac{1}{2}}(Y_{22} + Y_{2,-2})$, $xyz$;

$B_{2u}$: $\sqrt{\tfrac{1}{2}}\,i(Y_{22} - Y_{2,-2})$, $z(x^2 - y^2)$;

$E_u$: $\begin{cases} (1) & Y_{11}, \quad -ix_+; \\ (2) & Y_{1,-1}, \quad ix_-. \end{cases}$

The matrices are shown in Table AI.2.

## AII.3 $D_6$

The bases of the irreducible representations (Fig. AI.1) are as follows:

$A_1$: $Y_{20}$, $-(3z^2 - r^2)$;

$A_2$: $Y_{10}$, $iz$;

$B_1$: $\sqrt{\tfrac{1}{2}}\,i(Y_{33} - Y_{3,-3})$, $-x(x^2 - 3y^2)$;

$B_2$: $\sqrt{\tfrac{1}{2}}(Y_{33} + Y_{3,-3})$, $-y(3x^2 - y^2)$;

$E_1$: $\begin{cases} (1) & Y_{11}, \quad -ix_+; \\ (2) & Y_{1,-1}, \quad ix_-; \end{cases}$

$E_2$: $\begin{cases} (1) & Y_{22}, \quad -(x^2 - y^2 + 2ixy); \\ (2) & Y_{2,-2}, \quad -(x^2 - y^2 - 2ixy). \end{cases}$

The matrices are shown in Table AI.3.

**Table A1.2** Matrices for irreducible representations of $\mathbf{D}_{4h}$.

| $\mathbf{D}_{4h}$ | $\hat{E}$ | $\hat{C}_4^2(z)$ | $\hat{C}_4^3(z)$ | $\hat{C}_4(z)$ | $\hat{C}_2(x)$ | $\hat{C}_2(y)$ | $\hat{C}_2(u')$ | $\hat{C}_2(u)$ | $\hat{I}$ | $\hat{S}_4(z)$ | $\hat{S}_4^3(z)$ | $\hat{\sigma}_h$ | $\hat{\sigma}(xz)$ | $\hat{\sigma}(yz)$ | $\hat{\sigma}(u'z)$ | $\hat{\sigma}(uz)$ |
|---|---|---|---|---|---|---|---|---|---|---|---|---|---|---|---|---|
| $A_{1g}$ | 1 | 1 | 1 | 1 | 1 | 1 | 1 | 1 | 1 | 1 | 1 | 1 | 1 | 1 | 1 | 1 |
| $A_{2g}$ | 1 | 1 | 1 | 1 | −1 | −1 | −1 | −1 | 1 | 1 | 1 | 1 | −1 | −1 | −1 | −1 |
| $B_{1g}$ | 1 | 1 | −1 | −1 | 1 | 1 | −1 | −1 | 1 | −1 | −1 | 1 | 1 | 1 | −1 | −1 |
| $B_{2g}$ | 1 | 1 | −1 | −1 | −1 | −1 | 1 | 1 | 1 | −1 | −1 | 1 | −1 | −1 | 1 | 1 |
| $E_g$ | $\begin{bmatrix}1&0\\0&1\end{bmatrix}$ | $\begin{bmatrix}-1&0\\0&-1\end{bmatrix}$ | $\begin{bmatrix}i&0\\0&-i\end{bmatrix}$ | $\begin{bmatrix}-i&0\\0&i\end{bmatrix}$ | $\begin{bmatrix}0&-1\\-1&0\end{bmatrix}$ | $\begin{bmatrix}0&1\\1&0\end{bmatrix}$ | $\begin{bmatrix}0&-i\\i&0\end{bmatrix}$ | $\begin{bmatrix}0&i\\-i&0\end{bmatrix}$ | $\begin{bmatrix}1&0\\0&1\end{bmatrix}$ | $\begin{bmatrix}-i&0\\0&i\end{bmatrix}$ | $\begin{bmatrix}i&0\\0&-i\end{bmatrix}$ | $\begin{bmatrix}-1&0\\0&-1\end{bmatrix}$ | $\begin{bmatrix}0&-1\\-1&0\end{bmatrix}$ | $\begin{bmatrix}0&1\\1&0\end{bmatrix}$ | $\begin{bmatrix}0&-i\\i&0\end{bmatrix}$ | $\begin{bmatrix}0&i\\-i&0\end{bmatrix}$ |
| $A_{1u}$ | 1 | 1 | 1 | 1 | 1 | 1 | 1 | 1 | −1 | −1 | −1 | −1 | −1 | −1 | −1 | −1 |
| $A_{2u}$ | 1 | 1 | 1 | 1 | −1 | −1 | −1 | −1 | −1 | −1 | −1 | −1 | 1 | 1 | 1 | 1 |
| $B_{1u}$ | 1 | 1 | −1 | −1 | 1 | 1 | −1 | −1 | −1 | 1 | 1 | −1 | −1 | −1 | 1 | 1 |
| $B_{2u}$ | 1 | 1 | −1 | −1 | −1 | −1 | 1 | 1 | −1 | 1 | 1 | −1 | 1 | 1 | −1 | −1 |
| $E_u$ | $\begin{bmatrix}1&0\\0&1\end{bmatrix}$ | $\begin{bmatrix}-1&0\\0&-1\end{bmatrix}$ | $\begin{bmatrix}i&0\\0&i\end{bmatrix}$ | $\begin{bmatrix}-i&0\\0&i\end{bmatrix}$ | $\begin{bmatrix}0&-1\\-1&0\end{bmatrix}$ | $\begin{bmatrix}0&1\\1&0\end{bmatrix}$ | $\begin{bmatrix}0&-i\\i&0\end{bmatrix}$ | $\begin{bmatrix}0&i\\-i&0\end{bmatrix}$ | $\begin{bmatrix}-1&0\\0&-1\end{bmatrix}$ | $\begin{bmatrix}i&0\\0&-i\end{bmatrix}$ | $\begin{bmatrix}-i&0\\0&i\end{bmatrix}$ | $\begin{bmatrix}1&0\\0&1\end{bmatrix}$ | $\begin{bmatrix}0&1\\1&0\end{bmatrix}$ | $\begin{bmatrix}0&-1\\-1&0\end{bmatrix}$ | $\begin{bmatrix}0&i\\-i&0\end{bmatrix}$ | $\begin{bmatrix}0&-i\\i&0\end{bmatrix}$ |

**Table A1.3** Matrices for irreducible representations of $\mathbf{D}_6$.

| $\mathbf{D}_6$ | $\hat{E}$ | $\hat{C}_6^3(z)$ | $\hat{C}_6^4(z)$ | $\hat{C}_6^2(z)$ | $\hat{C}_6^5(z)$ | $\hat{C}_6(z)$ |
|---|---|---|---|---|---|---|
| $A_1$ | 1 | 1 | 1 | 1 | 1 | 1 |
| $A_2$ | 1 | 1 | 1 | 1 | 1 | 1 |
| $B_1$ | 1 | −1 | 1 | 1 | −1 | −1 |
| $B_2$ | 1 | −1 | 1 | 1 | −1 | −1 |
| $E_1$ | $\begin{bmatrix}1 & 0\\0 & 1\end{bmatrix}$ | $\begin{bmatrix}-1 & 0\\0 & -1\end{bmatrix}$ | $\begin{bmatrix}\epsilon & 0\\0 & \bar{\epsilon}\end{bmatrix}$ | $\begin{bmatrix}\bar{\epsilon} & 0\\0 & \epsilon\end{bmatrix}$ | $\begin{bmatrix}-\bar{\epsilon} & 0\\0 & -\epsilon\end{bmatrix}$ | $\begin{bmatrix}-\epsilon & 0\\0 & -\bar{\epsilon}\end{bmatrix}$ |
| $E_2$ | $\begin{bmatrix}1 & 0\\0 & 1\end{bmatrix}$ | $\begin{bmatrix}1 & 0\\0 & 1\end{bmatrix}$ | $\begin{bmatrix}\bar{\epsilon} & 0\\0 & \epsilon\end{bmatrix}$ | $\begin{bmatrix}\epsilon & 0\\0 & \bar{\epsilon}\end{bmatrix}$ | $\begin{bmatrix}\epsilon & 0\\0 & \bar{\epsilon}\end{bmatrix}$ | $\begin{bmatrix}\bar{\epsilon} & 0\\0 & \epsilon\end{bmatrix}$ |

| $\mathbf{D}_6$ | $\hat{C}_2(x)$ | $\hat{C}_2(b)$ | $\hat{C}_2(a)$ | $\hat{C}_2(d)$ | $\hat{C}_2(y)$ | $\hat{C}_2(c)$ |
|---|---|---|---|---|---|---|
| $A_1$ | 1 | 1 | 1 | 1 | 1 | 1 |
| $A_2$ | −1 | −1 | −1 | −1 | −1 | −1 |
| $B_1$ | 1 | 1 | 1 | −1 | −1 | −1 |
| $B_2$ | −1 | −1 | −1 | 1 | 1 | 1 |
| $E_1$ | $\begin{bmatrix}0 & -1\\-1 & 0\end{bmatrix}$ | $\begin{bmatrix}0 & -\epsilon\\-\bar{\epsilon} & 0\end{bmatrix}$ | $\begin{bmatrix}0 & -\bar{\epsilon}\\-\epsilon & 0\end{bmatrix}$ | $\begin{bmatrix}0 & \bar{\epsilon}\\\epsilon & 0\end{bmatrix}$ | $\begin{bmatrix}0 & 1\\1 & 0\end{bmatrix}$ | $\begin{bmatrix}0 & \epsilon\\\bar{\epsilon} & 0\end{bmatrix}$ |
| $E_2$ | $\begin{bmatrix}0 & 1\\1 & 0\end{bmatrix}$ | $\begin{bmatrix}0 & \bar{\epsilon}\\\epsilon & 0\end{bmatrix}$ | $\begin{bmatrix}0 & \epsilon\\\bar{\epsilon} & 0\end{bmatrix}$ | $\begin{bmatrix}0 & \epsilon\\\bar{\epsilon} & 0\end{bmatrix}$ | $\begin{bmatrix}0 & 1\\1 & 0\end{bmatrix}$ | $\begin{bmatrix}0 & \bar{\epsilon}\\\epsilon & 0\end{bmatrix}$ |

## AII.4 $D_{3h}$

The bases of the irreducible representations (Fig. 2.26(b)) are as follows:

*even representations*

$$A_1': \quad Y_{20}, \quad -(3z^2 - r^2);$$

$$A_2': \quad Y_{10}, \quad -ixy(3x^4 - 10x^2y^2 + 3y^4);$$

$$A_1'': \quad \sqrt{\tfrac{1}{2}}\,i(Y_{33} - Y_{3,-3}), \quad -yz(3x^2 - y^2);$$

$$A_2'': \quad \sqrt{\tfrac{1}{2}}(Y_{33} + Y_{3,-3}), \quad -zx(x^2 - 3y^2);$$

$$\mathrm{E}': \begin{cases} (1) & Y_{11}, \quad zx_+; \\ (2) & Y_{1,-1}, \quad zx_-; \end{cases} \quad \mathrm{E}'': \begin{cases} (1) & Y_{22}, \quad -(x^2-y^2+2\mathrm{i}xy); \\ (2) & Y_{2,-2}, \quad -(x^2-y^2-2\mathrm{i}xy); \end{cases}$$

*odd representations*

$$\mathrm{A}_1': \quad \sqrt{\tfrac{1}{2}}(Y_{33}+Y_{3,-3}), \quad -y(3x^2-y^2);$$

$$\mathrm{A}_2': \quad \sqrt{\tfrac{1}{2}}(Y_{33}-Y_{3,-3}), \quad \mathrm{i}x(x^2-3y^2);$$

$$\mathrm{A}_1'': \quad \mathrm{i}Y_{10}, \quad -z;$$

$$\mathrm{A}_2'': \quad Y_{20}, \quad xyz(3x^4-10x^2y^2+3y^4);$$

$$\mathrm{E}': \begin{cases} (1) & Y_{2,-2}, \quad \mathrm{i}z(x^2-y^2-2\mathrm{i}xy); \\ (2) & Y_{22}, \quad -\mathrm{i}z(x^2-y^2+2\mathrm{i}xy); \end{cases}$$

$$\mathrm{E}'': \begin{cases} (1) & Y_{1,-1}, \quad \mathrm{i}x_-; \\ (2) & Y_{11}, \quad -\mathrm{i}x_+. \end{cases}$$

(The functions $Y_{LM}$ are supposed to be multielectronic; therefore both even and odd basis sets include the same $L$ values, see Section 5.2.2.)

The matrices are shown in Table AI.4.

**Table A1.4** Matrices for irreducible representations of $\mathbf{D}_{3h}$.

| $\mathbf{D}_{3h}$ | $\hat{E}$ | $\hat{\sigma}$ | $\hat{C}_3(z)$ | $\hat{C}_3^2(z)$ | $\hat{S}_6(z)$ | $\hat{S}_6^5(z)$ |
|---|---|---|---|---|---|---|
| $A_1'$ | 1 | 1 | 1 | 1 | 1 | 1 |
| $A_2'$ | 1 | 1 | 1 | 1 | 1 | 1 |
| $A_1''$ | 1 | −1 | 1 | 1 | −1 | −1 |
| $A_2''$ | 1 | −1 | 1 | 1 | −1 | −1 |
| $E'$ | $\begin{bmatrix} 1 & 0 \\ 0 & 1 \end{bmatrix}$ | $\begin{bmatrix} -1 & 0 \\ 0 & -1 \end{bmatrix}$ | $\begin{bmatrix} \bar{\epsilon} & 0 \\ 0 & \epsilon \end{bmatrix}$ | $\begin{bmatrix} \epsilon & 0 \\ 0 & \bar{\epsilon} \end{bmatrix}$ | $\begin{bmatrix} -\bar{\epsilon} & 0 \\ 0 & -\epsilon \end{bmatrix}$ | $\begin{bmatrix} -\epsilon & 0 \\ 0 & -\bar{\epsilon} \end{bmatrix}$ |
| $E''$ | $\begin{bmatrix} 1 & 0 \\ 0 & 1 \end{bmatrix}$ | $\begin{bmatrix} 1 & 0 \\ 0 & 1 \end{bmatrix}$ | $\begin{bmatrix} \epsilon & 0 \\ 0 & \bar{\epsilon} \end{bmatrix}$ | $\begin{bmatrix} \bar{\epsilon} & 0 \\ 0 & \epsilon \end{bmatrix}$ | $\begin{bmatrix} \epsilon & 0 \\ 0 & \bar{\epsilon} \end{bmatrix}$ | $\begin{bmatrix} \bar{\epsilon} & 0 \\ 0 & \epsilon \end{bmatrix}$ |

| $\mathbf{D}_{3h}$ | $\hat{\sigma}(b)$ | $\hat{\sigma}(x)$ | $\hat{\sigma}(a)$ | $\hat{C}_2'''$ | $\hat{C}_2'$ | $\hat{C}_2''$ |
|---|---|---|---|---|---|---|
| $A_1'$ | 1 | 1 | 1 | 1 | 1 | 1 |
| $A_2'$ | −1 | −1 | −1 | −1 | −1 | −1 |
| $A_1''$ | 1 | 1 | 1 | −1 | −1 | −1 |
| $A_2''$ | −1 | −1 | −1 | 1 | 1 | 1 |
| $E'$ | $\begin{bmatrix} 0 & -\epsilon \\ -\bar{\epsilon} & 0 \end{bmatrix}$ | $\begin{bmatrix} 0 & -1 \\ -1 & 0 \end{bmatrix}$ | $\begin{bmatrix} 0 & -\bar{\epsilon} \\ -\epsilon & 0 \end{bmatrix}$ | $\begin{bmatrix} 0 & \epsilon \\ \bar{\epsilon} & 0 \end{bmatrix}$ | $\begin{bmatrix} 0 & 1 \\ 1 & 0 \end{bmatrix}$ | $\begin{bmatrix} 0 & \bar{\epsilon} \\ \epsilon & 0 \end{bmatrix}$ |
| $E''$ | $\begin{bmatrix} 0 & \bar{\epsilon} \\ \epsilon & 0 \end{bmatrix}$ | $\begin{bmatrix} 0 & 1 \\ 1 & 0 \end{bmatrix}$ | $\begin{bmatrix} 0 & \epsilon \\ \bar{\epsilon} & 0 \end{bmatrix}$ | $\begin{bmatrix} 0 & \bar{\epsilon} \\ \epsilon & 0 \end{bmatrix}$ | $\begin{bmatrix} 0 & 1 \\ 1 & 0 \end{bmatrix}$ | $\begin{bmatrix} 0 & \epsilon \\ \bar{\epsilon} & 0 \end{bmatrix}$ |

# APPENDIX III: Basis Functions of Irreducible Representations of Selected Point Groups†

## NOTATION

$S$ is the orbital ($l$, $L$), spin ($s$) or total ($J$) angular momentum (irreducible representation of rotation group). For the remaining notation see Section 6.3.5.

For the basis functions see Appendix II.

## AIII.1 O

$S = 1$; $\mathrm{D}^{(1)} = \mathrm{T}_1$:

$$\mathrm{T}_1: \quad \begin{cases} (1) & c_1 = 1; \\ (2) & c_0 = 1; \\ (3) & c_{-1} = 1. \end{cases}$$

$S = 2$; $\mathrm{D}^{(2)} = \mathrm{E} + \mathrm{T}_2$:

$$\mathrm{E}: \quad \begin{cases} (1) & c_0 = 1; \\ (2) & c_2 = c_{-2} = \tfrac{1}{2}; \end{cases}$$

$$\mathrm{T}_2: \quad \begin{cases} (1) & c_1 = 1; \\ (^*2) & c_2 = -c_{-2} = \tfrac{1}{2}; \\ (3) & c_{-1} = 1. \end{cases}$$

† After [5].

$S = 3; \quad \mathrm{D}^{(3)} = \mathrm{A}_2 + \mathrm{T}_1 + \mathrm{T}_2$:

$$\mathrm{A}_2: \quad ({}^*1) \quad c_2 = -c_{-2} = \tfrac{1}{2};$$

$$\mathrm{T}_1: \begin{cases} (1) & c_1 = \tfrac{3}{8}, \quad c_{-3} = \tfrac{5}{8}; \\ (2) & c_0 = {}^*1; \\ (3) & c_3 = \tfrac{5}{8}, \quad c_{-1} = \tfrac{3}{8}; \end{cases}$$

$$\mathrm{T}_2: \begin{cases} (1) & c_1 = {}^*\tfrac{5}{8}, \quad c_{-3} = \tfrac{3}{8}; \\ ({}^*2) & c_2 = c_{-2} = \tfrac{1}{2}; \\ (3) & c_3 = {}^*\tfrac{3}{8}, \quad c_{-1} = \tfrac{5}{8}. \end{cases}$$

$S = 4; \quad \mathrm{D}^{(4)} = \mathrm{A}_1 + \mathrm{E} + \mathrm{T}_1 + \mathrm{T}_2$:

$$\mathrm{A}_1: \quad (1) \quad c_4 = c_{-4} = \tfrac{5}{24}, \quad c_0 = \tfrac{7}{12};$$

$$\mathrm{E}: \begin{cases} (1) & c_4 = c_{-4} = \tfrac{7}{24}, \quad c_0 = {}^*\tfrac{5}{12}; \\ (2) & c_2 = c_{-2} = \tfrac{1}{2}; \end{cases}$$

$$\mathrm{T}_1: \begin{cases} (1) & c_1 = \tfrac{7}{8}, \quad c_{-3} = \tfrac{1}{8}; \\ (2) & c_4 = -c_{-4} = {}^*\tfrac{1}{2}; \\ (3) & c_3 = {}^*\tfrac{1}{8}, \quad c_{-1} = \tfrac{7}{8}; \end{cases}$$

$$\mathrm{T}_2: \begin{cases} (1) & c_1 = {}^*\tfrac{1}{8}, \quad c_{-3} = \tfrac{7}{8}; \\ ({}^*2) & c_2 = -c_{-2} = \tfrac{1}{2}; \\ (3) & c_3 = \tfrac{7}{8}, \quad c_{-1} = {}^*\tfrac{1}{8}. \end{cases}$$

$S = 5; \quad \mathrm{D}^{(5)} = \mathrm{E} + 2\mathrm{T}_1 + \mathrm{T}_2$:

$$\mathrm{E}_2: \begin{cases} (1) & c_4 = -c_{-4} = \tfrac{1}{2}; \\ (2) & c_2 = -c_{-2} = {}^*\tfrac{1}{2}; \end{cases}$$

$$\mathrm{T}_1: \begin{cases} (1) & c_5 = \tfrac{5}{128}, \quad c_1 = \tfrac{21}{64}, \quad c_{-3} = {}^*\tfrac{81}{128}; \\ (2) & c_4 = c_{-4} = \tfrac{1}{2}; \\ (3) & c_3 = {}^*\tfrac{81}{128}, \quad c_{-1} = \tfrac{21}{64}, \quad c_{-5} = \tfrac{5}{128}; \end{cases}$$

$$\mathrm{T}_1: \begin{cases} (1) & c_5 = \tfrac{63}{128}, \quad c_1 = \tfrac{15}{64}, \quad c_{-3} = \tfrac{35}{128}; \\ (2) & c_0 = 1; \\ (3) & c_3 = \tfrac{35}{128}, \quad c_{-1} = \tfrac{15}{64}, \quad c_{-5} = \tfrac{63}{128}; \end{cases}$$

$$\mathrm{T}_2: \begin{cases} (1) & c_5 = \tfrac{15}{32}, \quad c_1 = {}^*\tfrac{7}{16}, \quad c_{-3} = {}^*\tfrac{3}{32}; \\ ({}^*2) & c_2 = -c_{-2} = {}^*\tfrac{1}{2}; \\ (3) & c_3 = \tfrac{3}{32}, \quad c_{-1} = \tfrac{7}{16}, \quad c_{-5} = {}^*\tfrac{15}{32}. \end{cases}$$

$S = 6$; $\quad \mathrm{D}^{(6)} = \mathrm{A}_1 + \mathrm{A}_2 + \mathrm{E} + \mathrm{T}_1 + 2\mathrm{T}_2$:

$$\mathrm{A}_1:\ (1)\quad c_4 = c_{-4} = \tfrac{7}{16},\qquad c_0 = {}^*\tfrac{1}{8};$$

$$\mathrm{A}_2:\ ({}^*1)\quad c_6 = c_{-6} = \tfrac{5}{32},\qquad c_2 = c^*_{-2} = \tfrac{11}{32};$$

$$\mathrm{E}:\ \begin{cases} (1) & c_4 = -c_{-4} = \tfrac{1}{16},\qquad c_0 = \tfrac{7}{8}; \\ (2) & c_6 = c_{-6} = \tfrac{11}{32},\qquad c_2 = c_{-2} = \tfrac{5}{32}; \end{cases}$$

$$\mathrm{T}_1:\ \begin{cases} (1) & c_5 = \tfrac{11}{32},\qquad c_1 = {}^*\tfrac{3}{16},\qquad c_{-3} = \tfrac{15}{32}; \\ (2) & c_4 = -c_{-4} = {}^*\tfrac{1}{2}; \\ (3) & c_3 = {}^*\tfrac{15}{32},\qquad c_{-1} = \tfrac{3}{16},\qquad c_{-5} = {}^*\tfrac{11}{32}; \end{cases}$$

$$\mathrm{T}_2:\ \begin{cases} (1) & c_5 = \tfrac{165}{256},\qquad c_1 = \tfrac{5}{128},\qquad c_{-3} = {}^*\tfrac{81}{256}; \\ ({}^*2) & c_2 = -c_{-2} = \tfrac{1}{2}; \\ (3) & c_3 = {}^*\tfrac{81}{256},\qquad c_{-1} = \tfrac{5}{128},\qquad c_{-5} = \tfrac{165}{256}; \end{cases}$$

$$\mathrm{T}_2:\ \begin{cases} (1) & c_5 = \tfrac{3}{256},\qquad c_1 = \tfrac{99}{128},\qquad c_{-3} = \tfrac{55}{256}; \\ ({}^*2) & c_6 = -c_{-6} = \tfrac{1}{2}; \\ (3) & c_3 = \tfrac{55}{256},\qquad c_{-1} = \tfrac{99}{128},\qquad c_{-5} = \tfrac{3}{256}. \end{cases}$$

## AIII.2 $\mathrm{D}_4$

$S = 1$; $\quad \mathrm{D}^{(1)} = \mathrm{A}_2 + \mathrm{E}$:

$$\mathrm{A}_2:\ (1)\quad c_0 = 1;$$

$$\mathrm{E}:\ \begin{cases} (1) & c_1 = 1; \\ (2) & c_{-1} = 1. \end{cases}$$

$S = 2$; $\quad \mathrm{D}^{(2)} = \mathrm{A}_1 + \mathrm{B}_1 + \mathrm{B}_2 + \mathrm{E}$:

$$\mathrm{A}_1:\ (1)\quad c_0 = 1;$$

$$\mathrm{B}_1:\ (1)\quad c_2 = c_{-2} = \tfrac{1}{2};$$

$$\mathrm{B}_2:\ ({}^*1)\quad c_2 = -c_{-2} = \tfrac{1}{2};$$

$$\mathrm{E}:\ \begin{cases} (1) & c_1 = {}^*1; \\ (2) & c_{-1} = 1. \end{cases}$$

$S=3$; $\mathrm{D}^{(3)} = \mathrm{A}_2 + \mathrm{B}_1 + \mathrm{B}_2 + 2\mathrm{E}$:

$$\mathrm{B}_1\text{: } (1) \quad c_2 = -c_{-2} = \tfrac{1}{2};$$
$$\mathrm{A}_2\text{: } (1) \quad c_0 = {}^*1;$$
$$\mathrm{E}\text{: } \begin{cases} (1) & c_1 = \tfrac{3}{8}, \quad c_{-3} = \tfrac{5}{8}; \\ (2) & c_3 = \tfrac{5}{8}, \quad c_{-1} = \tfrac{3}{8}; \end{cases}$$
$$\mathrm{B}_2\text{: } ({}^*1) \quad c_2 = c_{-2} = \tfrac{1}{2};$$
$$\mathrm{E}\text{: } \begin{cases} (1) & c_1 = \tfrac{5}{8}, \quad c_{-3} = {}^*\tfrac{3}{8}; \\ (2) & c_3 = {}^*\tfrac{3}{8}, \quad c_{-1} = \tfrac{5}{8}. \end{cases}$$

$S=4$; $\mathrm{D}^{(4)} = 2\mathrm{A}_1 + \mathrm{B}_1 + \mathrm{A}_2 + 2\mathrm{E} + \mathrm{B}_2$:

$$\mathrm{A}_1\text{: } (1) \quad c_4 = c_{-4} = \tfrac{5}{24}, \quad c_0 = \tfrac{7}{12};$$
$$\mathrm{A}_1\text{: } (1) \quad c_4 = c_{-4} = \tfrac{7}{24}, \quad c_0 = {}^*\tfrac{5}{12};$$
$$\mathrm{B}_1\text{: } (1) \quad c_2 = c_{-2} = \tfrac{1}{2};$$
$$\mathrm{A}_2\text{: } (1) \quad c_4 = -c_{-4} = {}^*\tfrac{1}{2};$$
$$\mathrm{E}\text{: } \begin{cases} (1) & c_1 = \tfrac{7}{8}, \quad c_{-3} = \tfrac{1}{8}; \\ (2) & c_3 = {}^*\tfrac{1}{8}, \quad c_{-1} = {}^*\tfrac{7}{8}; \end{cases}$$
$$\mathrm{B}_2\text{: } (1) \quad c_2 = -c_{-2} = \tfrac{1}{2};$$
$$\mathrm{E}\text{: } \begin{cases} ({}^*1) & c_1 = \tfrac{1}{8}, \quad c_{-3} = {}^*\tfrac{7}{8}; \\ (2) & c_3 = \tfrac{7}{8}, \quad c_{-1} = {}^*\tfrac{1}{8}. \end{cases}$$

$S=5$; $\mathrm{D}^{(5)} = \mathrm{A}_1 + 2\mathrm{A}_2 + \mathrm{B}_1 + \mathrm{B}_2 + 3\mathrm{E}$:

$$\mathrm{A}_1\text{: } (1) \quad c_4 = -c_{-4} = \tfrac{1}{2};$$
$$\mathrm{B}_1\text{: } (1) \quad c_2 = -c_{-2} = {}^*\tfrac{1}{2};$$
$$\mathrm{A}_2\text{: } (1) \quad c_4 = c_{-4} = \tfrac{1}{2};$$
$$\mathrm{E}\text{: } \begin{cases} (1) & c_5 = \tfrac{5}{128}, \quad c_1 = \tfrac{21}{64}, \quad c_{-3} = {}^*\tfrac{81}{128}; \\ (2) & c_3 = {}^*\tfrac{81}{128}, \quad c_{-1} = \tfrac{21}{64}, \quad c_{-5} = \tfrac{5}{128}; \end{cases}$$
$$\mathrm{A}_2\text{: } (1) \quad c_0 = 1;$$
$$\mathrm{E}\text{: } \begin{cases} (1) & c_5 = \tfrac{63}{128}, \quad c_1 = \tfrac{15}{64}, \quad c_{-3} = \tfrac{35}{128}; \\ (2) & c_3 = \tfrac{35}{128}, \quad c_{-1} = \tfrac{15}{64}, \quad c_{-5} = \tfrac{63}{128}; \end{cases}$$
$$\mathrm{B}_2\text{: } ({}^*1) \quad c_2 = c_{-2} = {}^*\tfrac{1}{2};$$
$$\mathrm{E}\text{: } \begin{cases} (1) & c_5 = {}^*\tfrac{15}{32}, \quad c_1 = \tfrac{7}{16}, \quad c_{-3} = \tfrac{3}{32}; \\ (2) & c_3 = \tfrac{3}{32}, \quad c_{-1} = \tfrac{7}{16}, \quad c_{-5} = {}^*\tfrac{15}{32}. \end{cases}$$

$S = 6$; $D^{(6)} = 2A_1 + A_2 + 2B_1 + 2B_2 + 3E$:

$A_1$: (1) $c_4 = c_{-4} = \frac{7}{16}$, $c_0 = {}^*\frac{1}{8}$;

$B_1$: (1) $c_6 = c_{-6} = \frac{5}{32}$, $c_2 = c_{-2} = {}^*\frac{11}{32}$;

$A_1$: (1) $c_4 = c_{-4} = \frac{1}{16}$, $c_0 = \frac{7}{8}$;

$B_1$: (1) $c_6 = c_{-6} = \frac{11}{32}$, $c_2 = c_{-2} = \frac{5}{32}$;

$A_2$: (1) $c_4 = -c_{-4} = {}^*\frac{1}{2}$;

E: $\begin{cases} (1) & c_5 = \frac{11}{32}, \quad c_1 = {}^*\frac{3}{16}, \quad c_{-3} = \frac{15}{32}; \\ (2) & {}^*\frac{15}{32}, \quad c_{-1} = \frac{3}{16}, \quad c_{-5} = {}^*\frac{11}{32}; \end{cases}$

$B_2$: (*1) $c_2 = c_{-2} = \frac{1}{2}$;

E: $\begin{cases} (1) & c_5 = {}^*\frac{165}{256}, \quad c_1 = {}^*\frac{5}{128}, \quad c_{-3} = \frac{81}{256}; \\ (2) & c_3 = {}^*\frac{81}{256}, \quad c_{-1} = \frac{5}{128}, \quad c_{-5} = \frac{165}{256}; \end{cases}$

$B_2$: (*1) $c_6 = -c_{-6} = \frac{1}{2}$;

E: $\begin{cases} (1) & c_5 = {}^*\frac{3}{256}, \quad c_1 = {}^*\frac{99}{128}, \quad c_{-3} = \frac{55}{256} \\ (2) & c_3 = \frac{55}{256}, \quad c_{-1} = \frac{99}{128}, \quad c_{-5} = \frac{3}{256}. \end{cases}$

## AIII.3 $D_3$

$S = 1$; $D^{(1)} = A_2 + E$:

$A_2$: (1) $c_0 = 1$;

E: $\begin{cases} (1) & c_1 = 1; \\ (2) & c_{-1} = 1. \end{cases}$

$S = 2$; $D^{(2)} = A_1 + 2E$:

$A_1$: (1) $c_0 = 1$;

E: $\begin{cases} (1) & c_1 = 1; \\ (2) & c_{-1} = {}^*1; \end{cases}$

E: $\begin{cases} (1) & c_2 = 1; \\ (2) & c_{-2} = 1. \end{cases}$

$S = 3; \quad D^{(3)} = A_1 + 2A_2 + 2E$:

$$A_1: \quad (1) \quad c_3 = c_{-3} = \tfrac{1}{2};$$
$$A_2: \quad (1) \quad c_0 = 1;$$
$$A_2: \quad (1) \quad c_3 = c_{-3} = {}^*\tfrac{1}{2};$$
$$E: \begin{cases} (1) & c_1 = 1; \\ (2) & c_{-1} = 1; \end{cases}$$
$$E: \begin{cases} (1) & c_{-2} = {}^*1; \\ (2) & c_2 = 1. \end{cases}$$

$S = 4; \quad D^{(4)} = 2A_1 + A_2 + 3E$:

$$A_1: \quad (1) \quad c_0 = 1;$$
$$A_1: \quad (1) \quad c_3 = -c_{-3} = \tfrac{1}{2};$$
$$A_2: \quad (1) \quad c_3 = c_{-3} = {}^*\tfrac{1}{2};$$
$$E: \begin{cases} (1) & c_1 = 1; \\ (2) & c_{-1} = {}^*1; \end{cases}$$
$$E: \begin{cases} (1) & c_2 = 1; \\ (2) & c_{-2} = 1; \end{cases}$$
$$E: \begin{cases} (1) & c_4 = 1; \\ (2) & c_{-4} = 1. \end{cases}$$

$S = 5; \quad D^{(5)} = A_1 + 2A_2 + 4E$:

$$A_1: \quad (1) \quad c_3 = c_{-3} = \tfrac{1}{2};$$
$$A_2: \quad (1) \quad c_0 = 1;$$
$$A_2: \quad (1) \quad c_3 = -c_{-3} = {}^*\tfrac{1}{2};$$
$$E: \begin{cases} (1) & c_1 = 1; \\ (2) & c_{-1} = 1; \end{cases}$$
$$E: \begin{cases} (1) & c_5 = 1; \\ (2) & c_{-5} = 1; \end{cases}$$
$$E: \begin{cases} (1) & c_2 = {}^*1; \\ (2) & c_{-2} = 1; \end{cases}$$
$$E: \begin{cases} (1) & c_4 = {}^*1; \\ (2) & c_{-4} = 1. \end{cases}$$

$S = 6; \quad D^{(6)} = 3A_1 + 2A_2 + 4E:$

$$A_1: \quad (1) \quad c_0 = 1;$$

$$A_1: \quad (1) \quad c_6 = c_{-6} = \tfrac{1}{2};$$

$$A_1: \quad (1) \quad c_3 = -c_{-3} = \tfrac{1}{2};$$

$$A_2: \quad (1) \quad c_3 = c_{-3} = {}^*\tfrac{1}{2};$$

$$A_2: \quad (1) \quad c_6 = c_{-6} = \tfrac{1}{2};$$

$$E: \quad \begin{cases} (1) & c_1 = 1; \\ (2) & c_{-1} = {}^*1; \end{cases}$$

$$E: \quad \begin{cases} (1) & c_{-5} = 1; \\ (2) & c_5 = {}^*1; \end{cases}$$

$$E: \quad \begin{cases} (1) & c_{-2} = 1; \\ (2) & c_2 = 1; \end{cases}$$

$$E: \quad \begin{cases} (1) & c_4 = 1; \\ (2) & c_{-4} = 1. \end{cases}$$

# APPENDIX IV: Decomposition of Products of Representations

## AIV.1 DECOMPOSITION OF SYMMETRIC PRODUCTS OF DEGENERATE IRREDUCIBLE REPRESENTATIONS OF POINT GROUPS (SINGLE-VALUED REPRESENTATIONS)

| Group | Symmetric products |
|---|---|
| $\mathbf{C}_{\infty,v}$ | $[E_k^2] = A_1 + E_{2k} \quad (k = 1, 2, \ldots)$ |
| $\mathbf{D}_{\infty,h}$ | $[E_{kg}^2] = [E_{ku}^2] = A_{1g} + E_{2k,g} \quad (k = 1, 2, \ldots)$ |
| $\mathbf{C}_{2p+1}$ | $[E_k^2] = \begin{Bmatrix} A + E_{2k} & (k \leq \frac{1}{2}p) \\ A + E_{2p+1-2k} & (k > \frac{1}{2}p) \end{Bmatrix} \quad (k = 1, 2, \ldots, p)$ |
| $\mathbf{C}_{2p}$ | $[E_k^2] = \begin{Bmatrix} A + E_{2k} & (k < \frac{1}{2}p) \\ A + 2B & (k = \frac{1}{2}p) \\ A + E_{2p-2k} & (k > \frac{1}{2}p) \end{Bmatrix} \quad (k = 1, 2, \ldots, p-1)$ |
| $\mathbf{D}_{2p+1}$ and $\mathbf{C}_{2p+1,v}$ | $[E_k^2] = \begin{Bmatrix} A_1 + E_{2k} & (k \leq \frac{1}{2}p) \\ A_1 + E_{2p+1-2k} & (k > \frac{1}{2}p) \end{Bmatrix} \quad (k = 1, 2, \ldots, p)$ |
| $\mathbf{D}_{2p}$ and $\mathbf{C}_{2p,v}$ | $[E_k^2] = \begin{Bmatrix} A_1 + E_{2k} & (k < \frac{1}{2}p) \\ A_1 + B_1 + B_2 & (k = \frac{1}{2}p) \\ A_1 + E_{2p-2k} & (k > \frac{1}{2}p) \end{Bmatrix} \quad (k = 1, 2, \ldots, p-1)$ |
| $\mathbf{S}_{2(2p+1)}$ | $[E_{kg}^2] = [E_{ku}^2] = \begin{Bmatrix} A_g + E_{2k,g} & (k \leq \frac{1}{2}p) \\ A_g + E_{2p+1-2k,g} & (k > \frac{1}{2}p) \end{Bmatrix} \quad (k = 1, \ldots, p)$ |
| $\mathbf{C}_{2p,h}$ | $[E_{kg}^2] = [E_{ku}^2] = \begin{Bmatrix} A_g + E_{2k,g} & (k < \frac{1}{2}p) \\ A_g + 2B_g & (k = \frac{1}{2}p) \\ A_g + E_{2p-2k,g} & (k > \frac{1}{2}p) \end{Bmatrix} \quad (k = 1, \ldots, p-1)$ |

| Group | Symmetric products |
|---|---|
| $\mathbf{D}_{2p+1,d}$ | $[E_{kg}^2] = [E_{ku}^2] = \begin{cases} A_{1g} + E_{2k,g} & (k \le \frac{1}{2}p) \\ A_{1g} + E_{2p+1-2k,g} & (k > \frac{1}{2}p) \end{cases} \quad (k = 1, \ldots, p)$ |
| $\mathbf{D}_{2p,h}$ | $[E_{kg}^2] = [E_{ku}^2] = \begin{cases} A_{1g} + E_{2k,g} & (k < \frac{1}{2}p) \\ A_{1g} + B_{1g} + B_{2g} & (k = \frac{1}{2}p) \\ A_{1g} + E_{2p-2k,g} & (k > \frac{1}{2}p) \end{cases} \quad (k = 1, \ldots, p-1)$ |
| $\mathbf{S}_{2p+1}$ and $\mathbf{D}_{2p+1,h}$ | $[E_k'^2] = [E_k''^2] = \begin{cases} A' + E'_{2k} & (k \le \frac{1}{2}p) \\ A' + E'_{2p+1-2k} & (k > \frac{1}{2}p) \end{cases} \quad (k = 1, 2, \ldots, p)$ |
| $\mathbf{S}_{4p}$ | $[E_k^2] = \begin{cases} A + E_{2k} & (k < p) \\ A_1 + 2B & (k = p) \\ A + E_{4p-2k} & (k > p) \end{cases} \quad (k = 1, 2, \ldots, 2p-1)$ |
| $\mathbf{D}_{2p,d}$ | $[E_k^2] = \begin{cases} A_1 + E_{2k} & (k < p) \\ A_1 + B_1 + B_2 & (k = p) \\ A + E_{4p-2k} & (k > p) \end{cases} \quad (k = 1, 2, \ldots, 2p-1)$ |
| $\mathbf{T}$ | $[E^2] = A + E$<br>$[T^2] = A + E + T$ |
| $\mathbf{T}_d$ and $\mathbf{O}$ | $[E^2] = A_1 + E$<br>$[T_1^2] = [T_2^2] = A_1 + E + T_2$ |
| $\mathbf{T}_h$ | $[E_g^2] = [E_u^2] = A_g + E_g$<br>$[T_g^2] = [T_g^2] = A_g + E_g + T_g$ |
| $\mathbf{O}_h$ | $[E_g^2] = [E_u^2] = A_{1g} + E_g$<br>$[T_{1g}^2] = [T_{1u}^2] = [T_{2g}^2] = [T_{2u}^2] = A_{1g} + E_g + T_{2g}$ |

## AIV.2 DECOMPOSITION OF ANTISYMMETRIC PRODUCTS OF DEGENERATE IRREDUCIBLE REPRESENTATIONS OF POINT GROUPS (DOUBLE-VALUED REPRESENTATIONS)

| Group | Antisymmetric products |
|---|---|
| **T** | $\{\Gamma_8^2\} = A + E + T$ |
| $\mathbf{T}_h$ | $\{\Gamma_{8g}^2\} = \{\Gamma_{8u}^2\} = A_g + E_g + T_g$ |
| $\mathbf{T}_d$ and **O** | $\{\Gamma_8^2\} = A_1 + E + T_2$ |
| $\mathbf{O}_h$ | $\{\Gamma_{8g}^2\} = \{\Gamma_{8u}^2\} = A_{1g} + E_g + T_{2g}$ |

(For **T** and $\mathbf{T}_h$, $\Gamma_8 \equiv \Gamma_6 + \Gamma_7$ complex conjugate representations; see [1], pp. 79 and 86.)

# APPENDIX V: Effective Hamiltonians for Non-Kramers Doublets†

## AV.1 EFFECTIVE HAMILTONIAN LINEAR AND QUADRATIC IN MAGNETIC AND ELECTRIC FIELDS

| | |
|---|---|
| $g_{\parallel}\beta\mathscr{H}_z\boldsymbol{\sigma}_z$ | $D_{4h}$, $D_4$, $C_{4v}$, $D_{2d}$, $C_{4h}$, $C_4$, $S_4$; $D_{6h}$, $C_{6h}$, $C_{6v}$, $D_{3d}$, $D_{3h}$, $D_6$, $C_6$, $D_3$, $C_3$, $C_{3v}$, $C_{3i}$, $C_{3h}$ |
| $R\mathscr{E}_z\boldsymbol{\sigma}_z$ | $D_{2d}$ |
| $R\mathscr{E}_z\boldsymbol{\sigma}_y - R'\mathscr{E}_z\boldsymbol{\sigma}_x$ | $S_4$ |
| $R'(\mathscr{E}_x\boldsymbol{\sigma}_y + \mathscr{E}_y\boldsymbol{\sigma}_x)$ | $D_3$ |
| $R(\mathscr{E}_x\boldsymbol{\sigma}_x - \mathscr{E}_y\boldsymbol{\sigma}_y)$ | $D_{3h}$, $C_{3v}$ |
| $R(\mathscr{E}_x\boldsymbol{\sigma}_x - \mathscr{E}_y\boldsymbol{\sigma}_y) + R'(\mathscr{E}_x\boldsymbol{\sigma}_y + \mathscr{E}_y\boldsymbol{\sigma}_x)$ | $C_{3h}$, $C_3$ |
| $G_1(\mathscr{H}_x^2 - \mathscr{H}_y^2)\boldsymbol{\sigma}_x + 2G_2\mathscr{H}_x\mathscr{H}_y\boldsymbol{\sigma}_y$ | $D_{4h}$, $D_4$, $C_{4v}$, $D_{2d}$ |
| $G_1(\mathscr{H}_x^2 - \mathscr{H}_y^2)\boldsymbol{\sigma}_x + 2G_2\mathscr{H}_x\mathscr{H}_y\boldsymbol{\sigma}_y$ $+ G_1'(\mathscr{H}_x^2 - \mathscr{H}_y^2)\boldsymbol{\sigma}_x - 2G_2'\mathscr{H}_x\mathscr{H}_y\boldsymbol{\sigma}_x$ | $C_{4h}$, $C_4$, $S_4$ |
| $G_1[(\mathscr{H}_x^2 - \mathscr{H}_y^2)\boldsymbol{\sigma}_x + 2\mathscr{H}_x\mathscr{H}_y\boldsymbol{\sigma}_y]$ | $D_{6d}$, $D_6$, $C_{6v}$, $D_{3h}$ |
| $G_1[(\mathscr{H}_x^2 - \mathscr{H}_y^2)\boldsymbol{\sigma}_x + 2\mathscr{H}_x\mathscr{H}_y\boldsymbol{\sigma}_y]$ $+ G_2\mathscr{H}_z(\mathscr{H}_x\boldsymbol{\sigma}_x - \mathscr{H}_y\boldsymbol{\sigma}_y)$ | $D_{3d}$, $D_3$, $C_{3v}$ |
| $G_1[(\mathscr{H}_x^2 - \mathscr{H}_y^2)\boldsymbol{\sigma}_x + 2\mathscr{H}_x\mathscr{H}_y\boldsymbol{\sigma}_y]$ $+ G_1'[(\mathscr{H}_x^2 - \mathscr{H}_y^2)\boldsymbol{\sigma}_y + 2\mathscr{H}_x\mathscr{H}_y\boldsymbol{\sigma}_x]$ | $C_{6h}$, $C_6$, $C_{3h}$ |
| $G_1[(\mathscr{H}_x^2 - \mathscr{H}_y^2)\boldsymbol{\sigma}_x + 2\mathscr{H}_x\mathscr{H}_y\boldsymbol{\sigma}_y]$ $+ G_1'[(\mathscr{H}_x^2 - \mathscr{H}_y^2)\boldsymbol{\sigma}_y + 2\mathscr{H}_x\mathscr{H}_y\boldsymbol{\sigma}_x]$ $+ G_2\mathscr{H}_z(\mathscr{H}_x\boldsymbol{\sigma}_x - \mathscr{H}_y\boldsymbol{\sigma}_y)$ $+ G_2'\mathscr{H}_z(\mathscr{H}_x\boldsymbol{\sigma}_y + \mathscr{H}_y\boldsymbol{\sigma}_x)$ | $C_{3i}$, $C_3$ |
| $G_1[\sqrt{\tfrac{1}{3}}(\mathscr{H}_x^2 + \mathscr{H}_y^2 - 2\mathscr{H}_z^2)\boldsymbol{\sigma}_x + (\mathscr{H}_x^2 - \mathscr{H}_y^2)\boldsymbol{\sigma}_y]$ | $O_h$, $O$, $T_d$ |

† After [83].

| | |
|---|---|
| $G_1[\sqrt{\tfrac{1}{3}}(\mathcal{H}_x^2 + \mathcal{H}_y^2 - 2\mathcal{H}_z^2)\boldsymbol{\sigma}_x + (\mathcal{H}_x^2 - \mathcal{H}_y^2)\boldsymbol{\sigma}_y]$ $+ G_1'[\sqrt{\tfrac{1}{3}}(\mathcal{H}_x^2 + \mathcal{H}_y^2 - 2\mathcal{H}_z^2)\boldsymbol{\sigma}_y$ $- (\mathcal{H}_x^2 - \mathcal{H}_y^2)\boldsymbol{\sigma}_x]$ | $\mathbf{T_h}$, $\mathbf{T}$ |
| $\bar{G}_1(\mathcal{E}_x^2 - \mathcal{E}_y^2)\boldsymbol{\sigma}_x + 2\bar{G}_2\mathcal{E}_x\mathcal{E}_y\boldsymbol{\sigma}_y$ | $\mathbf{D_{4h}}$, $\mathbf{D_4}$, $\mathbf{C_{4v}}$, $\mathbf{D_{2d}}$ |
| $\bar{G}_1(\mathcal{E}_x^2 - \mathcal{E}_y^2)\boldsymbol{\sigma}_x + 2\bar{G}_2\mathcal{E}_x\mathcal{E}_y\boldsymbol{\sigma}_y$ $+ \bar{G}_1'(\mathcal{E}_x^2 - \mathcal{E}_y^2)\boldsymbol{\sigma}_y - 2\bar{G}_2'\mathcal{E}_x\mathcal{E}_y\boldsymbol{\sigma}_x$ | $\mathbf{C_{4h}}$, $\mathbf{C_4}$, $\mathbf{S_4}$ |
| $\bar{G}_1[(\mathcal{E}_x^2 - \mathcal{E}_y^2)\boldsymbol{\sigma}_x + 2\mathcal{E}_x\mathcal{E}_y\boldsymbol{\sigma}_y]$ | $\mathbf{D_{6d}}$, $\mathbf{D_6}$, $\mathbf{C_{6v}}$, $\mathbf{D_{3h}}$ |
| $\bar{G}_1[(\mathcal{E}_x^2 - \mathcal{E}_y^2)\boldsymbol{\sigma}_x + 2\mathcal{E}_x\mathcal{E}_y\boldsymbol{\sigma}_y]$ $+ \bar{G}_2\mathcal{E}_z(\mathcal{E}_x\boldsymbol{\sigma}_x - \mathcal{E}_y\boldsymbol{\sigma}_y)$ | $\mathbf{D_{3d}}$, $\mathbf{D_3}$, $\mathbf{C_{3v}}$ |
| $\bar{G}_1[(\mathcal{E}_x^2 - \mathcal{E}_y^2)\boldsymbol{\sigma}_x + 2\mathcal{E}_x\mathcal{E}_y\boldsymbol{\sigma}_y]$ $+ \bar{G}_1'[(\mathcal{E}_x^2 - \mathcal{E}_y^2)\boldsymbol{\sigma}_x + 2\mathcal{E}_x\mathcal{E}_y\boldsymbol{\sigma}_x]$ | $\mathbf{C_{6h}}$, $\mathbf{C_6}$, $\mathbf{C_{3h}}$ |
| $\bar{G}_1[(\mathcal{E}_x^2 - \mathcal{E}_y^2)\boldsymbol{\sigma}_y + 2\mathcal{E}_x\mathcal{E}_y\boldsymbol{\sigma}_y]$ $+ \bar{G}_1'[(\mathcal{E}_x^2 - \mathcal{E}_y^2)\boldsymbol{\sigma}_y + 2\mathcal{E}_x\mathcal{E}_y\boldsymbol{\sigma}_x]$ $+ \bar{G}_2\mathcal{E}_z(\mathcal{E}_x\boldsymbol{\sigma}_x - \mathcal{E}_y\boldsymbol{\sigma}_y) + \bar{G}_2'\mathcal{E}_z(\mathcal{E}_x\boldsymbol{\sigma}_y - \mathcal{E}_y\boldsymbol{\sigma}_x)$ | $\mathbf{C_{3i}}$, $\mathbf{C_3}$ |
| $G_1[\sqrt{\tfrac{1}{3}}(\mathcal{E}_x^2 + \mathcal{E}_y^2 - 2\mathcal{E}_z^2)\boldsymbol{\sigma}_x + (\mathcal{E}_x^2 - \mathcal{E}_y^2)\boldsymbol{\sigma}_y]$ | $\mathbf{O_h}$, $\mathbf{O}$, $\mathbf{T_d}$ |
| $G_1[\sqrt{\tfrac{1}{3}}(\mathcal{E}_x^2 + \mathcal{E}_y^2 - 2\mathcal{E}_z^2)\boldsymbol{\sigma}_x + (\mathcal{E}_x^2 - \mathcal{E}_y^2)\boldsymbol{\sigma}_y]$ $+ G_1'[\sqrt{\tfrac{1}{3}}(\mathcal{E}_x^2 + \mathcal{E}_y^2 - 2\mathcal{E}_z^2)\boldsymbol{\sigma}_y$ $- (\mathcal{E}_x^2 - \mathcal{E}_y^2)\boldsymbol{\sigma}_x]$ | $\mathbf{T_h}$, $\mathbf{T}$ |
| $\bar{g}_\parallel \beta \mathcal{H}_z\mathcal{E}_z\boldsymbol{\sigma}_z$ | $\mathbf{C_{4v}}$, $\mathbf{C_4}$; $\mathbf{C_{6v}}$, $\mathbf{C_6}$, $\mathbf{C_{3v}}$, $\mathbf{C_3}$ |
| $K(\mathcal{H}_x\mathcal{E}_x + \mathcal{H}_y\mathcal{E}_y)\boldsymbol{\sigma}_z$ | $\mathbf{C_{4v}}$; $\mathbf{C_{6v}}$, $\mathbf{C_6}$, $\mathbf{C_{3v}}$, $\mathbf{C_3}$ |
| $K(\mathcal{H}_x\mathcal{E}_y - \mathcal{H}_y\mathcal{E}_x)\boldsymbol{\sigma}_z$ | $\mathbf{D_4}$; $\mathbf{D_6}$, $\mathbf{C_6}$, $\mathbf{D_3}$, $\mathbf{C_3}$ |
| $K_1(\mathcal{H}_x\mathcal{E}_x + \mathcal{H}_y\mathcal{E}_y)\boldsymbol{\sigma}_z + K_2(\mathcal{H}_x\mathcal{E}_y - \mathcal{H}_y\mathcal{E}_x)\boldsymbol{\sigma}_z$ | $\mathbf{C_4}$ |
| $K(\mathcal{H}_x\mathcal{E}_y + \mathcal{H}_y\mathcal{E}_x)\boldsymbol{\sigma}_z$ | $\mathbf{D_{2d}}$ |
| $K_1(\mathcal{H}_x\mathcal{E}_x - \mathcal{H}_y\mathcal{E}_y)\boldsymbol{\sigma}_z + K_2(\mathcal{H}_x\mathcal{E}_y + \mathcal{H}_y\mathcal{E}_x)\boldsymbol{\sigma}_z$ | $\mathbf{S_4}$ |
| $K(\mathcal{H}_x\mathcal{E}_x + \mathcal{H}_y\mathcal{E}_y + \mathcal{H}_z\mathcal{E}_z)\boldsymbol{\sigma}_z$ | $\mathbf{T_d}$, $\mathbf{T}$ |
| $M_1\mathcal{H}_z^2\mathbf{1} + M_2(\mathcal{H}_x^2 + \mathcal{H}_y^2)\mathbf{1}$ | $\mathbf{D_{4h}}$, $\mathbf{D_4}$, $\mathbf{C_{4v}}$, $\mathbf{D_{2d}}$, $\mathbf{C_{4h}}$, $\mathbf{C_4}$, $\mathbf{S_4}$; $\mathbf{D_{6h}}$, $\mathbf{C_{6h}}$, $\mathbf{C_{6v}}$, $\mathbf{D_{3d}}$, $\mathbf{D_{3h}}$, $\mathbf{D_6}$, $\mathbf{C_6}$, $\mathbf{D_3}$, $\mathbf{C_3}$, $\mathbf{C_{3v}}$, $\mathbf{C_{3i}}$, $\mathbf{C_{3h}}$ |
| $M_0(\mathcal{H}_x^2 + \mathcal{H}_y^2 + \mathcal{H}_z^2)\mathbf{1}$ | $\mathbf{O_h}$, $\mathbf{O}$, $\mathbf{T_d}$, $\mathbf{T_h}$, $\mathbf{T}$ |
| $\bar{M}_1\mathcal{E}_z^2\mathbf{1} + \bar{M}_2(\mathcal{E}_x^2 + \mathcal{E}_y^2)\mathbf{1}$ | $\mathbf{D_{4h}}$, $\mathbf{D_4}$, $\mathbf{C_{4v}}$, $\mathbf{D_{2d}}$, $\mathbf{C_{4h}}$, $\mathbf{C_4}$, $\mathbf{S_4}$; $\mathbf{D_{6h}}$, $\mathbf{C_{6h}}$, $\mathbf{C_{6v}}$, $\mathbf{D_{3d}}$, $\mathbf{D_{3h}}$, $\mathbf{D_6}$, $\mathbf{C_6}$, $\mathbf{D_3}$, $\mathbf{C_3}$, $\mathbf{C_{3v}}$, $\mathbf{C_{3i}}$, $\mathbf{C_{3h}}$ |
| $\bar{M}_0(\mathcal{E}_x^2 + \mathcal{E}_y^2 + \mathcal{E}_z^2)\mathbf{1}$ | $\mathbf{O_h}$, $\mathbf{O}$, $\mathbf{T_d}$, $\mathbf{T_h}$, $\mathbf{T}$ |

## AV.2 EFFECTIVE HAMILTONIAN OF HYPERFINE INTERACTION

| | |
|---|---|
| $A_{\parallel}\hat{I}_z\boldsymbol{\sigma}_z + P\{3\hat{I}_z^2 - I(I+1)\}\mathbf{1}$ | $\mathbf{D}_{4h}$, $\mathbf{D}_4$, $\mathbf{C}_{4v}$, $\mathbf{D}_{2d}$, $\mathbf{C}_{4h}$, $\mathbf{C}_4$, $\mathbf{S}_4$; $\mathbf{D}_{6h}$, $\mathbf{C}_{6h}$, $\mathbf{C}_{6v}$, $\mathbf{D}_{3d}$, $\mathbf{D}_{3h}$, $\mathbf{D}_6$, $\mathbf{C}_6$, $\mathbf{D}_3$, $\mathbf{C}_3$, $\mathbf{C}_{3v}$, $\mathbf{C}_{3i}$, $\mathbf{C}_{3h}$ |
| $B_1(\hat{I}_x^2 - \hat{I}_y^2)\boldsymbol{\sigma}_x + B_2(\hat{I}_x\hat{I}_y + \hat{I}_y\hat{I}_x)\boldsymbol{\sigma}_y$ | $\mathbf{D}_{4h}$, $\mathbf{D}_4$, $\mathbf{C}_{4v}$, $\mathbf{D}_{2d}$ |
| $B_1(\hat{I}_x^2 - \hat{I}_y^2)\boldsymbol{\sigma}_x + B_2(\hat{I}_x\hat{I}_y + \hat{I}_y\hat{I}_x)\boldsymbol{\sigma}_y + B_1'(\hat{I}_x^2 - \hat{I}_y^2)\boldsymbol{\sigma}_y - B_2'(\hat{I}_x\hat{I}_y + \hat{I}_y\hat{I}_x)\boldsymbol{\sigma}_x$ | $\mathbf{C}_{4h}$, $\mathbf{C}_4$, $\mathbf{S}_4$ |
| $B_1[(\hat{I}_x^2 - \hat{I}_y^2)\boldsymbol{\sigma}_x + (\hat{I}_x\hat{I}_y + \hat{I}_y\hat{I}_x)\boldsymbol{\sigma}_y]$ | $\mathbf{D}_{6d}$, $\mathbf{D}_6$, $\mathbf{C}_{6v}$, $\mathbf{D}_{3h}$ |
| $B_1[(\hat{I}_x^2 - \hat{I}_y^2)\boldsymbol{\sigma}_x + (\hat{I}_x\hat{I}_y + \hat{I}_y\hat{I}_x)\boldsymbol{\sigma}_y] + B_2[(\hat{I}_z\hat{I}_x + \hat{I}_x\hat{I}_z)\boldsymbol{\sigma}_x - (\hat{I}_y\hat{I}_z + \hat{I}_z\hat{I}_y)\boldsymbol{\sigma}_y]$ | $\mathbf{D}_{3d}$, $\mathbf{D}_3$, $\mathbf{C}_{3v}$ |
| $B_1[(\hat{I}_x^2 - \hat{I}_y^2)\boldsymbol{\sigma}_x - (\hat{I}_x\hat{I}_y + \hat{I}_y\hat{I}_x)\boldsymbol{\sigma}_y] + B_1'[(\hat{I}_x^2 - \hat{I}_y^2)\boldsymbol{\sigma}_y - (\hat{I}_x\hat{I}_y + \hat{I}_y\hat{I}_x)\boldsymbol{\sigma}_x]$ | $\mathbf{C}_{6h}$, $\mathbf{D}_6$, $\mathbf{C}_{3h}$ |
| $B_1[(\hat{I}_x^2 - \hat{I}_y^2)\boldsymbol{\sigma}_x - (\hat{I}_x\hat{I}_y + \hat{I}_y\hat{I}_x)\boldsymbol{\sigma}_y] + B_1'[(\hat{I}_x^2 - \hat{I}_y^2)\boldsymbol{\sigma}_y - (\hat{I}_x\hat{I}_y + \hat{I}_y\hat{I}_x)\boldsymbol{\sigma}_x] + B_2[(\hat{I}_z\hat{I}_x + \hat{I}_x\hat{I}_z)\boldsymbol{\sigma}_x - (\hat{I}_y\hat{I}_z + \hat{I}_z\hat{I}_y)\boldsymbol{\sigma}_y] + B_2'[(\hat{I}_z\hat{I}_x + \hat{I}_x\hat{I}_z)\boldsymbol{\sigma}_y + (\hat{I}_y\hat{I}_z + \hat{I}_z\hat{I}_y)\boldsymbol{\sigma}_z]$ | $\mathbf{C}_{3i}$, $\mathbf{C}_3$ |
| $B_1\{\sqrt{\tfrac{1}{3}}[3\hat{I}_z^2 - I(I+1)]\boldsymbol{\sigma}_x - (\hat{I}_x^2 - \hat{I}_y^2)\boldsymbol{\sigma}_y\}$ | $\mathbf{O}_h$, $\mathbf{O}$, $\mathbf{T}_d$ |
| $B_1\{\sqrt{\tfrac{1}{3}}[3\hat{I}_z^2 - I(I+1)]\boldsymbol{\sigma}_x - (\hat{I}_x^2 - \hat{I}_y^2)\boldsymbol{\sigma}_y\} + B_1'\{\sqrt{\tfrac{1}{3}}[3\hat{I}_z^2 - I(I+1)]\boldsymbol{\sigma}_y - (\hat{I}_x^2 - \hat{I}_y^2)\boldsymbol{\sigma}_x\}$ | $\mathbf{T}_h$, $\mathbf{T}$ |

## AV.3 EFFECTIVE HAMILTONIAN INVOLVING MAGNETIC FIELD AND NUCLEAR SPIN

| | |
|---|---|
| $C_1(\mathscr{H}_x\hat{I}_x^2 - \mathscr{H}_y\hat{I}_y^2)\boldsymbol{\sigma}_x + C_2(\mathscr{H}_x\hat{I}_y + \mathscr{H}_y\hat{I}_x)\boldsymbol{\sigma}_y$ | $\mathbf{D}_{4h}$, $\mathbf{D}_4$, $\mathbf{C}_{4v}$, $\mathbf{D}_{2d}$ |
| $C_1(\mathscr{H}_x\hat{I}_x - \mathscr{H}_y\hat{I}_y)\boldsymbol{\sigma}_x + C_2(\mathscr{H}_x\hat{I}_y + \mathscr{H}_y\hat{I}_x)\boldsymbol{\sigma}_y$ $+ C_1'(\mathscr{H}_x\hat{I}_x - \mathscr{H}_y\hat{I}_y)\boldsymbol{\sigma}_y - C_2'(\mathscr{H}_x\hat{I}_y + \mathscr{H}_y\hat{I}_x)\boldsymbol{\sigma}_x$ | $\mathbf{C}_{4h}$, $\mathbf{C}_4$, $\mathbf{S}_4$ |
| $C_1[(\mathscr{H}_x\hat{I}_x - \mathscr{H}_y\hat{I}_y)\boldsymbol{\sigma}_x + (\mathscr{H}_x\hat{I}_y + \mathscr{H}_y\hat{I}_x)\boldsymbol{\sigma}_y]$ | $\mathbf{D}_{6d}$, $\mathbf{D}_6$, $\mathbf{C}_{6v}$, $\mathbf{D}_{3h}$ |
| $C_1[(\mathscr{H}_x\hat{I}_x - \mathscr{H}_y\hat{I}_y)\boldsymbol{\sigma}_x + (\mathscr{H}_x\hat{I}_y + \mathscr{H}_y\hat{I}_x)\boldsymbol{\sigma}_y]$ $+ C_2\mathscr{H}_z[(\hat{I}_z\boldsymbol{\sigma}_x + \hat{I}_y\boldsymbol{\sigma}_y) + C_3\hat{I}_z(\mathscr{H}_x\boldsymbol{\sigma}_x - \mathscr{H}_y\boldsymbol{\sigma}_y)$ | $\mathbf{D}_{3d}$, $\mathbf{D}_3$, $\mathbf{C}_{3v}$ |
| $C_1[(\mathscr{H}_x\hat{I}_x - \mathscr{H}_y\hat{I}_y)\boldsymbol{\sigma}_x + (\mathscr{H}_x\hat{I}_y + \mathscr{H}_y\hat{I}_x)\boldsymbol{\sigma}_y]$ $+ C_1'[(\mathscr{H}_x\hat{I}_x - \mathscr{H}_y\hat{I}_y)\boldsymbol{\sigma}_y - (\mathscr{H}_x\hat{I}_y - \mathscr{H}_y\hat{I}_x)\boldsymbol{\sigma}_x]$ | $\mathbf{C}_{6h}$, $\mathbf{D}_6$, $\mathbf{C}_{3h}$ |
| $C_1[(\mathscr{H}_x\hat{I}_x - \mathscr{H}_y\hat{I}_y)\boldsymbol{\sigma}_x + (\mathscr{H}_x\hat{I}_y + \mathscr{H}_y\hat{I}_x)\boldsymbol{\sigma}_y]$ $+ C_1'[(\mathscr{H}_x\hat{I}_x - \mathscr{H}_y\hat{I}_y)\boldsymbol{\sigma}_y - (\mathscr{H}_x\hat{I}_y - \mathscr{H}_y\hat{I}_x)\boldsymbol{\sigma}_x]$ $+ C_2\mathscr{H}_z(\hat{I}_x\boldsymbol{\sigma}_x - \hat{I}_y\boldsymbol{\sigma}_y) + C_3\hat{I}_z(\mathscr{H}_x\boldsymbol{\sigma}_x - \mathscr{H}_y\boldsymbol{\sigma}_y)$ $+ C_2'\mathscr{H}_z(\hat{I}_x\boldsymbol{\sigma}_y + \hat{I}_y\boldsymbol{\sigma}_x) + C_3'\hat{I}_z(\mathscr{H}_x\boldsymbol{\sigma}_y - \mathscr{H}_y\boldsymbol{\sigma}_x)$ | $\mathbf{C}_{3i}$, $\mathbf{C}_3$ |
| $C_1[\sqrt{\tfrac{1}{3}}(\mathscr{H}_x\hat{I}_x + \mathscr{H}_y\hat{I}_y - 2\mathscr{H}_z\hat{I}_z)\boldsymbol{\sigma}_x + (\mathscr{H}_x\hat{I}_x - \mathscr{H}_y\hat{I}_y)\boldsymbol{\sigma}_y]$ | $\mathbf{O}_h$, $\mathbf{O}$, $\mathbf{T}_d$ |
| $C_1[\sqrt{\tfrac{1}{3}}(\mathscr{H}_x\hat{I}_x + \mathscr{H}_y\hat{I}_y - 2\mathscr{H}_z\hat{I}_z)\boldsymbol{\sigma}_x + (\mathscr{H}_x\hat{I}_x - \mathscr{H}_y\hat{I}_y)\boldsymbol{\sigma}_y]$ $+ C_1'[\sqrt{\tfrac{1}{3}}(\mathscr{H}_x\hat{I}_x + \mathscr{H}_y\hat{I}_y - 2\mathscr{H}_z\hat{I}_z)\boldsymbol{\sigma}_y$ $- (\mathscr{H}_x\hat{I}_x - \mathscr{H}_y\hat{I}_y)\boldsymbol{\sigma}_x]$ | $\mathbf{T}_h$, $\mathbf{T}$ |
| $D_1\mathscr{H}_z\hat{I}_z\mathbf{1} + D_2(\mathscr{H}_x\hat{I}_x + \mathscr{H}_y\hat{I}_y)\mathbf{1}$ | $\mathbf{D}_{4h}$, $\mathbf{D}_4$, $\mathbf{C}_{4v}$, $\mathbf{D}_{2d}$, $\mathbf{D}_{6h}$, $\mathbf{D}_6$, $\mathbf{C}_{6v}$, $\mathbf{D}_{3d}$, $\mathbf{D}_{3h}$, $\mathbf{D}_3$, $\mathbf{C}_{3v}$ |
| $D_1\mathscr{H}_z\hat{I}_z\mathbf{1} + D_2(\mathscr{H}_x\hat{I}_x + \mathscr{H}_y\hat{I}_y)\mathbf{1} + D_3(\mathscr{H}_x\hat{I}_x + \mathscr{H}_y\hat{I}_y)\mathbf{1}$ | $\mathbf{C}_{4h}$, $\mathbf{C}_4$, $\mathbf{S}_4$; $\mathbf{D}_{6h}$, $\mathbf{C}_{6h}$, $\mathbf{C}_6$, $\mathbf{C}_{3i}$, $\mathbf{C}_{3h}$, $\mathbf{C}_3$ |
| $D_1(\mathscr{H}_x\hat{I}_x + \mathscr{H}_y\hat{I}_y + \mathscr{H}_z\hat{I}_z)\mathbf{1}$ | $\mathbf{O}_h$, $\mathbf{O}$, $\mathbf{T}_d$, $\mathbf{T}_h$, $\mathbf{T}$ |

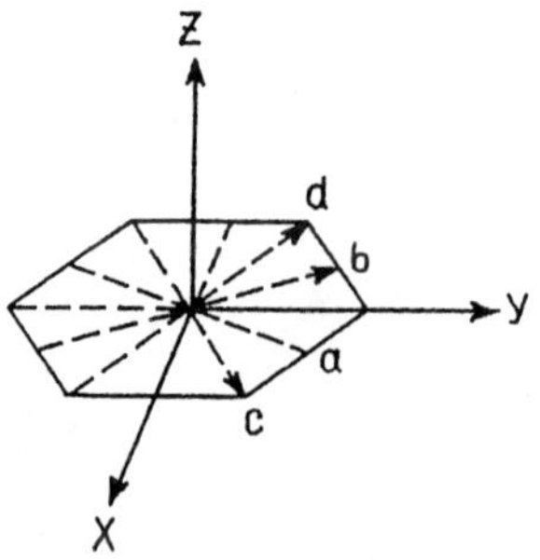


**Fig. AI.1** Symmetry axes of $\mathbf{D}_6$.

# References


[1] G. F. Koster, J. O. Dimmok, R. G. Wheeler and H. Statz, *Properties of the Thirty-Two Point Groups*. MIT Press, Cambridge, MA (1963).

[2] D. A. Varshalovich, A. N. Moskalev and V. K. Khersonskii, *Quantum Theory of Angular Momentum*. World Scientific, Singapore (1988).

[3] A. R. Edmonds, *Angular Momentum in Quantum Mechanics*. Princeton University Press, New Jersey (1956).

[4] M. Rotenberg, R. Bivins, N. Metropolis and F. K. Wooten, *The 3j- and 6j-Symbols*. MIT Press, Cambridge, MA (1959).

[5] A. M. Leushin, *Tables of Functions Transformed According to Irreducible Representations of Crystallographic Point Groups*, Nauka, Moscow (1968) (in Russian).

[6] D. G. Bell, *Rev. Mod. Phys.* **26**, 311 (1954).

[7] A. M. Leushin, *Tables of Matrices of Irreducible Representations of Crystallographic Point Groups. Collected Works*, Vol. 4, pp. 86–160. Kazan University (1968) (in Russian).

[8] O. V. Kovalev, *Irreducible Representations of the Space Groups*. Gordon and Breach, New York (1965).

[9] Y. Zak, A. Gasher, M. Gluck and Y. Gur, *The Irreducible Representations of Space Groups*. Benjamin, New York (1969).

[10] C. J. Bradley and A. D. Cracknell, *The Mathematical Theory of Symmetry in Solids: Representation Theory for Point Groups and Space Groups*. Clarendon Press, Oxford (1972).

[11] J. S. Griffith, *The Theory of Transition Metal Ions*. Cambridge University Press (1961).

[12] E. P. Wigner, *Group Theory and Its Application to the Quantum Mechanics of Atomic Spectra*. Academic Press, New York (1959).

[13] F. A. Cotton, *Chemical Applications of Group Theory*, 2nd edn. Interscience, New York (1971).

[14] S. Sugano, Y. Tanabe and H. Kamimura, *Multiplets of Transition Metal Ions in Crystals*. Academic Press, New York (1970).

[15] R. McWeeny, *Symmetry. An Introduction to Group Theory and Its Applications*. Pergamon Press, Oxford (1963).

[16] C. L. Ballhausen, *Introduction to the Ligand Field Theory*. McGraw-Hill, New York (1962).

[17] R. M. Hochstrasser, *Molecular Aspects of Symmetry*. Benjamin, New York (1966).

[18] C. D. Chisholm, *Group Theoretical Techniques in Quantum Chemistry*. Academic Press, London (1976).

[19] P. R. Bunker, *Molecular Symmetry and Spectroscopy*. Academic Press, New York (1979).

[20] M. Hamermesh, *Group Theory and Its Application to Physical Problems*. Addison-Wesley, Reading, MA (1964).

[21] R. L. Flurry, *Symmetry Groups. Theory and Chemical Applications*. Prentice-Hall, Englewood Cliffs, NJ (1980).
[22] I. G. Kaplan, *Symmetry of Multielectron Systems*, Nauka, Moscow (1969) (in Russian).
[23] L. H. Hall, *Group Theory and Symmetry in Chemistry*. McGraw-Hill, New York (1969).
[24] J. W. Leech and D. J. Newman, *How to Use Groups*. Meuthen, London (1969).
[25] D. B. Chestnut, *Finite Groups and Quantum Chemistry*. Wiley, New York (1974).
[26] D. M. Bishop, *Group Theory and Chemistry*. Clarendon Press, Oxford (1973).
[27] I. Bernal, W. C. Hamilton and J. S. Ricci, *Symmetry*. Freeman, San Francisco (1972).
[28] P. M. Zorkii and N. N. Afonina, *Symmetry of Molecules and Crystals*. Moscow University (1979).
[29] H. H. Jaffé and M. Orchin, *Symmetry in Chemistry*. Wiley, New York (1965).
[30] C. J. Hawkins, *Absolute Configuration of Metal Complexes*. Wiley-Interscience, New York (1971).
[31] R. F. Gillespie, *Molecular Geometry*. Van Nostrand-Reinhold, London (1972).
[32] V. I. Sokolov, *J. Mendeleyev All-Union Chem. Soc.*, **32** (1987) (in Russian).
[33] G. P. Kostikova and D. V. Korolkov, *Usp. Khim.* **53,** 591 (1985) (in Russian).
[34] H. A. Bethe and E. E. Salpeter, *Quantum Mechanics of One- and Two-Electron Atoms*. Springer, Berlin (1957).
[35] M. I. Petrashen and E. D. Trifonov, *Group Theory. Applications to Quantum Mechanics*. Nauka, Moscow (1967) (in Russian).
[36] H. Bethe, *Ann. Physik*. **3,** 133 (1929).
[37] A. B. Lever, *Inorganic Electronic Spectroscopy*. Elsevier, Amsterdam (1984).
[38] E. U. Condon and G. H. Shortley, *The Theory of Atomic Spectra*. Cambridge University Press (1935).
[39] L. D. Landau and E. M. Lifshitz, *Quantum Mechanics: Non-Relativistic Theory*, 3rd edn. Pergamon, Oxford (1977).
[40] I. I. Sobelman, *Introduction to the Theory of Atomic Spectra*. Fizmatgiz, Moscow (1963) (in Russian); *Atomic Spectra and Radiative Transitions*, Springer, Berlin. New York.
[41] L. S. Biedenharn and J. D. Louck, *Angular Momentum in Quantum Physics. Theory and Application*. Addison-Wesley, Reading, MA (1981).
[42] A. P. Jusys and A. J. Savukynas, *Mathematical Foundations of the Atomic Theory*. Vilnius (1973) (in Russian).
[43] E. El Baz and B. Castel, *Graphical Methods of Spin Algebra in Atomic, Nuclear and Particle Physics*. Marcel Dekker, New York (1972).
[44] V. I. Cherepanov and A. A. Shchetkov, *Zh. Eksp. Teor. Fiz.* **55,** 1805 (1968) [*Sov. Phys. JETP*. **XX**, XXX].
[45] I. G. Kaplan and O. B. Rodimova, *Zh. Eksp. Teor. Fiz.* **55,** 1881 (1968) [*Sov. Phys. JETP*. **XX**, XXX].
[46] S. V. Vonsovsky, S. V. Grum-Grzhimailo, V. I. Cherepanov, A. N. Men, D. T. Sviridov, Yu. F. Smirnov and A. E. Nikiforov, *Crystal Field Theory and Optical Spectra of Impurity Ions with Unfilled d-Shell*. Nauka, Moscow (1969) (in Russian).
[47] D. T. Sviridov and Yu. F. Smirnov, *The Theory of the Optical Spectra of Transition Metal Ions*. Nauka, Moscow (1977).
[48] E. König and S. Kremer, *Ligand Field Energy Diagrams*. Plenum, New York (1977).
[49] E. M. Ledovskaya and E. D. Trifonov, *J. Leningrad State Univ.* **10,** 21 (1962) (in Russian).
[50] M. Z. Balavichus, I. V. Radvilavichus and A. B. Bolotin, *Litovskii Fiz. Sb.* **3,** 389 (1963) (in Russian).
[51] K. R. Lea, M. J. Leask and W. P. Wolf, *J. Phys. Chem. Solids* **23,** 1381 (1962).
[52] A. G. McLellan, *J. Chem. Phys.* **34,** 1350 (1961).
[53] R. Pappalardo and D. L. Wood: *Molec. Spectrosc.* **10,** 81 (1963).
[54] F. C. Von der Lage and H. A. Bethe, *Phys. Rev.* **71,** 612 (1947).
[55] C. Kittel and J. M. Luttinger, *Phys. Rev.* **73,** 162 (1948).
[56] C. F. Ballhausen and H. B. Gray, *Molecular Orbital Theory*. Benjamin, New York (1964).
[57] J. N. Murrell, S. F. Kettle and F. M. Tedder, *Valence Theory*. Wiley, London (1965).

[58] J. E. Kimbal, *J. Chem. Phys.* **7**, 188 (1940).

[59] I. B. Bersuker, *Electronic Structure and Properties of Coordination Compounds. Introduction to the Theory*, 3rd edn. Khymiya, Leningrad (1986).

[60] R. McWeeny, *Methods of Molecular Quantum Mechanics* (2nd edition). Academic Press, Publishers (1989).

[61] E. M. Shustorovich and M. E. Dyatkina, *Dokl. Akad. Nauk SSSR* **128**, 1234 (1959) [*Sov. Phys. Dokl.* **XX**, XXX].

[62] B. S. Tsukerblat and M. I. Belinskii, *Magnetochemistry and Radiospectroscopy of Exchange Clusters*. Shtiintsa, Kishinev (1983).

[63] B. S. Tsukerblat, M. I. Belinskii and V. E. Fainzilberg, *Magnetochemistry and Spectroscopy of Transition Metal Exchange Clusters*. (*Soviet Scientific Reviews, Ser. B: Chemistry*, Vol. 9, pp. 337–481). Gordon and Breach, Reading, UK (1987).

[64] W. E. Hatfield, *Extended Interactions between Metal Ions in Transition Metal Complexes* (ed. V. Interrante), pp. 108–104 (1974).

[65] W. E. Hatfield, *Theory and Applications of Molecular Paramagnetism* (ed. E. A. Boudreax and L. N. Mulay), pp. 349–449. Wiley, New York (1976).

[66] V. T. Kalinnikov and Yu. V. Rakitin, *Introduction to Magnetochemistry*. Nauka, Moscow (1980) (in Russian).

[67] A. S. Davydov, *Quantum Mechanics*. Pergamon, Oxford (1976).

[68] P. P. Feofilov, *Polarized Luminescence of Atoms, Molecules and Crystals*. Fizmatgiz, Moscow (1959) (in Russian).

[69] A. A. Kaplyansky, *Opt. i Spektosk.* **16**, 1031 (1964) [*Sov. Phys. Opt. Spectrosc.* **XX**, XXX].

[70] M. Inoue and Y. Toyozava, *J. Phys. Soc. Japan* **20**, 363 (1965).

[71] T. R. Bader and A. Gold, *Phys. Rev.* **171** (1968).

[72] B. S. Tsukerblat and E. V. Vitiu, *Optical and Kinetic Effects in Strong Electromagnetic Fields*, pp. 48–61. Shtiintsa, Kishinev (1974) (in Russian).

[73] V. Heine, *Group Theory in Quantum Mechanics*. Pergamon, London (1960).

[74] R. S. Knox and A. Gold, *Symmetry in the Solid State*. Benjamin, New York (1964).

[75] A. Abragam and B. Bleaney, *Electron Paramagnetic Resonance of Transition Ions*. Clarendon Press, Oxford (1970).

[76] S. A. Altshuler and B. M. Kozyrev, *Electron Paramagnetic Resonance*. Nauka, Moscow (1972) (in Russian).

[77] A. B. Roytsin, *Some Applications of Symmetry Theory to Radiospectroscopic Problems*. Naukova Dumka, Kiev (1973) (in Russian).

[78] B. C. Grachev, *Zh. Eksp. Teor. Fiz.* **92**, 1834–1844 (1987) [*Sov. Phys. JETP* **XX**, XXX].

[79] R. McWeeny, *Spins in Chemistry*. Academic Press, New York (1970).

[80] A. B. Roytsin, *Usp. Fiz. Nauk* **5**, 677 (1971) [*Sov. Phys. Usp.* **XX**, XXX].

[81] W. B. Mims, *The Linear Electric Field Effect in Paramagnetic Resonance*. Clarendon Press, Oxford (1976).

[82] M. D. Glinchuk, V. G. Grachev, M. F. Deygen, A. B. Roytsin and M. A. Suslin, *Electric Effects in Radiospectroscopy*. Nauka, Moscow (1981) (in Russian).

[83] S. Washimia, K. Shinagawa and S. Sugano, *Phys. Rev.* **B1**, 2976 (1970).

[84] Brian L. Silver, *Irreducible Tensor Methods. An Introduciton for Chemists*. Academic Press, London (1976).

[85] K. Smith, *Table of Wigner 9j-symbols for Integral and Half-Integral Values of Parameters*. ANL-5860, Argonne National Lab., Chicago, Illinois (1958).

[86] K. M. Howell, *Revised Tables of 6j-symbols. Research Report 59-1*, University of Southampton, Southampton, England.

[87] A. Bencini and D. Gatteschi, *Electron Paramagnetic Resonance of Exchange Coupled Systems*. Springer, Berlin (1990).

[88] B. S. Tsukerblat, A. B. Ablov, V. M. Novotortsev, V. T. Kalinnikov, M. I. Belinskii and V. V. Kalmikov, *Dokl. Akad. Nauk SSSR* **210**, 1144 (19XX) [*Sov. Phys. Dokl.* **XX**, XXX].

[89] Yu. V. Rakitin, V. T. Kalinnikov and W. E. Hatfield, *Phys. Stat. Sol.* **(b)81**, 379 (1977).

[90] D. Gatteschi, L. Pardi and R. Sessoli, *Materials Sci.* **27**, 7 (1991).

[91] John C. Slater, *Quantum Theory of Molecules and Solids, v. I. Electronic Structure of Molecules*. McGraw-Hill Book company, Inc., NY (1963).

[92] R. Beckett, R. Colton, B. F. Hoskins, R. L. Matrin and D. G. Vince, *Aust. J. Chem.* **22**, 2527 (1969).

[93] B. S. Tsukerblat, B. Ya, Kuavskaya, M. I. Belinskii, A. V. Ablov, V. M. Novotortsev and V. T. Kalinnikov, *Theor. Chim. Acta*, **38**, 131 (1975); *Pisma Zh. Exp. Teor. Fiz.* **19**, 525 (1975).

[94] D. Gatteschi and B. S. Tsukerblat, *Mol. Phys.* **79**, 121 (1993).

[95] A. Ganeschi, D. Gatteschi, J. Laugier, P. Rey, R. Sessoli and C. Zanchini, *J. Am. Chem. Soc.* **110**, 2795 (1988).

[96] V. Ya. Mitrofanov, A. E. Nikiforov and V. I. Cherepanov, *Spectroscopy of Exchange-Coupled Complexes in Ionic Crystals*. Nauka, Moscow (1985) (in Russian).

[97] P. W. Anderson and H. Hasegawa, *Phys. Rev.* **100**, 675 (1955).

[98] P. G. de Gennes, *Phys. Rev.* **119**, 141 (1960).

[99] M. I. Belinskii and B. S. Tsukerblat, *Fiz. Tverd. Tela*, 4, 606 (1985).

[100] M. I. Belinskii, *Molec. Phys.* **60**, 793 (1987).

[101] G. Blondin and J.-J. Girerd, *Chem. Rev.*, **90**, 1359 (1990).

[102] M. I. Belinskii, B. S. Tsukerblat, S. A. Zaitsev and I. S. Belinskaya, *New J. Chem.*, **16**, 791 (1992).

[103] A. V. Palii, S. M. Ostrovskii and B. S. Tsukerblat, *New J. Chem.*, **16**, 943 (1992).

[104] S. M. Ostrovskii, A. V. Palii and B. S. Tsukerblat, *Zh. Struct. Khim.*, **34**, 33 (1993).

[105] V. P. Coropchanu, Sh. N. Gifeisman, V. Ya. Gamurar and B. S. Tsukerblat, *Zh. Struct. Khim.*, **34**, 14 (1993).

[106] A. L. Barra, D. Gatteschi, B. S. Tsukerblat, J. Döring, A. Müller and L.-C. Brunel, *Inorg. Chem.*, **31**, 5132 (1992).

[107] C. J. Gomez-Garcia, E. Coronado, R. Georges and G. Pourroy, *Physica B*, **182**, 18 (1992).

[108] J. J. Borras-Almenar, E. Coronado, R. Georges and C. J. Gomez-Carcia, *J. Magnet. and Magnetic Mat.*, **104–107**, 955 (1992).

[109] K. Nakamoto, *Infared and Raman Spectra of Inorganic and Coordination Compounds*. Wiley-Interscience, New York (1978).

[110] L. A. Gribov and W. J. Orville-Thomas, *Theory and Methods of Calculation of Molecular Spectra*. Wiley, New York (1988).

[111] R. Englman, *The Jahn–Teller Effect in Molecules and Crystals*. Wiley, New York (1972).

[112] I. B. Bersuker, *The Jahn–Teller Effect and Vibronic Interactions in Modern Chemistry*. Plenum, New York (1984).

[113] I. B. Bersuker and V. Z. Polinger, *Vibronic Interactions in Molecules and Crystals*. Springer, Berlin (1989).

[114] G. A. Gehring and K. A. Gehring, *Rep. Prog. Phys.* **38**, 1 (1975).

[115] A. A. Levin and P. N. D'yachkov, *Electronic Structure and Transformation of Heteroligand Molecules*. Nauka, Moscow (1990) (in Russian).

[116] Yu. E. Perlin and B. S. Tsukerblat, *The Effects of Electron Vibrational Interactions in Optical Spectra of Paramagnetic Impurity Ions*. Stiintsa, Kishinev (1974) (in Russian); Yu. E. Perlin, B. S. Tsukerblat, *Optical Bands and Polarization Dichroism of Jahn–Teller Effect in Localized Systems* (ed. Yu. E. Perlin and M. Wagner), pp. 251–346. Elsevier, Amsterdam (1984).

[117] N. N. Kristofel, *Theory of Small Radius Impurity Centers in Ionic Crystals*. Nauka, Moscow (1974).

[118] V. I. Cherepanov, V. N. Frolov and A. N. Frolov, *Opt. i Spektros.* **62**, 372 (1987) [*Sov. Phys. Opt. Spectros.* **XX**, XXX].

[119] S. B. Piepho, E. R. Krausz and P. N. Shatz, *J. Am. Chem. Soc.*, **100**, 2996 (1978).

[120] K. J. Wong, and P. N. Schatz, *Prog. Inorg. Chem.*, **28**, 369 (1981).

[121] B. S. Tsukerblat, A. V. Palii, V. Ya. Gamurar, A. S. Berengol'ts and H. M. Kishinevskii, *Phys. Lett. A*, **158**, 341 (1991).
[122] B. S. Tsukerblat, A. V. Palii, H. M. Kishinevskii, V. Ya. Gamurar and A. S. Berengol'ts, *Mol. Phys.* **76**, 1103 (1992).
[123] I. B. Bersuker and S. A. Borshch, In *Advances in Chemical Physics*, (Eds. I. Prigorine and S. A. Rice) vol. **81**, 703 (1992).
[124] S. A. Borshch, E. L. Bominaar, J.-J. Girerd, *New J. Chem.*, **17**, 39 (1993).
[125] A. J. Marks and K. Prassides, *New J. Chem.*, **17**, 59 (1993).
[126] V. Ya. Gamurar and B. S. Tsukerblat, *Zh. Struct. Khim.*, **31**, 16 (1990).
[127] V. ya. Gamurar, S. I. Boldirev, A. V. Palii and B. S. Tsukerblat, *Zh. Struct. Khim.*, **34**, 20 (1993).

## Additional references


[A.1] B.E. Douglas, S.A. Hollingsworth, *Symmetry in Bonding and Spectra*, Academic Press, San-Diego (1985).

[A.2] J.S. Griffith, *The Irreducible Tensor Method for Molecular Symmetry Groups*, Prentice-Hall Int. Inc., London-Tokyo-Sydney-Paris (1962).

[A.3] R.L. Carter, *Molecular Symmetry and Group Theory*, John Willey & Sons, Inc., New York (1998).

[A.4] A. Vincent, *Molecular Symmetry and Group Theory, A Programmed Introduction to Chemical Applications*, John Willey & Sons, Ltd, Chichester (2001).

[A.5] Brian R. Judd, *Operator Techniques in Atomic Spectroscopy*, Princeton University Press, Princeton-New-Jersey (1998).

[A.6] J.L. Birman, *Theory of Crystal Space Groups and Lattice Dynamics*, Springer-Verlag, Berlin-Heidelberg-New York-Tokyo (1984).

[A.7] A.S. Chakravarty, *Introduction to the Magnetic Properties of Solids*, John Wiley, New York (1980).

[A.8] S. Wolfram, *The Mathematica Book*, $5^{th}$ ed., Wolfram Media (2003), see: http://www.wolfram.com

[A.9] Per-Olov Lövdin, B. Pullman (eds), *Molecular Orbitals in Chemistry, Physics and Biology*, Academic Press, New York-London (1964).

[A.10] I.B. Bersuker, *Electronic Structure and Properties of Transition Metal Compounds, Introduction to the Theory*, Wiley, New York (1996).

[A.11] M.V. Nazarov, B.S. Tsukerblat, E.-J. Popovich, D.Y. Jeon, *Physics Letters* **A330**,291 (2004).

[A.12] K. R. Dunbar, E. J. Schelter, A.V. Palii, S. M. Ostrovsky, V. Yu. Mirovitskii, J.M.Hadson, M. A. Omary, S.I. Klokishner, B. S.Tsukerblat, *J. Phys. Chem.* **A 107**, 11102 (2003).

[A.13] E. Coronado, R. Georges, B. S. Tsukerblat, *Exchange Interactions: Mechanisms*, in: *Localized and Itinerant Molecular Magnetism: From Molecular Assemblies to the Devices*, NATO ASI Series, eds: E. Coronado, P. Delhaes, D. Gatteschi, J. Miller, Kluwer Academic Publishers pp. 65-84 (1996).

[A.14] J. M. Clemente, R. Georges, A. V. Palii, B. S. Tsukerblat, *Exchange Interactions: Spin Hamiltonians*, ibid., pp. 85-104.

[A.15] J. H. Van Vleck, *Revista de Matemática y Fisica Teó*rica, Universidad National de Tucumán, **14**, 189 (1962).

[A.16] P. M. Levy, *Exchange*, in: *Magnetic Oxides* (D. J. Craik, Ed.), John Wiley, pp. 181-232 (1975).

[A.17] P. M. Levy, *Phys. Rev. A.*, **155**, 135 (1964).

[A.18] P. M. Levy, *Phys. Rev. A.*, **177**, 509 (1969).

[A.19] K. I. Kugel, D. I. Khomskii, *Sov. Phys. JETP*, **37**, 725 (1973).

[A.20] K. I. Kugel, D. I. Khomskii, *Sov. Phys. Usp.*, **136**, 231 (1982).

[A.21] M.D. Kaplan, B.G.Vekhter, *Cooperative Phenomena in Jahn-Teller Crystals*, Plenum Press, New York-London (1995).

[A.22] M. V. Eremin, V. N. Kalinenkov, Y. V. Rakitin, *Phys. Status Solidi* B **89**, 503 (1978).

[A.23] M. Drillon, R. Georges, *Phys. Rev.*, **B 24**, 1278 (1981).

[A.24] M. Drillon, R. Georges, *Phys. Rev.*, **B 26**, 3882 (1982).

[A.25] B. Leuenberger, H. U. Güdel, *Mol. Phys.* **51**, 1 (1984).

[A.26] A. Ceulemans, G. A. Heylen, L. F. Chibotaru, T. L. Maes, K. Pierloot, C. Ribbing, L.G. Vanquickenborne, *Inorganica Chimica Acta*, **251**, 15 (1996).
[A.27] A. Ceulemans, L. F. Chibotaru, G. A. Heylen, K. Pierloot, *Mol. Phys*., **97**, 1197 (1999).
[A.28] A. Ceulemans, L. F. Chibotaru, G. A. Heylen, K. Pierloot, L. G. Vanquickenborne, *Chem. Rev*., **100**, 787 (2000).
[A.29] J.J. Borrás-Almenar, J.M. Clemente-Juan, E. Coronado, A.V. Palii, B.S. Tsukerblat, *J. Phys. Chem*., *A* **102**, 200 (1998).
[A.30] J.J. Borrás-Almenar, J. M. Clemente-Juan, E. Coronado, A. V. Palii, B.S. Tsukerblat, *Phys. Lett. A* **238**, 164 (1998).
[A.31] J.J. Borrás-Almenar, J.M. Clemente-Juan, E. Coronado, A.V. Palii, B.S. Tsukerblat, *J. Chem. Phys*., **114**, 1148 (2001).
[A.32] J.J. Borrás-Almenar, J.M. Clemente-Juan, E. Coronado, A.V. Palii, B.S. Tsukerblat, *Chem. Phys*. **274**, 131 (2001).
[A.33] J.J. Borrás-Almenar, J.M. Clemente-Juan, E. Coronado, A.V. Palii, B.S. Tsukerblat, *Chem.Phys*. **274**, 145 (2001).
[A.34] J.J. Borrás-Almenar, J.M. Clemente-Juan, E. Coronado, A.V. Palii, B.S. Tsukerblat, *J. Sol. State. Chem*. **159**, 268 (2001).
[A.35] J.J.Borras-Almenar, J.M.Clemente-Juan, E.Coronado, A.V.Palii, B.S.Tsukerblat, *J. Chem. Phys*., **114**, 1148 (2001).
[A.36] A.V.Palii , B.S. Tsukerblat, E. Coronado, J.M. Clemente-Juan, J.J. Borras-Almenar, *Polyhedron,* **22**(2003)2537.
[A.37] J.J. Borras-Almenar, J.M. Clemente-Juan, E. Coronado, A.V. Palii , B.S. Tsukerblat, *Polyhedron* , **22** (2003) 2551.
[A.38] A.V.Palii, B.S.Tsukerblat, E.Coronado, J.M. Clemente-Juan, J.J. Borrás-Almenar, *J. Chem. Phys*., **118** (2003) 5566.
[A.39] A.V. Palii, B.S. Tsukerblat, M. Verdaguer, *J. Chem. Phys*, **117** (2002) 7896.
[A.40] A.V. Palii, B.S. Tsukerblat, E. Coronado, J.M. Clemente-Juan, J.J. Borrás-Almenar, *Inorganic Chemistry*, **42** (2003) 2455.
[A.41] A.V. Palii, B.S.Tsukerblat , J.M. Clemente-Juan, E. Coronado, *Inorganic Chemistry*, **44** (2005) 3984.
[A.42] B. S. Tsukerblat, A.V. Palii, S.M. Ostrovsky, S. V. Kunitsky, S. I. Klokishner, K. R. Dunbar, *Journal of Chemical Theory and Computation,* **1** ( 2005) 668.
[A.43] A.V. Palii, S.M. Ostrovsky , S.I. Klokishner, B.S. Tsukerblat , K.R. Dunbar, *ChemPhysChem,* **7** (2006) 871.
[A.44] J.J. Borras-Almenar, J.M. Clemente, E. Coronado, B.S. Tsukerblat, *Inorganic Chemistry*, **38** (1999) 6081.
[A.45] J.J. Borras-Almenar, J.M. Clemente, E. Coronado, B.S. Tsukerblat, *Journal of Computational Chemistry*, **22**, (2001) 985
[A.46] O. Waldmann, *Phys.Rev*., **B61** (2000) 6138.
[A.47] D. Gatteschi, B.S. Tsukerblat, *Chemical Physics*, **202** (1996) 25.
[A.48] D. Gatteschi, B.S. Tsukerblat, V.E. Fainzilberg, *Applied Magnetic Resonance*,**10** (1996) 217.
[A.49] J.J. Borras-Almenar, J.M. Clemente-Juan, E. Coronado, B.S. Tsukerblat, *Chemical Physics*,**195** (1995) 1.
[A.50] J.J. Borras-Almenar, J.M. Clemente-Juan, E. Coronado, B.S. Tsukerblat, *Chemical Physics*,**195** (1995) 17.

[A.51] J.J. Borras-Almenar, J.M. Clemente-Juan, E. Coronado, B.S. Tsukerblat, Chemical Physics,**195** (1995) 29.

[A.52] E. Coronado, R. Georges, B. S. Tsukerblat, *Localization vs.Delocalization in Molecules and Clusters: Electronic and Vibronic Interactions*, in: *Localized and Itinerant Molecular Magnetism: From Molecular Assemblies to the Devices*, NATO ASI Series, eds: E. Coronado, P. Delhaes, D. Gatteschi, J. Miller, Kluwer Academic Publishers (1996), pp. 105-139.

[A.53] J.J.Borras-Almenar, J.M.Clemente-Juan, E.Coronado, A.V.Palii, B.S.Tsukerblat, *Magnetic Properties of Mixed-Valence Systems: Theoretical Approaches and Applications*, in: *Magnetoscience-From Molecules to Materials*, eds. J. Miller, M. Drillon, Willey-VCH, (2001) pp. 155-210.

[A.54] J.J. Borras-Almenar, J.M. Clemente-Juan, E. Coronado, R. Georges, B.S. Tsukerblat, *J. Magn. Magn. Mat.*, **140-144** (1995) 197.

[A.55] J. J. Borras-Almenar, E. Coronado, R. Georges, H. Kishinevskii, S. I. Klokishner, S. M. Ostrovskii, A. V. Palii, B. S. Tsukerblat, *J. Magn. Magn. Mat.*, **140-144** (1995) 1807.

[A.56] J. J. Borras-Almenar, J. M. Clemente-Juan, E. Coronado, R. Georges, B. S. Tsukerblat, *Mol. Cryst. Lig. Cryst.*, **274** (1995) 193.

[A.57]A. Bencini, A. V. Palii, S. M. Ostrovskii, B. S. Tsukerblat, M. G. Uytterhoeven, *Molecular Physics*, **86** (1995) 1085.

[A.58] J. J. Borras-Almenar, E. Coronado, R. Georges, A. V. Palii, B. S. Tsukerblat, *Chem. Phys. Lett.*, **249** (1996) 7.

[A.59] J. J. Borras-Almenar, E. Coronado, R. Georges, A. V. Palii, B. S. Tsukerblat, Phys. Lett. **A**, **220** (1996) 342.

[A.60] J. J. Borras-Almenar, J. M. Clemente-Juan, E. Coronado, R. Georges, A. V. Palii, B. S. Tsukerblat, *J. Chem. Phys.*, **105** (1996) 6892.

[A.61] I.B. Bersuker, *The Jahn-Teller Effect*, Cambridge University Press, Cambridge (2006).

[A.62] M.D. Kaplan, B.G. Vekhter, *Cooperative Phenomena in Jahn-Teller Crystals*, Plenum Press, New York-London (1995).

[A.63] J.J. Borras-Almenar, E. Coronado, S.M. Ostrovsky, A. V. Palii, B. S. Tsukerblat, *Chem. Phys.*, **240** (1999) 149.

[A.65] J.J. Borras-Almenar, J.M. Clemente, E. Coronado, S. M. Ostrovsky, A. V. Palii, B. S. Tsukerblat, *Vibronic Interactions in Mixed-Valence Clusters: General Overview and Applications, Proceedings of XIV International Symposium on Electron-Phonon Dynamics and Jahn-Teller Effect*, eds: G.Bevilacqua, L.Martinelli, N.Terzi , World Sci. Pub., Singapore, (1999), pp 302-310.

[A.66] J.J.Borras-Almenar, E.Coronado, S.M.Ostrovsky, A.V.Palii, B.S.Tsukerblat, Localisation vs. Delocalisation in the Dimeric Mixed-Valence Clusters in the Generalised Vibronic Model. Magnetic Manifestations, ibid, pp.293-301

# Index